Marines, Medals and Vietnam

William L. Myers

Lafayette, Louisiana

Copyright @ 2012 by William L. Myers
Library of Congress Catalog Number: 2012913723

All rights reserved. No part of this publication may be reproduced or used in any form or by any means - graphic, electronic or mechanical, including photocopying or information storage and retrieval systems - without written permission from the copyrights holder.

Format by Andre Andrepont and Carolyn Fontenot
Printed in the United States of America
by Andrepont Printing, Inc., Opelousas, Louisiana

ISBN: 978-0-9674365-1-7

Published by
Redoubt Press
4715 Woodlawn Road
Maurice, LA 70555
Phone: 337-898-0191
email: redoubt@cox.net

Preface

I have been at work on this project for almost four years and in that time became more of a chronicler and less an author. The process was intricate and difficult. I used documents which were based upon observations and opinions, perceptions, citations, and after–the-event oral histories. Many times this information did not always match. Many Marines experiencing the same event at the same time, many only a few feet apart, recall it in a different way. I have made an effort to resolve these conflicts, sometimes presenting both conflicting accounts. Often times I made a choice based on the preponderance of evidence rather than official documentation.

Where I have used previously published material I have tried to fill in the gaps, such as identifying casualties and others who participated in the action, and providing the names of those who earned medals.

In a work of this nature there are bound to be some inaccuracies and also a few errors. These I tried to limit and I apologize for my failings. I would appreciate your tolerance of them.

It was my intent to include as many names of Marine Corps and Navy veterans, attached to the Corps, as I could. Many are included but not nearly the number that I would have liked. All were magnificent in performing their duty. I hope that you enjoy reading about these heroes.

Semper Fidelis!
William L. Myers
Maurice, Louisiana
March 14, 2012

The following submission is a historical summary of the Vietnam War by a decorated veteran who wishes to remain anoymous. This Marine was a veteran of the fighting at Khe Sanh and earned the Silver Star. The summary:

The Vietnam veteran's duty was "police action": to halt the Communist takeover of South Vietnam. The Vietnam veteran, when serving in Vietnam, accomplished his duty. Moreover the Vietnam veteran won every battle he fought, including the battle for Khe Sanh. Thereafter the U.S. military withdrew its combat troops from South Vietnam, August 11, 1972. On January 27, 1973, a four party Peace Accord and Ceasefire was signed in Paris by all sides: U.S.A., South Vietnam, Viet Cong and North Vietnam. Our Last (non-combat) troops left South Vietnam March 29, 1973. When the ceasefire was signed more population and land were under Saigon control than any time previously. So in this regard, when we left we were winning.

Six months after our last troops left, the International Peace Commission reported North Vietnamese troops committing daily violations inside South Vietnam. While on monitoring patrol over South Vietnam in 1973, a Commission helicopter was shot down by the North Vietnamese Army. The Commission withdrew, July 1973. Full scale war returned. The Ceasefire's failure was North Vietnam's fault. December, 1974 the U.S. Congress ended U.S. Military hardware aid to the South Viatnamese military with the Case-Church Amendment.

In early 1975 the world press reported a massive, multi-division, tank and troop invasion launched from North Vietnam down into South Vietnam. Without continuing U.S. Military aid and replacement parts, South Vietnam alone surrended to North Vietnam, April 30, 1975. By then U.S. troops had been gone from South Vietnam's soil for over two years. The invasion ultimately reached Cambodia.

Unsuccessful foreign policy is different from battlefield defeat. Because the U.S.A. as a nation was not invaded, occupied or signatory to a surrender. Surrender is a defeat in war for a nation, ie; Germany and Japan in World War II. Not only is it a hasty generalization to state: The U.S.A. lost the war in South Vietnam," but also logically, The False Dilemma Fallacy: the "either/or situation"....such as the U.S.A. participating successfully until the 1973 Ceasefire; all sides signing that Ceasefire; The U.S.A. withdrawing; and then the last phase, or 3rd war counting the French, being fought between South Vietnam and North Vietnam from 1974 to 1975 resulting in South Vietnam's defeat and surrender on April 30, 1975.

The Vietnam veteran was the U.S.A. per se in Vietnam. There fore how may a nation be defeated in the war when its military troops are not actively participating at the time the defeat occurs? Militarily speaking then the South Vietnamese lost the war. Politically speaking some politicians lost the war. Unsuccessful foreign politics is different from 'defeat' in war.

This book is for all of the Marines who served in Vietnam and for those that I have known or served with. I wish that I could see all of their faces and speak with them one more time.

CONTENTS

Preface

1. Starlite 1
2. Operation Orange 24
3. Groucho Marx 47
4. A Hot Walk in the Sun 69
5. Getlins Corner 109
6. Attack on Echo 129
7. The Saga of the M-16 in Vietnam 148
8. Attack on the Dong Ba Tower 166
9. The Jacques Patrol 177
10. Preacher 199
11. Fix Bayonets! 217
12. "Follow Me!" 232
13. Hill 484 250
14. Dodge City 266
15. "Mo" Haas 278
16. Blood, Sweat and Tears 288
17. Medals 314

APPENDICES

Appendix I: Medal of Honor: Vietnam 317

Appendix II: Navy Cross: Vietnam 318

Appendix III: Silver Star: Vietnam 325

Photographs 369

Bibliography 378

Acknowledgements 379

Name Index 380

✶ ✶ ✶ ✶

By the summer of 1965, the Marines had become frustrated with fighting a guerilla war. Many hoped that the Viet Cong would tire of running and stand and fight.

Throughout July of 1965 the Third Marine Division's intelligence office had repeatedly turned up information regarding a large Viet Cong buildup south of Chu Lai.

On 15 August a soldier from the First Viet Cong Regiment surrendered to the ARVN and revealed that a regiment, numbering almost fifteen hundred men and supporting units, had set up a base in the Van Tuong village complex. The Viet Cong planned to attack the Marine garrison at Chu Lai which was about fifteen kilometers from there.

Lieutenant General Lewis W. Walt, the commander of Marine Forces in Vietnam, learning of the Viet Cong plan had to quickly make a choice. Two obvious courses of action were open to the Marines: they could remain within their defenses and wait for the enemy to attack, or they could strike the Viet Cong before the enemy was ready to move. Walt chose the latter for he had the authority and enough troops at his disposal to go on the offensive.

It would be a classic hammer and anvil operation with two battalions involved, Third Battalion, Third Marines and Second Battalion, Fourth Marines. The hammer would include three companies, India, Kilo and Lima, of 3/3 who would assault from the sea to the south and east of Van Tuong at Green Beach and would block the Viet Cong's southern route of escape. Mike Company would move by land from Chu Lai and set up a blocking position four kilometers northwest of the landing beaches.

The anvil would consist of three companies of Marines from 2/4 who would make an assault from the air into three landing zones that formed an arc southwest of Van Tuong. Hotel Company would land in LZ Blue which was about two thousand meters west of Green Beach. Echo Company would touch down at LZ White which was two thousand meters northwest. The Marines of Golf Company would land two thousand meters north in LZ Red. All three companies would then attack northeast toward Phase Line Banana, just short of Van Tuong.

The overall commander of the assault was Colonel Oscar F. Peatross, the commander of the Seventh Marines.

Operation Starlite was originally named Operation Satelite but a generator that supplied power to the operations clerks bunker failed and one of the clerks inadvertently typed in the name Starlite. Before the mistake was noticed the order had already been issued and Starlite was accepted as its official name.

"For twenty four hours it was continuous nonstop combat. Actually, it was not one large assault but just constantly fighting trying to find out who was behind you and who was in front of you. How anyone ever made any sense of the situation, I'll never understand."

> Private First Class Howard G. Miller
> I-3-3
> from *The First Battle*
> by Otto J. Lehrack

"Sergeant Jerry Tharp was a good friend. A Marine's Marine. One you could always depend on doing the job right. I was talking to him when the snipers bullet found him."

> First Lieutenant Homer K. Jenkins
> Hotel 2-4

This story concentrates on the actions of India Company, Third Battalion, Third Marines and Hotel Company, Second Battalion, Fourth Marines and Marines from other units associated with them. It also includes the helicopter squadron's HMM-361, HMM-261 and VMO-2.

On August 18, 1965 the commanding officer of 2/4 was Lieutenant Colonel Joseph R. "Bull" Fisher and commanding 3/3 was Lieutenant Colonel Joseph E. Muir. India 3/3 was led by Captain Bruce D. Webb while Hotel 2/4 was under the leadership of First Lieutenant Homer K. Jenkins. HMM-361 was led by Lieutenant Colonel Lloyd F. Childers., HMM-261 by Lieutenant Colonel Mervin B. Porter and VMO-2 by Lieutenant Colonel George F. Bauman.

References used in writing the following story were the books *No Shining Armor* and *The First Battle*. Both were written by Otto J. Lehrack. Also *U.S. Marines in Vietnam: The Landing and the Buildup, 1965* by Jack Shulimson and Major Charles M. Johnson, USMC

And an article that appeared in the September 2005 issue of Leatherneck Magazine. It was entitled *Starlite: "The Magnificant Bastards" Hit a Hot LZ* by B.L. Crumley.

Information was also obtained from the awards citations of many of the participants, the rosters and diaries of the participating units, and information obtained from the National Personnel Records Center.

1

Starlite

As the assault began Echo immediately ran into trouble as it moved off LZ White. The landing zone selected was located within the defensive perimeter of the 60th Viet Cong Battalion. On the first helicopter to land was First Lieutenant John A. Sullivan's second platoon that included Corporals Richard Tonucci and Ernie Wallace. Tonucci was a squad leader and Wallace was a machine gunner in that unit. His ammo carrier was seventeen year old Private First Class Jim Kehres. Since the Korean War seventeen year olds had not been allowed to serve in a combat zone. Somehow this fact had escaped those in charge.

From a ridgeline north and east of the Landing Zone, the Viet Cong used mortars and automatic weapons fire to stifle the companies advance. The Marines hit the deck and finding cover they fired their M-14 rifles and M-60 machine guns uphill. They killed three Viet Cong as they struggled to gain fire superiority. Maneuvering by squads, the company crept up the hill. Tonucci's squad had begun to move when Lance Corporal James Brooks was hit in the shoulder by a .50 caliber round that passed completely through his body, killing him. Working together, the older non-commissioned officers, and the young and green enlisted men destroyed one enemy redoubt after another. Finally after several hours of savage fighting, the Marines drove the Viet Cong from the hills.

During the initial moments of the operation other Marines died including Private First Class Henry C. Jordan of Hotel Company. As the helicopters brought in the last of the Marines into LZ Blue, Staff Sergeant Coy D. Overstreet, a helicopter crew chief, spotted well camouflaged enemy soldiers, disguised as bushes, advancing toward the helicopter. He swung his mounted M-60 machine gun to bear on them and fired a burst killing two and scattering the rest.

The machine gun that had killed Brooks was located, on Hill 43, directly in front of Hotel Company. The Marines were pinned down but Corporal Victor Nunez and others managed to return fire on the well concealed enemy.

Sullivan, firing a borrowed 3.5 inch rocket launcher, managed to bounce a round into a hooch from which the Marines were taking a lot of fire. The following explosion destroyed the enemy position.

By this time the Viet Cong had the range of the last helicopters to bring

Marines into the LZ. First Lieutenant's Stuart O. Kendall, Ramsey D. Myatt and Walter L. Sanders of HMM-361 were all wounded. Kendall was shot through the leg and the right hand. Myatt was first wounded by shrapnel but continued to fly and later was hit in the leg by small arms fire. Sanders suffered a fragmentation wound to his right lower leg. Staff Sergeant Dale O. Bredeson, Myatt's crew chief, kept his aircraft flying until it was too seriously damaged by enemy gunfire to continue.

First Sergeant Herschel D. Dorsett, an infantryman, flying as a gunner on Major Allan H. Bloom's aircraft managed to fire off a long burst that raked an entire enemy line of prone Viet Cong soldiers.

Bloom's co-pilot was Major Homer P. Jones a former jet pilot who now flew helicopters.

Sullivan's platoon, trying to take Hill 43, was stalled but big Ernie Wallace spotted a large group of Viet Cong moving down a trench line and closing in on the Marines from the rear. He aggressively attacked them and firing his M-60 from the hip and shoulder killed approximately twenty five enemy soldiers.

During this confusion Kehres and Wallace were separated. For the rest of the battle, Kehres helped out wherever needed until he was wounded.

Jenkins called a halt to the attack on Hill 43 and called for additional air strikes against it. In the meantime his other two platoons moved forward and continued the attack on Nam Yen 3. He chose to assault the village with the first platoon, led by First Lieutenant Charles C. Cooney, and let the third platoon, under First Lieutenant Robert S. Morrison, provide a base of fire. As they neared the village enemy fire suddenly poured into the Marines. When they passed the edge of the village and entered it the firing increased substantially.

The third platoon and the command group had to move through an open field as they approached the village. As the point men drew near Nam Yen 3 the rest of the Marines hugged a tree line that ran perpendicular to the open field and there waited to advance. The enemy let the point get close and then unleashed an intense volume of accurate small arms, recoilless rifle and rocket fire into the lead platoon. Quickly recognizing the gravity of the situation Private First Class Harry L. Kaus, Jr. and Lance Corporal Eddie L. Landry moved forward over thirty meters of open rice paddy swept by enemy fire. Kaus destroyed the crew of a rocket launcher and several of the enemy before he and Landry were shot down and instantly killed. Morrison also reacted aggressively and led his platoon in a savage attack on the hedgerow with his Marines firing as fast as they could from the hip.

Among the wounded was Lance Corporal Thomas J. Smith who suffered a gunshot wound to the back during this exchange and had to be evacuated.

Private First Class Richard M. Boggia and Lance Corporal Kenneth D. Stankiewicz were members of the machine gun section led by Corporal Jeffery R. Renfrow. With the increase in fire they dropped to the deck and set up their gun to cover the advance. Stankiewicz was the gunner and Boggia carried the ammunition. Suddenly Renfrow was hit by enemy fire and knocked down. Smoke and fire suddenly began to pour from his pack. An enemy round had ignited the flares that he carried there. Stankiewicz turned the gun over to Boggia and attempted to remove the pack from Renfrow. Although terrified Boggia got the gun going. The pack contained other explosives and Boggia expected it to explode at any moment. Stankiewicz, ignoring the danger, continued to work and finally removed the pack. Renfrow was pulled from the line of fire and evacuated with serious injuries. He would survive the battle.

Homer Jenkins contacted Bull Fisher and asked for help. Captain Bruce Webb in command of India Company 3/3 asked that he be allowed to cross over his boundary to attack An Cuong 2 which was out of his area of responsibility in an effort to take some of the pressure off of Hotel. Permission was quickly granted.

Jenkins pulled Cooney's platoon from the village choosing to consolidate his company and attack one objective at a time. Hill 43 was the highest piece of terrain and if captured Hotel 2/4 could control the entire area. With the help of a section of tanks and a flight of helicopter gunships Jenkins and his men were able to take the hill.

As Hotel Company evacuated their wounded the displaced Jim Kehres was shot through both buttocks and knocked to the deck as he was loading a wounded Marine onto a tank. It had been a busy day for him as he courageously crossed the fire swept rice paddies on several occasions to rescue wounded Marines and their equipment.

As soon as the Marines from India 3/3 received permission to attack the village of An Cuong 2 they took off running, with arms at high port, toward the enemy. Their aggressiveness surprised the enemy and they were initially able to take a few prisoners.

The Marines of India were soon busy trying to kill all of the Viet Cong that they could not capture. First Lieutenant Richard M. Purnell, the company executive officer spotted a number of enemy soldier's trying to escape down the riverbed. As he blazed away with his .45 caliber pistol he shouted "Get them!" The Marines cut them down in bunches as they fled.

Captain Webb called in a medevac for a Marine with a serious gunshot wound to the thigh and then ordered Private First Class Glenn Johnson to go back and bring up the others who were wounded.

Sergeant James Massey, the first platoon guide, was shot through the fleshy part of his upper arm by a bullet that ricochet's off of a shovel. He stayed in the fight and survived the operation.

Captain Webb assigned Corporal Robert E. O'Malley's first squad of the first platoon to ride a section of tanks that had managed to catch up with India Company.

O'Malley split the squad into three groups. He would ride the first tank along with Lance Corporal Christopher J. Buchs and Private First Class Robert L. Rimpson. Corporal Farrest Hayden and Lance Corporals James R. Aaron and Merlin E. Marquardt were assigned to the tank in the middle with the other members of the squad riding the third tank in the rear.

The tank commander aggressively led his vehicles down a trench line to the west and away from the rest of the company. With the first two tanks slightly ahead of the third they moved around a hedgerow and Marquardt was suddenly shot through the chest with several rounds fired from a Browning Automatic Rifle. The deadly rounds had come from the hedgerow. As the squad poured fired into the bamboo growth, Buchs raced to a corpsmen, who was treating another casualty, and asked for help but by the time he returned Marquardt was dead.

The Viet Cong were firing from trenches located on the other side of the bamboo. The Marines tried to dislodge them by throwing grenades but they were unable to penetrate the thick growth.

Buchs spotted a small opening in the hedgerow and he and O'Malley went through it and jumped into the trenches occupied by the Viet Cong. Firing furiously and accurately in close quarters O'Malley killed eight of the enemy and Buchs, employing hand grenades and his rifle, accounted for four others.

They soon ran out of ammunition, as other soldiers attacked, so they jumped out of the trench and quickly reloaded. This done, they jumped back in and continued the assault. O'Malley had Hayden check the bodies and positioned Rimpson, the grenadier, at the opening of the trench line. While Buchs provided cover O'Malley moved down the left side of the trench.

As Hayden went about his job an enemy soldier suddenly jumped up and threw a grenade at him. Fragments from the explosion hit Hayden in the hip and O'Malley in the foot. Buchs quickly shot the soldier dead. He then collected the enemy weapons and helped the wounded O'Malley and Hayden from the trench.

Rimpson jumped into the trench and killed another soldier at a distance of about fifteen yards with his grenade launcher. The other Marines were well within the blast radius of the exploding M-79 round but luckily none were hit by the flying fragments.

When one of the tanks with O'Malley's squad was hit with an anti-armor round the others directed their fire toward that weapon and silenced it. The Marines then loaded the tanks with their wounded. This proved to be a mistake as the enemy raked them with small arms and RPG fire and many of the injured were hit again.

Mortar rounds soon began to fall among the Marines and they again took cover in the trench. As O'Malley and his squad tried to exit the trench a round exploded nearby. He was wounded for a second time when a mortar fragment, from the blast, lodged in his forearm. The undaunted O'Malley picked up another wounded Marine and he and his squad continued to move back.

While O'Malley and his men were struggling with their fight, the rest of India Company assaulted An Cuong 2. The heavily fortified village contained about twenty five to thirty huts and was laced with log covered bunkers and trench lines.

As the Marines advanced the enemy fire from the village increased and one machine gun team, moving across an open rice paddy, was laced by automatic weapons fire. Three members of the team, Privates First Class Gilbert R. Nickerson, Walter L. Smith and James L. White, were shot down and killed. The only survivor, Private First Class Howard Miller, an ammunition carrier, slung his rifle over his back and then carrying his ammo box in one hand and the machine gun in the other kept moving. As he advanced and prepared to jump a trench a Chicom grenade exploded nearby and knocked him down. Miller was untouched by any fragments from the blast but it did slam him into the side of the trench and forced the air from his lungs. He lay there for a moment, caught his breath, and then continued to advance.

The Marines of India Company soon were forced to hit the deck by an intense enemy mortar attack. They dug in for a period of about twenty minutes and after they put some fire to their front the mortar attack stopped.

As they continued the assault the fighting broke down into a series of isolated small actions, as it usually does in war, with each man concerned only with the survival of him and the Marines in his small group.

Throughout the fight, Captain Webb appeared to be everywhere and was repeatedly exposed to enemy fire as he went about his duties as company commander. When direct communication broke down with the air support center the two Forward Air Controllers assigned to the company, Captain John D. Dalby and First Lieutenant Howard L. Schwend, spoke to the pilots directly and calling them in by their plane numbers brought in the needed support. This method was a little unorthodox but it got the job done.

They called for and controlled three separate air strikes that coincided with the companies attack and materially contributed to the defeat of the Viet Cong force.

The Marines of Colonel Muir's regiment had mastered the use of supporting arms and it allowed them to keep continuous pressure on the Viet Cong. Because of this the enemy suffered enormous casualties.

Private's First Class Charles D. Fink and John L. Jemison were members of Sergeant George A. Emerick's platoon that located a suspected enemy bunker complex. Before they were able to call in an air strike on it they were ordered forward. As soon as they passed it by the Viet Cong opened fire on them from the rear.

They took cover and Fink spotted a machine gun about fifty yards away that was putting out a heavy volume of fire into the ranks of the advancing Marines. At first Jemison, who carried the M-79, could not see the enemy position but under the direction of Fink he located it. He took careful aim and on the very first round that he ever fired in combat he hit the gunner on the head and took out part of the gun crew. As the excited Jemison rolled over, to reload the single shot weapon, he was seen. A Viet Cong soldier got behind the gun and fired a burst in his direction. One round hit Jemison in the head and killed him instantly. Fink shot the new gunner and as the other three members of the gun crew jumped up and ran he calmly shot them all to death.

A Viet Cong soldier fired an RPG at Fink. The round hit the paddy dike and exploded. Most of its power was absorbed by the dirt berm but some of its deadly force hit the stock of Finks rifle and shattered it, sending fragments into his forearms and face. The explosion rolled him onto his back and knocked his helmet and glasses askew.

Corporal Ronald R. Jones, the squad leader, came over and started to put a battle dressing on his arm. He was in a kneeling position as he worked on Fink. Jones suddenly flinched and said, "I think I just got shot in the shoulder." Fink checked and saw that he had indeed been hit in the shoulder and there was no exit wound. He used the battle dressing Jones carried and made a sling for him.

Emerick repeatedly exposed himself to hostile fire as he directed the fire of his men. Although badly outnumbered they repulsed the enemy and inflicted heavy casualties. With Fink and Jones wounded and Jemison dead Emerick sent the rest of the squad to find help while he remained to prevent the casualties from being captured. When he realized that help would not arrive in time he helped Fink and Jones to safety and then returned with two other Marines to recover the body of Jemison.

The ferocity of the India Company attack either killed or routed the entire enemy garrison in the village. The Marines were able to cross the streambed and then the trench line and had come up onto a flat spot where lay the bodies of many of the Vietcong. Staff Sergeant Jean Pinquet was making sure they were dead by putting rounds from his .45 caliber pistol into the heads of the dead enemy soldiers. Captain Webb ordered him to stop because he felt that it was inhumane.

Only seconds later one of the supposedly dead soldiers tossed a grenade into the company command group. Fragments from the blast killed Webb and another Marine and wounded several others.

Glenn Johnson was only a few feet from Webb, who was talking on the radio, when the grenade exploded. He jumped back and knocked both he and Private First Class Freddy Link into a trench. Gunnery Sergeant

Raymond B. Martin was in this group and was wounded in his right side and upper arm by grenade fragments.

With the death of Captain Webb command of India Company was placed on the shoulders of First Lieutenant Richard M. Purnell. To this point in the war he had seen very little action and this was his first major operation. Purnell would handle it well.

Johnson and First Sergeant Arthur Petty were two of the stretcher bearers that carried the body of Webb to the helicopter. Private First Class Gary N. Hammett lamented, "If it hadn't been for the captain telling the staff sergeant to quit what he was doing, Captain Webb might be alive today."

Petty supervised the loading of the badly wounded on the helicopters. When he tried to place one of India Companies NCO's, Corporal James P. Reed, on one of the choppers the pilot waved him off because his plane was already overloaded. Reed, a machine gun team leader and a company armorer, was shot full of holes but gamely lay there and got on the next bird. He survived the fight.

In the sector controlled by Hotel Company 2/4, Jenkins and his men were again moving against the village of Nam Yen 3. They were aided in the assault by two M48 tanks from Company C, Third Tank Battalion. Tonucci climbed onto the back of one of the vehicles and was trying to direct its fire unto a heavy .51 caliber enemy machine gun by using the tank-infantry phone. The tank was hit by a large anti-tank rocket and he was thrown into a rice paddy by the explosion.

The driver of the tank, Corporal Charles L. Denton, his platoon commander and the tank's gunner were wounded by the blast and the tank was set afire. Denton disregarded his own wound and the intense fire and pulled his platoon commander from the tank and assisted him to a place of relative safety. He then returned to the burning tank and guided it away from a large group of wounded Marines and to a protected position where the fire was then extinguished. The tank could not be repaired so engineers destroyed it in place.

Tonucci then moved down a trench in an attempt to eliminate the gun. Five Marines had been wounded and were lying in the open. Tonucci left Lance Corporal Joe C. Paul, who was also wounded, to protect them. Earlier in the day he had been placed on a helicopter for evacuation but before it took off he decided that his wound was not bad enough for him to leave the fight so he hopped off and rejoined his fellow Marines.

Tonucci was joined by a rifleman from his squad, Private First Class Ronald L. Centers, as he moved after the machine gun that was tormenting the Marines. The two Marines eliminated a grenade launcher in a bunker, the machine gun bunker, and still another bunker. Tonucci and Centers had no sooner killed the Viet Cong manning the gun than another crew quickly replaced them. They killed them also, and then took out another enemy bunker. The dynamic duo had killed fourteen of the enemy before they were finished.

As the two remaining tanks and three Ontos moved toward the village with the Marines of Hotel 2/4 they were unable to cross the flooded rice paddies. The vehicle commander of one of the Ontos, Corporal Robert G. Bousquet, took a round through the helmet, as he stood in the hatch, and was killed.

Jenkins and his Marines drove the Viet Cong from a trench they occupied. There they took a short break before continuing the assault. As they approached the village they were subjected to intense mortar and small arms fire from enemy soldiers who had tried to get behind the Marines. Corporal Edward L. Vaughn spotted an enemy mortar position to the rear and dashed across a rice paddy under intense Viet Cong fire so that he could fire from a more advantageous position. Once he got there, from a standing position, he shouldered his weapon and fired three hundred rounds and killed the entire seven man mortar crew. When a new crew moved toward the mortar he repeated his actions and killed them too.

With all of the yelling and confusion Boggia became separated from his gunner, Stankiewicz. He and several other Marines found themselves pinned against a dike by enemy sniper fire.

Corporal Ronnie Spurrier, a fire team leader was in this group. He found himself in an exposed position without ammunition. Private First Class Robert L. Stipes, an automatic rifleman in Spurrier's team, ran out into the paddy in front of his fire team leader and delivered a ferocious volume of accurate fire that allowed Spurrier and the others to withdraw to safety.

Sergeant Jerry D. Tharp was shot and killed by a sniper who was located in a tree. He had been talking to Homer Jenkins and when he paused and raised his head to look over a small berm and was hit.

Hotel lost its company Gunny about the same time. Gunnery Sergeant Albert H. Raitt was standing on the edge of a ditch, firing his M14 at the Viet Cong, when he was shot and killed.

Private First Class Carlos R. Hanley was hit in the left arm by fragments from an exploding mortar round.

Tonucci and diminutive Private First Class John E. Slaughter, a grenadier, went after the sniper. Slaughter weighed about a hundred pounds and started the day carrying seventy two rounds for his single shot, crack barrel, grenade launcher. In total his ammunition weighed about thirty six pounds. Slaughter was resupplied with the same amount of rounds three to four times during the fight. He had been loading and firing the weapon as fast as he could since Hotel Company had hit the landing zone.

Slaughter was wounded by the fragments from an exploding enemy round. But he now had so many targets that his wounds were of little concern. With Tonucci spotting, Slaughter popped rounds into the tree tops one at a time. They killed the sniper and by the end of the day the barrel of his M79 was completely worn.

The seemingly inexhaustible Wallace was also busy at this time and as he fearlessly exposed himself to enemy fire the bipod was shot from his weapon. A well camouflaged enemy force tried to out flank Jenkins troops as they neared the village, and although the Viet Cong looked like trees and bushes, Wallace was not deterred or fooled and was able to separate them from the terrain. He continued to fire and killed another fifteen of the enemy. That brought his total for the day to forty killed.

When Boggia finally located Stankiewicz he discovered that his friend had been seriously wounded in the arm. The injured Marine gave the gun to Boggia and he was placed on a tank to be evacuated. While there Stankiewicz was hit again and killed.

When the action got hot and the Marines were in danger of having their assault stopped, Private Samuel J. Badnek, no garrison Marine, decided to do something about it and removed all combat equipment from his body. He then dashed forty five yards through heavy fire to reach the Viet Cong positions. Once there he hurled several grenades into their redoubt and personally killed eight of the guerillas. Stunned by his assault the enemy became disorganized. The fire was so intense that those who witnessed Badnek's mad dash toward the Viet Cong could not believe that he had survived. Although wounded in the head he continued to ferociously engage the enemy, directing accurate and effective rifle fire into their ranks.

Meanwhile, Paul came under attack as he tried to protect the wounded. They were subjected to devastating mortar, recoilless rifle, automatic weapons and rifle fire and were pinned down. He fearlessly placed himself between the wounded men and the enemy. Making Paul's situation more difficult, the Viet Cong added a couple of white phosphorous rifle grenades rounds. When they exploded some of the burning pieces landed on him. Paul was hit several times but refused to quit and continued to fire his weapon until all of the wounded were evacuated. He, too, was eventually evacuated but died on the way to a medical facility.

The attack by Hotel 2/4 was again repulsed by the enemy defending the village at a high cost to each side. They returned to LZ Blue and dug in. At that point Jenkins only had 28 Marines in his in his company that were able to continue the fight.

Among the Hotel Company wounded was Private First Class Donald L. Harris who received a gunshot wound to the neck.

In the India 3/3 sector O'Malley was wounded for a third time. A mortar round exploded nearby and a shell fragment hit him in the chest and punctured his lung. It caused O'Malley's lung to collapse but he continued until all of his dead and wounded were loaded onto the tanks. Rimpson was also hit by a shell fragment near his eye. When the eye popped out of its socket, he quickly pushed it back in, and although suffering from impaired vision, continued.

O'Malley and his men moved to the helicopter landing zone for evacuation. Once there, Buchs had to knock out an enemy machine gun, with an M-79, that was firing on the helicopters as they landed. There were fifteen casualties including O'Malley to be evacuated. He waited until all were aboard before he got on the second helicopter. This one was piloted by First Lieutenant Richard J. Hooten.

His co-pilot was First Lieutenant Kenneth L. Slowey and the crew chief was Lance Corporal Richard E. Ely. Later in the day Slowey would be superficially wounded in the upper right arm, and Ely would lose a toe to enemy fire.

Company A, First Amphibian Tractor Battalion commanded by Second Lieutenant Robert F. Cochran, Jr. was moving inland on a resupply mission to the troops on the front line. This armored column consisted of five amphibious tractors and two tanks. When they stopped in order to check their positions they were hit by small arms fire from the left flank. As the enemy fire subsided they again proceeded down the road in the direction of India Company. The lead vehicles had to slow down when they reached a sharp curve that entered a wooded area. This forced those following to stop. At this point the lead tractor on which Cochran was riding was hit by recoilless rifle fire and knocked out.

The ambush site was bordered by a rice paddy on one side and a hedgerow covered by a dense thicket on the other. The site was well chosen and left the Marines with little room to maneuver. Several of the vehicles were immediately hit and disabled.

Cochran evacuated his crew and as the enemy attacked and tried to enter his disabled vehicle he removed all of the machine gun ammunition so that the Viet Cong could not make use of it. He then directed all of the dazed and wounded men to the safety of two amphibian tractors that had gained hull defilade positions behind the dike of the rice paddy. This done he moved toward a vehicle that he thought would provide a good view of the ambush site. Cochran was severely wounded as he moved toward this tractor. In an effort to not expose the men in the vehicle to enemy fire he tried to enter it through the top hatch. There he was hit once more and killed.

Staff Sergeant Jack Marino, Jr. was riding in the last tractor when the column stopped. Thinking that they had they had made contact with India Company he dismounted and started to move to the front. The first vehicle was hit and the ambush was sprung. When Cochran was killed Marino, although wounded by fragments from and exploding mortar round in the right leg and right hand, assumed command and set up a protective perimeter of fire. For hours it was a desperate standoff but he and his crewman used a machine gun, rifles, pistols and hand grenades to good effect.

Most of the nine Marines remaining with Moreno were wounded, some more than once, but they continued to fight effectively and repulsed every attack. When the enemy attempted to attack from a tree line to the rear they fired at the Viet Cong with an intensity that also stopped that attack. It was estimated that over several hundred of the enemy had converged on the column.

Sergeant Robert F. Batson was forced from his amtrack armed only with his combat knife and he was quickly shot down and killed. When his body was found he still had the knife clutched tightly in his hand.

When Sergeant James E. Mulloy, Jr., a supply administrative man, tried to maneuver through the Viet Cong fire his tractor became bogged down in a rice paddy. He immediately took charge and rallied the Marines and organized the defense of their stricken vehicle. Mulloy then administered first aid to the wounded and inspired them with his calmness. He soon realized that if he remained in the tractor he would have limited visibility so he decided to move and take up a better position in the paddy. The position he chose was a good one and for twenty hours he conducted a virtual one man defense.

He had a good view of the stranded column and as the Viet Cong tried to assault it he was able to pick them off, sometimes individually and at other times in small groups. Several of the enemy boarded one of the abandoned vehicles by walking up the ramp. One emerged with a soap dish in his hand and turned to show his prize to another. Taking careful aim, Mulloy shot them both down.

In time the enemy realized that he was a major deterrent to achieving their objective, of wiping out the entire Marine force, and they attempted to eliminate him. Mulloy repeatedly repulsed all attempts by the Viet Cong to kill him and, in the process, inflicted extremely heavy casualties on the enemy. When help arrived the next morning Mulloy, although wounded by shell fragments in the right wrist and hand, made certain that all of the wounded were evacuated before he sought relief.

Captain Allan W. Lamb, a tank officer attached to Regimental Landing Team 7, led a relief column consisting of a mounted rifle company (India 3/3) on five LVTP5'S to aid the stricken column. When his tank was hit directly in its front by a well placed round from an enemy 57mm recoilless rifle Lamb continued on undaunted and after more than eight hours of continuous action, the Viet Cong broke contact and the trapped Marines and their equipment were rescued.

During the rescue attempt Emerick and two other Marines were blown from the top of the tractor they were riding. Emerick was dazed and confused but otherwise uninjured. He had lost his rifle and, now weaponless, helped tend to the other Marines who had been slightly wounded. The next morning a piece of the barrel and receiver group of a rifle was found. The serial number on the receiver identified what remained of the weapon as Emerick's.

On his own initiative Schwend called for and controlled several helicopter evacuations of dead and wounded Marines.

On the first evacuation the helicopters of HMM-361 were called in to take out the wounded. The pilot of the lead aircraft was First Lieutenant Paul W. Bronson and his copilot was First Lieutenant Roger W. Cederholm. Believing that they were going into a hot LZ, Bronson asked his wing to fly high while he went down. This plane was flown by First Lieutenants Daniel T. Armstrong and Richard G. Adams. A Marine in an exposed position guided them in but as they neared the ground the Viet Cong blasted them with a withering fire. Armstrong immediately dropped down and the gunners on his aircraft worked the enemy over with their M-60's. With his controls badly shot up Bronson managed to level off and finally crash land, about a mile away, amongst a bunch of Sampans on the beach.

They were again subjected to heavy enemy fire from a nearby thicket. Lance Corporal Jimmie R. Clouse, the crew chief, was hit in the chest and abdomen by enemy rounds. The wound to the abdomen was so severe that it eviscerated him. Bronson moved to him quickly and was able to push his intestines back in. He held them in place and carried Clouse to the other aircraft. The badly wounded crew chief would survive and after several weeks of hospitalization eventually be returned to duty.

Another courageous pilot was Captain Benny D. Rinehart. He made three flights into the fully exposed landing zone and on each one remained until his helicopter was fully loaded with the wounded. Rinehart's first helicopter was so shot up that he returned in a different one and took out additional wounded. This bird also suffered similar severe damage, so he switched to another plane and returned for a third time. He rescued a total of 44 casualties in a period of about ten hours.

NOTES:
Marines killed on Operation Starlite are listed as follows:
Members of India-3-3:
Batson, Robert F. Sgt. 1812065; 27; Dallas, Texas
Brand, Thomas R. HN 6946202; I-3-3; 19; Allen Park, Michigan
Cotten, Ollie R. Pfc. 2042487; 19; Salt Lake, Michigan
Delmark, Francis J. D. LCpl. 2100716; 22, Salt Lake City, Utah
Jemison, John L. Pfc. 2056778; 20; Pittsburgh, Pennsylvania
Kopec, Edward Pfc. 2129987; 19; Chicago, Illinois
Marquardt, Merlin E. LCpl. 2097254; 19; Fort Wayne, Indiana
Nickerson, Gilbert R. Pfc. 2909031; 20; Racine, Wisconsin
Smith, Walter L. Pfc. 2084015; 19; Memphis, Tennessee
Thomas, David J. Cpl. 2057965; 23; Dallas, Texas
Towne, Peter C. Sgt. 1808614; 27; Morris, Connecticut
Vannatta, Jon D. LCpl. 2043921; 20; Richmond, Indiana
Webb, Bruce D. Capt. 068857; 31; Wheaton, Illinois
White, James L. LCpl. 2022472; 21; Chicago, Illinois

Member of Company A, First Amphibian Tractor Battalion:
Cochran, Robert F. Jr. 2dLt. 089648; 23; Poplarville, Mississippi

Members of Hotel-2-4:
Bousquet, Robert G. Cpl.; 24; Holyoke, Massachusetts
Brooks, James R. Jr. Pfc. 2061602; 20; Anderson, South Carolina
Dewitt, James P. LCpl.. 2058275; 21; Fruita, Colorado
Duran, Steve G. HN 5977969; 20; Deming, New Mexico
Henrich, Bruce J. Pfc. 2134183; 18; Detroit, Michigan
Jordan, Henry C. Pfc. 2075834; 20; New York, New York
Kaus, Harry L. Jr. Pfc. 2077042; 18; Dunkirk, New York
Landry, Eddie L. LCpl. 2035493; 20; Gonzales, Louisiana
Nickerson, William W. Cpl. 2055960; 20; Racine, Wisconsin
Raitt, Albert H. GySgt. 580671; 38; Neptune City, New Jersey
Sawyer, James H. Pfc. 1987561; 20, Morgantown, West Virginia
Smith, James A. SSgt. 1420044; 30; Warsaw, Kentucky
Short, Mitchell C. LCpl. 2072644; 18; Canoga Park, California
Stankiewicz, Kenneth D. LCpl. 2063870; 19; Buffalo, New York
Tette, John B. Pfc. 2127060; 19; Rochester, New York
Tharp, Jerry D. Sgt. 1559204; 28; Kemp, Texas

On this day there were many heroes. The Medal of Honor was earned by Lance Corporal Joe C. Paul and Corporal Robert E. O'Malley. Paul's award was posthumous.

The Navy Cross was awarded to Private Samuel J. Badnek, Second Lieutenant Robert F. Cochran, Jr., Lieutenant Colonel Joseph E. Muir, Sergeant James E. Mulloy, Jr., Corporal Ernie W. Wallace and Captain Bruce D. Webb. The awards to Cochran and Webb were posthumous.

The Silver Star was earned by Lance Corporals Christopher J. Buchs, Richard J. Hoeck and Eddie L. Landry, Privates First Class Ronald L. Centers and Harry L. Kaus, Jr., Corporals Charles L. Denton, Richard L. Tonucci and Edward L. Vaughn, Sergeant George A. Emerick, Lieutenant Colonel Joseph R. Fisher, First Lieutenants Homer K. Jenkins, John A. Kelly, Robert S. Morrison and Richard M. Purnell, Captains Allan W. Lamb and Benny D. Rinehart, Staff Sergeant Jack Marino, Jr., Colonel Oscar F. Peatross. The awards to Kaus and Landry were posthumous.

The Bronze Star was awarded to Captain's Paul W. Bronson, Jay A. Doub and David A. Ramsey, First Lieutenants Charles C. Cooney and Howard L. Schwend and Private First Class Robert L. Stipes.

Sergeant James E. Mulloy, Jr. first joined the Marine Corps in 1952 and served two years on active duty before going into the reserves for four years. He was a fire team leader in the Seventh Infantry Company out of Louisville, Kentucky. In 1958 he returned to the ranks of the regular Marine Corps and served until 1984 when he retired as a Chief Warrant Officer-4.

Staff Sergeant Jack Marino, Jr. enlisted in the Marine Corps in 1944 and served with George Company, Third Battalion, Twenty Ninth Marines, as part of the newly formed Sixth Marine Division, during the battle at Okinawa. After the Japanese surrender, his unit sailed to Tsingtao, China to accept the surrender of the Japanese there, and remained on occupation duty until April 1, 1946. Discharged when he returned to the United States, Marino joined the Marine Reserve unit in Mobile, Alabama in 1950. They were soon called to active duty and he served, in Korea, with Baker Company, Seventh Engineer Battalion. Marino also served a second tour in Vietnam as a member of Bravo Company, First Amphibian Tractor Battalion. He retired as a Master Sergeant in 1972.

Captain Benny D. Rinehart enlisted in the Marine Corps in 1955 and served in the infantry until he was selected for flight school in 1956. He remained in the Corps until April of 1977 when he retired with the rank of Lieutenant Colonel. During the course of the war in Vietnam, on a second tour in 1969, he earned a Bronze Star for heroism.

Lieutenant Colonel Lloyd Childers enlisted in the Navy at the beginning of World War II and served as an aviation radioman. He flew as an aerial flight gunner in action against the Japanese forces in the Battle of Midway. While participating in a torpedo plane attack as a member of Torpedo Squadron Three, Childers aircraft was exposed to intense anti-aircraft fire and overwhelming fighter plane opposition. Childers vigorously used his machine gun in an effort to repel the attacking Japanese aircraft. When he was seriously wounded an unable to use his machine gun he fired at the planes with his .45 caliber pistol.

Immediately after the war Childers joined the Marine Corps Reserve and returned to active duty in 1946. He later went through flight training and became a pilot in 1950. Childers served in Korea with VMF(N) 513 during the early stages of that war. He retired in 1968 as a Lieutenant Colonel.

Captain Jay A. Doub was the commanding officer of Kilo Company during Operation Starlite. His company came under withering hostile fire on at least five occasions. Each time, Doub personally led the assault on the objective and captured the position with a minimum of Marine Casualties.

He enlisted as a private in 1948 and was a member of Fox Company, Second Battalion, Fifth Marines in Korea during the tough fighting of November and December of 1950. Doub worked his way up to the rank of Staff Sergeant before being commissioned a second Lieutenant in March of 1953.

In combat, Doub did not think that officers needed to carry weapons. He felt that they would be too busy directing their troops to fire weapons.

Despite the severe injuries he suffered when Joe Muir was killed, Doub remained on active duty until his retirement in 1969 with the rank of Major.

Lieutenant Colonel Joseph E. Muir was a former enlisted man. He enlisted at the end of World War II and worked his way up from the ranks. He was appointed a Second Lieutenant in 1948. He served in Korea near the end of the war with Weapons Company, Third Battalion, First Marines.

He was a slightly built, soft spoken individual who was greatly respected by the men of the battalion. Muir seemed to inspire confidence in others wherever he went. The peripatetic Muir was held in awe by his Marines and many thought that he would eventually be the commandant. He was fondly called "The Grasshopper" because during a combat situation he moved from place to place and seemed to be involved in everything at the same time.

Muir earned the Bronze Star Medal for the period of January 18 to May 7, 1965. He was instrumental in the planning and execution of operations against the Viet Cong in the first few weeks of the war.

He believed in readiness and was a staunch believer in the use of supporting arms. Once he took command of the third battalion it did not take long before all of the officers, staff NCO's and many of the lesser ranks could call for artillery, air, or naval gunfire. He drilled his men at all hours, night and day, to perfect all of the infantry skills.

Muir always looked out for the welfare of his men and was determined that none would be killed unnecessarily. He was a master in the use of supporting arms and could use them effectively either simultaneously or in sequence.

During Operation Golden Fleece on September 11, 1965 Colonel Muir was killed when he stepped on a 155mm howitzer shell rigged as a land mine.

The Marines of 3/3 were moved to Hill 55 in the Danang area and were there to provide security so that the South Vietnamese could harvest their rice. They were flown in by helicopter on 10 September and quickly settled in.

Upon arrival, Captain David A. Ramsey and his company gunnery sergeant had walked all around the top of the hill avoiding what appeared to be a trail. They had even issued C-rations to the troops from there, all without incident.

Around midnight Muir, and five other Marines, came up to the top of the hill to meet with Ramsey. It was a bright moonlit night and they discussed what they were to do the next day. As daybreak approached, Muir and the others in his party, which included Captain Jay Doub and two radiomen, decided to return to the command post.

Muir hadn't walked fifty yards when he disappeared in a huge explosion. He had stepped on what turned out to be a 155mm artillery shell, buried with an antipersonnel fuse on the nose. Parts of the colonel splashed over Gunny Versimato, Top Petty and Ramsey. They ran over there to see what had happened and discovered that the colonel had been killed. The explosion had blown off both of his legs and most of both arms. Muir's body was missing from about the rib cage on down and there was absolutely no possibility of him being alive. One of the other Marines, a radioman, Private First Class Paul W. Mansir, was also killed. Doub was badly wounded and had lost one of his eyes. He was bleeding profusely and was in bad shape but would survive.

The other radioman had also been badly mutilated but was alive and conscious. All of one leg, part of the other, his genitals and part of one arm had been blown away. He was bleeding profusely from a cut on his forehead and thinking that it was his only injury kept asking the corpsman, "Will it leave a scar?"

The other two Marines walked off without a scratch.

Ramsey was the Operations Officer of the Third Battalion, Third Marines during Operation Starlite. Working tirelessly and with superb efficiency under a twenty-four hour deadline, he prepared an effective operation plan. When the Battalion Command Group was subjected to heavy enemy fire during the conduct of the operation Ramsey quickly organized administrative personnel and radiomen into a mobile infantry unit which drove through the enemy held area and reestablished contact with friendly units. He retired, with the rank of Lieutenant Colonel in December of 1980.

Sergeant George A. Emerick, a Marine since 1959, would remain in the Marine Corps for twenty six years and retire with the rank of Sergeant Major in 1985.

Colonel Oscar F. Peatross had previously earned the Navy Cross in World War II as a member of Carlson's Raiders. This action occurred on August 17, 1942 during the famous raid on Makin Island. He later authored a book entitled *Bless 'Em All: the Raider Marines of World War II*.

Peatross graduated from the University of North Carolina in 1939 and joined the Marine Corps the following year. He served as an officer and attained the rank of Major during World War II and was a battalion commander during the Korean War. Peatross was in command of the Marine Corps Recruit Depot at Parris Island when he retired as a Major General in 1971.

Lieutenant Colonel Joseph R. Fisher enlisted in the Marine Corps in 1942. He was twice wounded, in his right shoulder and chest, on Iwo Jima where he earned the Silver Star on February 21, 1945 as a member of Company C, First Battalion, Twenty Sixth Marines. Fisher had entered the battle as a platoon sergeant but due to attrition was acting as the platoon leader. Although sustaining painful machine gun bullet wounds to his right shoulder and chest he refused to quit the fight and led his men until he collapsed from the effects of his injuries and was evacuated.

While fighting in the Korean War, Fisher earned the Navy Cross during the First Marine Division's famous breakout from the Chosin Reservoir. At that time he was the Commanding Officer of Company I, Third Battalion, First Marines. As a First Lieutenant he inspired his men to make a heroic effort in repulsing several vicious attacks by the Chinese on November 28 and 29, 1950. Fighting with the same unit, Fisher also earned Bronze Stars on September 20, 1950 and March 23, 1951.

Lance Corporal Ernie Wallace of Hotel Company 2/4 recalls: We knew that the operation was going to be tough, but we had no idea how many of the enemy we would encounter. On our prior patrols we had had absolutely no contact.

The landing zone was hot as we landed. We were the second wave coming in, and the VC had already made contact. We tried to set up a 360 defensive perimeter, but we were taking a lot of fire from all directions and were finding it difficult to do. We were facing open rice paddies and there was a ville to the left, Nam Yen 3, with some high ground, Hill 43, to the right. We were taking fire from everywhere.

From the start we were working to link up with Mike Company, but we were in a firefight the entire day; we'd move a little bit, then a little bit more. We got separated as a company. We were one platoon here, one platoon here, and another here, and it ended up we were facing a heavy weapons company, something we were not aware of.

We got to an open ditch near the bottom of a hill. From the time we hit the landing zone we were completely surrounded by the enemy. We bypassed Hill 43. We did not take it because in order to do so we would have had to cross a ditch that ran from Nam Yen 3 to the base of the hill.

I spotted a large number of Viet Cong coming into the trench line from our rear. I think that there were about twenty five of them. I was giving covering fire for the assault from the high ground. I fired into them until I ran out of ammunition and then I went across the ditch.

As we moved forward I noticed that the little pine trees ahead were actually well-camouflaged Viet Cong so I started to yell for the others to start shooting the trees.

Up until Starlite, although we had been probed, we were in a defensive mode. We didn't have our weapons loaded. We could carry magazines but we couldn't have a round chambered, but that all changed during Starlite. After that, we were on the offensive.

A three war Marine, Arthur O. Petty first enlisted in the Marine Corps in 1941. He retired in 1973 with the rank of Sergeant Major.

Lance Corporal Phillip F. O'Donnell remained in the Marine Corps and retired as a Sergeant Major with thirty years of service in 1993.

Corporal Ronnie Spurrier enlisted in the Marine Corps in 1959. He remained on active duty for twenty years retiring with the rank of Gunnery Sergeant in 1979.

During the war in Vietnam 57 Marines earned the Medal of Honor. Defying the odds, two of these went to a pair of friends who grew up together.

Robert E. O'Malley and Thomas P. Noonan were born in 1943 in the same neighborhood in Woodside, New York. They entered kindergarten together in 1949 and were lifelong friends. In 1961 they went their separate ways, O'Malley, eighteen, followed his three brothers into the Marine Corps and Noonan tried the seminary. Not finding that to his liking Noonan, a champion wrestler, then went to Hunter College in the Bronx and earned a degree in physical education.

In late December of 1967, two years after O'Malley earned the Medal of Honor, and now twenty four, Noonan joined the Marine Corps as an enlisted man. He was sent to Vietnam in July of 1968 and in a very short time was promoted to lance corporal and made a fire team leader with Golf Company, Second Battalion, Ninth Marines.

On February 5, 1969, during Operation Dewey Canyon, Noonan was killed as he attempted to rescue a wounded Marine exposed to intense enemy fire. For his actions that day he was posthumously awarded the Medal of Honor.

After the war O'Malley kept in touch with the Noonan family. He visited Florence Noonan, Tommy's mother, every Memorial Day until her death. Mrs. Noonan proudly kept a photo of both Marines on her living room wall.

A Street on the Marine Corps Air Station at Yuma, Arizona is named after Noonan.

Others who were either mentioned in the chapter or were wounded that day are as follows:

Aaron, James R. LCpl. 2058940; I-3-3; 20; Detroit, Michigan
Adams, Charles W. LCpl. 2043706; H-2-4; 20; Indianapolis, Indiana
Adams, Richard G. 1stLt. 085218; HMM-361; 24; Cambridge, Mass.
Anderson, George E. Cpl. 1947641; I-3-3; 21; Austin, Texas
Anderson, Jesse R. Jr. Pfc. 2091316; I-3-3; 19; Denver, Colorado
Armstrong, Daniel T. 1stLt. 084144; HMM-361; 24; Los Angeles, Cal.
Bianchini, Michael L. Pfc. 2080675; H-2-4; 19; San Francisco, Cal.

Barrett, John P. Pfc. 2088428; H-2-4; 19; Detroit, Michigan
Bell, Robert L. Sgt. 1653652; I-3-3; 26; Richmond, Virginia
Billings, Howard R. Cpl. 2035397; I-3-3; 19; Little Rock, Arkansas
Bloom, Allan H. Maj. 061041; HMM-361; 34; Philadelphia, Penn.
Boggia, Richard M. Pfc. 2104902; H-2-4; 19; New York, New York
Bredeson, Dale O. SSgt. 1526842; HMM-361; 30; Des Moines, Iowa
Bronson, Paul W. 1stLt. 077326; HMM-361; 28; Los Angeles, California
Buchs, Christopher J. LCpl. 2107157; I-3-3; 20; Fraser, Michigan
Carter, Tommie L. Cpl. 1975128; H-2-4; 23; Richmond, Virginia
Casabar, Lorenzo Sgt. 1964591; H-2-4; 25; Pearl Harbor, Hawaii
Cederholm, Roger W. 1stLt. 088160; HMM-361; 28; Chicago, Illinois
Centers, Ronald L. Pfc. 2128537; I-3-3; 19; Cincinnati, Ohio
Chiles, Charles R. 1stLt. 090334; C-1-12; 24; Los Angeles, California
Clouse, Jimmie R. LCpl. 2060906; HMM-361; 20; Cincinnati, Ohio
Cochrane, William T. LCpl. 2114069; I-3-3; 20; Jacksonville, Florida
Comer, Andrew G. Maj. 057653; H&S-3-3; 36; Minneapolis, Minnesota
Cooney, Charles C. 1stLt. 088368; H-2-4; 23; Erie, Pennsylvania
Cortez, Joe A. Pfc. 2058283; H-2-4; 20; Denver, Colorado
Costa, James T. LCpl. 2080636; H&S-3-3; 19; San Francisco, California
Dalby, John D. Capt. 081155; I-3-3; 27; Houston, Texas
Denton, Charles L. Cpl. 2073195; C-3-Tanks; 20; Detroit, Michigan
Doub, Jay A. Capt. 060072; K-3-3; 35; South Bend, Indiana
Dorsett, Herschel D. 1stSgt. 305488; HMM-361; 43; Cincinnati, Ohio
Edgell, Joseph C. LCpl. 2050958; H-2-4; 20; Philadelphia, Pennsylvania
Ely, Richard E. LCpl. 2029522; HMM-361; 21; New York, New York
Emerick George A. Sgt. 1863759; I-3-3; 23; Chesterfield, Michigan
Enos, Charles R. Cpl. 2022517; I-3-3; 20; Chicago, Illinois
Etheridge, James E. Sgt. 1687992; I-3-3; 26; McAlester, Oklahoma
Field, Robert E. Pfc. 2030307; I-3-3; 20; New York, New York
Fink, Charles D. LCpl. 2114187; I-3-3; 19; Jacksonville, Florida
Fisher, Joseph R. LtCol. 045857; CO-2-4; 43; Westwood, Massachusetts
Ford, David A. Pfc. 1955109; I-3-3; 22; Omaha, Nebraska
Foster, Joe Jr. Pfc. 2124426; I-3-3; 19; Waycross, Georgia
Galloway, Edward R. LCpl. 2073318; H-2-4; 20; Detroit, Michigan
Gregg, Raymond E. Cpl. 2036628; I-3-3; 20; Kansas City, Missouri
Gregory, John G HN 5876334; I-3-3; 23; Portland, Oregon
Hall, George Sgt. 1413134; H-2-4; 30; Philadelphia, Pennsylvania
Hammett, Gary N. Pfc. 2098853; I-3-3; 19; Baltimore, Maryland
Hanley, Carlos R. Pfc. 1843587; H-2-4; 23; Owensboro, Kentucky
Harris, Donald L. Pfc. 2036393; H-2-4; 20; Buffalo, New York
Hayden, Farrest LCpl. 2062881; I-3-3; 20; Los Angeles, California
Hoeck, Richard J. LCpl. 1960766; H-2-4; 22; Wyoming, Minnesota
Hogstad, Edward B. LCpl. 2041689; H-2-4; 21; Los Angeles, California
Homer, Lee E. Pfc. 2107055; I-3-3; 19; Detroit, Michigan
Hooton, Richard J.J. 1stLt. 088161; HMM-363; 24; Pensacola, Florida
Hoppes, Michael L. LCpl. 2069918; I-3-3; 19; Indianapolis, Indiana
Horne, David L. Sgt. 1560601; I-3-3; 27; Unknown
Hults, Daniel E. Pfc. 2046360; H&S-3-3; 20; Seattle, Washington
Inks, Richard L. Cpl. 2057661; I-3-3; 20; Seattle, Washington
Irish, Michael W. Cpl. 2107286; I-3-3; 19; Detroit, Michigan
Johnson, Clifford J. Cpl. 2046329; I-3-3; 20; Seattle, Washington
Jones, Homer P. Maj. 067787; HMM-361; 32; Unknown
Jones, Ronald R. Cpl. 2058623; I-3-3; 21; Detroit, Michigan
Jones, Ronnie F. Cpl. 2062668; H-2-4; 21; Los Angeles, California
Kehres, James W. Pfc. 2122336; H-2-4; 17; Detroit, Michigan

Kendall, Stuart O. 1stLt. 085049; HMM-361; 26; Corvallis, Oregon
Kiger, Richard W. Cpl. 2042586; H-2-4; 20; Detroit, Michigan
Kolb, Bruce A. LCpl. 2078702; I-3-3; 19; Philadelphia, Pennsylvania
Kremer, James H. Cpl. 2097274; I-3-3; 19; Indianapolis, Indiana
Laizure, Virgil E. Cpl. 1876793; I-3-3; 24; San Francisco, California
Lamb, Allan W. Capt. 065948; RLT-7; 37; Ellensburg, Washington
Lane, Bobby R. SSgt. 1448517; I-3-3; 30; Macon, Georgia
Larock, Robert E. HN 6820504; I-3-3; 20; Ogdensburg, New York
Layman, Dwight R. Cpl. 2042429; I-3-3; 20; Detroit, Michigan
Lehrack, Otto J. Capt. 082134; I-3-3; 23; Flushing, New York
Lewis, Harold W. Sgt. 1884120; I-3-3; 23; St. Albans, New York
Lightfoot, Clarence W. Cpl. 2076602; I-3-3; 19; New York, New York
Machado, Joseph P. LCpl. 2049581; I-3-3; 20, Los Angeles, California
Manning, James F. Cpl. 2009426; H-2-4; 22; Albany, New York
Mansir, Paul W. Pfc. 2024820; H&S-3-3; 22; Fort Worth, Texas
Marino, Jack Jr. SSgt. 991615; A-1-Amphib; 41; Mobile, Alabama
Martin, Raymond B. Jr. GySgt. 624893; I-3-3; 43; Philadelphia, Pennsylvania
Martinez, Jimmy M. Cpl. 2049910; I-3-3; 20; Phoenix, Arizona
Matton, Gregory L. Cpl. 2042695; H&S-3-3; 20; Detroit, Michigan
McKnight, James J. Cpl. 2022985; I-3-3; 20; St. Paul, Minnesota
Miller, Larry T. Pfc. 2094783; H&S-3-3; 19; Dallas, Texas
Morris, Gary W. LCpl. 2054752; I-3-3; 19; El Paso, Texas
Morrison, Robert S. 1stLt. 087345; H-2-4; 23; Lake Forest, Illinois
Morrow, Dennis B. Pfc: 2108590; H-2-4; 19; Los Angeles, California
Muir, Joseph E. LtCol. 049816; CO-3-3; 37; Meadow Bridge, W.V.
Mulloy, James E. Sgt. 1288230; H&S-3-3; 30; Jeffersontown, Kentucky
Myatt, Ramsey D. 1stLt. 088389; HMM-361; 24; Oklahoma City, Okla
Noel, Robert L. SSgt. 1937369; I-3-3; 22; Richmond, Virginia
Noonan, Thomas P. Jr. LCpl. 2292900; G-2-9; 25; Maspeth, New York
O'Donnell, Phillip F. LCpl. 2050829; H-2-4; 21; Philadelphia, Pennsylvania
O'Malley, Robert E. Cpl. 1972161; I-3-3; 22; Brooklyn, New York
Overstreet, Coy D. SSgt. 1507995; HMM-361; 30; Dallas, Texas
Pacheco, Jose A. LCpl. 2049535; I-3-3; 22; Los Angeles, California
Partridge, James S. HN 5422714; I-3-3; 22; Seattle, Washington
Pasley, David C. LCpl. 2022548; I-3-3; 20; Chicago, Illinois
Pass, Richard S. LCpl. 2021983; H&S-3-3; 20; Chicago, Illinois
Paulson, Albert L. Sgt. 1443491; I-3-3; 30; Atlantic, Iowa
Petoskey, Regonel L. Pfc. 2097282; H&S-3-3; 19; Indianapolis, Indiana
Petty, Arthur O. 1stSgt. 305727; I-3-3; 42; Detroit, Michigan
Pinguet, Jean SSgt. 1316822; I-3-3; 31; Baltimore, Maryland
Pooley, Ronald J. Pfc. 2110002; H-2-4; 19; Minneapolis, Minnesota
Puhlick, Peter S. LCpl. 2016634; H-2-4; 21; Miami, Florida
Ramsey, David A. Capt. 070095; OO-3-3; 34; San Francisco, California
Reed, Franklin M. Sgt. 1492334; H-2-4; 29; Dayton, Tennessee
Reed, James P. Cpl. 2059737; I-3-3; 20; San Francisco, California
Reeff, James M. Pfc. 2108073; H&S-3-3; 19; Seattle, Washington
Renfrow, Jeffery R. Cpl. 1690036; H-2-4; 26; Chicago, Illinois
Reynolds, Paul R.LCpl. 2090099; I-3-3; 19; Baltimore, Maryland
Rhoden, Paul E. LCpl. 2060711; H-2-4; 20; St. Louis, Missouri
Rimpson, Robert L. Jr. Pfc. 2036627; I-3-3; 20; Kansas City, Missouri
Rinehart, Benny D. Capt. 072921; HMM-361; 29; Vandalia, Illinois
Rodriques, Jesse R. PFC. 2127768; H-2-4; 19; Los Angeles, California
Sanders, Walter L. 1stLt. 085068; HMM-361; 29; Norfolk, Virginia

Schmidt, James W. Pfc. 2129785; I-3-3; 19; Chicago, Illinois
Schwend, Howard L. 1stLt. 082552; I-3-3; 26; Salt Lake City, Utah
Slaughter, John E. Pfc. 2078141; H-2-4; 19; Philadelphia, Pennsylvania
Slowey, Kenneth L. 1stLt. 087648; HMM-363; 24; Dubuque, Iowa
Smith, Thomas J. LCpl. 2075007; H-2-4; 20; New York, New York
Spurrier, Ronnie Cpl. 1858630; H-2-4; 24; Louisville, Kentucky
Stanton, Edward R. II Pfc. 2122331; I-3-3; 19; Detroit, Michigan
Stipes, Robert L. Pfc. 2034556; H-2-4; 20; South Charleston, West Virginia
Stubbs, Reuben Jr. 2075062; H-2-4; 20; New York, New York
Taylor, Roger R. Pfc. 2099840; H-2-4; 19; Pittsburgh, Pennsylvania
Tonucci, Richard L. Cpl. 2011754; H-2-4; 21; Derby, Connecticut
Tramel, Charles T. Cpl. 2057927; I-3-3; 20; Dallas, Texas
Vaughn, Edward L. Cpl. 1946372; H-2-4; 23; Beaumont, Texas
Waddell, Lyle K. Pfc. 2080320; I-3-3; 19; Dallas, Texas
Wagner, Walter W. Sgt. 1958680; H-2-4; 22; Chicago, Illinois
Walker, Michael F. Cpl. 2043832; I-3-3; 20; Indianapolis, Indiana
Wallace, Ernie W. Cpl. 2034491 ; H-2-4; 22; Wayne, West Virginia
Walt, Lewis W. LtGen. 05436; CO-MFV; 49; Ft. Collins, Colorado
Watson, Robert B. Cpl. 2035642; H&S-3-3; 20; Oil City, Louisiana
Wireman, Charles W. Pfc. 2049724; I-3-3; 20; Los Angeles, California
Woodson, Ralph A. Jr. Pfc. 2078709; H-2-4; 20; Crystal, West Virginia
York, Roger D. HN 6871444; H&S-3-3; 20; Odessa, Texas

✷ ✷ ✷ ✷

In his fine memoir, *A Rumor of War*, Philip Caputo wrote: "In April 1966, One-Three suffered heavy casualties in an operation in the Vu Gia Valley. I had been transferred from the battalion to regimental HQ, where I was assigned as an assistant operations officer. There, I saw the incompetent staff work that had turned the operation into a minor disaster. Part of the battalion was needlessly sent into a trap, and one company alone lost over one hundred of its one hundred and eighty men. Vietnamese civilians suffered too. I recall seeing the smoke rising from a dozen bombed villages while our artillery pounded positions in the hills and our planes darted through the smoke to drop more bombs. And I recall seeing our own casualties at the division hospital. Captain Greer, the intelligence officer, and I were sent there from the field to interview the survivors and find out what had gone wrong. We knew what had gone wrong—the staff had fouled up—but we went along with the charade anyway. I can still see the charnel house, crammed with wounded, groaning men, their dressings encrusted in filth, the cots pushed one up against the other to make room for the new wounded coming in, the smell of blood, the stunned faces, one young platoon leader wrapped up like a mummy with plastic tubes inserted in his kidneys, and an eighteen-year-old private, blinded by shellfire, a bandage wrapped around his eyes as he groped down the aisle between the cots. "I can't find my rack," he called. "Can somebody help me find my rack?"

Caputo continues, "The men who fought in Vietnam at this time had joined the service in peacetime, before the death toll built up to a daily announcement. One-Three's losses between March and August of 1965 amounted to one hundred and ten killed and wounded, or ten percent. During this battle Bravo Company lost one hundred and eight men in only one hour. In this early period the Marines in this battalion had been together for years and were very close to one another. They assumed that they they would be together until the end of their enlistments."

The following story takes what happened during the first few days of April 1966, out of the darkness, and puts it where it belongs, as a part of the history of the Marine Corps in Vietnam. Until now Operation Orange has been treated as a shameful chapter in an unpopular war. That it has been denied a place in history weighs especially heavy on those who participated in the operation. It's as if the sacrifices made during that viscous fight had been rendered meaningless. I believe this to be the first recorded version of what happened on Operation Orange.

> "Women wear pants, Marines wear trousers"
>
> Staff Sergeant Marvin L. Paxton
> Senior Drill Instructor
> Platoon 363
> San Diego, California
> Late summer, 1960

The area of operations for Orange was such that the confluence of the Song Vu Gia River and its tributary, the Song Con River, effectively divided the area into three distinctly separate sectors, two to the north of the Vu Gia and one to the south. Because of the location of the air field and a proposed camp for the Special Forces, control of, and operations in each of the sectors was required. At least one rifle company would be needed in each sector.

The airfield was located to the north of the Vu Gia and it was deemed necessary that the assault begin there. At 0700 on April 1, 1966 Delta Company, First Battalion, Third Marines and Delta Company, First Battalion, First Marines landed there. During the initial assault events progressed as anticipated and the two companies that landed there quickly secured the two northern sectors. These sectors included the airfield and the area adjacent to it. At that time one company had been retained as reserve should that requirement be needed.

On 2 April, Bravo Company of the First Battalion, Third Marine Regiment, the reserve company, landed to sweep the sector south of the river. It was here that the only organized enemy force of any strength was encountered. Based on available intelligence, the strength of this force was considerably larger than had been anticipated, and the size of the force employed was insufficient. They could not rapidly overcome the enemy resistance and there was no reserve force available to help.

Bravo Company had walked into a cleverly designed trap and would soon be decimated.

In compiling this story information was gleaned from interviews with the participants, the awards citations of many of the participants, the rosters and diaries of the participating units, the command chronology of the First Battalion, Third Marines for April 1966, and information obtained from the National Personnel Records Center.

2

Operation Orange

The First Battalion, Third Marines was under the leadership of Lieutenant Colonel Robert R. Dickey, III. The Operations Officer for the battalion was Major John J. Tolnay who supervised the preparation for Operation Orange. Major Tolnay oversaw the placement of personnel and their supporting elements for the operation which began on the morning of April 1, 1966 when Delta Company made a helicopter landing on the north side of the Vu Gia River.

Bravo Company, the reserve company, had moved from its position on Iron Bridge Ridge to the Battalion rear the day before, and prepared to land on the second of April. They were to conduct search and destroy operations on the south bank of the Song Vu Gia to the southwest.

First Lieutenant Willis Wilson, the commander of the third platoon, remembers that, "The battalion had been part of a defensive perimeter around Danang since we landed there on a troop ship from Okinawa. We had the northern segment. Having settled in and operated among the locals for about five months we were selected to move out of our position and become part of the assaulting force in Operation Orange. We were temporarily replaced by another unit so we could move out. The night before the operation we moved to Battalion Headquarters and were briefed on the operation and given our orders. Though we had an opportunity to sleep in relative comfort in the big tents, Staff Sergeant Thurman Owen and I decided we would sleep in the field with our troops. It was a relaxing environment and we didn't get to sleep as early as we probably should have."

On that morning, Bravo Company ate steak and eggs for breakfast before moving to the landing zone where the choppers were to pick them up. The first platoon, under First Lieutenant Dennis Peterson, was the assault platoon and would be the first to land in the H-34s. They would be followed by the second and third platoons, respectively in CH-36s.

Wilson recalls, "The next morning - the morning of the operation - we had a good breakfast and boarded helicopters to be carried to the combat area."

The weather during the operation was hot and humid with temperatures up to 105 degrees during the day to 75 degrees at night with occasional thunderstorms at night.

Lance Corporal John Hagler remembers, "On the morning of April 2, 1966 reveille was held at 0430. After chow we were issued additional ammo and grenades. We were also given a one hundred round box of belted machinegun ammo, and one 60mm mortar round. Some Marines were also given LAAWS or rifle grenades to carry. With our packs, weapons, and cartridge belts this brought our gear weight to over 70 pounds. Of course we bitched about the extra weight having no idea how valuable that extra gear would be in the next two hours. Had we known we would have gladly carried double.

I remember that morning being very hot. In fact it was so hot that the lift capability of the choppers, especially the 34s, was diminished. We were lined up in sticks of six or eight Marines but, with all our gear, and because of the heat, the 34s were not able to lift off carrying that many Marines. My fire team, which was the first team of the first squad, was put aboard a 36 with a squad from the second platoon.'

"The flight to the landing zone was very quite except for the whop of the rotor blades of the chopper. As I looked at my fellow passengers I could see that each Marine was very much into himself. All were firmly gripping their weapons; most were staring at the deck. Some were smoking, some were praying, their lips visibly moving. When we crossed the Line of Departure, the word was passed to lock and load. It was then that adrenaline started running through my veins.'

"Our landing zone was a sandy area along the south bank of the Vu Gia River. We approached the landing zone from a high altitude and when we were positioned over it we started a corkscrew decent. I could see the landing zone out of the starboard hatch of the chopper. There were tracers crisscrossing it. From across the river I could see softball size tracers arcing up towards our chopper. The landing was hot.'

"I sat on a cloth bench on the port side of the chopper. The fear, the anxiety, the adrenaline rush were overwhelmed by a feeling of dread and claustrophobia. We were trapped on this chopper. We were being shot at and there was nothing we could do. The only place in the world I wanted to be was off of that chopper. Should I sit on my flak jacket and helmet? It wouldn't even slow down a bullet. There was nothing, absolutely nothing that I could do. It was the most helpless feeling in the world.'

"We got off of the chopper on the perimeter of the landing zone and dropped into the prone position among the second platoon. I wanted to get my fire team back with the first platoon but I didn't know where around the perimeter the first platoon was. I saw Sergeant David D. Patten of the third platoon standing and directing the fire of his squad. The firing slacked, then ceased. I found my platoon and joined my squad."

They landed at about 0720 and the second platoon, led by First Lieutenant Kenneth E. Martin, attacked west along the river. The third platoon, under First Lieutenant Wilson, attacked south and inland, and were followed by company headquarters, and then the first platoon.

The third platoon made contact soon after the landing and the firefight lasted several minutes. They had run into resistance in the hamlet of Ngoc Kinh (1).

"There was no enemy fire at the landing zone," Wilson recalls. "We assembled in squads, went over the orders, and moved out toward the objective. I had the squads spread out so we could cover each other in a 'fire and manouver' sequence. Very soon we encountered some enemy rifle fire. The center squad returned fire riddling the bodies of a couple of enemy riflemen. I had the cautious feeling that these were sacrificial scouts positioned to warn the main force of our movement."

In these hamlets the enemy was much stronger than anticipated and in addition to the Q17 Local Force Company, with a strength estimated at sixty to seventy men, was two companies of the 321st Viet Cong Battalion. This force occupied positions in the hamlets that provided excellent cover and concealment. These positions were heavily entrenched with a network of obstacles consisting of booby traps and barbed wire. Their weapons included small arms, automatic weapons and 60mm mortars.

Wilson continues, "I remember one of the platoon's men being wounded badly enough that he could not proceed with us. We radioed company headquarters they should schedule an evacuation for him. The company first sergeant wanted to know the service number of the wounded man. We were not to use names because sometimes the enemy had captured this information and contacted next of kin back home to demoralize them. I didn't want to take the time to dig my platoon leader's notebook out of my pack and look up the information. The man had a very unique first name so I told the radio man to give just the first name, and we moved out toward the first rallying point."

Several walking wounded were sent to the rear to be medevaced. Among them was the company First Sergeant, James R. Berheide, who had been shot in the arm. He walked along holding his bandaged arm complaining, "I've been through World War II and Korea and it took these little bastards to get me."

When the Company Command post was pinned down by heavy enemy fire, Berheide had deliberately exposed himself to gain a better observation site from which to locate the enemy positions and had been hit in the left arm. Despite his wound, he ensured that the administrative chief was briefed upon assuming the duties of First Sergeant.

The second platoon proceeded through Tam Hiep towards Ngoc Kinh (2). They encountered no resistance in Tam Hiep and actually crossed 500 meters of open rice paddies to reconnoiter the near edge of Ngoc Kinh (2) with one squad. When they were within 30 to 50 meters from the hamlet the squad received heavy automatic weapons fire from the Viet Cong entrenched behind the wood line. In front of their position were barbed wire and bamboo fence obstacles. The platoons other two squads who were waiting in Tam Hiep attempted to envelop Ngoc Kinh (2) from the opposite flank but sustained heavy casualties from effective sniper and heavy automatic weapons fire.

When the second platoon came under heavy fire from automatic weapons and mortars, Martin was wounded in the hand in the initial burst of fire, but remained in command and kept the platoon moving. When they were pinned down behind a small dike in the open field, he continually exposed himself to call in effective close air support and was able to reorganize his platoon and see to the evacuation of its wounded.

One of Martin's men, Private First Class Claude G. Lebas, a rifleman, spotted the enemy mortar position and withdrew to the rear where he picked up three assault weapons. He then returned alone through an open area that was covered by heavy fire from both directions. Using these weapons he destroyed the Viet Cong mortar position just as it was adjusting fire on the company command group. When Lebas returned to his platoon, he discovered that a machine gunner had been killed and that he and the weapon were one hundred meters forward of their secured position. He shouted to his fellow Marines, and asked for covering fire as he rushed out to recover the machine gun. Enemy fire was so intense that his covering force was driven back to a more secure area. Undaunted, Lebas grabbed the gun and returned to safety amid a hail of bullets. It was later discovered that two enemy rounds were imbedded in his haversack, stopped only by the rations he was carrying.

Private First Class Dale E. Lawson, another rifleman with the second platoon, moved to a wounded Marine who was down in front of the platoon lines. He suffered a painful gunshot wound to his left hand while doing so, but was still able to administer first aid to the casualty, and was able to move him to a position of relative safety. Lawson realized that he would be unable to return to his original position because of the murderous incoming fire so he remained with the wounded man. He soon sustained a second and more serious wound when he was shot in the left side of the abdomen. Despite his wounds, Lawson maintained his position for seven hours while providing covering fire that prevented the Viet Cong from advancing in his area.

Another Marine whose actions stood out on this day was Corporal Manuel Avalos, Jr., a fire team leader with the second platoon, pinned down in that open field. On four separate occasions he advanced through the murderous Viet Cong fire and brought back wounded Marines to an area where they could receive medical attention. When both his squad leader and another fire team leader were wounded, Avalos assumed command of the squad, and with keen judgment and daring initiative maneuvered the fire teams in a manner which allowed his squad to be the only Marine unit to reach a tree line surrounding the village. From this position they were able to direct extremely effective fire on the deeply entrenched enemy.

Lance Corporal Alexander J. Menzies, an Automatic Rifleman, realized that one of the machine guns was running low of ammunition and crawled through twenty meters of withering enemy small arms fire to obtain the needed ammunition and then returned with it. The gun continued to fire and when it required a change of barrels Menzies again crawled through enemy fire to obtain the barrel and bring it back. When his squad leader was hit and went down in front of the squad's position, he crawled to him and pulled him to safety. This time, when Menzies returned to the machine gun position he was fataly wounded by enemy fire.

During the heat of the action Hospitalman Third Class David C. Boos dashed into the withering fire to aid Marine casualties. While attending to a stricken man, he was seriously wounded. He ignored his own suffering and stalwartly lay in the blazing sun for three hours, all the while instructing others on the proper care of the wounded.

Another corpsman, Irwin Gordon, a husky Canadian, moved into the rice paddy and treated many Marines who were already wounded. As he worked on them under the blazing sun, four more were also hit and he moved to them, in turn, and administered first aid.

Serving as a messenger with the second platoon was Lance Corporal Gerold L. McDougall. He quickly realized that first aid was needed for the numerous casualties. McDougall took the medical pack from Boos, and following his instructions, braved the withering enemy fire to treat the stricken men. This done, he then exposed himself to the violent action as he returned to the disabled Boos and pulled him across more than two hundred meters of fire swept terrain to a position of relative safety.

Hagler recalls, "Small arms fire, automatic weapons and machineguns, mortar rounds and grenades were going off. I could hear the distinct thunk, thunk, thunk, of a 50 caliber machinegun."

The embattled second platoon soon requested assistance and at 0902 the third platoon was ordered to proceed immediately to support them. They moved out but soon encountered sniper fire and a water obstacle that halted their advance and delayed any assistance that they could provide for their embattled fellow Marines. These hamlets had not been prepped by artillery initially, because a policy existed that would not allow preparatory fire on a hamlet unless hostile fire had been received from it. Artillery fire could not now be used because both of the platoons involved were too close to the enemy positions.

At 0945 an air strike was requested on Ngoc Kinh (2) so that the second platoon could move it's dead and wounded back and permit the employment of supporting arms on the hamlet. They marked their front lines with air panels and the Artillery Observer delivered smoke rockets as a further identification of the strike area for the attacking aircraft.

Also at 0945, the first platoon and the command post group continued to move southwest to Ngoc Kinh (1) toward Lap Thuan through rice fields and would occasionally receive sniper fire. On these occasions they would hit the deck.

Lance Corporal Roy F. Parr, one of a two man combat engineer team attached to the first platoon, remembers taking cover behind a dirt berm as enemy rounds kicked up dirt all around them. "Lance Corporal Jim Skemanski, the other member of our team, a slightly built Marine who wore glasses, lay close to me and all that I could see of him were his helmet and eyeglasses. He says, 'Just like in the movies.' At that moment I could have killed him."

Hagler recalls, "We were spread out along a village trail in the prone position. There was a vicious battle raging to our front and right flank. I could hear the sound of something dropping to my rear. I heard it again and then again. I looked back near my feet and saw the tail fins of three 60 mm rounds sticking up out of the dirt trail. One round was just a foot from my boot; the other two were within a radius of five feet. I was stunned. Had any of the three rounds gone off I would have been flayed. To this day I wonder how three mortar rounds, which are not that accurate, could land in such close proximity to each other? But, even more so, I wonder, what are the chances that all three would be duds?"

The airstrike delivered on Ngoc Kinh (2), particularly the 20mm strafing runs, stopped the enemy fire long enough for the second platoon to move its casualties toward Tam Hiep. Every time the aircraft approached the target, the Marines from the second platoon would move toward Tam Hiep carrying their wounded with them. This system of removing the wounded to a place of relative safety continued until early evening before all of the wounded were moved to Tam Hiep.

The second platoon and the Bravo Company command group came on line south of the third platoon at 1135. The third platoon continued from the western section of Ngoc Kinh (1) towards the hamlet Lap Thuan. The command group moved to the high ground so that they would be able to observe and control the battle.

At 1135 the third platoon came under intense enemy mortar fire, close range small arms fire, and barbed wire obstacles covered by automatic weapons. This prevented the forward movement of the platoon.

"After emerging from a small bit of woods," Wilson recalls, "we spread out and moved across an open field toward our next obstacle, a barbed wire fence. The squad I was with in the center, one squad was in the river bank, and the other squad was in reserve to the rear. The barbed wire fence is where the enemy hit us with small arms fire and explosives - either mortars or mines. I wasn't sure which. My response was to ask for small arms support from the squad on the right and get us though that fence."

Private First Class Anthony K. Kelichner remembered that, "As we cleared the village and entered a series of rice paddies, mortar rounds started exploding around us. My squad started running toward the nearest rice paddy dike. I was the automatic rifleman in our fire team and I had rifle magazines criss-crossing my upper body with hand grenades hanging all over my pack harness. I was also carrying extra bandoliers of M-79 rounds and an extra can of M-60 machine gun ammo. I was really loaded down."

"My squad was advancing on line on the right flank walking on the shore of the river which was approximately eight to ten feet below the plane of the battlefield," recalls Private First Class Daniel R. Kaylor. "When the mortars started we took cover against the river bank. The rest of the third platoon was on line crossing the open field and moving toward the tree line. They were led by Sergeant Owen and Lieutenant Wilson."

Wilson courageously moved up and down his platoon's positions, directing his men and judiciously ordering the deployed squads to bypass the barbered wire in an attempt to eliminate the Viet Cong threat.

When his platoon sergeant, Staff Sergeant Owen, suffered multiple fragmentation wounds and became entangled in the barbed wire, Wilson attempted to cross seventy five meters of open terrain in an effort to retrieve the sergeant, who had been mortally wounded. The body of Owen was still being hit by small arms fire. Before, Wilson could reach the sergeants body; he was caught in a hail of small arms fire and pinned down.

Wilson recalls that Owen was not hung up on the barbed wire. "We were close to a barbed wire fence when the explosions hit and he was killed instantly. I pulled him back ... out of the line of fire I hoped. I took his shotgun and fired toward the enemy and tried to keep them back so we could regroup and secure an area so we could evacuate wounded Marines."

No longer able to help Owen, Wilson resumed direction of the platoon and was able to establish a strong base of fire to provide cover for a deployed squad. When a radio operator was wounded in another sudden flurry of Viet Cong fire, Lieutenant Wilson again braved the withering fire to assist the downed Marine. He was then hit by enemy fire sustaining a serious gunshot wound to the upper right chest.

"As I moved up and down the line," Wilson remembers, "I felt a punch in my stomach and assumed I was hit there in the gut. When I looked down I saw no wound. Then I saw a small, clean hole in my right chest. That is where the bullet entered. The feeling in my gut must have resulted from the shock wave. Later I recalled that I should have tried to tape my Geneva Convention Card over the hole ... the way we had been taught to deal with a 'sucking chest wound.'" But, Wilson was not done yet. With unbelievable determination and presence of mind, he continued to maintain direction of the platoon. Increasing the rate of fire from his base squads, he launched an attack by the enveloping squad which silenced the enemy fire. "I kept moving and trying to secure the area until I started to feel dizzy. Then I yelled to the radio operator to radio to the company headquarters that one of the squad leaders was in command of the platoon."

Kelichner continues, "When we took off running, I tripped and fell flat on my face in the paddy. I lay there with bullets buzzing past my head and impacting all around me. Mortar rounds were exploding all around us. I raised my head and saw that I was the only Marine still out there. I jumped up, grabbed all of my gear, and started running again. I ran about ten feet and got "smacked" by something that sent me flying. I soon realized that I was again lying flat on my face in the muck and mud with bullets still impacting around me and mortar rounds exploding all around. I started to rise and began to gather all my gear when I realized that another Marine and a Navy corpsman had come to my aid. With one on either side of me they got me on my feet. They grabbed my gear and we started running. Bullets were flying everywhere. As we neared the end of the rice paddy and approached the paddy dike wall we came upon a hedgerow that was about four foot tall. We didn't slow down and dove right through the bushes and landed on the other side, protected from the enemy fire and on drier ground."

"The corpsman examined me and told me that I was shot in the shoulder. He put a pressure bandage on to stop the bleeding. There were three or four other wounded on the other side of the bushes with me including a Lieutenant with a "sucking chest wound", and he wan't doing very well.. We were medevaced out in a CH-46 chopper and when the chopper landed we boarded. They were still shooting at the helicopter when we took off."

Kaylor continues, "Across the river there was a Viet Cong soldier shadowing us and taking shots at us every so often, although we were not allowed to return fire on him. He had me in his sights and was landing hits all around me when Corporal Michael E. Cirrilo, our squad leader, led us into the area where Lieutenant Wilson was lying wounded. It was in a depression below the line of fire. Sergeant Owen was right there in front of of us hanging entangled in the barbed wire fence. We were returning fire into the tree line when another Lieutenant, I don't know who he was, came over and assessed Wilson's condition. He told us that we had to get him to the medevac area or he would die right there. About that time another Marine from my squad made a dash for our position when his right shoulder exploded from an enemy round. He collapsed in our midst and was taken to the evacuation area along with Lieutenant Wilson."

Wilson remembers that, "I kept fading in and out of consciousness. But I do recall someone nearby on a radio bringing in support. There were a few men from the platoon who were preparing to lay me down and carry me to a helicopter. I asked them to carry me sitting up so I could breathe. Lying down pretty much shut down my breathing. They did very well, and I was soon being lifted off in the chopper. I don't think there were any more Marines aboard besides the crew."

With Owen dead and Wilson wounded, and unable to continue, Corporal Gwinn A. Henry, Jr., a fire team leader, took over the third platoon. In doing so he exhibited outstanding leadership as he directed the fire of his Marines, set up security and kept a constant check on his personnel and equipment.

Sergeant David D. Patten, a squad leader, with the third platoon was leading his squad near the village of Lap Thuan when one of his fire teams was pinned down by a heavy volume of fire from Viet Cong automatic weapons. Without hesitation he moved to the team's position and with enemy rounds kicking up dirt all around him, stood beside his prone comrades to gain better observation and fire his M-79 grenade launcher into the enemy trench lines. Patten's fire was so effective that the Viet Cong action subsided and they were able to overrun and eliminate the enemy position.

They continued on their original mission and again made contact with the Viet Cong. The point rifleman was shot down and became the target of concentrated enemy fire. Patten raced across about twenty meters of open ground to reach the wounded Marine where he quickly administered first aid and pulled the Marine to safety. This undoubtedly saved the man from receiving additional wounds.

As Patten moved to rescue the wounded pointman, Lance Corporal Daniel J. Piotrowski quickly moved to an exposed position and set in with his automatic rifle and began to lay down a base of covering fire for the two exposed Marines. Even though taken under fire himself he remained at his position and continued to provide protective fire until they reached safety. In doing so he was shot and killed.

Private First Class Webster Bernhardt recalls, "I was right next to Piotrowski when he was killed by an AK-47 head shot."

At this time the first platoon was ordered to reinforce them.

"We moved up the trail a short distance to where the third platoon was engaged." Hagler remembers. "Along the trail were dead and wounded Marines. The fire was heavy. I saw Lance Corporal Gregory C. Love off the trail standing up being treated by a corpsman. His utilities had been cut off of him and he was covered from head to toe in his own blood."

Love had been hit in the left arm and left leg by shell fragments.

The third platoon continued to advance by fire and maneuver until they came to a hedgerow where they were subjected to heavy mortar and automatic weapons fire. They suffered heavy casualties and were forced to withdraw to the east and south.

Hagler describes the situation, "The trail ran into a dried up rice paddy and, this is where the third platoons advance had been stopped. Thirty yards to the front of the paddy was a barbed wire fence. Beyond the fence was a wooded area from which the Viet Cong were firing from bunkers and a trench line. To the right, or north, the paddy sloped off and down to a small stream."

'The Viet Cong were covering this approach with a machinegun. To the left, or south, the paddy opened up into a large field of dried up rice paddies that extended about 300 yards to a range of foothills. At the base of these foothills was the company headquarters."

'The first platoon moved, under fire, into the paddy along the north side, where it sloped down to the stream, and along the barbed wire fence, where the incoming fire was merciless. Like the third platoon, the first platoons advance was stopped. We hugged the deck and returned fire, but could not suppress the heavy volume of incoming fire."

The first platoon provided a base of fire and the 4.2's, 81's and 105's were used to quiet the fire coming from around Lap Thuan. The two embattled platoons set up a defense at the western edge of Ngoc Kinh (1). Enemy sniper and mortar fire from Lap Thuan continued until darkness on these positions.

"We got as many of the dead and wounded out of the field of fire and back down the trail as possible." Hagler continued. "The body of Staff Sergeant Owen, the platoon sergeant of the third platoon, was hung-up on the barbed wire. Several attempts were made to retrieve him, but to no avail."

Lance Corporal Milton Pittman of the first platoon recalls, "When we came in the third platoon was taking heat, it was ugly. By the time we fought our way to help, Owen was already on the wire. He was hung on the wire gate that led to the river. They tied a rope around my waist and Diez and I crawled out to get him, but he was already gone. We cut him down, he was cold and clammy, rigor had set in, so he was still. I took the rope off me and tied it around him, so they drug him in."

Private Dennis R. Hooyman, of the first platoon, courageously went to the aid of several fellow Marines who were wounded and still exposed to the withering enemy fire. He had already been painfully wounded in the left hand and the left side of his face by shell fragments, but he refused medical attention and moved from one casualty to the next, helping four of his wounded comrades to an area of relative safety.

John Hagler continues, "We were pinned down in the open. Air strikes were called in on the enemy positions to our front. The F-4 Phantoms come in so close and low that I could make out the design on the pilots helmets. The shrapnel from the bombs ripped through the air above our heads and we could feel the intense heat of the napalm."

"After the air strikes, artillery was called in, 'danger close'. The big 155 mm shells impacted into the wooded area and shrapnel tore into the hard earth of the paddy. Luckily no Marines were hit but we were still receiving enemy fire. HM3 Irwin "Butch" Gordon, one of the first platoon corpsmen, was everywhere braving the heavy enemy fire, as he moved from one wounded Marine to the next giving care and comfort."

"One of the most unusual incidents of the day occurred about this time. A machine gunner shouted, 'Ammo up!' A Marine from the gun team, carrying two cans of of gun ammo, one in each hand, broke from cover and dashed to the gun position. He had not taken five steps when he was shot through the calf muscle and went flying 'ass over teakettle'. Everyone who saw this, including me, broke out in hysterical laughter, like it was the funniest thing we had ever seen. Tears were rolling down my cheeks, I could not stop laughing. This went on for several minutes before we could regain control of ourselves."

Corporal Gary D. Nail, the Company driver, was responsible for the evacuation of four wounded Marines to the safety area. Nail, accompanied by two other Marines, crossed nearly three hundred meters of terrain that was exposed to enemy fire, and carried one of the casualties back to the secured area. This act was repeated three more times, each instance drawing fire from the enemy. Later, when the headquarters group came under the accurate fire of an enemy sniper, Nail exposed himself completely, and drew fire from the sniper who was located by the muzzle blast. He then directed effective return fire that silenced the sniper.

The walking wounded and litter cases from the two platoons were moved to a landing zone for medical evacuation. The bodies of those killed in action could not be recovered until after darkness because the Viet Cong had centered their fire on the location of the bodies. That night Gwinn Henry led the third platoon back into the battlefield, which was still under mortar attack, to retrieve the bodies of the fallen.

Hagler recalls that, "Lance Corporal Hugh G. Ventus, from the first squad, was overcome by the extreme heat and was in danger of dying, and a decision was made to try and get him over to the Company Headquarters because that would be the quickest way to get him evacuated. Sergeant Donald R. Burton, my squad leader, called me over to his position and told me the situation and plan. I was to take my fire team and carry Ventus on a poncho across to Company Headquarters. We carried him, staying as close to the deck as possible. We would move a few yards then hit the deck. When we arrived, Corporal Gary D. Nail, the Company radio operator, informed me that Sergeant Burton had been killed. Refusing to believe it, I said, 'No he hasn't, I just left him.' Nail just looked at me."

Pittman remembers what happened to Burton. "He was hit behind the left ear when he stood up to evacuate the perimeter and set us up to retrieve Sergeant Owen's body off the wire. He fell forward and dropped the baby pictures that he had received from home. His child had just been born. He was a proud father. One sad thing after another, the day before he was hit he had written a letter home, which meant that the final letter from him to his wife would come after the telegram informing them, he was gone."

Hagler recalls, "With Burton dead I was now the squad leader and I wanted to get back to my squad and platoon as soon as possible. Before I could take my fire team back to the first platoon Captain Richard V. Hunt, the Company Commander, told me he needed my fire team to stay at the

Company Headquarters to increase security. They had been coming under increasingly heavier fire."

'My fire team formed a make shift perimeter around a paddy that was being used as a landing zone for resupply and medivac choppers. Every helicopter that came in received heavy fire. Enemy snipers were taking shots at anyone that moved and Private Randolph Redmond was shot in the stomach."

"Redmond was wounded, but that was to be expected, for he was clumsy and would not listen." Pittman added.

Helicopters arrived and evacuated the wounded. Several squadrons were involved. Among them were Marine Medium Helicopter Squadrons 163, 164, 263, and 361. The gunships of Marine Observation Squadron 2 also provided support. They were drawing a heavy volume of fire from Lap Thuan. One helicopter was unable to lift off and was abandoned by the crew.

"The emergency medivac chopper that was called in for him was hit by heavy ground fire and crash landed in the landing zone." Hagler continues, "Another helicopter, which had landed just before the downed one, quickly picked up the air crew. The severely wounded Redmond could not be not be evacuated until the next day."

First Lieutenant Harry R. Ramsburg, a pilot with HMM-263, adds, "I was the pilot of the aircraft that landed in the zone and was burned that night. The one they burned had Snoopy painted on the door. We were on a med-evac and I was flying wing on Major Jerry Kettering. They were shooting at him as we entered the landing zone, but not leading him enough, and hit our bird and knocked out the engine so we left it. We thought we were on fire but they just hit an oil line. Ken Reed was my co-pilot but I can't remember the rest of the crew. We ran to Kettering's plane and Doc Rundle, the squadron surgeon, was flying on it. The zone was a company CP -- it was very hot -- so we planned to retrieve the bird the next day."

Captain Terry L. Wilcox, another pilot with HMM-263, wrote, "All of the H-34's I flew got us back. Some were so badly damaged they were put in the junk pile and scavenged for parts. None of my crew was ever hurt. A copilot was hit in the leg. When he grabbed his leg the slug fell to the deck. It left one big black and blue bruise. I had medevacs re-wounded, photographers and journalists wounded, but no crew members."

First Lieutenant Norman J. Mahalick, flying with VMO-2 recalls, "Norm Ehlert and I were "short-timers" and due to rotate home on April 30th, so we were assigned as the medevac ship's escort gunship for the regular 24-hour assignment that day and "Wimpy" Norton was the assigned medevac pilot for the same 24 hour period. That night the medevac aircraft was scrambled and we chased Wimpy to heavy fighting and a very hot landing zone. There were mortars exchanging shots from both sides, and a myriad of friendly and enemy tracers flying back and forth.

Wimpy turned all his exterior lights off and was successful in getting into the zone while receiving heavy ground fire. Norm and I attacked what we could confirm as the enemy. Wimpy got airborne and informed Norm and I that they had too many wounded for the one aircraft to fly out and we were needed to go in and pickup as many of the remaining wounded as we could.

We had Wimpy turn his lights back on at 1,000 feet for proper separation and turned our lights off as we started our approach. We were receiving heavy fire from several directions, but mostly from the west and figured they could see our red instrument lights and turned them down to the extremely dim position.

Upon landing we were immediately surrounded by Marines who assumed the prone position and fired at the nearby enemy. The wounded were loaded and as we were lifting off a mortar round hit and exploded near the front side of the aircraft. I felt the concussion as it exploded and threw mud and shrapnel across our windshield.

As we proceeded to the new Naval Hospital with our lights still out we passed between two large shadows. We had actually flown through a flight of H-34's that had been scrambled and also had their lights off.

The next day I went for a swim and some sun and out come's my good friend and future roommate, Dave Owens. As it turned out it was Dave's formation that we nearly collided with."

At approximately 2130, using the cover of darkness, the first platoon moved to the command post. They carried with them the dead and wounded from both platoons. They set in on Bravo Company's western flank. The snipers assigned to them from the headquarters company proved to be very effective.

Hagler recounts this event. "The day moved on. The first, second and third platoons were heavily engaged, and air strikes and artillery continued to pound the enemy positions. At the same time, our causalities continued to mount. There was smoke and carnage everywhere. At dusk my fire team moved to a position just up the hill in order to protect the rear of the Company Headquarters."

All the while, Captain Hunt continually exposed himself to the intense enemy fire in order to evaluate the situation, issue appropriate orders and inspire his Marines despite their having sustained heavy casualties. With most of the communications destroyed by the Viet Cong fire he was able to maintain contact with elements of the company by using runners or by personal contact as he moved from one critical part of the battlefield to another. By doing this he was able to maintain the momentum of his unit's assault.

'We dug in for the night in a rocky area, which made for lousy protection." Hagler recalls. "It was difficult to dig in the rocky soil, but we did the best we could. I made sure that my team cleaned their weapons and, we did so, one at a time. I scrounged ammo, grenades, and anything else I could find from a stockpile of munitions taken off of the already evacuated dead and wounded. I gathered as much as I could and made sure that this ammunition was distributed among the members of my team."

'I returned to my position and placed my grenades so they could be easily reached. I checked my magazines and made sure that they were fully loaded and clear of dirt and debris. I broke down my rifle into the barrel and receiver group, firing mechanism, and stock and cleaned these parts with an oily cloth and a small paintbrush. This done, I took a little container of lubreplate and smeared the greasy contents on every place on my rifle where metal rubbed against metal. I reassembled the rifle, inserted a magazine, chambered a round and clicked on the safety. I fully believed that we were all going to die that night and I wanted the enemy to pay a dear price."

"When I realized that I hadn't eaten, I took out a can of beefsteak and potatoes and opened it. I took a bite and threw up. The beefsteak and potatoes tasted like rotting flesh. Somehow I had not realized that the putrid smell of decomposing bodies permeated the air. How could I have not noticed that smell?"

Despite their concerns there was very little enemy activity during the evening of April 2 and the morning of April 3, 1966. At about 0355 the downed helicopter was destroyed by fire and numerous explosions from an unknown cause. Two Marines were wounded and one was killed. The survivors were pulled from the helicopter by the ever present corpsman, Gordon. Other Marines were there to help carry them to safety. The burns on the wounded were severe. The flames were so intense and hot that no one was able to get near the helicopter.

To Hagler the memory is vivid. "Sometime after midnight there was an explosion to my rear and the night sky was illuminated by fire. I heard screams and my name was called telling me to get my fire team down there. I rushed my fire team down into the company area. The downed chopper was blazing. I did not know what happened or why. I did not know if it was due to enemy action or happenstance, but the helicopter had exploded and burned, killing Nail and badly burning another Marine."

One of the badly burned Marines was Corporal Larry M. Anderson, the company radio operator. He had been burned on both legs, his face, neck and both hands.

Pittman remembers, "Nail left the hill to climb into that downed chopper to get out of the rain. I saw the blast but there was no way to reach it."

The pilot of the stricken craft, Ramsburg, speculates, "We heard that the Marines who were inside the downed bird tried to heat up their C-rations which caused the explosion. The aircraft was full of holes and oil and gas and the resultant fumes were everywhere."

Nail's body was removed from the helicopter at 0630 and by 0900 all remaining wounded and the bodies of those killed were evacuated by helicopter.

Hagler remembers, "My team went back to our position and waited for the final assault, which never came. For some unknown reason, the enemy had passed up a chance destroy a badly crippled Company Headquarters. Maybe we had hurt them worse then they had hurt us."

'That night, what was left of the first, second and third platoons were able extract themselves from the terrible situation of that day. They carried out all their causalities."

Pittman recalls, "Williams's body was bloated so badly that it was unrecognizable. Richards, the crazy Indian, was hit twice and the list goes on."

Included on this list are the following Marines who received gunshot wounds. Corporal Raymond C. Allen through the right elbow, Sergeant Curtis Baggett through the right side of his chest, Corporal David Bell in the left shoulder, Private First Class Michael Bledsoe through the lower left leg, Lance Corporal Alvin Bryant to the right arm, Lance Corporal Raymond Buckalew through the left leg, Private First Class James Burkes to the left thigh, Sergeant Robert Colbert in the head and right arm, Sergeant Floyd Darrh to the right leg, Private Lee Davis in the left thigh, Private Ysidore Garza to the chin, Lance Corporal Larry Hunter in the left thigh, Private William Hutson to the right side of the chest, Private Anthony Kelichner in the left shoulder,Corporal Timothy Kendall in the left leg and right buttocks, Corporal Everette Kratzer in the left leg, Lance Corporal Felipe Longoria was in the right hip and groin, Staff Sergeant James Martin in the right buttocks, First Lieutenant Kenneth Martin through the right wrist, Private Johnny Militello in the left hand, Corporal Warren Millidge in the left thigh, Sergeant Joseph Mitch through the left forearm, Lance Corporal John Paratore in the chest, Sergeant David Platt to the right forearm, Lance Corporal Duane Pounder in the left forearm, Private Paul Richards in the right hand, Lance Corporal Robert Salas in the neck, right arm and right shoulder, Private First Class Fernand Szabo in the right shoulder, Lance Corporal Thomas Williams through the right hand, Lance Corporal Roy Wilson a superficial wound to the chest, and Private First Class Carroll Young in the right thigh.

Others were wounded by shrapnel. Corporal Barney Cushing in the back, Lance Corporal Ronald Glover in the chest and right leg, Corporal Richard Kinnard with superficial wounds to the head, Sergeant Howard Perry in the chest, both arms and both legs, Lance Corporal Oliver Renfro in both arms, left side of the abdomen and chest, and fractures of both legs, Lance Corporal Anthony Santiago in the left foot and right ear, Corporal

Thomas Strobel in the head and left hand, and Lance Corporal Bobby Wilson in the right leg.

At 1400, Bravo Company, without its second platoon moved east about 600 meters and set up a defensive position.

John Hagler continues, "The next morning the first and third platoons were combined into a single under strength platoon. I do not remember what happened that day. I know we moved east along the foothills and set in looking down on the area of the fight of the day before. India Company of the Third Battalion, Third Marines, commanded by Captain G.V. Gardner, which had been sent in to support us, was down in the west end of the village.

'Towards dusk that company got into a heavy firefight, which we could see. Some years later I talked to, by then, Major Gardner, and he told me that his company lost several Marines killed and wounded, in that firefight.

That evening we received orders to sweep Ngoc Kink (1) and Lap Thuan the next morning.

The morning of 4 April we were joined by Delta Company and the combined force started sweeping on the east end of Ngoc Kinh (1), and swept west through the west end of Lap Thuan. The sweep began early and took all day. Contact was light and we sustained few casualties."

"On 4 April," Parr remembers, "Skemanski and I placed the demolition charges that destroyed the village and its bunker network in the valley."

Hagler continues. "Late that afternoon Bravo Company was lifted by helicopter to Battalion Forward which was on a hill farther into the Vu Gia river valley.

We continued to sweep west into the valley until the end of the operation on the tenth or eleventh.

We did make contact a few more times, and took a few more casualties, but the worst was over after 3 April."

NOTES:
Marines killed on Operation Orange are listed as follows:
Burton, Donald R. Sgt. 1915979; B-1-3; 26; Coatesville, Pennsylvania
Cook, Michael D. Pvt. 2135876; H&S-1-3; 18; Rubidoux, California
Courchane, Dale L. Cpl. 2012981; B-1-3; 20; Wenatchee, Washington
Cuellar, Julian C. Cpl. 2081158; B-1-3; 20; San Antonio, Texas
Fuller, Eugene E. Pfc. 2112318 19; H&S-1-3; Minneapolis, Minnesota
Haines, Paul A. LCpl. 2064998; 19; B-1-3; Canton, Ohio
Hasty, William D. LCpl. 2091249; H&S-1-3; 22; Birmingham, Alabama
Menzies, Alexander J. N. LCpl. 2133733; B-1-3; 20; Walworth, New York
Moss, Weldon D. Pfc. 2135169 B-1-3; 22; Ethete, Wyoming
Nail, Gary D. Cpl. 2083004; B-1-3; 20; Salina, Kansas
Owen, Thurman W. SSgt. 1403627; B-1-3; 31; Wichita Falls, Texas
Piotrowski, Daniel J. LCpl. 2122153; B-1-3; 20; Jackson, Michigan
Russell, David G. LCpl. 2130604; B-1-3; 19; Salem, Oregon
Villa, Feliberto Pfc. 2142407; B-1-3; 19; Ingleside, Texas

Williams, Ralph M. LCpl. 2042407 B-1-3; 20; Chicago, Illinois
Wright, Richard H. Pfc. 2145963; B-1-3; 21; Halstead, Kansas
Young, Donald E. LCpl. 2090141; C-3-Eng; 20; Baltimore, Maryland

On this terrible day there were many heroes and far too few were recognized for their heroism. The Navy Cross was earned by Lance Corporal Claude G. Lebas and First Lieutenant Willis C. Wilson.

The Silver Star was awarded to Corporal Manuel Avalos, Jr., Sergeant David D. Patten, Lance Corporal Daniel J. Piotrowski, and Private First Class Alexander J.N. Menzies. The awards to Menzies and Piotrowski were posthumous. Sergeant Floyd J. Darrh, who was wounded that day, would later earn the Silver Star during a second tour of duty in Vietnam. Major Alvah J. Kettering had previously earned the Silver Star in December of 1965.

The Bronze Star was earned by First Sergeant James R. Berheide, Hospitalman Third Class David C. Boos, First Lieutenants Richard V. Hunt and Kenneth E. Martin, Corporals Gwinn A. Henry, Jr. and Gary D. Nail, and Lance Corporal Gerold L. McDougall, Private First Class Dale E. Lawson, Private Dennis R. Hooyman, and Major John J. Tolnay.

Roy Parr, who served a second tour in Vietnam as a grunt with Mike Company, Third Battalion, Twenty Seventh Marines, remembers Donald Young. "Don and I were friends and he went in with the second platoon and was killed shortly after they landed. He had met somebody from his high school and asked if our two teams could switch so that he could go in with the second platoon. I told him it didn't matter to me and we switched. Don was a good kid and I have often thought about the exchange that took place that day. The other engineer with Don was wounded by rifle fire in the shoulder if I remember correctly."

John Hagler describes the members of his fire team: "Michael Wancea was from New Jersey. Maybe 5'10." Slim build. I think 19 years old, but not older. He was bright, learned fast, worked hard and did a good job on any task that he was assigned. When we were on Okinawa, on paydays, he didn't hit the bars like most of us. He would go to the PX and buy a pack of Swisher Sweet cigars, the kind with the wooden tip, and a jar of Tang. He didn't smoke other than that, but he loved his Swisher Sweets. Because he stayed out of the bars and skivvy houses he always had money. Don Allen and I, both of whom thought that if you weren't in the field there should be a party going on, never had any money. We were constantly going to him with our hands out. Sometimes he would go into convulsions, but he never said no. If he had charged us interest he would have made a fortune."

'Frederick McCoy, I believe was from B' Mo (Baltimore). He was 5'7" or 5'8" with a slim build. He was eighteen years old and was smart and street wise. Just cocky enough so that you knew that he knew what he was doing. He was open and friendly. I knew from the first time I met him that he could be trusted and counted on."

'I don't remember where John Dutton was from but he was 5'9" with a medium build. I think 19 years old. He was a salt of the earth kind of Marine. Intelligent and hard working, I couldn't ask for a better member of my fire team."

'Sergeant Burton joined Kilo Company shortly after I did in the early summer of 1965. At that time we were located on Camp Pendleton at Camp Margarita and were about to begin Lock-on (intensive training to prepare to ship out to Okinawa). He was assigned to the first platoon and became my squad leader. Sergeant Burton was about 5' 10" tall with a slim build. He was light skinned and had a small mustache. Burton was a natural leader; born with that ability to inspire not only the Marines who served under him, but those who served with and around him. He was bright, intelligent and could communicate his orders in a terse and cogent manner. He was sharp, always squared away, and led by example. I learned much from him.'

'When we were on Okinawa I asked my parents to send me a case of Garrett Sweet Snuff. When it arrived I introduced the squad, including Sergeant Burton, to the pleasures of sweet finely ground tobacco. One night, as we were going by truck from the field back to Camp Hanson, I pulled out a tin of snuff and placed some between my cheek and gum. I handed the tin to Sgt Burton and he started raising it to his mouth. He stopped just before he got it to his lips and, without lowering his hand, he looked me in the eyes and said, "Goddamn you Hagler, if my wife knew I was doing this she would kill me."

In Viet Nam he was rock solid. Even during the worst of times he was able to provide that small amount of inspiration, even if it was just a kick in the ass that would keep you going. When he was killed it deeply affected me, and still does to this day."

'They had joined the Marine Corps before the first combat troops landed in Vietnam. The war was on the horizon but it was not yet considered a quagmire that swallowed up young Americans. Their outlook and attitude was different from those who would join the service even a year later. They were bright, intelligent and motivated. I didn't have to stand over them 24/7 to make sure they completed their assigned duties. They worked hard training in the States and they worked hard training on Okinawa. They were good at their jobs. When we landed in Vietnam I laid my life in their hands and never looked back. I trusted them completely. These three Marines were in my fire team or in my squad the entire time I was with Bravo Company we went on over one hundred combat patrols and participated in numerous operations. They never let me down. I could not have handpicked three better Marines."

On the website *The Vietam Veterans Memorial*, First Lieutenant Curtis B. Bruce remembers Staff Sergeant Owen and Sergeant Burton: "He had been my platoon sergeant in Camp Pendleton for two years before we went to Vietnam. I became the XO of Bravo Company shortly after we arrived. During our time together at Camp Pendleton in 3/5 he was my platoon sergeant in the 106 platoon and then in third platoon of Kilo Company. That was my first assignment out of The Basic School and he helped me learn how to be a Marine officer. He was a fine man, always ready with a smile and a joke, and a persuasive leader for his Marines. I was the Battalion Intelligence Officer during the battle when he was killed and to this day I feel guilt because I wasn't with him. He, however, had one of the best platoon commanders in the Corps, Lieutenant Bill Wilson, who won a Navy Cross that day, If Bill couldn't have kept Staff Sergeant Owen alive, no one could."

"Burton was a fine Marine and I have fond memories of him. He was always squared away and kept his Marines the same way. I was his platoon commander at Camp Pendleton, in Okinawa, and for a short time in Vietnam."

Years later Willis Wilson was to write, "As I lay wounded on the battlefield that day I was in a state of semi-consciousness It was a strange state of calmness and shame ... what can I do to make up for my failure? I was trying to make up for what I felt was my failure to protect 'my Marines.' I kept assuming that I had issued an ineffective 5-paragraph order going into the battlefield. It was too complicated and based on very limited information about the terrain. I was feeling very guilty and inadequate as a platoon leader and kept thinking about my wife and 1-year-old son and hoping they knew I loved them.

After evacuation I remember being in a field hospital sitting on a canvas operating table. A doctor was putting a drainage tube in two places on my chest to take the pressure off of my lungs. The tube was feeding what I recall as a gallon jug. I'm not sure how he found the bullet, but he did find it just under the skin in my lower right back. He made an incision, and I felt and heard the bullet hit the canvas near my butt. I asked if I could keep that bullet. I believe it was a 7.62 mm steel jacketed bullet which was slightly deformed, I think, from hitting my rib. Maybe it was a ricochet. It served as a reminder of how lucky I was. I made a key chain fob out of that bullet by heating it up melting the lead inside the steel jacket then putting a screw-eye in the melted lead. I carried it with me at all times, and each time I was tempted to get upset or dejected I would pull out that bullet and relax with the truth that I was a lucky man, and things could be much worse ... life is good; thank you, God. Some years later I was on a trip with the family, and we stopped in a filling station. I went inside and bought some snacks. After getting back on the road and stopping later, I discovered the bullet was missing from the keys. Sometimes I think back and wish I had made more of an effort to track that bullet down. I still am able to remember that bullet and thank God for a good life.

Within what I recall as a few days I was being loaded onto a large airplane and flown to Clark Air force Base in the Philippines. I spent some days there in a ward with other chest wound patients. We were watching TV when a little incident came on that made us laugh. We all shared chest pains as a result and that made it even funnier."

Others who were either mentioned in the chapter or were wounded that day are listed as follows:
Aaron, Raymond C. Jr. Cpl. 1998791; B-1-3; 22; Montgomery, Alabama
Anderson, Larry M. Cpl. 2059535; B-1-3; 21; Minneapolis, Minnesota
Baggett, Curtis E. Sgt. 1653834; B-1-3; 27; Richmond, Virginia
Bell, David L. Cpl. 2015852; B-1-3; 22; Wyoming, Pennsylvania
Bennett, Ernest L. Pfc. 2091618; B-1-3; 20; Albany, New York
Berheide, James R. 1stSgt. 570830; B-1-3; 39; Kalida, Ohio
Bernhardt, Webster W. III LCpl. 2131748; B-1-3; 19; Los Angeles, California
Binnebose, Paul E. LCpl. 2135825; B-1-3; 19; Santa Barbara, California
Bledsoe, Michael A. Pfc. 2156880; B-1-3; B-1-3; Salt Lake City, Utah
Boos, David C. HM3 ; B-1-3; 21; Philadelphia, Pennsylvania
Bruce, Curtis B. 1stLt. 087878; H&S-1-3; 24; West Lafayette, Indiana
Bryant, Alvin LCpl. 2091928; B-1-3; 19; Jacksonville, Florida
Buckalew, Raymond W. LCpl. 2082266; B-1-3; 21; Los Angeles, California
Burke, William LCpl. 2075397; B-1-3; 19; Chicago, Illinois
Burkes, James Jr. Pfc. 2168897; B-1-3; 19 Los Angeles, California
Callaway, James M. Pfc. 0000000; B-1-3; 20; Tacoma, Washington
Caputo, Philip J.1stLt. 089046; C-1-3; 23; Chicago, Illinois
Cerillo, Michael E. Sgt. 2031797; B-1-3; 21; New York, New York
Colbert, Robert O. Sgt. 1882342; B-1-3; 24; Vallejo, California
Cushing, Barney A Cpl. 1984364; B-1-3; 22; Albuquerque, New Mexico
Darrh, Floyd J. Sgt. 1297336; B-1-3; 32; Constantia, New York
Davis, Lee R. Pvt. 2085320; B-1-3; 20; Macon, Georgia
Dickey, Robert R. III LtCol. 049621; CO-1-3; 40; Huntly, Virginia
Dickmann, Wayne E. Pfc. 2109245; B-1-3; 19; Milwaukee, Wisconsin
Diez, Larry W. Pvt. 2151635; B-1-3; 18; Los Angeles, California
Dowdy, Ralph W. Pvt. 2135746; B-1-3; 19; Los Angeles, California
Ehlert, Norman E. 1stLt. 083255; VMO-2; 26; Harvey, Illinois
Eubanks, D. L. Pvt. 2157027; B-1-3; 19; Pensacola, Florida
Foreman, Walter D. HM3 6872721; B-1-3; 21; Houston, Texas
Forgacs, Terrance S. Pfc. 2103810; B-1-3; 20; Ravenna, Ohio
Gardner, Grady V. Capt. 063874; I-3-3; 37; St. Louis, Missouri
Garza, Ysidore Jr. Pvt. 2185076; B-1-3; 19; San Antonio, Texas
Gordon, Irwin W. HN 77803503; B-1-3; 20; Halifax, Nova Scotia
Glover, Ronald W. LCpl. 2114901; B-1-3; 19; Kansas City, Missouri
Hagler, John LCpl 2048689; B-1-3; 21; Pasadena, Texas
Henry, Gwinn A. Jr. Cpl. 1912372; B-1-3; 24; South Charleston, West Virginia
Hooyman, Dennis R. Pvt. 2138298; B-1-3; 19; Milwaukee, Wisconsin
Hunter, Larry E. LCpl. 1993868; B-1-3; 19; Macon, Georgia
Hunt, Richard V. Capt. 073960; B-1-3; 28; San Francisco, California
Hutson, William A. Pvt. 2046116; B-1-3; 21; Macon, Georgia
Jordan, Charles G. 1stLt. 087417; B-1-3; 25; Macon, Georgia

Kaylor, Daniel R. Pfc. 2146827; B-1-3; 19; Allegan, Michigan
Kelichner, Anthony K. Pfc. 2133671; B-1-3; 19; Buffalo, New York
Kendall, Timothy Cpl. 1814759; B-1-3; 26; Chicago, Illinois
Kettering, Alvah J. Major 067506; HMM-263; 32; Riverside, California
Kinnard, Richard E. Jr. Cpl. 2061331; B-1-3; 20; Cincinnati, Ohio
Kumrow, Lawrence L. HM3 6979031; B-1-3; 20; New Castle, Pennsylvania
Kratzer, Everette C. Cpl. 2007570; B-1-3; 21; New Orleans, Louisiana
Lawson, Dale E. Pfc. 2141790; B-1-3; 19; Los Angeles, California
Lebas, Claude G. LCpl. 2141933; B-1-3; 22; Long Beach, California
Longoria, Felipe LCpl. 2110526; B-1-3; 19; San Antonio, Texas
Love, Gregory C. LCpl. 2135740; B-1-3; 20; Los Angeles, California
Mahalich, Norman J. 1stLt. 088540; VMO-2; 23; Los Alamitos, California
Martin, James M. SSgt. 1388050; B-1-3; 31; Omaha, Nebraska
Martin, Kenneth E. 1stLt. 090343; B-1-3; 23; East Williston, New York
Matthews, Don J. Cpl. 2007543; B-1-3; 22; New Orleans, Louisiana
McCoy, Frederick D. Pvt. 2122878; B-1-3; 19; Indianapolis, Indiana
McDougall, Gerold L. LCpl. 2054092; B-1-3; 20; El Paso, Texas.
Militello, Johnny T. Pfc. 2134307; B-1-3; 20; Detroit, Michigan
Milledge, Warren E. Cpl. 2006103; B-1-3; 22; Cincinnati, Ohio
Mitch, Joseph E. Sgt. 586509; B-1-3; 39; Philadelphia, Pennsylvania
Mosley, Richard J. Pfc. 2135717; B-1-3; 19; Los Angeles, California
Norton, Robert L. 1stLt. 067718; VMO-2; 32; West Hartford, Connecticut
Paratore, John M. LCpl. 2072800; B-1-3; 19; Memphis, Tennessee
Parr, Roy F. LCpl. 2094989; C-3-Eng; 19; Dallas, Texas
Paxton, Marvin L. SSgt. 584570; D-1-3; 33; Bryan, Ohio
Perry, Howard D. Jr. Sgt. 1877236; B-1-3; 25; Dallas, Texas
Peterson, Dennis A. 1stLt. 087897; B-1-3; 24; Los Angeles, California
Pittman, Milton, LCpl. 2082287; B-1-3; 20; Los Angeles, California
Platt, David L. Sgt. 1854790; B-1-3; 26; Minneapolis, Minnesota
Pounder, Duane E. LCpl. 2083229; B-1-3; 21; Boston, Massachusetts
Ramsburg, Harry R. Capt. 090389; HMM-263; 23; Westminster, Maryland
Redmond, Randolph Pvt. 2077129; B-1-3; 21; Seattle, Washington
Reed, Kenneth R. 1stLt. 091622; HMM-263; 22; Tucson, Arizona
Renfro, Oliver R. LCpl.. 2094539; B-1-3; 20; Shreveport, Louisiana
Richards, Paul Pvt. 2077148; B-1-3; 21; Seattle, Washington
Salas, Robert M. LCpl. 2128019; B-1-9; 19; Los Angeles, California
Santiago, Anthony LCpl. 2127916; B-1-3; 20; Los Angeles, California
Strobel. Thomas W. Cpl. 2021178; B-1-3; 22; Milwaukee, Wisconsin
Szabo, Fernand Pfc. 2105056; B-1-3; 19; New York, New York
Sylva, Valentin Z. Jr. Pfc. 2158261; B-1-3; 19; Dallas, Texas
Tolnay, John J. Maj. 058602; OO-1-3; 34; St. Augustine, Florida
Toney, Charles L. LCpl. 2122668; B-1-3; 20; Indianapolis, Indiana
Vasquez, Roy SSgt. 1614894: B-1-3; 28; San Antonio, Texas
Ventus, Hugh G. LCpl. 2128548; B-1-3; 19; Cincinnati, Ohio
Wancea, Michael LCpl. 2030573; B-1-3; 19; Winfield Park, New Jersey
Wilcox, Terry L. Capt. 089013; HMM-263; 24; Syracuse, New York
Williams, Thomas C. LCpl. 2107378; B-1-3; 20; Portland, Oregon
Wilson, Bobby G. LCpl. 2124827; B-1-3; 19; Dallas, Texas
Wilson, Roy R. LCpl. 2073086; B-1-3; 20; Charlotte, North Carolina
Wilson, Willis C. 1stLt. 087454; B-1-3; 25; Sharpsville, Pennsylvania
Young, Carroll W. Pfc. 2151647; B-1-3; 19; Norwalk, California
Young, James C. Pfc. 2125763; B-1-3; 19; Amarillo, Texas
Young, Robert L. Pfc. 2130196; B-1-3; 19; Chicago, Illinois

✯ ✯ ✯ ✯

On August 6, 1966 in northern Quang Tri province just below the DMZ the valley just to the east of the Razorback was infested with company sized units of the North Vietnamese Army. The command group of the Marine Corps decided to use artillery to reduce the size of these enemy units, and to avoid losing an unnecessary number of Marines on the ground. In order to do so small reconnaissance teams, made up of four to five men were inserted by helicopter into enemy territory. Their job was not to fight but to set up observation posts that were used to locate enemy concentrations. Once they were located these recon teams would call in artillery fire by using battery powered radios.

Team 61 of the First Force Recon Battalion consisted of four men led by Staff Sergeant Billy M. Donaldson. The team went by the name "Groucho Marx", their radio call sign. The only equipment they carried in addition to their weapons was two radios and a set of 7 x 50 power binoculars.

They were inserted by helicopter into a valley about twelve miles west of Dong Ha. The patrols position was three miles north of the Rockpile, a steep rock formation that jutted about 700 feet straight up in the air. Their mission was to observe and maintain surveillance of the main valley running east and west through their area of operation, capture one North Vietnamese prisoner, and be prepared to call in supporting arms on enemy targets of opportunity.

> "The grenade that wounded me hit about two feet away.
> It exploded, but there was no pain."
>
> Captain Howard V. Lee
> Echo-2-4
> Medal of Honor
> From *Bonnie Sue*

Information for this story was obtained, with permission, primarily from Marion F. Sturkey's fine book entitled *Bonnie-Sue: A Marine Helicopter Squadron in Viet Nam*.

Also used were articles that appeared in newspapers including, the *Pacific Stars and Stripes* and *The Jacksonville Daily News*, the awards citations of many of the participants, personal interviews, the rosters and diaries of the participating units, and information obtained from the National Personnel Records Center.

3

Groucho Marx

Team Groucho Marx was inserted at around 1900 and they moved easily through the elephant grass and set up for the night. They detected the movement of enemy soldiers just before 2400.

Around 1100 the next day they spotted the cooking fires of the North Vietnamese along a stream bed and were able to pinpoint their location. Donaldson, and his radioman Corporal Thomas E. Bachta, made radio contact with Cam Lo and called in artillery fire and gunships on the enemy.

Realizing that the North Vietnamese now knew that Marine spotters must be in the valley Donaldson had the patrol move slowly to a vantage point about 100 meters away. No one was able to sleep as enemy patrols searching for them passed nearby many times during the night.

During the daylight hours of 8 August the enemy intensified their efforts to locate the patrol. By mid-morning the North Vietnamese came within 50 meters as they conducted an on line search. All the while Donaldson continued to call in artillery fire.

As the enemy intensified their search efforts Donaldson radioed in and asked that reinforcements be sent in. Four H-46 helicopters from HMM-265, call sign Bonnie Sue, carrying a "Sparrow Hawk" reaction force of 44 Marines from Echo Company, Second Battalion, Fourth Marines under the command of Second Lieutenant Andrew W. Sherman took off. They were escorted by gunships from VMO-2, call sign Deadlock.

The insertion was unopposed and the reaction force quickly linked up with the recon team on top of a small, steep knoll. It was only about 30 feet high but stood out in the surrounding terrain. Sherman was a former enlisted man and had served time as a drill instructor. He wisely had his men set up a defensive perimeter and had them dig fighting holes. After a careful search of the area it was discovered that the enemy had vanished and by 1500 a decision was made to extract the entire reaction force.

This time eight H-34 helicopters from HMM-161, call sign Barrelhouse, again escorted by the gunships of VMO-2, flew in for the extraction. The zone was supposed to be clear but as a precaution only the lead aircraft went in. The North Vietnamese opened up with heavy machine gun fire and one .50 caliber round passed through the cabin. It wounded two passengers and mortally wounded, Corporal Ronald L. Belknap, the gunner. The missile penetrated the right side of his neck and then hit the back of his head, killing him instantly. Belknap, a precision measurement equipment chief, volunteered for the mission.

Four more of the aircraft landed amidst the heavy enemy fire and twenty of the Marines made it into the helicopters and were flown out. As the rest of the H-34's started down to retrieve the rest of the reaction force they were waved off by Sherman.

He and the remaining 23 men hastily withdrew toward the top of the knoll where their fighting holes awaited them. Donaldson declined to be withdrawn with the initial airlift, as he felt he could assist Sherman in the evacuation of the remainder of the force. Bachta, although suffering from heat exhaustion, remained in order to assist with communications. They now occupied a good defensive position with clear fields of fire on all sides.

As the North Vietnamese prepared to assault the position, Corporal Douglas H. Van observed a disabled Marine lying only ten feet from an enemy position. He left his covered position and braving the hostile fire, reached the man and carried him to the safety of his own lines.

About an hour later nearly 250 North Vietnamese soldiers attacked Sherman and his 23 Marines. The Marines used all of the weapons at their disposal and blazed away at the onrushing enemy.

Private First Class Robert Callaway recalls, "They hit us all of a sudden. I saw this one guy yelling and waving his arms around. He looked like a squad leader or platoon commander. I shot him. We were shooting them with pistols and throwing grenades like they were going out of style."

The North Vietnamese retreated to the jungle and regrouped. There they added reinforcements and attacked again. The Marines also repulsed this attempt on their redoubt. But they had been hurt. The erstwhile Sherman was magnificent as he rallied his men against the vicious assault on the knoll. When they were forced to give a little ground he moved about in full view of the enemy and directed the defenders and led a counterattack that drove the North Vietnamese from the Marine positions.

During the counter attack Lance Corporal George R. Gibson leaped from protective cover and charged into the enemy fire. Although painfully wounded by the fragments of an exploding grenade he continued to press the attack until the original positions had been retaken. He then crossed twenty five meters of fire swept terrain to retrieve badly needed ammunition.

The peripatetic Gibson then moved over to an unmanned machine gun and put it into action. He courageously stood his ground as the North Vietnamese tried to take out his position with hand grenades and small arms fire. As the fight continued he was hit again, this time by machine gun fire. Despite suffering a painful gunshot wound to the right shoulder he killed six more of the enemy soldiers. When the North Vietnamese set up a machine gun that threatened to hit the Marines with sweeping gunfire Gibson moved across the slope of the knoll, exposed to enemy fire, and flanked the weapon. His one man effort succeeded in knocking it out.

Sherman was shot through the head and killed. Private First Class

Benjamin Hamrick was killed by multiple fragmentation wounds penetrating his entire body. Corporal Dennis R. Schmidt died when he was hit by gunshot wounds that penetrated his right shoulder and lower chest. Also killed was Private Vernal S. Martin.

Many others were wounded by exploding enemy ordinance. Among the casualties were Privates First Class Richard Corson and James J. Swartz, and Lance Corporals Sterling A. Fletcher and Richard L. Iverson. Corson was hit by a shell fragment that removed the third finger on his right hand. Swartz was hit by shrapnel in the left wrist and left shoulder. Fletcher suffered fragment wounds in both feet and legs and multiple fragment wounds to the right side of his face and both buttocks. Iverson received a fragmentation wound to the left knee.

Sergeant Robert L. Pace, a platoon sergeant from Echo Company, was next in command but did not make it through the next enemy assault. The North Vietnamese charged for a third time and were again pushed back but Pace was down with blood gushing from fragmentation wounds to the face, neck and chest. He was out of the fight but would survive.

Although severally wounded himself. Donaldson had to take command of the group. He had been badly wounded in the arm and head. When he saw an enemy grenade land in a fighting hole that contained a wounded Marine he jumped in and threw it out. It exploded as it left his hand and he was hit by the fragments but saved the life of the Marine. Bachta took charge of one side of the perimeter and simultaneously radioed the situation to his superiors and supporting aircraft.

Lance Corporal Timothy A. Roberts, a fire team leader with Echo Company, stood out during the night long battle as he employed the fire power of his team against their foe. In doing so, he repeatedly exposed himself to the enemy. During the course of the action he suffered a laceration wound to his left hand and was hit in the left side and left leg with shrapnel.

They were running out of ammunition and it would be dark soon. As time passed more reinforcements filled the ranks of the North Vietnamese. The Marines were in a desperate situation and Bachta radioed in, asked for ammunition, and added, "You better hurry!"

Captain Howard V. Lee knew that without ammunition the men on the knoll were doomed. There was no place for the helicopters to land but he decided to try anyway. Lee asked for six volunteers and all of the remaining members of Echo Company asked to go. Lee had to choose.

Two helicopters from HMM-161 were loaded with all of the grenades and ammunition that they could carry. Lee and three of the men boarded one of the birds. The other three Marines climbed into the other plane. As they took off and headed for the embattled Marines they were escorted by two gunships from VMO-2.

The first helicopter landed in a clear, flat area between the knoll and the enemy. Lee and the three troopers with him jumped from the aircraft, and weighed down by the ammunition crates that they carried, ran as fast as they could toward the Marine perimeter. Under fire they reached the knoll and dived into nearby fighting holes.

The other aircraft landed on a narrow ridge about 100 meters from the others. The Marines exited the bird and found themselves stranded. They were cut off and had no radio. Lee realized their predicament and contacted Major Vincil W. Hazelbaker, who was flying one of the VMO-2 gunships high overhead, and asked for assistance. As darkness approached he had trouble locating the Marines until they popped a green smoke grenade.

Hazelbaker and his copilot, First Lieutenant Anthony B. Costa, dove down toward the floor of the valley. They made two passes on which the door gunners worked over the enemy positions with machine gun fire. On the third pass, Hazelbaker landed the craft. The three marines piled in and the Major quickly lifted the helicopter into the air and dived behind the ridge and used it for protection from the enemy guns.

HMM-265 was located at Marble Mountain which was about 80 miles from the embattled Marines near the Razorback. They received the call for a medevac mission. A flight of four aircraft were assigned the mission. The number three aircraft was flown by Captain Richard O. Harper. First Lieutenant Marion F. Sturkey was his copilot. Manning the starboard gun was the crew chief, Sergeant Herbert S. Murff. The door gunner on the other side was Lance Corporal Luke A Stephen, an aircraft electronics systems technician.

The number one aircraft, piloted by Captain Dale Tinsley and First Lieutenant Joseph T. Roberts, was downed temporarily by a hydraulics malfunction. Their wingman flying helicopter number two would also be left behind. This aircraft piloted by First Lieutenants George G. Richey, Jr. and Paul J. Albano would stay with the other while mechanics fixed the problem.

Harpers plane was now in the number one position and on his wing was a helicopter flown by Captain Leo J. Farrell and First Lieutenant Charles D. Joyner. On this aircraft Corporal John A. Hedger was the crew chief and Staff Sergeant Edward R. Dusman, an aircraft communication/navigation system man, was the door gunner.

Tinsley's bird would now carry twelve Marines from Echo-2-4, who upon landing would exit the aircraft, and fight off the North Vietnamese while the dead and wounded were loaded aboard. They were to land in a relatively clear bombed out area to the southwest of the knoll. Once they landed they would have to sit there exposed to enemy fire.

As they descended, two aircraft from VMO-2 stayed with them to provide guidance and protection. When they were about twenty feet above the ground the North Vietnamese cut loose on them. Enemy rounds ripped through the cockpit. Captain Harper was shot, once in the right hand and once in the chest. The round that hit him in the chest ripped through his flak jacket about three inches below his heart and exited out of his back. Instinctively, Sturkey and Harper, working together at the controls, managed to take off. An enemy round hit Sturkey in the foot and slammed his shin against the instrument panel. Both men had suffered severe wounds but continued to fly and they gained speed and altitude.

They then assessed the damages incurred. Stephen had a lacerated right forearm but could still man the machine gun. Murff had been hit in the abdomen by one of the enemy rounds and was in bad shape.

None of the twelve grunts aboard had been wounded and two of them were trying to help the badly wounded Murff. In a span of about ten seconds the helicopter had been hit by 34 enemy rounds, all in the area of the cockpit.

Although critically wounded and in great pain, Harper managed to stay lucid, and in control as he and Sturkey kept the stricken aircraft airborne and flew it the ten miles to Don Ha.

During this time Tinsley and his wingman Richey arrived over the valley and found it in total darkness. He radioed to Lee that he would attempt to land on the knoll. The North Vietnamese poured fire into the helicopter as it neared the ground. Tinsley and his copilot, Roberts, although hit in the arm by shell fragments, kept the aircraft under control.

In the cabin of the helicopter a round slammed through the helmet of the gunner, Dusman, and ripped into his face. Hedger had both of his legs shattered by a 50. caliber round that passed through his thighs, just above the knee. As he crumpled to the cabin floor, Dusman rushed to his side and desperately tried to stop the massive flow of blood.

Tinsley and Joe Roberts also managed to fly their stricken craft back to Dong Ha.

Staff Sergeant Conrad L. Ortego, a wounded veteran of the Korean War, and a platoon sergeant with Echo Company, was a part of the reaction force and recalls. "I was on two different helicopters that tried to land on the knoll and reinforce the Marines fighting there. On both occasions the pilots were forced to turn back by the heavy enemy fire."

Back on the knoll only 16 men were capable of fighting. Lance Corporal Danny R.Vance was hit in the right knee by shrapnel during a rescue mission in which he left his fighting hole and dashed to a wounded Marine who had been struck in the chest and lay in the path of the onrushing North Vietnamese. He dragged the wounded man to a protected position and rendered first aid. From there he continued to fight and played an integral role in repulsing the enemy attacks on the knoll.

Also distinguishing himself was Lance Corporal Gary N. Butler, a radio operator with Echo Company, who although painfully wounded in the right arm by enemy small arms fire, remained at his radio and relayed all command post traffic as the helicopters attempted to extract the trapped

Marines. He maintained his position for fourteen hours until relieved. During each savage enemy attack Butler transmitted accurate and timely information, exhibiting exceptional knowledge of the situation in which the Marines found themselves.

During the fight Sergeant Tommy E. Stevens spotted an enemy soldier in a fighting hole with a wounded Marine. He dashed through the enemy fire, killed the North Vietnamese soldier, and then manned the position and protected the fallen Marine.

Although added to the Echo Company casualty list many Marines continued to fight. This list now included Lance Corporals Nathaniel Calvin, Jesus Daniels, Jr., Roger A. Davis, Edward McDermott and Leroy Williams, Corporals David A. Smith and Richard A. Strock, and Private First Class Frederick Rode.

Daniels received a fragmentation wound to the left buttocks and a severe concussion from and exploding round. McDermott was hit by shrapnel in the left thigh and left ankle. Strock suffered a fragmentation wound to the left elbow. A shell fragment hit Rode in the left elbow. On 19 August he received a second purple heart when he was struck by a missile near the coccyx area. Both Vance and Williams would be killed later in the war, Williams, a month later, on September 9 and Vance on January 7, 1968 while serving on a second tour and as a member of a CAP unit.

Lee, also, was now among the wounded. A grenade landed about two feet from him and exploded. Fragments from the explosion tore into his body. One hit his right eye and blinded him. Other fragments hit Lee in the right arm and lower body with one metal shard passing completely through his buttocks and exiting on the inside of his thigh. The force of the explosion ripped off his pistol belt and canteen.

Hospitalman Third Class Nicholas C. Tarzia had repeatedly moved through withering enemy fire to aid the wounded. On this occasion, he raced twenty meters across open ground to assist the wounded Lee. When his first aid pack was blown away by enemy fire, Tarzia used his own clothing for bandages as he treated and comforted the many casualties. After he had completely exhausted all available first aid materials and could be of no further assistance to the wounded, Tarzia gallantly helped defend his disabled comrades by manning a position on the defensive perimeter.

Although blind in one eye and badly injured, Lee remained in command but by midnight could only periodically direct the battle. Bachta alertly maintained the continuity of command by directing fire, controlling air cover and radioing situation reports. He called in and asked for more ammunition. The battalion commander, requested to talk with the "6 actual", Lee. Bachta informed him that Captain Lee was so badly injured that he was not able to do anything. The Commanding Officer then asked, "Well, then, who is in charge?" Bachta replied, "I am."

A supply aircraft, unable to land in the zone, attempted to drop ammunition to the beleaguered ground force but the supplies landed outside of the unit's defensive perimeter. Lance Corporal Roger J. Baca left his covered position to retrieve the vital munitions. He dashed across an open area and returned with a portion of the desperately needed ammunition. Baca then made two more trips into the fire swept area to retrieve more of the munitions.

The situation was getting desperate for the defenders on the knoll and Hazelbaker, flying overhead in his gunship, decided to do something about it. He daringly maneuvered his aircraft through the darkness and hostile fire and, coming up the backside of the small hill, landed near the Marines position. He resupplied the ground troops with three thousand rounds of his own ammunition. Ammunition meant to be used by the helicopter. This would keep them going for a little while.

Hazelbaker knew that the grunts needed additional ammunition, and he believed that they could do it again. He and Costa asked for and received permission to fly the resupply mission and they, in turn, asked for a volunteer crew. When many stepped forward they selected, Corporal Eppie Ortiz and Lance Corporal James E. McKay. Ortiz would fly as the crew chief and McKay, a former grunt, as the gunner.

Three other aircraft accompanied Hazelbaker on the mission. The second plane was flown by First Lieutenant Shepard C. Spink. Staff Sergeant Stanley L. Brewczynski, a career Marine, who was to retire as a Master Sergeant in 1975 and die shortly thereafter, was the co-pilot. Lance Corporal Michael C. Martin was the crew chief. Captain Ben A. Meharg was the pilot of the third helicopter in the flight. His copilot was First Lieutenant Beasley Willis with Sergeant Allen M. Shepherd flying as crew chief and Lance Corporal John F. Bornemann as aerial gunner. The fourth helicopter in the flight was flown by Captain John O. Enockson with Staff Sergeant William T. Cummings, Jr. as crew chief and Sergeant David E. Warrick, a Navigation Systems Technician, the aerial gunner.

They exchanged the gunship for a cargo hauler and it was loaded with eight hundred pounds of rifle ammunition, M-79 rounds and grenades. The plan was to use the same approach to the knoll as before, unload the ammunition, and take out as many of the wounded as possible. Everything went well and Hazelbaker landed the aircraft and Ortiz and McKay threw out the ammo crates but before any wounded could be loaded aboard the helicopter was disabled.

The aircraft received a crippling round when an enemy rocket, fired at short range, hit its rotor head and fragments from the explosion wounded three of the crew members. Costa was hit in the leg by a steel fragment and Ortiz had his helmet shattered by the blast and fragments penetrated his skull. Three fragments of hot steel tore completely through McKay's flak jacket and wounded him in the right side of his back. He crawled from the helicopter and as he started to run toward a nearby fighting hole an enemy projectile struck him in the left knee and knocked him down. He tried to rise but couldn't and tumbled down a ravine and lost consciousness. The bullet that ripped through his knee had severed the tendons and his lower leg flopped around like a loose hinge. McKay could not stand or walk. One of the Marines from Echo Company saw him fall and crawled down and together they returned to the perimeter.

The grunts on the knoll added to their firepower by removing the M-60 machine guns, used by the door gunners, from the helicopter.

Corporal James Pace left his covered position and retrieved the badly needed ammunition from the downed helicopter. Although painfully wounded he crawled across an open area under the intense enemy fire to the aircraft and returned with the vital ammunition. Pace then crawled from position to position on the knoll and distributed the ammunition to his fellow Marines.

The fearlessly determined Van again stood out, as he moved through enemy fire and delivered ammunition to one Marine position that had almost exhausted its supply.

Weakened by an excessive loss of blood and unable to get out of his fighting hole Lee had to turn over the command of his troops to Hazelbaker. The gunbird pilot was up to the task and provided firm leadership and reorganized the defense. During the rest of the night, Hazelbaker and Costa distinguished themselves on the knoll by organizing their defenses and providing exceptional leadership to the surrounding force. They supervised the distribution of the ammunition, ensured that the casualties were taken care of, and controlled supporting air strikes.

The North Vietnamese apparently realized that the Marines were again well supplied with ammunition, and they relented in their attempts to overrun the position. They however, kept as close to the Marine perimeter as possible. By doing this they were able to avoid the murderous fire from the gunships and Spooky. Major Hazelbaker commanded operations until daylight brought relief in the form of Marines from Foxtrot Company.

The fighting was over.

The helicopters flew in and took out the most seriously wounded Marines first. They flew them to Dong Ha and then returned to pick up the wounded that could still walk. They returned once more to retrieve the bodies of those who had been killed in the fight. For them time no longer mattered.

The bodies of dead North Vietnamese soldiers littered the area around the top of the knoll. The enemy had fought fiercely and bravely and as they retreated they took with them the bodies and the equipment of many of their comrades killed by the Marines. Wisely they had not attempted to retrieve those close to the top of the knoll. The Marines of 2/4 counted 37 bodies, already swarming with flies, lying within fifteen meters of the perimeter. It was estimated that an additional 100 were probably killed out of an attacking force of around 250.

Five Marines had been killed in the fight. One was from HMM-161, and the others were from Echo Company. Twenty seven were wounded; fifteen were from Echo, one from the recon patrol, and eleven from the helicopter crews.

NOTES:
The following Marines were killed on August 8, 1966:
Belknap Ronald L. Cpl. 2001400; HMM-161; 22; Daly City, California
Hamrick, Benjamin N. Pfc. 2139760; E-2-4; 19; Kingmont, West Virginia
Martin, Vernal G. Pvt. 2053431; E-2-4; 21; Marshfield, Wisconsin
Schmidt, Dennis R. Cpl. 2104609; E-2-4; 21; North Plainfield, New Jersey
Sherman, Andrew M. 2dLt. 097870; E-2-4; 32; Doylestown, Ohio

Many Marines were decorated for heroism on this day. Among them was Captain Howard V. Lee who earned the Medal of Honor.

The Navy Cross was awarded to Staff Sergeant Billy M. Donaldson, Lance Corporal George R. Gibson, Major Vincil W. Hazelbaker and Second Lieutenant Andrew M. Sherman. The award to Sherman was posthumous.

The Silver Star was earned by Corporal Thomas E. Bachta and Hospitalman Third Class Nicholas C. Tarzia.

Corporals John A. Hedger and James Pace, Lance Corporals Roger J. Baca and James E. McKay, Captain Ben A. Meharg and Sergeant Tommy E. Stevens earned the Bronze Star.

Captain James W. Rider, a pilot with VMO-2 writes: "During my first tour in I Corps I was involved with the formation and execution of the "Sparrow Hawk" mission. Sparrow Hawk was an airborne reaction force that consisted of a rifle platoon, transported by UH-34's and later CH-46's, escorted by two armed UH-1E's, with a Command and Control Huey piloted by a Tactical Air Controller Pilot (TACP), overhead. As the mission evolved there was always a rifle platoon on standby as the Third Marine Division, Sparrow Hawk Reaction Force."

"Sparrow Hawk was very successful as a reaction force. There was one notable tragedy. In May 1966 while flying med evac missions with a UH-1E slick and a gunship, our flight evacuated 14 or more casualties from a Sparrow Hawk platoon that had been lifted into a rice paddy surrounded by a concealed VC Company. "Doc" Mayton, our corpsman and Corporal Bob Abshire, my crew chief were both awarded the Navy Cross for unbelievable courage picking up casualties from the paddy, within point blank range of the surrounding VC. The company commander of that Platoon, Captain "Al" Christy wrote several articles for the Marine Corps Gazette, describing this and other actions in the areas surrounding Hill 55."

Donaldson, a career Marine, first enlisted in 1954. In addition to the

Navy Cross he also earned the Bronze Star Medal in Vietnam. Donaldson retired in 1973 having attained the rank of First Sergeant. Years later, when asked about the other members of the patrol, Donaldson replied. "I didn't know any members of the team. They were assigned to me the day of the patrol. I now know of Bachta because of him receiving the Silver Star."

Thomas E. Bachta, who first enlisted in 1963, went on to serve thirty years in the Marine Corps and retired with the rank of Sergeant Major in 1993. For his heroic actions on the night of August 8 and 9, 1966 he was originally recommended for a Navy Cross. Bachta was sent to DaNang for a rest, and while there, stole a jeep. As a punishment this paladin's award recommendation was degraded from the Navy Cross to a Silver Star.

Richard O. Harper, Who went on active duty in 1955, made a career of the Marine Corps and retired from active duty in 1978 as a lieutenant colonel. He was originally recommended for a Silver Star for his actions on this day but it was eventually downgraded to a lesser award.

Vincil W. Hazelbaker served on active duty in the Marine Corps from 1953 until his retirement as a Colonel in 1979. He flew jets until 1963 when he underwent training to convert to becoming a helicopter pilot.

Herbert S. Murff recovered from his serious wounds and remained in the Marine Corps retiring in 1978 with the rank of Master Sergeant. He first enlisted in 1956. He died in 1979 from an illness complicated by the after effects of his wound. His name was added to the Wall in 1997.

Jim McKay had joined the Marine Corps in June of 1964. He was short and slightly built and his soft spoken demeanor belied his inner toughness. Trained as an armorer, McKay had come to Vietnam in early 1965 as a member of the Third Shore Party Battalion. It was not a front line unit and he wanted more action, so in December of the same year he requested a transfer to Force Recon. His request was granted and he spent the next couple of months in the field with a recon unit. In February of 1966 with his thirteen month tour in Vietnam getting short McKay was still not satisfied.

McKay wanted to be a gunner on a Marine Corps gunship so he and the Marine Corps arraigned an unusual deal. He would extend his combat tour in Vietnam for an additional six months and become a door gunner with VMO-2. He started his new assignment in March. He was to be a gunner when airborne and a loadmaster on the ground. During the following six months he flew on many dangerous missions. On 3 August he was the gunner flying with First Lieutenant Richard L. Drury when the lieutenant was severely wounded in the right hand, arm, and leg by an enemy .50 caliber round.

August 8, 1966 was McKay's last scheduled day to fly in Vietnam. When his afternoon flight was finished he was supposed to turn in his gear, fly to Da Nang, and then to Singapore where he was to spend a ten day R&R as part of the extension pact that he had made with the Corps. He would then fly directly to the United States from Singapore.

McKay turned in all of his gear and then was made aware that some of his old recon buddies were part of the Groucho Marx patrol. He knew that they were trapped and fighting for their lives on the knoll, so he refused to catch many of the available flights to Da Nang. As soon as the opportunity arrived he volunteered to fly with Major Hazelbaker.

Instead of flying to Singapore the badly injured McKay ended up in the Oakland Navy Hospital and after a lengthy stay, recovered from his wounds and was assigned to the Weapons Training Battalion at Edson Range. He then transferred to Headquarters Battery, First Battalion, Twelfth Marines. There McKay finished his four year commitment to the Corps and was honorably discharged on June 12, 1968.

John F. Bornemann completed his tour in Vietnam and returned to the United States. He was one of 22 Marines killed in the fiery crash of a CH-53 helicopter at New River, North Carolina on June 23, 1967. The crash occurred on a Friday morning at 0901 as the helicopter, from HMH-461, piloted by Lieutenant Colonel Joseph L. Davis and Captain Peter F. Janss, and loaded with 30 Marines was landing. The Marines were members of various squadrons on the base and were wearing full combat gear. They were taking part in what was called General Military Subjects, or GMS school. It was a training exercise in which the aviation personnel were being reacquainted with their basic responsibilities as infantryman. Each month every squadron and support unit on the base would assign one or two people to the school.

In this instance they were practicing the tactic known as vertical envelopment --- dropping right down on the enemy using helicopters. They took off from a pickup point and headed for the landing zone. Half way there the occupants felt a *thump* from below. A UH-1 Huey, from VMO-1, piloted by Captain William L. Buchanan and First Lieutenant William W. Storbeck, had been given clearance to take off and had flown directly into the underside of the CH-53. The pilot, Buchanan, a veteran of the war in Vietnam, had already earned the Bronze Star Medal while flying with VMO-6 in December of 1965. He was instructing the inexperienced copilot, Storbeck, on tactical maneuvers. Neither plane saw the other.

The official investigation of the collision determined that the helicopters were in contact with different air traffic controllers working on different radio frequencies.

The Huey exploded and crashed, killing its occupants instantly, and the control cables of the CH-53 were severed. The pilot, Davis, tried to autorotate and almost succeeded but the aircraft hit hard, bounced, and its tail section snapped off. Without the tail rotor the plane started to spin and some of the Marines were thrown free. It came down hard again and this time finally stopped spinning but the fuel lines had been broken and the helicopter burst into flames.

The spinning had caused numerous injuries and broken bones among the Marines inside.

Sergeant Thomas L. Williams, a member of the helicopter crew in the '53 was forcefully thrown through a plexiglass covered hatch at the rear of the aircraft. His hurtling body tore out the glass and its supporting metal structure as it left the plane about 300 feet from the ground. Williams's body was later found in the trees.

Lance Corporal Jon B. Rittler of VMO-1, who survived the crash recalls, "I can remember that when the Huey hit us a big, black Sergeant was standing, unrestrained. I thought he was the crew chief. He had this look of terror on his face with a questioning look like 'What do we do now?'" It was probably Williams.

The copilot, Janss, recalls, "We never saw the other helicopter. I felt the impact and our plane started to porpoise. We tried to work the controls but there was nothing that we could do. When we hit the ground I was knocked unconscious. Someone pulled me from the wreckage. "I had no injuries, no blood, or anything. I was just knocked unconscious." Janss had already served a tour in Vietnam with HMM-161 and would later return for another.

"I was sitting in the tail section close to 'Frenchy' Lafountaine." Rittler remembers. "When we hit the ground the tail section snapped off and I can remember flying backwards through the air. I hit on my back and when I sat up I was facing the aircraft and it was burning. Frenchy was lying close to the fire so I pulled him to safety but I couldn't help any of the others because the fire by that time was too hot."

Joe Davis escaped unscathed but wandered off in a daze and left Janss trapped, unconscious in the cockpit. First Lieutenant James M. "Duffy" DuFriend of H&MS-26, one of the lucky ones, was thrown free and suffered only minor injuries. He kicked out the plexiglass windows of the cockpit and pulled Janss to safety. DuFriend, a former enlisted man, had been trained as a parachute rigger.

The following information was gleaned from articles, concerning the collision that appeared in the *Jacksonville Daily News* on June 23 and June 26, 1967.

Included among the dead were the crew chief, Sergeant Kenneth G. Ross, Williams, and Private First Class Jerry W. Scurlock. All were members of HMH-461. Also killed were Corporals Franklin L. Bell,

George S. Bondarewicz, Phillip E. Dennon and Private First Class Frank E. Stodelmaier of HMM-261, and Corporal Milton J. Parkerson, Lance Corporals Wayne E. Montgomery, Kenneth R. Neill and David G. Newell, and Private First Class Bruce E. DeRolf of H&MS-26, and Corporal Kenneth L. Wassan and Lance Corporals Robert N. Smith and Private First Class Michael J. Vicknus of HMM-161 and Lance Corporals Larry J. Bailey and Robert K. Naugle of HMM-365 and Corporal Joseph Howard and Private First Class Clifford C. Tressler of MABS-26.

Ross and Williams had already served tours in Vietnam in 1966, Ross with HMM-364 and Williams with HMM-362. In addition, Ross had been wounded on July 30, 1966.

Lance Corporal James J. Spence and Corporal Gary P. Linick of VMO-1 and Hospitalman Third Class Charles E. Nightingale of MABS-26 were among the injured who survived the crash. Their injuries were critical. Lance Corporal Lafountaine of HMM-162 suffered a broken femur in his right leg and was hospitalized for several weeks.

Along with Rittler and DuFriend others escaped the crash with minor injuries. They included Sergeant Richard C. Harrelson of HMH-461 who had also done a tour in Vietnam with HMM-361. The others were Corporals Jimmy Noble and Garry J. Ryan of HMM-261 and Sergeant Rudolph H. Talbert and Private First Class Andrew R. Brown of MABS-26.

Rittler remembers being taken to sick bay where he was checked out and released. "I was walking back to the barracks and Gunny White from my squadron picked me up in his car and brought me there. He wanted to know if I was alright."

Ronald Winter, in his fine book *Masters of the Art*, wrote:

That night, long after the fires had been extinguished and the bodies taken to sick bay for identification, a Marine entered our barracks, looking for some friends. He had been in the '53 and survived the crash. We gathered around him asking questions.

"I was sitting right next to the tail section when it broke apart," he said. "I had my safety belt on and I didn't get tossed around like the others. Then when it hit the second time I was thrown clear. I landed on my pack and passed out for a few seconds. When I came to I ran over and started pulling some of the others out. I carried three to safety." And then he stopped his throat constricting and his voice cracking just a bit. "But they were piled up near the exits, with broken legs and arms. They couldn't get out the escape hatches and the fuel was spraying over them.

And then it exploded. I couldn't get near it. For 20 seconds all I could hear was screaming. Then nothing. I'll never forget those screams as long as I live."

He never told us his name, and no one wanted to intrude at that point. He went on his way, taking his memories and leaving us with ours.

Rittler thinks that he may be the Marine mentioned in the Winter book. "We shared a three story barracks with HMM-161. They were on the first floor and my squadron was on the third floor. We shared a recreation room on the first floor and I knew most of those guys. They knew that I had been in the crash and I probably stopped and talked with them awhile. I don't recall the conversation but it almost has to be me. I was one of the few survivors that were uninjured and I lived in the same barracks as Winter."

In Masters of the Art, Winter writes: Mike Vicknus, a metalsmith from my squadron, was a good friend of mine with whom I had swooped north to Connecticut a few months earlier, was in the GMS class. We'd had fun that weekend. He had a Plymouth Cyclone, with a big engine and four on the floor, and I remember him blasting through Richmond on I-95 at something over 90 miles per hour. I remember telling him that the sign on the side of the road was the route number, not the speed limit, but he laughed and kept on going.

Linick remembers, "John Borneman was my best friend. My buddy. He was short and chubby but all Marine. When he was promoted to Sergeant I was promoted to Corporal and he gave me his chevrons. I still have them today. On the day of the crash we were practicing infantry tactics and Borneman was my squad leader. I had the first fire team but the Gunny caught Borneman and I fucking around and switched me to the third fire team. We were in the tail section of the '53 when it crashed. Borneman and the first fire team were up front and everyone in that section burned to death.

We were in, what we called 'Grunt School", and were playing war games. We circled the field and came in to land where we were supposed to exit the helicopter and assault this position. As we started to land we felt the impact and I knew that we were in trouble. I can still remember hearing the exclamations, 'Oh Shit!', 'Oh God!' 'Oh Fuck!" I remember later one of the survivors telling me that he thought it was all part of the exercise. Of course, I knew better. I flew every day. That's what I did. Sergeant Williams was standing and I could tell from the look in his face that he also knew that we were crashing. The helicopter started to auto rotate and there was a terrible vibration with dust flying everywhere. When it hit the tail section broke off and I don't remember anything else. I do remember that the Marine sitting across from me was thrown on top of me and I pushed him away. He was killed in the crash and I cannot recall his name. I try but I just can't. I still see his face every day.

I was really messed up bad in the crash. The tail section, at first, didn't burn. The Marines in the tail section got busted up but there was time to pull the injured and dead from that section before the entire plane exploded. The Marines in the front of the aircraft, with the exception of the pilots, burned to death. Of course, Borneman was up there and he didn't have a chance. They thought that I was dead so they left me, for a while, in the wreckage and worked on others. I was told that the Gunny who separated John and I was the one who noticed that I was still alive and pulled me from the wreckage. My injuries were so severe that I was in the hospital for seven months. I suffer from survivor guilt and I am still a little angry. I wanted to go to Vietnam and I was unable to. The Marine Corps considered me to be permanently damaged and they got rid of me. Frenchy didn't get hurt real bad and he got to go to Vietnam, as did Rittler who escaped the crash virtually uninjured. Jim Spence, who was the automatic rifleman in my fire team, broke something like 42 bones. The other two Marines in my team who I did not know well were killed. I cannot recall their names.

I suffered a lot of joint injuries and had to have my right knee and right hip replaced. I also had to have surgery on the joint of my right shoulder. I also suffered extensive head injuries and had at least a hundred stitches in my face alone. I was fortunate that the guy who sewed me up was a plastic surgeon and he did a fantastic job. I hardly notice the scars anymore. With the head injuries, I wasn't in a coma, but I was delirious, and when I talked with anyone none of what I said made any sense. I was in and out like that for about a month. I also broke my left wrist and I have had to have back surgery since then. Of course, I was a Marine and I did exercise and all of the therapy required. Like many others I tried to drown my pain in alcohol and I had problems with that. I quit drinking on the 23rd, which was the day of the crash. When that day would come around on any month I would just get drunk, now I just go to church. I am clean and sober now and have been for some time. I am a lifetime member of Alcoholics Anonymous.

I was the crew chief of the Huey that hit us. In fact, Borneman and I had both been its crew chief at one time or the other. We had just recently worked on that aircraft and had gotten it air worthy."

Brown was interviewed by a reporter from the *Jackonsonville Daily News* a few days after the crash and he said, "The only thing I don't seem to recall is the impact when we hit the ground. Except that I really didn't hear anything, but I remember the violent shaking. I remember seeing the helicopter coming apart and spinning. My first thought was that it was an exercise of some sort. I didn't think that we would crash. When I looked at Sergeant Williams, the expression on his face gave it away."

Brown suffered a broken foot in the crash. "I started to walk away from the crash, an instructor helped me walk part of the way, I walked about 50 yards from the crash and then I fainted. When I came to I saw an ambulance parked near the wreckage scene and I walked over to it."

Brown was then taken to the MCAF dispensary where X-Rays revealed the condition of his foot. He was then brought over to the Naval Hospital for treatment.

When asked if he would go up in a helicopter again he replied, "Last Friday I swore I wouldn't, but really, I would do it. He smiled, "I'm glad that I got out."

In a written witness statement for the Judge Advocate Corps investigation of the crash Norman Lafountaine recalled, "The first time I realized we were in trouble was when I felt a jar, but I didn't think too much of it until I saw fragments come through the roof and the fiber from the sound proofing was tearing up and it was coming down in little flakes. The tail ramp operator, Sergeant Williams, was standing up. He looked out the window and when he turned around he looked like he had seen a ghost. His eyes were big. I think what he saw was this Huey that hit, but still as to myself, I didn't know what was going on. I just thought that maybe we started to have a little bit of trouble. Then everything started to happen. The fragments started coming through, the floor started shaking. The noise I heard was sort of a clanking. It sounded to me like guide lines and metal contact and what I figured, being a mechanic and all that the rotor being out of balance, it probably tore up and was conflicting with the balance of the guide lines and that is what that is probably from. We started a downward decent nose down. We were bouncing and clanking.

I was sitting in the aircraft on the right hand side in approximately the sixth seat. I had my cartridge belt with all my gear on it, magazines and all. It was quiet a bulky load, so I had my seatbelt strapped on the outside of it, and when we started to spin, I pulled it tighter and I was held in pretty good, and I put my rifle alongside my leg as I was instructed because the spinning around wouldn't allow me to put it between my legs and hold on to it. I didn't take too much notice of the people around me. PFC Brown was sitting on my left and the person on my right was Lance Corporal John Rittler from VMO-1 and he got out very lucky. He got just a scratch on his nose. To the right of Rittler, who was the fire team leader, was a Corporal Ryan and he got out pretty good too.

As we started spinning and we were completely out of control as far as that goes, without a tail rotor an all, and I got a quick glance out of the window one time during a spin. I couldn't really estimate as we flashed by the window so fast, I couldn't pick up any details to determine how high we were. I just thought it was all over right there. I braced myself. And it seemed forever that we were spinning. There were people yelling and then we finally hit. We were still spinning when we hit. And I'm pretty sure when we hit, the way we were spinning and the way we dug in, that the rear of the helicopter must have broke just about where we were sitting because I remember I was thrown out.

The immense heat was all fiery. Dirt was being dug up. I went through it and I was all by myself. When I began to pick myself up, I was 15 to 20 feet from the plane and lying on my side. My leg all buckled up under me. I got my wits right away and I looked at the plane and it was burning real fast with a yellow-orange flame and it looked like fuel. I could smell fuel and what I figured was that when we hit it might have ruptured the fuel cell. The explosion is probably what sent us out. There was no trace of my seat belt after that. There was some sheet metal around me where I fell. There was a colored boy to my right, maybe a few paces from me. I'm not sure if he was dead or in shock or what. I don't know what his name is. I rolled over on my back and noticed that my leg was all twisted up and I figured that it was broken. I tried to keep my head. I didn't want to go to pieces. I knew if I did I would be in worse trouble, so I grabbed the foot, threw my leg straight out in front of me and backed up so that I left a trail behind me and I managed to get my cartridge belt and everything off, gas mask, and I got up on my elbows and my rear and I started going backwards to get away from the fire because it was burning. I had a few blisters on my face from getting so close to it and that's when I bumped into a body and I couldn't get around it to well. There was shrubbery and I couldn't get over that. John Rittler yelled at me and I looked up and told him that I couldn't move, so he came over and he grabbed hold of me, and he tried to drag me, but I was too heavy for him. He only had me a short distance from the fire, but it was good enough for the moment, and he went and got Sergeant Rudolfo Escotto, a GMSP instructor, who was at the scene at the time. Corporal Noble put me in a pick-up truck. I think it was an MP truck, and Noble stayed with me."

Staff Sergeant Escotto was on the ground and witnessed the collision. "The Huey just took off and flew right into the underside of the '53. It immediately exploded into flames and fell straight to the ground. All aboard must have died instantly. It looked like the pilot of the '53 had it under control but it bounced and fell over on its side. It was beginning to burn as we approached it and the body of one Marine was hanging from the wreakage. It looked like he was dead but I didn't want his body to burn so we pulled him out. The aircraft really began to burn with an intense heat and we had to back away. I heard someone calling for help from the other side of the wreckage and I remember thinking, 'How did he end up on that side.' I ran over there and I saw the Marine (Lafountaine). It looked like he had a broken leg and couldn't move. If we had left him where he was he would have either burned to death or suffocated. Another Marine (Rittler) and I had to crawl to him because the fire was so hot. We dragged him to safety. The pilot appeared to be uninjured but was walking around in circles in a daze and talking to himself. I had one of my Marines escort him from the area. Lieutenant DuFriend rescued the copilot."

Escotto continues, "I was the oldest member of the GMSP instructor staff and I was 32 years old that day. I remember that DuFriend was a former enlisted man and a hell of a nice guy. I stayed in the Corps for 31 years and retired as a Master Gunnery Sergeant. I was a veteran of the war in Korea and I did two tours in Vietnam."

Lance Corporal Victor H. Ridlon, a crash crewman, was on runway watch when the collision took place. "We had just come on duty and were sitting in a crash truck on the "hot spot." I can't remember the entire crew but my buddy Lance Corporal Michael Haggerty was there. Suddenly we heard an explosion. I looked up and knew that they were in trouble. The Huey was on fire and came down and hit the ground about 20 yards from where we were sitting. Our driver, Sergeant Reising, immediately responded and got a good spot on the burning Huey. Haggerty, working the turret gun, covered the downed helicopter with foam and extinguished the fire. I was supposed to work the handline but since the fire was out I took a fire extinguisher and ran up to the cock pit with the intention of rescuing the pilots but when I looked in I could tell that they were both dead. Both bodies were charred and were burnt beyond recognition. As instructed we stayed with the Huey. The '53 had gone down near the end of the runway in the area where we held practice fires. After a while we moved over to the '53 and helped pull bodies away from the aircraft. Everything was covered with foam and there were bodies and parts of bodies everywhere. There was even a torso up in the trees. The smell of burnt flesh was permeating the air. I can still recall that smell to this day."

In the crash barn was Lance Corporal Lawson R. Davis. "I was assigned to a truck that day. It was an MB-5, a truck that had only one turret. I was the turret man. When the crash bell went off we knew that it was a bad one. They announced that we have a midair collision between two helicopters with a number of Marines aboard….Roll! Roll! Roll! All available personnel and trucks were needed. We ran to the truck, jumped aboard, and took off across the tarmac and when we came up from behind the hangers I remember two columns of smoke rising up. I said, 'Oh boy! This ain't a drill.'

Davis continues by describing the crash site, "We had a crash crew training area that was down on the west end of the runway and at the far end of that at the edge of the woods there was a simulated bunker area. The '53 was loaded with a bunch of Marines and it was going to set down and they were going to attack this bunker complex as part of their infantry training. The Huey had come up under the '53 and when he hit it, it busted off the rotors. At that point it was just a bunch of dead weight and it inverted and went straight down killing everybody on board. It blew up and there was a fire. The 'Hot Spot' went to the first fire. That's what we were trained to do. They were fighting the fire with the turret when we went by them on the runway. The driver just floor boarded it and we hauled ass the rest of the way. When we got to the end of the runway he slowed down so that he wouldn't bump us all out of there. There was a little ditch to go over. We cut right into the training area where the '53 had gone down. The adrenalin was super high. As soon as we spotted the training kicked in. The driver slapped my leg and I stepped down on the pedal and started shooting foam. As I worked the turret I looked up and saw the two pilots standing on the edge of the woods. I wasn't looking for any survivors. I was busy. I had a job to do.

The pilot and the co-pilot were alive. They had escaped through the front window and were standing over near the edge of the woods. They looked like ghosts just standing over there watching the plane burn. The crew chief had tried to go out through the front. When we found him his arms were up and his hands were wrapped around the fire partition that separated the cockpit from the belly of the helicopter. He was trying, I guess, to go through that door when it exploded because that's where we found his body with his hands still hooked there. A couple of guys were thrown out of the chopper and their bodies were lying out in the open when we turned them over one of thems face was gone. After we knocked the fire down we took the handlines, it was just a bunch of tangled piping and wires, and put out little fires that existed here and there. We then started the recovery process, removing the bodies from the wreckage. This took most of the day."

Davis continues by describing the recovery process. "It was awful. We would pick the poor guys up and put them in a body bag. Whatever position they were laying in when the fire started was the way they were when we picked them up. They were just burnt to a crisp. If an arm was stuck up in the air, it didn't move, it stayed that way which made it difficult for us to get them in the body bag. It was a mess. Some people did survive although I don't know how many. Other Hueys began to land and started taking people off and bring them to the hospital at Camp Lejuene. Then we lined the bodies up and they also were flown to Camp Lejuene. The only two people that I saw alive were the ones who I thought were the pilots.

As we continued we were instructed to find every weapon and get those serial numbers recorded. We had to work real slow for they were taking pictures and documenting everything. There was a lot of stuff that had to be done, mainly getting the dog tags on the guys. Identifying them then getting the weapons. Someone was writing down all of the serial numbers so that we could account for those. There was a lot of stuff going on. At the end of the day our clothes smelled pretty bad. We were rank. The smell of burnt flesh is terrible. It's a smell that you don't forget."

Another member of the crash crew, Sergeant Floyd E. Miller, remembers the crash. "I was in the crash barn when it occurred. I hurried and put on my bunker gear and drove out to the crash site and when I got there the fire was out. I worked out their all day and participated in the removal of the bodies. It was not a very pleasant job, in fact, even now on very hot days I can remember the smell of the burnt bodies. Most were burnt beyond recognition. It was in June and it was a very hot day so I took off my fire coat and helmet and continued to work in my boots, trousers and gloves."

"I had just gotten off duty and was at the PX when I saw black smoke coming from the area where we held practice fires", recalls Lance Corporal Louis Roman. "I knew that we didn't hold practice on weekdays so I knew something was wrong. I decided to help but by the time I reached the crash site the fire had already been suppressed so I assisted with the removal of the bodies. It was a mess. All of the bodies were completely burned and the smell was unbelievable. All clothing and skin was burned from the ones I saw. One blackened Marine spoke to us as we approached him with a body bag. He said, 'You're not going to put me in that bag' and then he died.

Others mentioned in the chapter are as follows:
Abshire, Bobby W. Cpl. 1979928; VMO-2; 22; Fort Worth, Texas
Albano, Paul J. 1stLt. 088756; HMM-265; 25; Buffalo, New York
Baca, Roger J. LCpl. 2142068; E-2-4; 19; Phoenix, Arizona
Bachta, Thomas E. Cpl. 2051749; 1-F-Recon; 21; Chicago, Illinois
Bailey, Larry J. LCpl. 2203412; HMM-365; 19; Spartansburg, Pennsylvania
Bell, Franklin L. Cpl. 2168482; HMM-261; 20; Los Angeles, California
Bondarewicz, George S. Cpl. 2173484; HMM-261; 20; Buffalo, New York
Bornemann, John F. LCpl. 2056765; VMO-2; 21; Pittsburgh, Pennsylvania
Brewczynski, Stanley L. SSgt. 1503352; VMO-2; 30; Pulaski, Wisconsin
Brown, Andrew R. Pfc. 2295943; MABS-26; 20; Milwaukee, Wisconsin
Buchanan, William L. Capt. 085553; VMO-1; 27; Daytona Beach, Florida
Butler, Gary N. LCpl. 2147570; E-2-4; 19; Oklahoma City, Oklahoma
Callaway, Robert Pfc. 2156844; E-2-4; 19; Butte, Montana
Calvin, Nathaniel Jr. LCpl. 2149612; E-2-4; 19; St. Louis, Missouri
Christy, Howard A. Capt. 069814; A-1-9; 29; Provo, Utah
Corson, Richard III Pfc. 2136761; E-2-4; 19; Hartford, Connecticut
Cummings, William T. Jr. SSgt. 1558314; VMO-2; Memphis, Tennessee
Daniels, Jesus Jr. LCpl. 2140484; E-2-4; 19; San Francisco, California
Davis, Joseph L. LtCol. 048137; HMH-461; 41; New Orleans, Louisiana
Davis, Lawson R. Jr. LCpl. 2198767; MABS-26; 20; Lexington, North Carolina
Davis, Roger A. LCpl. 2101492; E-2-4; 20; Birmingham, Alabama
Dennon, Phillip E. Cpl. 2151461; HMM-262; 20; Sharonville, Ohio
DeRolf, Bruce E. Pfc. 2102223; H&MS-26; 21; Michigan City, Indiana
Donaldson, Billy M. SSgt. 1461338; 1-F-Recon; 30; Valiant, Oklahoma
Drury, Richard L. 1stLt. 088238; VMO-2; 24; Phoenix, Arizona
DuFriend, James M. 1stLt. 089180; H&MS-26; 29; Newton, Kansas
Dusman, Edward R. SSgt. 1555282; HMM-265; 29; New York, New York
Enockson, John O. 1stLt. 085563; VMO-2; 29; Minneapolis, Minnesota
Escotto, Rudolfo SSgt. 1223936; H&MS-26; 32; Stockton, California
Farrell, Leo J. Capt. 085103; HMM-265; 28; Des Moines, Iowa
Fletcher, Sterling A. LCpl. 2141384; E-2-4; 19; Baltimore, Maryland
Gibson, George R. LCpl. 2010487; E-2-4; 20; Pasadena, Texas
Haggerty, Michael LCpl. ; MABS-26; Lacrosse, Wisconsin
Harper, Richard O. Capt. 071918; HMM-265; 30; Beeville, Texas
Harrelson, Richard C. Sgt. 1906461; HMH-461; 25: Raleigh, North Carolina
Hazelbaker, Vincil W. Maj. 063157; VMO-2; 39; Grangeville, Illinois
Hedger, John A. Cpl. 2043471; HMM-265; 22; St. Louis, Missouri
Hinkel, Russell C. LCpl. 2098995; 1-F-Recon; 20; Baltimore, Maryland
Howard, Joseph Cpl. 2130147; MABS-26; 20; Chicago, Illinois
Iverson, Richard L. LCpl. 2159622; E-2-4; 19; Denver, Colorado
Janss, Peter F. Capt. 075407; HMH-461; 32; Des Moines, Iowa
Joyner, Charles D. 1stLt. 087794; HMM-265; 26; Philadelphia, Pennsylvania
Lafountaine, Norman S. LCpl. 2155239; HMM-162; 18; Dartmouth, Massachusetts
Lee, Howard V. Capt. 069961; E-2-4; 33; Dumfries, Virginia
Linick, Gary P. Cpl. 2150529; VMO-1; 20; Chicago, Illinois
Martin, Michael C. LCpl. 2058808; VMO-2; 21; Detroit, Michigan
Mayton, James A. HM1 2973461; VMO-2; 33; Manchester, Tennessee

McDermott, Edward J. Jr. LCpl. 2092855; E-2-4; 19; Hartford, Connecticut
McKay, James E. LCpl. 2095581; VMO-2; 20, San Francisco, California
Meharg, Ben A. Capt. 078224; VMO-2; 30; Austin, Texas
Miller, Floyd E. Sgt. 1674897; MABS-26; 27;
Montgomery, Wayne E. LCpl. 2294480; H&MS-26; 19; Saline, Kansas
Murff, Herbert S. Sgt. 1607851; HMM-265; 28; Pemiscot, Missouri
Naugle, Robert K. LCpl. 2207372; HMM-365; 20; Elizabeth, New Jersey
Neill, Kenneth R. LCpl. 2250964; H&MS-26; 19; Metropolis, Illinois
Newell, David G. LCpl. 2182146; H&MS-26; 18; Weymouth, Massachusetts
Nightingale, Charles E. HM3 9189312; MABS-26; 20; St. Paul, Minnesota
Noble, Jimmy Cpl. 2138682; HMM-261; 20;
Ortego, Conrad L. SSgt. 1165457; E-2-4; 33; Opelousas, Louisiana
Ortiz, Eppie Cpl. 2029702; VMO-2; 21; New York, New York
Pace, James Cpl. 2034336; E-2-4; 21; South Charleston, Virginia
Pace, Robert L. Sgt. 1885775; E-2-4; 25; Seattle, Washington
Parkerson, Milton J. Cpl. 2231213; H&MS-26; 20; Springfield, Oregon
Richey, George G. Jr. 1stLt. 088665; HMM-265; 27; Washington, D.C.
Rider, James W. Capt. 077451; VMO-2; 30; West Seneca, New York
Ridlon, Victor H. LCpl. 2249241; MABS-26; 19; Lynn, Massachusetts
Rittler, Jon B. LCpl. 2181606; VMO-1; 20; San Leandro, California
Roberts, Joseph T. III Capt. 084240; HMM-265; 27; Annapolis, Maryland
Roberts, Timothy A. Cpl. 2138986; E-2-4; 19; Portland, Oregon
Rode, Frederick W. Pfc. 2218812; E-2-4; 18; Philadelphia, Pennsylvania
Roman, Louis LCpl. 2220822; MABS-26;   Palmer, Massachusetts
Ross, Kenneth G. Sgt. 1873293; HMH-461; 25; Peachland, North Carolina
Ryan, Garry J. Cpl. 2154228; HMM-261; 20; Jacksonville, Florida
Scurlock, Jerry W. Pfc. 2260951; HMH-461; 19; Memphis, Tennessee
Shepherd, Allen M. Sgt. 1554080; VMO-2; 26; Philadelphia, Pennsylvania
Smith, David A. Cpl. 2083824; E-2-4; 20; Knoxville, Tennessee
Smith, Robert N. LCpl. HMM-161;
Spence, James J. LCpl. 2274127; VMO-1; 19; Kissimee, Florida
Spink, Shepard C.1stLt. 086010; VMO-2; 25; New York, New York
Stephen, Luke A LCpl. 2037354; HMM-265; 21; Denver, Colorado
Stevens, Tommy E. Sgt. 1912220; E-2-4; 24; South Charleston, West Virginia
Stodelmaier, Frank E. Pfc. 2173643; HMM-261; 20; Elmira, New York
Storbeck, William W. 1stLt. 092242; VMO-1; 24; Highland, Illinois
Sturkey, Marion F. 1stLt. 091161; HMM-265; 23; Plum Branch, South Carolina
Strock, Richard A. Cpl. 2099101; E-2-4; 20; Baltimore, Maryland
Swartz, James J. Pfc. 2151204; E-2-4; 19; Cincinnati, Ohio
Talbert, Rudolph H. Sgt. 2072804; MABS-26; 22; Little Rock, Arkansas
Tarzia, Nicholas C. HM3 6941016; E-2-4; 20; Westbury, New York
Tinsley, Dale L. Capt. 070187; HMM-265; 32; San Francisco, California
Tressler, Clifford C. Pfc. 2300742; MABS-26; 19; Essex Junction, New York
Van, Douglas H. Cpl. 2181505; E-2-4; 19; San Francisco, California
Vance, Danny R. LCpl. 2154364; E-2-4; 19; Cleveland, Ohio
Vicknus, Michael J. Pfc. 2203370; HMM-161; 19; Cleveland, Ohio
Warrick David E. Sgt. 1846579; VMO-2; 26; Pittsburgh, Pennsylvania
Wassan, Kenneth L. Cpl. 2232975; HMM-161; 19; Holden, Louisiana
Williams, Leroy LCpl 2151180; E-2-4; 18; Cincinnati, Ohio
Williams, Thomas L. Sgt. 1658422; HMH-461; 28; Wilmington, North Carolina
Willis, Beasley 1stLt. 088058; VMO-2; 23; Nashville, Tennessee
Winter, Ronald E. Cpl. 2208381; HMM-161; 19; Wynantskill, New York
Young, Ralph M. Cpl. 2214025; MABS-26; 21; San Diego, California

★ ★ ★ ★

Captain Francis J. West, Jr., a Marine reserve officer, was invited to apply for assignment to active duty during the summer of 1966 to research and write small unit action stories. Returning to active duty in May of 1966, he wrote the highly acclaimed *Small Unit Action in Vietnam: Summer 1966*.

The following story is about an action that took place on August 10, 1966 involving members of the First Battalion, Fifth Marines during Operation Colorado. The battalion made initial contact at about 1100 with elements of two North Vietnamese Army battalions that developed into a full-fledged battle fought in a driving rainstorm which did not clear until about 1730.

"If I go down, your job is to get me and the crew out.
If my radios go bad, you've got the lead."

Captain James M. Perryman
VMO-6
Silver Star
From *Gunbird Driver*
By David Ballentine

Small Unit Action in Vietnam: Summer 1966 was published by the government of the United States and is therefore in the public domain.

At dawn on August 11, 1966 the author arrived by helicopter in 1/5's perimeter, some 20 miles northwest of Chu Lai and 6 miles west of Tam Ky, a district headquarters near the South China Sea. On that perimeter 10 hours earlier, the battalion had fought the only major battle of Operation Colorado. The author was well acquainted with the officers and men of the battalion and so, gathering in small groups, they told him in detail what had occurred and pointed out the exact positions they had held. He wrote the somber aftermath from personal observation.

I took the liberty to make some small changes in the text and to add additional information to Captain West's excellent account of the action that occurred that day. His story remains essentially the same.

Also used were the awards citations of many of the participants, the rosters and diaries of the participating units, and information obtained from the National Personnel Records Center.

4

A Hot Walk in the Sun

On August 7, 1966 the First Battalion of the Fifth Marines was dropped by helicopters into assigned objectives. It was a composite lift led by Marine Medium Helicopter Squadron-362 (HMM-362), which was commanded by Lieutenant Colonel Alfred F. Garrotto, which placed the battalion in the vicinity of Hiep Duc. Approximately 90 aircraft were involved in the strike, making it by far the largest in the Chu Lai area. On this day as on every day thereafter, operations were made more hazardous by afternoon thundershowers which brought low ceilings of visibility.

The battalion's area of operations lay some 3,000 meters to the east of the valley where the second battalion was evacuating refugees. Charlie Company was lifted by helicopter into a small valley that had been hit the previous night by a B-52 strike.

As usual the landing zone was hit by air and artillery strikes before the 160 man company landed. The helicopters whirled down and the Marines jumped into the waist deep rice paddies and waded toward the surrounding tree lines, staying in their helicopter teams. Once out of the paddies, the platoon sergeants sorted out their respective platoons, while the platoon commanders oriented themselves on their maps, which was not an easy process. The company was twisted around the paddies in a jagged circle.

The Marines set up in a tight defensive perimeter within three minutes. They followed a well established procedure that they had done dozens of times. This time, like most others, they encountered no resistance.

"Hell," growled one Marine, "it's just going to be another hot walk in the sun."

First Lieutenant Marshall "Buck" Darling studied his map carefully. Satisfied he had located his position exactly; he reported by radio to battalion headquarters and called a meeting of his platoon commanders. The company would sweep east up the mile long valley with a platoon on either side of the main trail and one in reserve. He gave the first platoon two scout dog teams and kept an engineer attachment with company headquarters.

Before the company could move, the engineers had to destroy two booby traps on the trail at the edge of the landing zone. Both traps had been plainly marked to warn the villagers and the markers were still in place when the Marines arrived. One was an explosive charge buried under a pile of loose earth—the other a scooped—out section of the trail studded with bamboo stakes and cleverly camouflaged. With demolitions the engineers quickly disposed of both obstacles.

First Lieutenant Arthur Blades attached a scout dog and handler team to both his first and second squads. The platoon moved forward to search the scattered huts. The area was poor. Most dwellings were small one room huts with hard dirt floors, mud and bamboo walls and straw roofs. Nearly all contained deep bomb shelters. From past experience the Marines knew that there was little chance that enemy soldiers would hide there. When Marines were on large offensive operations, the Viet Cong, unless cornered, fled rather than fought. Only stragglers would go to ground in exposed areas. The German shepherd dogs enabled the Marines to move swiftly. The villagers rarely emerged from their hiding places when the Marines or even the interpreter yelled at them. Bur one low growl worked wonders. Cave after cave was emptied in seconds. Still the search yielded nothing— only frightened women and children.

Blades and the platoon were disappointed. They were spoiling for a fight and thoroughly exasperated with the situation. Nevertheless, the platoon commander did not allow his private opinions to influence his tactical decisions. Throughout the long and empty afternoon he yelled at his squad leaders to keep contact with each other, scolded his troops for bunching up, and insisted his flankers beat through the underbrush and not drift into a single column. The sun sapped the Marines and gradually the pace slackened. After a few hours the dogs showed signs of fatigue and overheating. Blades prodded his men to stay alert. To an observer, he pointed out with particular pride the leadership his squad leaders were showing.

"Look at them," he said, "two are lance corporals and one just made corporal. But I wouldn't want anybody else. They know their people and work hard. They are real hard noses."

In the third hour of the search, the Marines found a house hidden in a tree grove which contained Viet Cong khaki uniforms, medical supplies and U.S water cans. The material judged of intelligence value was saved; the rest, as well as the house was burned.

In the late afternoon, Darling reported to battalion that the valley contained no enemy force. He requested a helicopter pickup. Battalion concurred.

While waiting, the Marines sat down and cooked C-rations. Few felt hungry enough to eat hot canned meat under a hot sun. In small huddles, the Vietnamese children had edged forward to peep at the Marines. A rifleman enticed one little boy to overcome his fear and venture forward to gulp a mouthful of food. Other children followed suit, timidly at first, then with gathering confidence. For the last hour the troops were in the valley, they played with and fed the children.

The helicopters came in and the Marines walked out into the rice paddies to board them.

"You know," Blades commented as he led his platoon to the landing zone, "I've been in this country for 30 days and I've never heard a shot fired in anger. I'm beginning to wonder if there really is an enemy here at all."

The children waved shyly. The adults stood and watched without expression or movement. Charlie Company flew to another objective.

Three days later, they found their fight, a savage, slugging encounter which made them wonder if they would ever again gripe about a lack of action.

After his companies, searching separately for the elusive enemy during the first few days of Operation Colorado, Had met no hard resistance, Lieutenant Colonel Harold L. Coffman had consolidated his 583-man battalion and was sweeping toward the sea, some seven miles to the east. For three consecutive days, the route of the battalion lay along a dirt road which wound through valleys out of the foothills of scrub covered mountains and east across monotonous expanses of flat land stretching to the sea in an unbroken succession of rice paddies, tree lines and hamlets. The troops had uncovered little evidence to indicate the presence of a large enemy force, but each day it seemed they saw fewer villagers, while the intensity of sniper fire increased.

As HMM-363 continued to fly at maximum support of the operation two pilots, Major Peter C. Scaglione and First Lieutenant Walter W. Smith, were wounded in separate incidents on August 9. Scaglione was wounded by shrapnel in the lower right leg while Smith received a fragmentation wound to the right eye.

On the morning of 10 August, the enemy snipers were unusually persistent. All three rifle companies—Alpha, Bravo, and Charlie—encountered small groups of snipers every few hundred meters along the route. Enemy snipers in Vietnam are like hornets, if ignored entirely, they can sting. But if reaction is swift and aggressive, they can be swatted aside. Responding aggressively, the Marines poured out a large volume of fire each time they were fired upon. The snipers, however, carefully kept their distance, rarely firing at ranges closer than 500 yards. The previous day a few North Vietnamese had waited until the Marine point squad was within 200 meters before firing. Those enemy soldiers had been pinned down, enveloped and dispatched.

Coffman and his company commanders did not like the situation for the troops were expending ammunition at a rapid rate with no telling effect upon the enemy. Toward noon, they ordered the squad leaders to supervise very selective return of fire in order to conserve rounds. Marching under a clear sky and a searing sun, Coffman knew the helicopters could re-supply his battalion but disliked making that request if not solidly engaged.

At approximately 1100, the battalion arrived at the hamlet of Ky Phu. Coffman called a halt and the men settled down in what little shade they could find and began opening cans of c rations. Soon word was received for the battalion to remain in position pending the arrival of the regimental commander, Colonel Charles A. Widdecke.

After a conference with Colonel Widdecke, Coffman issued the order to push forward again. At 1400, the battalion resumed the march with the hamlet of Thon Bay as its objective. All indications were that the battalion would reach Thon Bay about 1600. Coffman liked to allow himself ample time to set in before dark. It took a few hours to tie in the lines of a battalion properly and on previous days he had allowed several hours for the task

The companies guided on the main road which led to the sea just as they had done on previous days. In front of the Marines lay acres of rice paddies gridded by thick tree lines and tangles of scrub brush, familiar enough landscapes. Groves of palm trees and patches of wooden huts dotted the roadside. Storm clouds were billowing over the mountains to the west behind the Marines.

Charlie and Alpha Companies, forming a dual point, struck off together, covering respectively the right and left flanks of the road. Both companies spread out far across the paddies. In trace along the road followed the battalion command group, consisting of the 81mm mortar platoon, the battalion headquarters, the 106mm recoilless rifle platoon (sans anti-tank guns), the logistics support personnel, and others. Bravo Company brought up the rear. The battalion was thus spread in a wide "Y" formation, the stem anchored on the road and the prongs pushed well out in the paddies.

Upon resuming a march, a battalion commander can generally expect a time lag of several minutes caused by a few false starts as squads, platoons, and companies bump and jerk along before sorting themselves out and hitting a smooth, steady pace. This did not happen on the afternoon of 10 August. The battalion moved swiftly. The platoons at point fanned out on both sided of the road. Slogging through paddies and twisting through tree lines, they covered more than a mile in the first 20 minutes. Rain was washing away their sweat and impeding their vision when they arrived on the outskirts of the tiny hamlet of Cam Khe at 1510. They noticed that the huts they passed were empty and there were no farmers working in the rice paddies. Giving cursory glances into dugout shelters and caves, the Marines saw that they were packed with villagers.

Because of this fact the men were alert and wary when they passed through and around the hamlet. As the second platoon, Alpha Company, pushed through the scrub growth on the left flank of the hamlet, the men saw to their front a group of about 30 enemy soldiers cutting across a paddy from left to right. The platoon reacted instinctively. They did not wait to be told what to do. Throwing their rifles to their shoulders, they immediately cut down on the enemy. Their initial burst of fire was low, short and furious. Caught in the open and moving awkwardly through the water and slime the enemy could not escape. Shooting from a distance of less than 150 yards, the second platoon wiped them out in seconds. Farther back in the column, men thought a squad was just returning fire on a sniper. Although they did not yet know it, the battle that the Marines had sought was joined. The Marines had struck the first blow and it was a painful one.

The North Vietnamese, however, counter punched hard. From a small hedgerow behind the fallen enemy, several semiautomatic weapons opened up and rounds cracked by high over the heads of the Marines. The troops were now keyed for battle. Excited and stirred by their swift, sharp success, the platoon shifted its direction of advance and splashed into the paddy. The volume of enemy fire increased and bullets sprouted in the water around the Marines. The platoon momentum slowed as the Marines started flopping down into the water to avoid the fire. But no one had yet been hit, and the platoon quickly built up a base of fire and continued the movement by short individual rushes.

The company commander, Captain James R. Furleigh, came up, bringing with him the first platoon. That unit in turn rushed into the paddy on the left flank of the second platoon. The volume of enemy fire was swelling. With seventy bulky, slow moving targets to hit at close range, the enemy gunners improved their aim as well.

Almost in the same second, a man in each platoon was struck by machine gun bullets. Other Marines stopped firing to help the wounded or merely look. The rate of outgoing fire dropped appreciably. Encouraged by this, the North Vietnamese redoubled their rate of fire. No longer forced to stay down themselves they aimed more carefully and bullets hit more Marines lying in the water. The machine gun in the tree line in front of the second platoon chattered insistently as it traversed back and forth in low sweeping bursts over the Marines heads. Two more Marines were hit. In an attempt to avoid the murderous fire the men ducked low and fired even less in return. The attack had bogged down.

The second platoon had advanced 40 meters across the paddy; the first platoon no more than 20. In the tree line 100 meters away, they could see the North Vietnamese moving into better firing positions. Most of them were wearing camouflaged helmets some had on flak vests. The Marines could find no cover or concealment in the paddies and time was running against them. The machine gun had them pinned. The rain, and mud and their heavy gear prohibited a quick, wild, surging assault. Furleigh, a sharp eyed, quick minded West Point graduate assessed the situation. As he saw it there were two alternatives. They could either go forward or back. What he could not do was let the company stay where it was. He knew that if he urged the men they would go forward in bounds until they advanced into the tree line. Casualties from a frontal assault against an effectively placed machine gun would be heavy. As he considered his options three more men were hit. That clinched it for him and he decided to pull back. By retreating his men would escape the heavy fire and there would be time for the situation to clear. Battalion could then issue specific orders.

What Furleigh did not realize at that particular moment was that heavy fighting was raging in half a dozen other places, including battalion headquarters. While Company A was attacking in the paddy mortar shells had fallen along the road, just missing the battalion command group. The headquarters element was quite distinguishable with its fence of radio antennas had hastily sought the concealment of the bushes and houses to the left of the road. The NCOs shouted to their sections to disperse yet stay close and the radio operators tried to copy incoming messages and transmit at the same time. The officers were busy trying to pinpoint their position and decide on a course of action, whenever one was taken under small arms fire coming from all directions. Reports filtered in by runner and radio that Alpha Company to the northeast was pinned down and withdrawing and that to the west, at the rear of the battalion, Bravo Company was battling. To the east, on the right side of the road, Charlie Company reported that it too was engaged.

In that situation, the battalion commander could not determine precisely the size or the nature of the engagement. Unbeknownst to Coffman was that the enemy had also been surprised by the sudden engagement. The headquarters group was busy defending itself. It was raining in such heavy sheets that at times figures only yards could not be seen. The visibility ceiling for aircraft had dropped to 50 feet so that no jet or helicopter support was available.

Coffman stayed calm. His was a seasoned battalion that he had commanded for 12 months. He knew his company commanders. Faced with a battle that denied tight central control he let his junior officers direct the fighting while he concentrated on consolidating the battalion perimeter as a whole and shifting forces as the need arose.

The situation was terribly confused. Although the battalion was on the defensive, its individual units were on the offensive. Platoons from each company were attacking separate enemy fortified positions. Caught in an ambush at his left flank, Coffman refused to draw the battalion in tight. In attempting to consolidate the companies had to fight through enemy groups. To relieve pressure on units particularly hard pressed Marines not personally under fire moved to envelope the flanks of the North Vietnamese. The extraction of wounded comrades from the fields of fire is a tradition more sacred than life. It was best accomplished by destroying the positions of the North Vietnamese which had the Marine casualties under fire. So isolated were the fragments of the fight that each action is best described as it happened—as a separate event. Fitted together these pictures form the total picture of a good, simple plan which was aggressively executed, with instances of brilliant tactical maneuvers occurring at crucial moments.

By reason of extremity, Furleigh's Alpha Company played the key role in the fight during the initial hour. They were in the thick of the fight from the start. Private First Class Larry Bailey, a mortar man assigned to company headquarters, had moved up with his company commander. He described it in this way: 'The VC were everywhere. They were in the banana trees; they were behind the hedgerows, in the trenches, behind the dikes and in the rice paddies."

Bailey soon became a casualty as he suffered a gunshot wound to his right arm.

Pulling back out of the paddy had not proved easy, several of the Marines had been wounded and one more was killed, bringing to a total of five the number of American dead in the paddy. The dead included Sergeants Ernest B. Amador and John D. Smith, Privates First Class' Lawrence J. Kindred and Everette A. Thompson and Private Richard A. Skinner. Smith, an Army veteran, died as a result of a fragmentation wound to the abdomen. Displaying excellent fire discipline, the North Vietnamese singled out targets and concentrated a score of weapons on one man at a time. That unified enemy fire altered the exposed positions of some Marines from dangerous to doomed. Those already rendered immobile by wounds were most vulnerable to sustained sniping. These casualties, which included Bailey, had to be immediately moved from the beaten zone of the bullets.

It was an arduous moment. No man could stand erect in that storm of steel and survive. So the wounded were dragged along through the flooded paddies by their buddies, much like an exhausted swimmer is towed through the water by a lifeguard. Fire was returned on the enemy by those Marines who were not dragging or being dragged. No on later felt that their fire had inflicted casualties on the enemy. Though not lethal this fire was delivered in such steady volume that it forced the enemy gunners to shoot without aiming properly. Had the platoons not reestablished a steady stream of return fire, it is doubtful they could have escaped. By doing this they maintained unit integrity, keeping their squads and fire teams intact and bringing their wounded with them. By repeated exhortations, curses and orders, Furleigh provided the guidance necessary to steady the men and prevent any slackening of fire in the moments of confusion.

Once the two platoons had reached the hedgerow, they spread out to form a horseshoe perimeter with the first platoon to the left of the second. The open end of the horseshoe faced southwest, toward the battalion command group, and the closed end faced the enemy to the north. Furleigh tried to call artillery fire down on the enemy. In the full fury of the thunder and lightning storm the adjusting rounds could not be seen or heard. Nor did anyone in Alpha Company have a clear idea where the front lines of the other companies were. For fear of hitting friendly troops Furleigh cancelled the mission after the first adjusting rounds had gone astray.

Conspicuously absent from the battle at this crescendo was the mighty fire power of supporting weapons, proclaimed by some critics as the saving factor for Americans in encounters with the enemy. The North Vietnamese had numerical and fire superiority. Initially, it was they who freely employed supporting arms, namely mortars and recoilless rifles. What had developed for Alpha Company – and for the battalion—was a test of its rifleman.

They responded magnificently. Once tied in, the two platoons of Alpha Company needed no urging to keep fire on the enemy. At first there were an abundance of targets to shoot at. The North Vietnamese kept leaping up and darting about from position to position. The Marines, lying prone and partially concealed in the undergrowth put out a withering fire. The men shouted back and forth, identifying targets and exclaiming at their hits. They were getting back for the frustration in the paddy. Losing in a contest of aimed marksmanship, the North Vietnamese pulled a 60mm mortar into plain view and aimed it at the opposite hedgerow. While they were dropping shells down the tube the Marines who could see the weapon were screaming: "Give us a couple of LAAWs—LAAWs up!" Several of the short fiberglass tubes were passed forward, thrown up from man to man. Some, after long immersion, in the water, failed to function but others did and a direct hit was scored on the mortar.

While their attention was concentrated to the front the second platoon came under heavy fire from the right. So low and steady were the bursts from the automatic weapons that the platoon was unable to move against them. When the bullets came tearing in they carried with them the sound of the weapon. Had the fire come from across the paddy the rounds would have passed before the weapon could be heard firing. Thus, Furleigh judged there was a dug in force within the hedgerow not over 60 meters from his right flank. He called battalion and asked that Charlie Company be committed to attack along his right flank. Coffman concurred and ordered Lieutenant Buck Darling to the assault.

Struck to earth any time they stood up the North Vietnamese opposite Furleigh had ceased their jack-in-the-box tactics and were staying low. The machine gun which had stopped the enemy cold in the first attack swept wide steel swaths over the Marines heads. Attempts to knock out the gun had been unsuccessful and had cost the company lives.

Lance Corporal Richard P. Donathan had been the first to try. Donathan was lying near Furleigh when the machine gun first opened up and killed some Marines. Known throughout the battalion for his aggressive actions in fire fights, he was not cowed by the near presence of death. He Asked Furleigh if he could work his way around the right flank "to get the gun." Furleigh told him to go ahead and he had set off. Several other Marines just got up and followed him. H moved rapidly up a trail on the right of the hedgerow, his swift foray catching some enemy soldiers by surprise These his small band cut down but the sound of the firing alerted the machine gun crew. The gun swung toward them. Caught in the open, the raiding party was at the mercy of the enemy. Behind Donathan a Marine went down. The men on the line heard Donathan shout, "Corpsman!"

Hospitalman Third Class Terry C. Long hurried forward. He found the wounded man lying on the trail in front of the hedgerow. While he was bandaging the man he heard from up the trail Donathan shout again, "Corpsman!" Long left the first casualty, having assured him that he would return, and ran on. Several yards farther, he came across another Marine, hit in the leg. The casualty told him that Donathan had gone on alone. Long went forward to look for him.

Both men showed singular fortitude and determination. To go forward alone against the enemy who has struck down all others takes raw courage. A deliberate, conscious act of the will was made by each man when he went on alone, knowing that he did not have to do so. Donathan went forward driven by his determination to eliminate the machine gun nest. Long went forward sensing that Donathan might need him.

He worked his way carefully, bent over to present a smaller target. Occasional clusters of bullets whizzed past him. He saw a pack lying near some bushes and identified it as Donathan's. He dropped his own pack beside it and continued on, armed with a pistol and clutching his medical kit. A few yards farther on he saw an M14 and a bandolier of ammunition lying on the trail. He knew Donathan could not be far away. He looked into the bushes growing on the side of a bank next to the trail.

There was Donathan, wounded but still conscious. Long slipped down to him and began dressing the wound. He had almost finished the task when he was hit. He cried out and pitched over Donathan who then sat up and reached for him.

"Where you hit, T.C.", He asked?

"Back of the knee," Long replied. "The right one. Went right through—maybe shattered."

Despite his own wounds, Donathan managed to inject Long with morphine. He was trying to bandage the knee when two bullets tore into his back. He fell on top of Long, still conscious but unable to move. Pinned by Donathan's weight and weak from the morphine and his wound, Long could not wiggle free. Lying in each other's arms they talked back and forth and tried to comfort one another. It was mostly just idle talk, like many previous conversations they had in rear areas. After a while Donathan's voice just trailed off. Death claimed him quietly.

Long lay, in the mud, under the body, while rain pelted his face. Despite the morphine he felt a terrible stinging in the back of his right knee. Time passed. But no one came.

It was not that they didn't try to find him. Although the trail was raked by fire, Marines crawled out by ones and twos from the hedgerow to pull back the others who had accompanied Donathan. There were five bodies sprawled in plain view of the enemy. Four were retrieved by Marines who crept through the bushes to the edge of the trail, then reached out and pulled the wounded men back into the concealment of the hedgerow. The fifth casualty, Sergeant Sam Baker, lay in the corner of a rice paddy. Each time a Marine left the undergrowth to edge toward him a fusillade of shots would force him back. Finally, Private First Class Robert English, a man light on his feet and agile in his movements, sprinted from the hedgerow, grabbed Baker and ploughed back into the brush before the North Vietnamese found the range.

Furleigh then had his platoons intact and accounted for with the exception of Long and Donathan. Private First Class William Bielecki, the company radioman who had formerly been with the third platoon, went off to look for them—a lone man in search of two lost friends. He found their packs, and the rifle and ammunition. Bielecki must have stood within twenty feet of them when he retrieved the rifle but he did not see or hear them. He looked down into the bottom of the ditch. Bielecki might have looked right over them but they were shielded from his view. As he scrambled back to the lines he became because fatigued because in addition to his own rifle he carried the packs and the rifle of the lost Marines. His movements were awkward. Nevertheless he struggled on until it suddenly occurred to him that salvaging nonessential gear under heavy fire really was not necessary or wise. He threw away the packs. Bielecki entered the lines carrying two rifles and reported to his company commander.

From Bielecki's report, Furleigh guessed that both Donathan and Long had been killed and their bodies dragged away by the North Vietnamese. The machine gun was still firing whenever a Marine exposed himself. But the rain and wind had slackened and two armed Hueys whirled over the battlefield. So entangled were the battle lines that neither Furleigh nor the other company commanders were able to direct the pilots on targets. The North Vietnamese inadvertently solved the problem by firing at the helicopters, the troublesome machine gun 150 yards to Furleighs front being one of the first enemy weapons to do so. The Hueys responded viciously, diving to pump hundreds of rounds into the tree line, turning low and tightly, and then raking the area from the opposite direction. Incoming on Alpha Company dropped abruptly as the enemy ducked into holes. Furleigh took full advantage of the respite to move his wounded to the battalion aid station located to his rear.

With Charlie Company driving on the right and Alpha Company holding steady to the front the North Vietnamese began pulling their forces westward, which was to the left, in an attempt to outflank Furleigh and drive toward the battalion command group. The men of Alpha's first platoon were becoming worried about their left flank, having seen several of the enemy scurrying in that direction behind the paddy dikes. Furleigh radioed to the third platoon, which had been in the rear near the battalion command group skirmishing with snipers. He told the acting platoon commander, Staff Sergeant Albert J. Ellis, to bring the platoon up and refuse the left flank of the first platoon.

The 37 men of the platoon moved forward to the north edge of the hedgerow. They were immediately engaged by the enemy who were running left along a scrub covered paddy dike 100 meters to their front. The volume of fire was intense, preventing the Marines from slipping farther right to tie in with Furleigh's group. Guided by the experience of combat the platoon members fanned out and flopped down to form a semi-circular perimeter. They were not in visible contact with the first platoon but could hear the sounds of American weapons about 60 meters to their right. To the front and left were the enemy soldiers. The third platoons fight was the Marine rifleman's dream: an engagement in which the enemy clearly showed themselves and tried to sweep the field by superior marksmanship. The third platoon had waited long for the enemy to make the mistake of choosing to stand and fight.

With dozens of visible targets the platoon at first ignored the basics of fire discipline and everybody just blazed away. The impact of the heavy 7.62mm bullets knocked some of the enemy completely off the dike and sent them spinning and thrashing into the paddy. The noise was deafening. The platoon commander, Sergeant Ellis, a veteran of the fighting in Korea and a Marine since 1952, was furious. His men simply didn't have enough cartridges to expend at such a fast rate and he doubted they would receive a resupply as long as the wind, rain and lightning continued. Ellis almost went hoarse shouting, "Knock it off! Knock it off! We don't have enough ammo. You squad leaders get on your people!" Slowly the volume of outgoing fire dropped.

The platoon settled into a routine set more by reflex action than design. Made acutely aware by Ellis that they might fight indefinitely, the men, only minutes before profligate, became absolutely miserly in their use of bullets. They would fire only when a distinct target appeared and then generally but one round per man.

The enemy, pushed to the earth, started building up their base of fire. Soon automatic weapons were rattling all along the dike and the Marines felt the sharp blast of 60mm mortar shells slamming into their perimeter. The North Vietnamese had better cover and firing positions than the Marines. They could steady their weapons on the mud dike and expose only their heads and shoulders while firing. They had ample ammunition and outnumbered the Marines perhaps five to one. One out of every four Marines was hit in the fight but only two were killed. The enemy used aimed fire, attested by the fact that every member of the platoon later recounted seeing the splashes of rounds hitting near him.

Near misses to Marines were common and one was even comical. Lance Corporal Robert Matthews, a fire team leader in the first squad, was firing from the prone position when a bullet hit his pack and knocked him sideways. He lay quite still, feeling a hot, sticky substance spread over his back. He yelled, "I've been hit!" Another rifleman crawled to him and gently slid off the damaged pack. Then the rifleman laughed and said: "That's not blood." Matthews' 'wound' had been caused by the bursting of a can of shaving cream.

Corporal Rodney Kohlbuss second squad took several casualties in quick succession. He ordered his men to pick up the wounded and move to more protected positions. The men found the shift difficult but so strong are the habits of training that they tried to take all of their equipment with them. Kohlbuss yelled to them to drop the excess gear and move. This they did while the other two squads provided covering fire.

In addition to those killed or wounded in action, Ellis had one man, Private First Class George Fudge, missing from the platoon during the first hour. When the fight first began Fudge was walking well ahead of his platoon in order to maintain contact with the first platoon to their front. During the initial burst of firing he thought he heard a strange sounding machine gun to his right front. The first platoon seemed not hear it, for they veered toward the swelling sounds of the fight to the left. Still thinking the point was just brushing off snipers Fudge was reluctant to alert the third platoon by voicing his suspicions. Instead He decided to investigate the noise he had heard. Avoiding the main trail he cut between two huts and proceeded to pick his way carefully through a thin screen of underbrush. When he was abreast of the back yard of another house he stopped to look and listen. He flicked an indifferent glance at the yard which was studded with stumpy banana trees and was about to proceed when he looked again in disbelief. The trees were walking. Fudge was not inexperienced. A trained sniper he had spotted and shot several well camouflaged Viet Cong in previous battles. But never had he seen such perfect concealment. Had the North Vietnamese not moved he would have walked right past them and probably been shot in the back.

With their backs toward him the North Vietnamese were clustered around a machine gun set up on a paddy dike. Fudge did not hesitate. It never did occur to him to go back and get help. Standing 50 meters from the enemy soldiers, he raised his rifle to his shoulder, sighted in carefully and fired twice. Two of the enemy fell and the others, obviously stunned, turned and just gaped at Fudge. He fired two more times and two more enemy soldiers went down. Before he could fire again the fifth enemy soldier reacted like a stuntman in a war movie. Pushing off from his heels he flipped backwards over the dike in a somersault and came up blazing away with a submachine gun. That alerted other North Vietnamese that there was an enemy in the midst of their positions.

Bullets whipped by Fudge from all directions. He fell flat and lay perfectly still for a moment. He was startled by the savage, if belated, onslaught and angered that he had missed a perfect score. The unmanned machine gun attracted his attention next. He threw a grenade and it landed squarely on target. Satisfied with himself on that account he crawled back toward his platoon belatedly aware that the company had engaged more than a few snipers.

En route, he bumped into a party of men from the first platoon who were moving the wounded to the rear. He joined them and helped carry the poncho liners. Fudge made several round trips and added to his tally when an enemy soldier stepped out from behind a bush 150 meters away. He dropped him with one round.

When Fudge finally rejoined the third platoon Ellis was so glad to see the deadly sharpshooter that he didn't chew him out for being gone so long. He just sent Fudge into the fray and told him to get busy. He did not disappoint his platoon commander. Before dark his rifle brought down five more of the enemy.

The platoon was armed with LAAWs, grenade launchers, machine guns and rifles. The men who had LAAWs and M79s engaged the fortified positions from which the enemy was laying down a web of cross fires. Lance Corporal Robert Goodner proved most effective with the LAAWs. With one shot he blasted an automatic weapons emplacement 150 meters away. The back blast from the recoilless weapon whipped up a gust of spray which marked his position and under a hail of bullets he half crawled half swam away.

Others had noted his success and he was asked to try for another gun which had a group of Marines pinned. Goodner wormed his way to a vantage point, waited until the gun fired and sighted in. The range this time was 250 meters. The rocket hit the target squarely and pieces of the gun flew into the air.

To conserve ammunition the machine gunners kept their bursts extremely short but with so many targets the barrels were soon steaming. Having hastily set up the gunners found that their fields of fire were extremely limited. Attempts to shift positions for delivery of enfilade fire were thwarted by the special attention given them by enemy gunners. The weapons squad, Lance Corporal Ronald Moreland, was one Marine who did not curse the rain for it kept his guns from overheating and malfunctioning.

The performance of the rifleman was a study in marksmanship. The leaders of the platoon had been known to walk the line in a firefight urging the men to "hold them and squeeze them, hold steady and shoot low." The men had overcome their initial desire to fire frantically and were putting out rounds one at a time, firing sparsely and carefully. Ammunition had been replenished slightly by taking the bandoliers of the casualties and redistributing them to those still firing. The Marines had a clear view of the dike. At 100 yards the North Vietnamese were in serious trouble dueling with rifleman trained to hit a 20 inch bull's-eye at 500 yards.

Failing to break the Marine perimeter by frontal fire, the enemy again tried to shift their forces and turn the Marines left flank. Corporal Carl Sorenson held that flank with his third squad. His men shouted to him that they could see large groups of the North Vietnamese crawling and darting to their left. He passed the word to his platoon commander, Ellis, who then told Furleigh, who in turn notified battalion. Lieutenant Colonel Coffman called Bravo Company, already fighting their way forward, and told them to get a unit up on the double to tie in to the left flank of the third platoon of Alpha Company.

Kohlbuss' squad exacted a terrible toll when the enemy lifted their base of fire and tried to slip past. In moving, the enemy soldiers exposed part of their bodies over the top of the dike. Every time an enemy soldier rose up the entire squad would fire together. They developed a rhythm to their volleys. It was like knocking down ducks in a shooting gallery. A figure would pop up behind the dike, a dozen rifles would crack and the figure would pitch sideways and disappear from sight.

But while the battle to the front was going well for the third platoon the pressure on the left flank was increasing. More and more of the North Vietnamese were coming from the northeast and trying to cut wide around the platoon. The men of the third squad estimated that between 75 and 100 enemy soldiers were seeking to skirt around them. The platoon did not have enough ammunition to beat off a determined attack by a group that size. They needed help so Ellis sent Sorenson to the left rear with instructions to find Bravo Company and guide forward a relief force.

Bravo was moving up but not without difficulty. When the fight had first begun the company was spread out far to the rear of the battalion command group. The march order went: first platoon, second platoon and third platoon with the first platoon dispersed through both the paddies and the tree lines on the left flank.

The first platoon ran into trouble shortly after Alpha Company became engaged. To their flank they saw about 40 of the enemy with bushes tied to their backs trotting north across a wide field. They were seeking the concealment of a tree line. The platoon fanned out and gave chase. The second squad surged ahead and swept through the same field the North Vietnamese had deserted. The first and third squads were slightly to the rear and keeping to the edge of the hedgerows.

Twenty feet from the tree line the second squad was lashed by a blaze of automatic weapons fire. Trapped in the open the squad was hit hard and men began yelling for help. Three of the nine men had been badly wounded and one had been killed. The Marines who could still fire did so and the sound of their weapons brought help. One of the wounded was Private First Class Terrance P. Bourgett who suffered a penetrating wound to the right foot.

The second squad was lying flat in the grass and the other two squads, staying in the hedgerows to the rear, could not see them. By this time they were under fire and being kept busy. But Sergeant Darwin R. Pilson, the right guide of the platoon, worked his way forward to the sound of the M14s. The second squad had been expending rounds feverishly trying to smother with fire an enemy machine gun and then move out of the open into the cover of the tree line. The North Vietnamese, however, had returned round for round from a deep trench and the squad had made no progress. When the Marines' fire became particularly intense, the machine gun would stop firing, only to begin a short time later from a different section of the trench. The men shot several enemy soldiers who incautiously poked their heads through the underbrush but they could not knock out the gun which was delivering fire not a foot over their heads. When Pilson reached them, Private First Class Eugene Calogne had just killed a sniper with his last bullet.

Pilson dumped his ammunition on the ground beside them and told the squad leader, Corporal Catarino V. Nuncio Jr. that he would bring help, and crawled away. He reached the company headquarters and reported to the company commander, Captain John P.T. Sullivan that the squad was pinned down and had taken several casualties. He grabbed a grenade launcher and was about to set off again when Sullivan told him, "Slow up, I'll get you some help."

Sullivan now had three distinct problems to solve. In addition to the second squad's predicament he had just received word from the battalion commander to send men forward to block the left flank of Alpha Company which was in danger of being enveloped. And the company command group itself was being subjected to intense fire from a village 500 yards to their left rear.

The first platoon was fighting on the flank and the third platoon was guarding the rear, under fire but not pressed. That left the second platoon to commit. Sullivan split the platoon, sending the second and third squads up the road to find and help Alpha Company, while Sergeant Ronald Lee Vogel took the first squad and set out to relieve Nuncio's squad.

Nuncio would soon receive gunshot wounds to his face, legs and right arm.

Pilson had gone ahead laden with ammunition and spoiling for a fight. Marines engaged in a dozen places saw him go by, moving steadily into the thick of it, stopping only to fire or reload or throw a grenade. Perhaps the gods of war favored the dauntless that day, since he never got scratched though men fell on both sides of him. He reached the second squad, distributed more ammunition, and joined the fray.

Vogel's squad slugged its way forward. The rain was falling in sheets and the North Vietnamese held many of the intermingled hedgerows. It was impossible to identify a man at 70 yards. Vogel lost a man before they had gone a hundred yards when a figure in utilities and a Marine helmet loomed up out of the dusk across a paddy. The squad paid him no attention until he fired and killed a Marine then ducked to the undergrowth. The slain man's friend, Lance Corporal Robert Monroe, jerked a grenade from his cartridge belt shouting, "I'm going to kill that bastard." Vogel told him to keep low and stay in the hedgerow but Monroe, beside himself with fury, started to move into the open anyway. Vogel reared up and hit him with the force of the blow knocking Monroe flat. Other Marines held him fast until he calmed down and agreed to follow orders.

Vogel's men reached the field without taking any other casualties. They split up and crawled through the grass to search for the dead and wounded of the second squad. None presumed to assault the trench line only a few feet away. Their orders were to recover the casualties and it was this task that they set about doing.

But in a fascinating testimony to the thoroughness of the training they had received, the unwounded Marines of the second squad had continued the attack. Nuncio and some others crept forward trying to penetrate the enemy lines. The grass and their closeness to the earth impeded their vision so the squad members could not see one another, yet they all moved in the same direction—forward.

Vogel had to split his own squad to find them. Some men dragged the casualties back, while others inched forward listening for the sound of M14s. Monroe found Private First Class Gregory Pope lying under a bush a few yards from the tree line. Pope was in the rifleman's classic prone position, legs spread, elbows up and in, cheek resting along the stock of his rifle. So intent was his concentration that he ignored Monroe's presence at first. From the constant cracklings overhead it was obvious to Monroe that the enemy was equally intent on disposing of Pope and had a lot more firepower. Monroe, fully recovered from his irrational rage, now in turn became exasperated with Pope.

"Hey," he yelled, "what are you doing? You're all alone out here."

Startled, Pope replied, "Is that right? Then let's get the hell out of here."

Monroe wasn't exactly right although he had no way of knowing otherwise. Three Marines had almost succeeded in storming the trench. This group was aggressively led by Private First Class Lolesio Peko who had reckoned that if they were able to sneak close enough, they could rush the crew of the machine gun before the enemy moved to another position. So thinking, Privates First Class' Peko, Eugene Calogne and Carmelo Millian edged toward the sound of the gun and right into its field of fire before they realized that they were trapped. Peko and Millian were hit as they moved between two trees not over 20 feet from the trench. Peko suffered shrapnel and missile wounds to the back, head and left leg while Millian had been shot in the abdomen. The trees were about 13 feet apart and were used as aiming points by the gun crew. Calogne helped the two men crawl to shelter behind the tree on the right and there the three lay, listening to the bullets fly by and pitching grenade after grenade into the trench with no noticeable effect. They had run out of grenades and were firing carefully spaced single shots when some men from Vogel's squad heard them and came up. These Marines stayed to the left of the machine gun's fire lane and protected the flank. Hospitalman Third Class Harold Lewis reached Peko and Millian and he and Calogne pulled them back.

This completed the extraction of the second squad, a unit of resolute men.

The wounded were brought back to Bravo Company's command center, a position at that moment almost as perilous as the ones where they had been hit. It was an easy target to mark since the Marines were constantly bringing in wounded and ducking out with ammunition and instructions. Among the wounded were Private First Class Joseph Brown, who had a fragmentation wound in the right wrist and a complete loss of hearing, and Lance Corporal James R. Jackson with shrapnel wounds in the right cheek and right leg and arm.

The command group was always in motion with Gunnery Sergeant Thomas Beandette, Corporal Theodore Smith and Hospitalman Second Class Robert Feerick sallying forth several times to bear in the wounded. Privates First Class Patrick Scullin and James Henderson wandered in carrying a wounded Marine and a prisoner whom they had captured by knocking him unconscious. They had become separated from the machine gun section with the second platoon and together had fought up and down the left flank before making contact with some other Marines.

From a village 400 meters to the northwest the North Vietnamese brought heavy weapons to bear on the command center. They tried to hit the Marines clustered there with a 3.5 inch rocket launcher and missed. They then tried with 57mm recoilless rifles and missed. They also tried with .50 caliber machine guns and the rounds went high.

They tried with a 60mm mortar and succeeded. They almost blew up the wounded of the second squad at their moment of deliverance. The casualties were being carried across the paddy in front of the hedgerow when a mortar round plunged down behind them. The litter bearers hastened their steps and gained the concealment of the bushes just as a second shell burst behind them. Lance Corporal Van Futch, a company radio operator sitting in the hedgerow, had been watching the mortars chase the wounded across the paddy, and thought: "Uh-oh, here it comes now. The next one will be right in here." He was correct. The next shell dropped in the middle of the hedgerow and struck down two more Marines. Flak jackets were hastily thrown over the wounded as the men prepared to receive more incoming.

None came. A rifleman had located the enemy mortar pit and Sergeant

Peter Rowell quickly fired his own 60mm mortar. The counter mortar fire silenced the enemy weapon. Rowell would later be wounded in the face by shrapnel and shot in the left arm.

During that exchange Sullivan took shrapnel in the leg. It slowed him down but did not impede his effectiveness. He was more worried about communications. The North Vietnamese had come up on the battalion's radio frequency and were jamming radio contact between the companies. Over the air the enemy played music, jabbered at a fast rate and whistled shrilly.

It worked but the Marine communicators didn't let the enemy know it. It had happened to the battalion on a previous operation and the battalion communications officer, Captain Milton L. Harman, had profited from that experience. His communicators were instructed to ignore the interference and continue transmitting as if nothing were wrong. After a while the enemy gave up the jamming. But Sullivan was uneasy over the prospect that they could resume the tactic at any time.

And after pulling back the casualties and straightening out his left flank he needed to contact his third platoon. The second platoon was fighting with Alpha Company and the men around him, from the first platoon, were near exhaustion from their efforts. He had to have fresh troops to carry the wounded to the battalion evacuation point and he could use more firepower if the enemy persisted in probing a route for envelopment.

He managed to contact the third platoon commander, First Lieutenant Woody F. Gilliland, and that staunch individual lost no time in bulling his way forward. His platoon arrived fresh and intact. The company commander and his gunnery sergeant made no effort to hide their feelings at the sight of the ex-football player jogging toward them along the hedgerow.

"Brother, I could kiss you!" exclaimed the gunny, momentarily forgetting rank and sex.

Sullivan turned responsibility for the casualties over to Gilliland, whose platoon carried them to a rice paddy marked as a landing zone. Suitably, a landing zone must be secure—free from hostile fire—before the helicopters can land. But large, lumbering craft though the H-34 troop helicopter is, it can be surprisingly difficult to destroy, as events were to show.

It was still raining but the ceiling had raised enough for the medical evacuation helicopters to come in, provided they were not shot out of the air. The troops on the ground tried to clear an area but it just couldn't be done. The battalion command group was still using rifles and the 106mm recoilless rifle platoon was shooting snipers out of trees. The three rifle companies were fighting tooth and nail. There was no respite. The Hueys were striking all around the perimeter. The Marines marked their lines or targets by popping smoke grenades, only to have the enemy follow suit. A Marine would pitch a yellow smoke grenade and no sooner would it billow than a half dozen clouds of yellow would filter from surrounding hedgerows.

Sullivan resorted to the standing operating procedure established for such emergencies. The troops would heave a combination of different colored grenades and the pilots would identify over the radio the color schemes. Notified when they had seen the right combination, the Hueys could bear in. Their presence suppressed enemy fire but the minute they flew off; the North Vietnamese emerged from their holes and resumed the battle. As Gilliland gathered the casualties, he knew their only chance of being flown out depended on the skill of the H-34 pilots. If they came in they would do so virtually unprotected.

Of the four H-34s, from Marine Medium Helicopter Squadron-362, which conducted the medical mission, two were shot down, neither over the battlefield itself. The first craft in, piloted by Captain Gregory W. Lee, had been wracked by fire. Crippled after running a gauntlet of crossfire's, it fluttered back to base headquarters two miles south at Tam Ky, where it sputtered out altogether. The second craft was luckier. First Lieutenant Ellis Laitala dropped his bird down to 200 feet and still could not see the nose of the helicopter. He tried twice more to find a break in the cloud cover and finally succeeded, only to run into fire. The enemy had had ample time to prepare for his arrival after he had clattered over the landing zone a few times and when he did cut through the rainy mist, they had a 30. caliber machine gun talking. Laitala's copilot, First Lieutenant Richard Moser, saw a burst of tracers zip by his right window chest high. Turning to tell Laitala that the enemy was zeroed in, he saw another burst streak by the left window.

"It's a good thing that guy didn't hold one long burst," he said.

Laitala made ten trips to bring in ammunition and first aid dressings and to evacuate casualties. On each approach and take off he received fire. He put his helicopter through a series of desperate gyrations each time to shake off the streams of tracers, pitting flying skills against marksmanship.

The third Pilot to land shared Lee's fate. Major Raymond Duvall's craft was hit repeatedly. During the two hours he was flying in the area he flew through more concentrated fire than he had seen in his eleven previous months in Vietnam. Despite the intensity of that fire Duvall refused to allow his gunners, Sergeant Charles M. Olson, the crew chief, and Lance Corporal Alexander Winters, Jr., to open up. In this area, and at dusk, it was difficult to distinguish the Marine positions from those of the enemy. A wild machine gun burst, if the helicopter suddenly rocked, could kill Marines just as quickly as North Vietnamese. What finally forced him down was a hit in the rotor blade. The torn hole caused a terrible shrieking noise with every revolution of the blade and the troops on the ground were sure that it would crash. But, like Lee, Duvall and his copilot First Lieutenant Joel G. Vignere managed to wobble back to Tam Ky.

Among the helicopters, though, the one most memorable to Gilliland and the troops of Bravo Company was YL-54. "I'll never forget that one,"

Gilliland said. "I don't know how he did it. He should have been nailed a dozen times."

Captain Robert J. Sheehan was flying YL-54 in an exceptional manner. Ordinarily, a helicopter is travelling through the air at a speed of 80 to 90 knots when it approaches a landing zone. Sheehan hit the landing zone doing 115 knots—to a layman this difference may not seem like much but Sheehan's copilot, First Lieutenant Marshall Morris, explained:

"They had our altitude pegged. I'd say if we were going 5 knots slower, they'd have had us. Captain Sheehan really revved it up and just plain outran the tracers. It was a speed I know I couldn't do."

In a conversation later, however, Sheehan himself was quick to point out that landing an H-34 helicopter could not be a one man show.

"It's a team effort," he said, "like a rifle squad. The crew chief checks out the side door to make sure the tail is clear of obstructions when we come in. The gunner has to suppress hostile fire. The copilot backs up the pilot at all times. The copilot doesn't grab the controls but he palms them, like kid gloves. If the pilot is hit on landing and the copilot is daydreaming, the bird would probably crash."

On his first trip in, Sheehan picked up eight wounded and headed out south at treetop speed. He flew straight into a wall of bullets, one of which hit the carburetor. Sheehan quickly pulled right and the tracers fell behind. The hostile fire was like that on each of the nine trips he made and the helicopter was struck on three separate occasions. On the second trip, his gunner, Sergeant Peter B. Jensen an aircraft structures mechanic, was hit but the round ricocheted off his thick pilot's flak jacket. Sheehan allowed his gunners to fire and he could actually see their rounds finding targets. Jensen, a former drill instructor, spun two enemy soldiers completely around with one long burst of his M60 machine gun while the crew chief, Lance Corporal Edward M. Baker, dropped another that was crouched in a trench.

Altogether, Sheehan flew in 2,400 pounds of ammunition 400 pounds of battle dressings and took out 20 casualties. The last evacuation proved to be the most difficult. Coming in, Sheehan attracted fire from all directions. Some of the enemy soldiers were hidden not more than fifty yards from the helicopter, whose occupants could see the hostile positions much more clearly than could the Marines on the ground. But all the linked cartridges for the machine guns had been used up. Their plight seemed so bad Baker swung himself out of the helicopter door on to the steel lift step and returned fire with a .38 caliber pistol. A Navy corpsman named King, along to attend to the casualties saw this and said, "Fuck it, I'd better get out there too."

With that, he leaned out and began to fire his .45.

Sheehan put down and the infantryman brought up a casualty. They shouted, "Two more are coming!" Sheehan jerked his thumb up in the air to signal that he understood and would wait. And wait he did, for a full five minutes while the North Vietnamese tried frantically to destroy YL-54. Tracers were whining by at all angles, like a swarm of angry bees. From the village outside Sullivan's perimeter came the fire helicopter pilots hate most, that of .50 caliber machine guns speaking in tandem. The tracers rushed by in streams. Sheehan watched a paddy dike to his left front shred away. The foliage on a nearby hedgerow fell away like leaves in an October wind.

Across a paddy, a group of Marines struggled forward, half-dragging, half-lugging two of the wounded wrapped in ponchos. Sheehan remembered thinking that it would be a good idea to carry a number of stretchers in his helicopter when he went to the assistance of Marines in the future, if there were a future for YL-54. He was not going to leave without those two Marines but he thought the furious fires would reach him before they did. Gilliland shared that belief and stared at the stubborn helicopter in amazement. It sat, until the two wounded had reached it. Then Sheehan whirled away and Gilliland vowed to remember that helicopter.

Once airborne and headed south, Sheehan called over the intercom to check on his men.

"Hey," he said, "how are you guys doing back there?"

"Hell, Captain," came the cheeky reply, "we're having a ball."

After 10 trips into the hot landing zone Sheehan took advantage of a lull to inspect his aircraft where he discovered that a direct hit through the carburetor was one of four that he had received during the course of his flight.

When the medevac was finally completed, Laitala joined Captains Robert C. Willeumier and Robert A. Bracy and joined two aircraft of another squadron in lifting a company of reinforcements for 1/5. This mission was led by Major Vincent J. Guinee, Jr. of HMM-361 and totaled five aircraft. On the first pass, the gunner in the number three aircraft YN-21, Staff Sergeant Robert T. Walsh, was mortally wounded. He lived a short time and was kept alive by artificial respiration but failed to make it to an aid station alive. Two of the aircraft were hit while 78 troops were being lifted. A factor lending difficulty to the situation was that the lift occurred just after dark.

The weather and the situation were clearing all around the perimeter. The Hueys were fluttering back and forth and had pinpointed the sections of the village to the left front of Bravo Company from which the heaviest concentration of hostile fire was pouring, including the .50 caliber machine guns which had unsuccessfully searched for Sheehan. One helicopter pilot acting as the Tactical Air Controller Airborne (TACA) had the responsibility for selecting and designating targets for fixed wing aircraft. He called down the jets, A4D attack planes specially designed for close air support. The Huey pilot could see the tracers of a .50 caliber machine gun winking from the side of a hill above the village. Captain David Y. Healy, flying with Marine Attack Squadron-311 dropped down on the target, his jet ducking up and out over the village before the hill reverberated from the shock of a 2,000 pound bomb. The bomb had landed directly in the middle of three weapons positions and its explosion silenced all of the heavy weapons on the hilltop. As he leveled off, the Huey pilot informed him he had received a heavy volume of fire from the village as he went in and when he pulled out of his run. Healy had been unaware of this.

Circling above the battlefield, Healy's wingman, First Lieutenant John F. Schneider, Jr., could see the village clearly. The TACA told him to come in. Schneider had begun his dive when the TACA radioed him to pull out; he wanted more time to spot the exact source of the fire. He did this by flying over the village and drawing fire, a tactic not recommended for the faint of heart. Satisfied he had designated the target area properly, he told Schneider to come in again.

Schneider entered his dive doing 300 knots. He concentrated on his target—the northeast end of the village—released the 2,000 pounder and pulled out doing 450 knots.

To the men of Bravo Company who watched his dive, it had been a marvelous spectacle. In the growing dusk and gloom they had seen the jet slide down a thick red stream of tracers, then pull out, leaving behind a shattering splash of light and dirt. The infantrymen actually cheered.

In his gathering speed and approach angle, Schneider had been completely unaware his jet was the object of such concentrated fire. His plane had not been touched.

The recoilless rifles and .50 caliber machine guns did not speak again from the village. Following the strike, Bravo Company received only desultory sniper fire and Sullivan consolidated his lines with remarkable ease. Coffman then directed him to bring his company up the road. Help was needed in bringing out the casualties from Charlie Company, which had fought the hardest battle of all.

It was the premonition of a combat rifleman which kept Charlie Company from walking into a bad situation even before the fight had started. Charlie Company had swept east through the village on the right side of the road and arrived at a large open rice paddy bordered by thick hedgerows. There the road split, one trail angling northeast off to the left, flush against a tree line, the other running due east across a paddy, 75 meters removed from the undergrowth.

Corporal Frank Parks was leading the point squad. He was worried by the absence of the villagers and the lack of cows in the fields. He believed the company was going to be hit. While he was hesitating, a fusillade broke out to his left rear, where Alpha Company was. He thought someone had flushed a few snipers. But faced by two trails, he chose to bring the lead element of the company out across the paddy away from the hedgerow to the left. He reasoned that, if they were hit from that flank, they could take cover behind the road dike and build up a superior volume of fire. He did not want to be hit from positions two feet away.

Parks gestured to his point man, Private First Class Tyrone Cutrer, and Cutrer parted the bushes and walked into the open. Other Marines followed and the company bore to the right, leaving the hedgerow on the left flank for Alpha Company to prod. Cutrer's platoon, the third, was well into the paddy when they began taking fire from the hedgerow. The rounds were passing high and didn't bother the men. The Marines estimated that not more than five or six enemy soldiers were shooting at them.

Their reaction was immediate; they wheeled left and rushed the tree line. They screamed and shouted as they slogged across the paddy, a tradition that had become a habit in the company over many months and many fire fights. They could hear answering shouts and cries to the rear where the 81mm mortar platoon was marching. Between the rifle company and the heavy mortar platoon a bond of friendship had been struck and, hearing Charlie Company go into action, the mortar men were lending them all the verbal support they could. The air was filled with rifle shots, wild shrieks, and loud cries of "Go get 'em Charlie!" "Whomp up on those bastards!", "Do some dinging Charlie!" "Kill them dead!"

Park's point squad had a jump on the rest of the company and had almost closed on the hedgerow when a man was hit and went down. The others slowed their momentum, hesitated, then flopped down no more than 15 meters from the bushes.

This slack period while they tended to the casualty gave the North Vietnamese time to recover and build up an effective base of fire. Before the Marines could resume their push, heavy automatic fire was pouring above their heads. Still, they were very close, so close that Cutrer yelled "Let's go up on the bush line!" and bounded forward the few remaining yards. He was thrown back out by the blast of a grenade and for a few seconds stood erect in front of the bushes, deaf and dazed. Recovering his senses, he picked up the rifle that had been blasted from his hands and rushed forward again. This time he was joined by two more Marines and all three ran full into another grenade. Cutrer's luck held and he was the only one not injured. He dragged his two companions down into the shelter of a drainage ditch outside the hedgerow and put battle dressings on their wounds. Finishing that task, he picked up a grenade launcher and pumped several shells into the bushes in quick succession. Strung out along the ditch, the squad lay flat and covered the hedgerow with area fire. Their attack had been stopped cold.

First Lieutenant Buck Darling, commanding Charlie Company, later expressed dissatisfaction with the tendency of the troops at precisely the worst moment to turn aside from attacking the enemy to care for the wounded.

"Once a person gets hit," he said, "and your fire and maneuver stops in a paddy, your momentum is dead. It gives the enemy a chance to sight in. When the next man gets up, he'll get dinged—then nobody wants to get up. So you might as well have them crawl back across the paddies. If you could get them up on a line and charge, you might carry the position—with casualties, of course. But you'll probably not get the men to do that all at once together."

"If I'd made it in that first half hour," he added ruefully, "I'd have squeezed them in."

At about this time, Darling received a call from Lieutenant Colonel Coffman. Coffman explained that Alpha Company was being hit from a trench line to their right flank and he wanted Darling to attack it and relieve part of the pressure on Alpha. Darling thought that Coffman must have read his mind, since that trench line was the enemy position which had just repulsed the third platoon and Darling at that precise moment was preparing to assault it in force.

Darling was a seasoned commander and a master of small unit tactics. He had been with the battalion for 30 months, longer than any other man, and had extended his tour in Vietnam to keep his company. Unruffled by fire and at his best when actively engaged, Darling took his time to gauge the measure of the enemy which confronted him. His third platoon was engaged on the left flank, his second had encountered no enemy on the right, and his first platoon was holding fast to the rear in reserve. Before further committing his forces, Darling turned control of the company over to his executive officer, First Lieutenant Ronald Benigo, and went forward to assess the situation. Benigo was a former enlisted man and a 1964 graduate of the Naval Academy. Upon graduation he was ranked fourteenth out of a class of over 900.

In this action Darling was motivated not by bravado but by his knowledge of close in combat.

"A small unit leader," he said, "in thick brush can do nothing talking over the radio. He has to go see, which means you have to leave somebody back to coordinate things while you go up to decide on a tactical maneuver."

What Darling saw prompted him to employ the classic small unit maneuver: lay down a frontal base of fire and envelop from the flank. It is a simple, direct solution but very hard to repulse if the defenders have left the end of a flank dangling. And the North Vietnamese had done exactly that.

Darling brought up the second platoon and dispersed them along the dike road. From there they could deliver fire on the hedgerow and be protected themselves. They moved far enough out into the paddy to shoot past the right flank of the third platoon and the machine gun crews set up their guns in pairs, with excellent fields of fire. Darling thus had over a hundred weapons massed to rake a tree line not 200 meters long.

Next he called up the first platoon and told their platoon commander, First Lieutenant Arthur Blades, to take the hedgerow by assault. It was an under strength platoon even at the beginning, numbering only 37 men, including attachments. The second and third squads held but six men each. It was agreed Blades would mark his progress by smoke as he went, so that the base of fire could be shifted and kept ahead of him.

Blades deployed his platoon on line to the left rear of the third platoon. He pushed straight north through the underbrush with his three squads abreast, the third squad on the right nearest the third platoon, the second in the center, the first on the left. The platoon moved up abreast of the third platoon without opposition.

The third squad was guiding on a deep, narrow trench line cut under the bushes just at the edge of the paddy. The rest of the platoon was strung out left for 60 meters. The men could not see farther than 20 meters through the maze of undergrowth, palm and banana trees, and thatched houses.

There came one of those odd lulls in a firefight when everyone stopped firing at the same time. That was the moment Lance Corporal Palmer Atkins chose to move his squad, the third, into a small clearing. From less than 30 feet away a brace of automatic weapons withered the Marines skirmish line. Four of the six riflemen were struck down. The other two fell flat and returned fire.

Blades called Darling, asking for additional men so he could protect his flanks. While he was on the radio, the last two members of the third squad, including Adkins, who suffered a gunshot wound to the left hand, were hit by small arms fire. Blades had lost a whole squad—three of the six casualties, Lance Corporal Gregory M. Howard and Privates First Class David L. Faught and Paul E. Sudsbury, were dead—and had not struck a blow at the enemy. He had no idea how many enemy soldiers opposed him or how well they were armed. He did not know, nor would he have cared if he had known, that his platoon faced the contest which characterized Marine Operations at places like Tarawa, Peleliu, Iwo Jima and Seoul: an assault against a determined, entrenched and well disciplined enemy.

Responding to Blades' call for help, Darling gathered a group of Marines and a machine gun crew and sent them forward.

As the action had increased in intensity, control of the company had fragmented and a distinct separation of responsibilities within the command group had occurred. This was as it should be but rarely is. Darling controlled the overall tactics and the commitments of his platoons. Gunnery Sergeant Steve Jimenez functioned as general supervisor and foreman. He pointed lost Marines in the direction of their units, organized special details to carry ammunition or casualties, and ensured that the spread of outgoing fire along the long two platoon base stayed steady and even. The company first sergeant, Thomas J. Dockery, saw to the evacuation of the wounded. Dockery set up an aid station and evacuation point to the rear originally to handle only Charlie Company's casualties. But Lieutenant Colonel Coffman, seeing that the top sergeant had organized a system, directed that the battalion aid station be set up alongside him. Soon Dockery found himself keeping a record of all the casualties, allotting spaces on helicopters according to the corpsman's recommendations, and keeping the battalion commander informed as the number of wounded grew. These chores he handled well.

"But my biggest problem," he said, "was holding back forward observers, logistics support people, 81mm mortar men, engineers, company and battalion headquarters personnel, and the radiomen who wanted to quit their usual job and go up to the front. Even the corpsmen who were supposed to stay in the battalion aid station were heading out with grenades and bandages."

Within the perimeter of the battalion command group, the S-3, Major Bayard "Scotty" Pickett, had the same problem. He had to physically restrain Marines from leaving what they considered unnecessary jobs and rushing to the front.

"But, hell," the ex-All American football player grinned, "I really wasn't too mad at them when I hauled them back. I couldn't be—I did the same thing myself."

Dockery, a smiling Irishman and a ready talker, kept the wounded talking to the corpsmen and to each other—talking about anything to keep their thoughts away from their wounds and their bodies away from lapsing into shock. Still, one Marine died of shock before Dockery could get him evacuated. "What got me about that one," he said, "was that his death wasn't necessary. He was shot in the elbow but a lot of guys were hit worse and made it. He just clammed up inside himself and we couldn't snap him out of it, not even by slapping him. He said he was going to die and he did."

Darling's executive officer, Lieutenant Benigo, handled the job of getting the wounded off the field and back to Dockery. Altogether, that amounted to 38 men, but Benigo was not around for the final tally. While carrying a wounded man out of the paddy, he was struck on the back of the skull by a round which spun his helmet off and threw him flat.

"My God," he thought, "I'm dead."

Part of his scalp had been laid bare and he was bleeding hard. As if in conformation of his own belief, he heard a voice yell, "The lieutenant's dead."

Then he thought, "No, I'm not dead."

He scrambled to his feet, picked up the wounded man again, and staggered back to the aid station. He made two more round trips, weaving in a drunken fashion and ignoring all suggestions that he get on a helicopter himself. When he returned a third time to the aid station, he was set upon and forcibly evacuated, to the last protesting he had only been scratched and the bleeding would stop at any moment.

Most of Charlie Company's wounded came from the first platoon. When the reinforcements Darling had dispatched reached Blades, he placed them on his left flank, thereby freeing his remaining two squads to clear the trench line. The third squad on the extreme right having been wiped out, he sent his second squad forward parallel to the trench but shielded by two houses. In this movement, two more Marines went down. Blades felt sick. Although a hard, driving man, he was close to his men and had argued insistently to keep his young squad leaders and make rank within his own platoon, not to bring in leaders from other platoons. Now fully one third of his organic unit was down and the enemy force seemed unhurt.

Instead of falling back, the platoon redoubled its efforts. From behind the houses, the Marines lobbed grenades into the trench, while men back a few yards with Blades blasted away with automatic weapons. The platoon commander was hit in the back by grenade fragments which ripped his flak jacket. He flung the jacket off and continued to throw grenades. After a series of quick throws, his men held their fire for a moment to gauge the strength of the enemy. The opposing fire had definitely slackened, so Lance Corporal Irwin Brazzel led his three men in a dash forward to the next house in an attempt to out flank the North Vietnamese. Two automatic weapons opened up again and Brazzel and another Marine were cut down. That left two fighting men in the second squad. Blades was in anguish. Brazzel, hit in the shoulder, crawled behind a house. The other Marine lay exposed and motionless. A corpsman, Hospitalman Lawrence T. Steiner, rushed forward to help him and was killed.

Blades wasn't sure that they had killed a single enemy. But the Marines were bombarding the trench with grenades and three grenade launchers and a dozen rifles. Still, two—and only two—automatic weapons replied after each fusillade.

Brazzel called to his lieutenant.

"Sir, the sniper is on the other side of the fence. I can't shoot through it but I think you can work around it alright."

Blades moved up with the first squad—his last squad. The squad leader, Corporal Christopher Cushman, deliberately stood erect for a second, and then dropped flat. The sniper Brazzel had warned about sprang up to fire and was shot by Lance Corporal Walter McDonald, a combat photographer who had swapped his camera for a rifle.

Next Blades had his party provide covering fire for a Marine lying wounded in a clearing. The Marine crawled into a house next to the trench and started kicking out the back wall so he could throw grenades into the trench. Hearing the commotion just above his head, an enemy soldier riddled the wall, wounding the man again. While the enemy soldiers' attention was so diverted, McDonald darted around the house and dropped a grenade right on him.

The Marine assault was now gaining momentum. At the enemy fell back, three engineers attached to the platoon worked their way up the trench itself, keeping the pressure on the rear of the enemy while Blades' party pounded them from the right. When the enemy, retreating to concealed positions around a tall haystack, pinned down Blades and his men, the engineers crept up the trench. Once close in Lance Corporals Clifford Butts and William Miller rose up and fired furiously at the foliage around the haystack, While Private First Class William Joy hurled grenades as fast as he could. They blew the haystack, most of the surrounding foliage, and some of the enemy apart and forced the others to abandon the position and pull back.

The engineers, who had worked together on other operations, kept their trio intact under the intense fire. But that was the exception. The other Marines had naturally split into pairs. Clearing the trench line became a team effort. One Marine would throw grenades while the other covered him by rifle fire. In this way, Lance Corporal William Cox and Private First Class Michael Stevenson worked their way to Brazzel and dragged him back.

Blades kept urging the men forward. He directed and the men responded. Even those who did not belong to the platoon, but who came to fight, took their cue from his leadership; such as McDonald, who later said, "I just did what the Lieutenant told me to do."

There were two Marines who crawled up the trench and asked the platoon commander if they could help. Blades responded, "Yes, with grenades." At which they extracted several grenades from their pouches and held them out to him.

"Who the hell do you think I am," Blades roared, "John Wayne? Get out of that trench and go throw your own grenades!"

They did.

The platoon sergeant, Orrin W. Spahn, kept an eye on the massive base of fire as the platoon advanced, threw smoke grenades ahead to shift the fire. He stuck close to Blades. Both used grenade launchers, a weapon they found particularly effective in rooting out the tenacious enemy.

The Marines were pushing the defenders back but they still weren't sure how many there were or how many they had killed. The men could see occasional targets, however. A head or a back would poke up here or there from the trench for an instant and the Marines would cut loose. Fighting at less than 15 yards, they were sure they were dropping some but the two automatic weapons continued to blaze at them from successive positions up the trench line. Grimly, the men dogged after the enemy. The end came suddenly when the North Vietnamese ran out of the trench line at the point of the hedgerow. Blades was grinding forward on the left. Darling's base of fire was sweeping the open paddies to the right. The Marines sensed victory when some of the enemy broke and ran. Cushman saw a figure in gray khaki hop out of the trench and duck into the bushes. The squad leader waited until he moved then shot him. McDonald nailed two more in a similar manner.

But that was all. The rest stayed and died in a roar of exploding grenades and automatic rifle fire. Blades radioed to Darling. He had to put the call through himself. In the closing minutes of the battle both his platoon sergeant, Spahn, and his plucky radio operator, Private First Class William Brown, had been hit. Spahn had been shot through the left ankle and tibia. Brown was hit in the right arm by shrapnel and Cox also suffered a grazing wound to the head from a round fired by a North Vietnamese sniper.

"We've taken the objective," he said.

The platoon commander limped over to the last section of the trench and peered down. It was clogged with bodies pressed side by side or laying in heaps, smashed and torn by bullets and grenades. The Marines counted 19 bodies, most packed within 15 meters of trench line. They picked up 17 new automatic weapons and packs crammed with stick grenades and linked ammunition smeared with Vaseline.

The discipline of the North Vietnamese in firing just two weapons at a time had been excellent. Their positions were deep, covered and camouflaged. A detailed map found on the body of their company commander indicated the care with which he had prepared his fire plans and drilled his men. Yet, instead of ambushing an annihilating the lead Marine platoon, they were overrun and killed.

The factors contributed to the success of the Marines' grinding assault— Darling's plan, Blades' leadership and the troops' aggressiveness, especially the latter. In the opening minutes of the attack, Blades lost 10 out of the 12 men in the two squads first to engage the enemy, including both squad leaders. The assault could have crumbled then and there. It didn't. The men went on in. They weren't perfect. They made mistakes and Blades was the first to point them out. In particular, he noted that men were wounded or killed because they stood erect when they should have crawled. They did so because they were tired and it was easier to move by standing. The weight and bulk of their gear contributed greatly to this fatigue. Still, they adapted to two man teams and waded in slugging, and kept slugging, until they destroyed the enemy force.

It would be nice to end the story here, with the Marines holding the field of battle and the North Vietnamese, beaten at every turn, slipping away in the growing dusk, never to return. But Vietnam isn't like that. It doesn't just end decisively, nor did this engagement, really.

The North Vietnamese pulled back at dark and Kilo Company, Third Battalion, Fifth Marines, was flown in to lend a hand, but the fight had passed. The battalion buttoned up tightly in a circular perimeter. Flare ships kept the area lighted and massive artillery fires ringed the battalion. Not even snipers harassed the lines. The companies passed a quiet night, noticeable for its lack of activity.

But for one Marine it was a night of terror. It had taken corpsman T. C. Long an hour and a half to crawl out from under Donathan's body. When at last he had freed himself, it was dusk and he hadn't the strength to move any more. He lay in the mud with the stinging in his kneecap where the ants were feeding on the raw flesh and waited and dozed and prayed. Sometime during the night, two North Vietnamese walked past the body and tripped over him. They stopped and stripped both Donathan's body and Long of gear. He played dead until they walked away. At times Long blanked out. Once he awoke with a terrible thirst and to a puddle close by. As he drank, he heard footsteps approaching. He turned his head to look and was blinded by a bright light. Long blinked dazedly into the beam of the flashlight for a few seconds, then it went out and he heard the footsteps receding. "Why didn't he kill me?" he thought.

At first light Captain Furleigh sent out a strong patrol to find the bodies of the missing men. This time Bielecki saw Long lying in a rice paddy beside the trail. They carried him and Donathan's body back.

While Lieutenant Coffman sent out patrols to police the battle area and pick off enemy stragglers, the press came in to get the story. The men had little to say. To each other they talked long and fully and eagerly. But to strangers they were reluctant to speak.

By midmorning, the patrols and outposts were engaged in desultory exchanges with enemy skirmishers and snipers. The men walked warily when they left the perimeter. It was obvious there were still many of the enemy in the area.

That was why the Marines didn't quite believe it even when they saw the helicopter land and the officers in short sleeve utilities jump out.

"Is it?" a private first class asked his sergeant.

"Sure looks like it," the sergeant replied. "I don't know anyone else in the Marine Corps who wear four stars."

General Wallace M. Greene, Jr., the commandant of the Marine Corps, had come to the battlefield. With him walked Lieutenant General Lewis W. Walt, commander of the Marine forces in Vietnam. General Walt had a habit of dropping in unexpectedly in unsecure areas and most of the men had seen him before and were not surprised to see him again. The feeling among the troops was that, while it was alright for General Walt to expose himself, the Commandant shouldn't do so. The generals walked the trench line Blades' platoon had cleared and asked pointed questions about the tactics and weapons used.

Ordinarily, Buck Darling is a talkative person who can go on for hours when asked about tactics operations. But he wasn't used to talking with generals and his cryptic words to the commandant might be the clearest code of the Marine in combat. General Greene asked him what happened.

"Well, General," he replied, "we got into a fight with the enemy."

The Commandant then asked what he did.

"General," he answered, "we killed them."

The general officers left and the battalion passed the afternoon burying enemy dead, patrolling and resting. They were going to spend the night they were going to spend the night there and they weren't happy about it. They thought the enemy had the area plotted perfectly. The battalion commander issued to his company commanders the march order for the next morning and the hour of stand to alert for that night.

At dusk the Marines were manning the lines in force. The sky to the west was still red. Two snipers were silhouetted perfectly against the red background as they climbed up palm trees and were dropped to the ground in a burst of automatic rifle fire by men from Bravo Company. The Marine patrols went out and one from Alpha Company, minutes after leaving the lines, killed two more of the enemy. After this quick contact, the other patrols around the perimeter were pulled in and the battalion sat defensively, waiting for probes.

None materialized and the hours dragged by. Then, at 0350, the North Vietnamese struck. With sharp suddenness, the first 82mm mortar round exploded right on the edge of the trench where the generals had stood that afternoon. It was exact firing. Other shells dropped in, striking near the command post of Alpha Company. In the blackness a Marine cried: "My God, somebody help me. I'm hit bad. Please get me a doctor. I'm dying." Corpsmen from both Alpha and Charlie Companies raced to the man, but he died.

The battalion command center was hit hardest by both mortars and recoilless rifles. Two more men died there and several others were wounded. The Bravo Company command post was established in a store house near the battalion command post. A 57mm recoilless shell struck the ground just in front of it and bounced into the side of the building. The explosion collapsed the inside of the shelter yet dealt the Marines trapped within only ringing ear drums and multiple scratches. Private First Class Edward H. Simonson, a radioman, was wounded in the left side and right heel.

The hut where Darling, Dockery and Jimenez were sleeping fared better. The enemy fired at least five shells at it and all passed high. Darling lay flat and listened to the shells flutter past, each sounding like a bird trying to fly with a broken wing. In the din, he just barely heard another sound and shouted: "Shut up, everybody, lie still and listen. Try to get a fix on the sound of their weapons.

They could hear in the distance the slight but unmistakable pop of a mortar and the much louder bang of the recoilless rifle.

Battalion was way ahead of them. Hueys had been called to fly over, spot the weapons by their flashes and destroy them. With the noise of their arrival the hostile weapons stopped firing.

In a much less effective manner the enemy had simultaneously hit the perimeter with a ground attack. About a squad of infantry firing automatic weapons moved toward Bravo Company's positions. The Marines on the line laid down devastating blanket of fire and the enemy fell back and did not return.

The next morning, the battalion set out to walk the final four miles into task force headquarters. For the first two miles they would follow the same road they had taken for the past four days. Coffman again set out a double point, with Alpha Company on the left of Bravo Company, which guided on the road. The companies moved across the rice paddies and through the hedgerows and encountered only scattered sniper fire.

In midmorning the battalion took its first casualty. The Point of Alpha Company stumbled over the tripwire of a grenade and went down with shrapnel in both legs. It was Private First Class English, the man who, during the battle, had moved so swiftly to rescue a wounded Marine. While waiting for a helicopter to evacuate English, Captain Furleigh told his radioman: "Pass the word to all platoons to watch where they walk. Keep an eye out for mines and booby traps."

Less than ten minutes later, Furleigh crossed through a back yard at the head of his command group to get a better glimpse of his lead platoon. He saw them spread out in a paddy on the other side of a bush line. The captain headed for the nearest opening and pushed aside the brush in his way. A grenade went off under him and blew him back into his radio operator. Both collapsed with multiple wounds with Furleigh suffering shrapnel wounds to his right arm, hand and both legs He was a resolute, intelligent captain who deserved a better finish to his tour in Vietnam than medical evacuation.

Around this same time Private First Class Charles J. Corso and Hospitalman Jerry C. Hayes were both hit in the abdomen by shots fired by a sniper. Also wounded was Lance Corporal Conrad J. Gaywont. Private First Class Harlie L. Millis was shot in the head and chest.

Coffman sent Lieutenant Blades forward from Charlie Company and the march was resumed. The men trudged under the hot sun across the paddies and thought of nothing in particular and said very little. They were tired and the muck of the paddies slowed their pace. Bravo Company on the right had easier going along the road and began to out distance them. Blades, incredibly fresh, spurred them on by shouting, "Come on! What's the matter with you? Square away and walk tall, Marines. Put some pride in that step!"

That was the way the battalion walked in to the task force area, jaunty and yet tired; glad to be back and proud of themselves. One rifleman actually starts whistling the Marine Corps Hymn as they neared the battalion area. "Knock that off," growled his buddy, "where do you think you are—on the grinder back at boot camp?"

"No, man," came the reply, "but I can dream can't I?"

So they came back for a few days rest and replenishment before going out again.

And again.

NOTES:

There were 16 Marines and 3 Navy corpsmen killed in action with another 65 wounded. The Marines killed more than 100 North Vietnamese before the battle ended on the August 11.

The Men who died on Operation Colorado are listed as follows:
Amador, Ernest B. Sgt. 1912852; A-1-5; 25; San Antonio, Texas
Donathan, Richard P. LCPL. 2035921; H&S-1-5; 22; Grand Junction, Colorado
Elrod, James T. HM3 6952120; B-1-5; 19; Moultrie, Georgia
Faught, David L. Pfc. 2226693; C-1-5; 20; Grand Rapids, Michigan
Haddix, Douglas B. Cpl. 2128498; C-1-5; 19; Springfield, Ohio
Higbee, Robert D. LCpl. 2120952; C-1-5; 20; Drexel Hill, Pennsylvania
Howard, Gregory M. LCpl. 2116184; C-1-5; 22; Greenway, Virginia
Jackson, Walter P. HM3 3912853; A-1-5; 23; Danville, Illinois
Kindred, Lawrence J. Pfc. 2176590; A-1-5; 18; Overland, Missouri
Rolle, Melvin Pfc. 2142256; C-1-5; 19; Miami, Florida
Skinner, Richard A. Pvt. 2032817; A-1-5; 20; Greenbelt, Maryland
Smith, John D. Sgt. 2023911; A-1-5; 24; Camden, New Jersey
Steiner, Lawrence T. HN 6997503; C-1-5; 21; Houston, Texas
Sudbury, Paul E. Pfc. 2160268; C-1-5; 21; Corinna, Maine
Tasker, Kenneth E. Pfc. 2227365; B-1-5; 19; Deer Park, Maryland
Thompson, Everette A. Pfc. 2245514; A-1-5; 19; Grand Rapids, Michigan
Walsh, Robert T. SSgt. 1399730; HMM-361; 30; La Crosse, Wisconsin
White, Theodore G. Jr. Sgt. 1921078; 24; Baltimore, Maryland
Williamson, Richard W. LCpl. 2135312; B-1-5; 19; Fife, Virginia

For heroism on Operation Colorado the Silver Star Medal was earned by Captain John P.T. Sullivan, First Lieutenants Ronald Benigo and Arthur C. Blades, Hospitalman Third Class James T. Elrod, Major Vincent J. Guinee, Jr., Private First Class Lolesio Peko, Sergeant Darwin R. Pilson, Captain Gregory W. Lee and Staff Sergeant Robert T. Walsh. The awards to Elrod and Walsh were posthumous.

The Bronze Star was awarded to First Lieutenant Woody F. Gilliland and HM3 Harold F. Lewis.

Staff Sergeant Robert T. Walsh was a former Drill Instructor who had just returned from Vietnamese language school, a requirement for becoming a load master. When he rejoined HMM-361 he was in charge of the hydraulic shop and put back on flight pay for he often volunteered to fly as a door gunner. Corporal Lawrence C. Isham, a fellow squadron member and a qualified parachute rigger, recalls that day:

"The squadron was resupplying the ground troops during Operation Colorado. Corporal Don Wilmot, who was the door gunner and crew chief for YN-21, and I were flying all day until about 1800 when our pilot set down and told us to go to the mess tent and get something to eat. We just sat down to eat when the lead platoon for the ground troops were ambushed about three miles out from the main base of operations. There was an almost immediate call for medevacs. Don and I jumped up and ran back to YN-21, but by the time we got there, Walsh and another crew chief that were on standby had already signed the flight sheet so Don and I couldn't go. Walsh was killed on the first or second medevac. I believe to this day that I wouldn't be alive if it wasn't for Walsh taking my place."

Walsh was actually involved in a night reinforcement flight. When his flight was menaced by Viet Cong automatic weapons fire he silenced the enemy guns with accurate bursts of fire of his own which enabled his flight to continue. After delivering the reaction force and picking up an injured Marine the helicopter lifted off and as it cleared the trees it was engulfed with streams of enemy tracers. As Walsh put fire on the most menacing enemy positions the helicopter was viciously raked from all directions and he was wounded. He disregarded these wounds and continued to fire until he lost consciousness.

Major Vincent J. Guinee, Jr. had earned a Silver Star only about a month before Operation Colorado while flying with HMM-361. He earned this medal for heroism on July 9-10, 1966.

Colonel Charles F. Widdecke joined the Marine Corps in 1941 and earned the Silver Star during the fighting in the Marshall Islands during World War II. He earned the Navy Cross as a Captain while leading Company C, First Battalion, Twenty Second Marine regiment during the invasion of Guam. Widdecke retired from active duty as a Major General in 1971. He died in 1973 at the age of 54.

David A. Ballantine, a helicopter pilot with VMO-6 from March of 1966 to April of 1967 made this observation of door gunners in his Vietnam memoir, *Gunbird Driver*:

Door gunners were an impressive lot. These guys volunteered and they ran the gamut of squadron jobs from administration to ordnance, from intelligence to hydraulics. They were hidden tigers. I was amazed by the spectrum of guys who flew as door gunners. After signing for the bird, I'd head out to preflight and there was the crew chief, the man who flew daily, took care of his helicopter, and was well known to me and to all the pilots. With him was a door gunner, a trained volunteer from some other, nonflight-related job in the squadron.

I'd brief them on what we were up to and my useful specifics. We'd strap the gunbird to our asses and leap off the Ky Ha cliff. These gunners, and of course the crew chiefs manned swivel-mounted M-60s in the cabin just behind the pilots. They dished it out the helicopter's sides and saved my ass on more occasions than I will ever know. They stood in the rear doors, gunner's belt securing them to the helicopter, slightly crouched over their guns, and hammering away. God bless 'em!

Lieutenant Colonel Harold Coffman enlisted in the Marine Corps in 1945 and after reaching the rank of sergeant was commissioned a second lieutenant in 1948. He earned the Bronze Star Medal as a platoon commander with Antitank Company, First Marine Division, in Korea. While in command of Battalion Landing Team 1/5 in Vietnam he earned a second Bronze Star and the Purple Heart. He retired from the Marine Corps as a Brigadier General in 1977 after 33 years of service.

Captain James R. Furleigh recovered from his wounds and served a second tour in Vietnam in 1969. He stayed in the Marine Corps until 1990 and retired with the rank of Colonel.

Donald Alexander, on the website *The Vietnam Veterans Memorial*, shares this recollection of Richard Donathan:

"I knew Donathan as the "Animal" because he was so strong and looked like a weightlifter. On August 10 as we moved towards our ambush he humped by Sgt. Sam Baker and I carrying two spools of com wire on his back and Sam began to tease him about being hard and he said, "You don't know what hard is man!" and grabbed a black bird off a paddy dike, that for some reason had not flown away, and bit its head off. He spit out the head and threw the flopping corpse away. I was later told that he and 'Doc" Terry C. Long were hit at the same time and left on the field over night."

When Major Duvall was called into action on August 10, 1966 he was the leader of a flight of helicopters from HMM-362 that were part of a resupply and troop lift mission and was waiting for a thunderstorm to pass so that the mission could be completed. He was ordered to scramble an aircraft to assist two medevac helicopters that were receiving intense enemy fire. Quickly unloading his helicopter he flew out to the med evac pickup zone where the Marine's from 1/5 were heavily engaged. Spotting the smoke grenade marking the zone he started an approach. Due to heavy automatic weapons fire he had to break off his approach. He climbed back out, turned and approached from a different direction. Faced by lighter fire he landed, picked up the wounded, and flew them to Tam Ky. He then returned for two more loads of wounded personnel. On the first trip he brought 500 pounds of medical supplies and grenades and on the second 1000 pounds of ammunition.

As he prepared to pick up a fourth load he received a call for help from another ground unit. They smoked the new landing zone and as he approached his craft was taken under heavy automatic weapons fire. Duvall, and his co-pilot Vignere, veered to the east as the crew chief, Olson, marked the location of the enemy positions with a red smoke grenade. He called for suppressive fire and fixed wing aircraft began hitting the area.

Duvall then proceeded to the original medevac zone where he was again met by intense automatic weapons fire. He received a major hit in the main rotor system but continued into the zone and loaded the waiting wounded. Duvall then flew through the enemy fire back to the medical facility. Upon landing for refueling he discovered that the aircrafts electrical system was damaged and further flight was impossible. For his devotion to duty, Duvall received the Navy Commendation Medal.

First Lieutenant Marshall Buckingham Darling, III had previously distinguished himself in combat during a rescue operation in the vicinity of Tam Ky on June 16, 1966. This was the night of the famous Howard's Hill fight. A recon unit led by Staff Sergeant Jimmie E. Howard had been trapped and were taking a pounding from the Viet Cong. Darling was in Command of Company C during the rescue operation. He earned a Bronze Star Medal for recovering the body of his second platoon commander, Second Lieutenant Ronald W. Meyer. While trying to maneuver toward the enemy, Meyer had been mortally wounded. Accompanied by Lance Corporal James E. Brown and halted several times by intense enemy fire Darling persisted and was finally able to retrieve the body.

He earned a second Bronze Star as the commander of Golf Company, Second Battalion, Fifth Marines for heroic actions and superb leadership between March 21, 1970 to February 21, 1971.

Darling, then a colonel, also distinguished himself for outstanding meritorious service as Commanding Officer, Marine Air Ground Task Force 4-90 while deployed to U.S. Facility, Subic Bay from June to August 1991 during the eruption of Mount Pinatubo, and the simultaneous onslaught of Typhoon Yunya.

He served 30 years in the Marine Corps and retired in 1993 as a Colonel.

First Lieutenant Arthur Blades transitioned to aviation, and returned to Vietnam for his second tour as a helicopter pilot with HMM-364 in 1970. He went on to serve 30 years in the Marine Corps and retired with the rank of Lieutenant General in 1996.

Blades shared his observations of Buck Darling and others: "Buck was a one of a kind individual who was outspoken, unconventional, unpretentious, intelligent and direct in his problem solving. He had a sharp sense of humor, was well read on many subjects, enjoyed military history and loved working with wood.

In a traditionally clean shaven Marine Corps, he always wore a mustache and kept his head clean shaven. This never changed over the years.

Buck was a "natural" troop leader in the field. He inspired confidence, loyalty, trust and initiative in his troops. They loved him.

In the field Buck was confident in his abilities and calm under duress. He had superior situational awareness and was always anticipating the possible.

Always one to worry about his troops and the enemy, Buck usually dispersed the command section with its radio operators to avoid being targeted. He always ensured that units were covered by heavy weapons during open field movement. Buck always had his fire support officer plotting an artillery "wall of steel" around the company as it moved and ensured that the artillery was prepared to assist.

His walk was very distinctive and in a column of 200 men he could always be picked out because of it.

During the night Buck would check the lines and would often spend time with the men. He was rarely challenged because all knew him even in the dark – they just knew him.

During my second tour in Vietnam the Seventh Marines had their headquarters south of Da Nang near the Que Son Mountains (LZ Baldy). Helicopters stationed at Marble Mountain would fly unit supply out of the Headquarters LZ. It was during one of those resupply operations that I recognized a familiar directness to the orders and his voice on the radio. I had transitioned to helicopters and was flying with HMM-364, The Purple Foxes. After shutting down we spent about an hour recalling events in 1/5.

Captain Francis "Bing" West was a reserve officer called to active duty during the summer of 1966. Buck Darling referred to him as the "Red Leopard." Following the rescue operation on Howards Hill he referred to Charlie Company as "Suicide Charlie." It was a name that stuck with the troops and his company. C/1/5 was the "go to" company when difficult situations arose.

Platoon Sergeant Oren Spahn was reportedly the most competent Sergeant with the company. It was my good fortune to have him as the Platoon Sergeant. Experienced and competent he effectively led and instilled confidence in the platoon. Liked and trusted in the field.

Buck vouched for him as I was being assigned to the platoon. It is possible that he may have been involved with Buck at Howard's Hill.

Spahn was badly wounded in the leg early in the fight on 10 August. His position was near the rice paddy area and not readily visible to me when he was wounded. He was subsequently medevaced."

Others mentioned in the chapter or wounded in action are as follows:
Adkins, Palmer R. LCpl. 2128731; C-1-5; 20; Jenkins, Kentucky
Alexander, Donald W. Cpl. 1617926; A-1-5; 26; Springfield, Missouri
Bailey, Gordon L. Sgt. 1610206; H&S-1-5; Los Angeles, California
Bailey, Larry D. Pfc. 2134062; A-1-5; 20; Melvindale, Michigan

Baker, Edward M. Pfc. 2134405; HMM-362; 20; Detroit, Michigan
Baker, Sam R. II Sgt. 1973607; A-1-5; 22; Bronx, New York
Beandette, Thomas GySgt. 1364972; B-1-5; 36; Montpelier, Vermont
Benigo, Ronald 1stLt. 089391; C-1-5; 26; Detroit, Michigan
Benoit, David W. Pfc. 2136671; A-1-5; 19; Hartford, Connecticut
Bielecki, William Pfc. 2130067; A-1-5; 20; Chicago, Illinois
Bourgett, Terrance P. Pfc. 2110185; B-1-5; 20; Minneapolis, Minnesota
Bracy, Robert A. Capt. 084294; HMM-362; 26; Nashville, Tennessee
Brazzel, Ervin L. LCpl. 2113369; C-1-5; 20; San Francisco, California
Brown, Joseph Pfc. 2220264; B-1-5; 19; Philadelphia, Pennsylvania
Brown, James E. LCpl. 2059877; C-1-5; 21; Oakland, California
Brown, William H. Jr. Pfc. 2112993; C-1-5; 20; Charlotte, North Carolina
Bush, Richard H. Jr. LCpl. 2109483; B-1-5; 20; Nashville, Tennessee
Calonge, Eugene E. Pfc. 2105524; A-1-5; 19; New York, New York
Coffman, Harold L. LtCol. 049702; CO-1-5; 39; Huntington, Penn
Corso, Charles J. Pfc. 2088114; A-1-5; 20; Oklahoma City, Oklahoma
Cox, William J. LCpl. 2133910; C-1-5; 19; Kansas City, Missouri
Cushman, Christopher S. Cpl. 2141189; C-1-5; 20; Albany, New York
Cutrer, Tyrone E. Pfc. 2109624; C-1-5; 20, Nashville, Tennessee
Darling, Marshall B III 1stLt. 087880; C-1-5; 25; Lancaster, California
Dockery, Thomas J. 1stSgt. 1170512; C-1-5; 33; Philadelphia, Pennsylvania
Duvall, Raymond L. Jr. Maj. 054440; HMM-362; 33; Escambia, Florida
Ellis, Albert J. SSgt. 1352191; A-1-5; 32; Albany, New York
English, Robert T. LCpl. 2134156; A-1-5; 20; Detroit, Michigan
Feerick, Robert M. HM2 4544231; B-1-5; 30; New York, New York
Fudge, George B. Jr. Pfc. 2131000; A-1-5; 20; St. Louis, Missouri
Furleigh, James R. Capt 085376; A-1-5; 25; Manchester, New Hampshire
Futch, Van M. LCpl. 2114051; H&S-1-5; 20; Jacksonville, Florida
Garr, Raymond G. LCpl. 2132535; B-1-5; 20; Jamestown, Kentucky
Garrotto, Alfred F. LtCol.. 025722; HMM-362; 43; Kansas City, Missouri
Gaywont, Conrad J. LCpl. 2109165; A-1-5; 20; Milwaukee, Wisconsin
Gilliland, Woody F. 1stLt. 092092; B-1-5; 23; Amarillo, Texas
Goodner, Robert W. LCpl. 2094847; A-1-5; 20; Midland, Texas
Gorman, William M. SSgt. 1642965; B-1-5; 26; Unknown
Guinee, Vincent J. Jr. Maj. 064735; HMM-361; 35; New York, New York
Haga, Gene H. SSgt. 1818836; H&S-1-5; 25; Jacksonville, Beach, Florida
Hansen, Dennis G. Cpl. 2059126; A-1-5; 21; Minneapolis, Minnesota
Hairston, Clifford O. Pfc. 2237826; B-1-5; 19; Roanoke, Virginia
Harman, Milton L. Capt. 083355; H&S-1-5; 26; Syracuse, New York
Hayes, Jerry C. HN 7727137; H&S-1-5; 20; Raleigh, North Carolina
Healy, David Y. Capt. 074133; VMA-311; 31; New York, New York
Henderson, James W. Pfc. 2078016; B-1-5; 21; Philadelphia, Pennsylvania
Howard, Jimmie E. SSgt. 1130610; C-1-Recon; 37; Burlington, Iowa
Isham, Lawrence C. Cpl. 2009513; HMM-361; 22; Albany, New York
Jackson, James R. LCpl. 2095317; B-1-5; 19; Memphis, Tennessee
Jensen, Peter B. Sgt. 1377825; HMM-362; 31; Grand Rapids, Michigan
Jiminez, Steve M. GySgt. 1641509; C-1-5; 30; San Antonio, Texas
Kelley, Joseph P. LCpl. 2061309; C-1-5; 21; Cincinnati, Ohio
Kohlbuss, Rodney Cpl. 2130029; A-1-5; 20; Chicago, Illinois
Lantry, Thomas H. HM3 3912970; H&S-1-5; 21; Spokane, Washington
Laitala, Ellis E. 1stLt. 085480; HMM-362; 25; Columbia South Carolina
Lasich, Patrick J. Pfc. 2226750; B-1-5; 19; Detroit, Michigan
Lee, Gregory W. Capt. 082477; HMM-362; 28; King of Prussia, Pennsylvania

Lego, George M. LCpl. 2099938; B-1-7; 20; Lanse, Pennsylvania
Lewis, Harold HM3 6866350; H&S-1-5; 21; Denver, Colorado
Lewis, Richard A. Jr. Pfc. 2150462; A-1-5; 19; Chicago, Illinois
Long, Terry C. HN 6859938; H&S-1-5; 22; Siletz, Oregon
Matthews, Robert A. LCpl. 2100037; A-1-5; 20; Pittsburgh, Pennsylvania
Meyer, Ronald W. 2dLt. 091945; C-1-5; 23; Dubuque, Iowa
Miller, Robert E. Pvt. 2092826; B-1-7; 20; Louisville, Kentucky
Miller, William E. LCpl. 2116784; H&S-1-5; 20; Pittsburgh, Pennsylvania
Milian, Carmelo LCpl. 2074942; B-1-5; 21; New York, New York
Millis, Harlie L. Pfc. 2112990; A-1-5; 19; Raleigh, North Carolina
Monroe, Robert W. LCpl. 2122739; B-1-5; 19; Indianapolis, Indiana
Moreland, Ronald LCpl. 2088198; A-1-5; 20; Oklahoma City, Oklahoma
Morris, Marshall L. Jr. 1stLt. 089716; HMM-362; 23; Clemson, South Carolina
Moser, Richard 1stLt. 087346; HMM-362; 24; New Haven, Connecticut
Mowan, John E. Pfc. 2152417; B-1-5; 19; Indianapolis, Indiana
Nasalroad, Lonnie W. Pfc. 2131003; A-1-5; 20; St. Louis, Missouri
Nuncio, Catarino V. Jr. Cpl. 2045749; B-1-7; 21; Houston, Texas
Olson, Charles M. Sgt. 2022633; HMM-362; 22; Minneapolis, Minnesota
Parks, Frank E. Cpl. 2044103; C-1-5; 21; Indianapolis, Indiana
Peko, Lolesio Pfc. 2142244; B-1-5; 22; Albany, New York
Pickett, Bayard S. Maj. 058478; H&S-1-5; 38; Charleston, South Carolina
Pilcher, William G. LCpl. 2114851; B-1-5; 20; Kansas City, Missouri
Pilson, Darwin R. Sgt. 2050793; B-1-5; 24; Philadelphia, Pennsylvania
Pope, Gregory A. Pfc. 2134329; B-1-5; 19; Detroit, Michigan
Rodriguez, Henry Y. Pfc. 2164611; B-1-5; 19; San Antonio, Texas
Rowell, Peter M. Sgt. 1866864; B-1-5; 25; Kittery York, Maine
Scaglione, Peter C. Jr. Maj. 066767; HMM-362; 33; Salisbury, Connecticut
Scullin, Patrick Pfc. 2104821; B-1-5; 20; Staten Island, New York
Sheehan, Robert J. Capt. 082647; HMM-362; 26; Cambria, Pennsylvania
Simonson, Edward H. Pfc. 2141661; B-1-5; 19; Los Angeles, California
Smith, Theodore G. Cpl. 2043258; B-1-5; 21; Akron, Ohio
Smith, Walter W. 1stLt. 088056; HMM-363; 25; New Orleans, Louisiana
Sorenson, Carl M. Cpl. 2079462; A-1-5; 20; Chicago, Illinois
Spahn, Oren W. Sgt. 1635090; C-1-5; 26; Holidaysburg, Pennsylvania
Stevenson, Michael Pfc. 2231209; C-1-5; 19; Unknown
Sullivan, John P.T. Capt. 084248; B-1-5; 29; Chicago, Illinois
Vidovich, Matthew Cpl. 2030513; B-1-5; 20; New York, New York
Vignere, Joel G. 1stLt. 089778; HMM-362; 25; Buffalo, New York
Vogel, Ronald L. Sgt. 2012912; B-1-5; 22; Seattle, Washington
West, Francis J. Capt. 084938; HQMC; 26; Milton, Massachusetts
Widdecke, Charles A. Col. 08547; CO-5; 47; Dallas, Texas
Willeumier, Robert C. Capt. 084298; HMM-362; 27; Evanston, Illinois
Wilmot, Donald W. Cpl. 2023615; HMM-361; 21; Philadelphia, Pennsylvania
Winters, Alexander Jr. LCpl. 2075193; HMM-362; 20; New York, New York

✯ ✯ ✯ ✯

On March 30, 1967 the infantrymen of Company I, Third Battalion, Ninth Marine Regiment were at full strength and involved in Operation Prairie III They had been on the move under unrelenting heat and an unmerciful sun for close to three days. They were out of water and miserable but had been warned that there would be no evacuation for heat casualties on this operation by by their commanding officer Captain Michael P. Getlin. The Third Battalion was on a sweep southwest of Con Thien and north of Cam Lo and it was hoped that India Company would draw attention from the North Vietnamese Army.

Most of the men in India Company were very young, with many still in their teens but almost all of them were combat veterans.

"I'll guarantee you we're going to get into something."

Captain Michael P. Getlin
I-3-9
Spoken to HM3 Kenneth R. Braun
from The "Flaming I" At Getlin's Corner by R.R. Keene.
February 2009 edition of Leatherneck

References used to write this story include the magazine articles, The "Flaming I" At Getlin's Corner by R.R. Keene from the February 2009 edition of Leatherneck Magazine. Forever Proud by Malcolm McConnell from the November 1988 edition of Readers Digest. And Missoula Man Honored with Navy Cross a newspaper article from the Missoulian newspaper on June 20, 1969

Also used were the following books, Keith William Nolan's Operation Buffalo, Marion Sturkey Bonnie Sue: A Marine Corps Helicopter Squadron in Vietnam, Edward f. Murphy's Semper Fi Vietnam and The Navy Cross edited by Paul Drew Stevens.

Additional information was also gleaned from the website The Vietnam Veteran's Memorial, the awards citations of many of the participants, witness statements for these awards, command chronologies, the rosters and diaries of the participating units, and information obtained from the National Personnel Records Center.

So isolated were the fragments of the fight that each action is best described as it happened—as a separate event. Fitted together these acts of heroism form the total picture as it occurred at crucial moments.

5

Getlin's Corner

On March 30, 1967, India Company, under the command of Captain Michael P. Getlin, was establishing night ambush sites when all elements became engaged simultaneously and the command group and a small security element were attacked by a reinforced North Vietnamese company supported by heavy automatic weapons and mortar fire. Captain Getlin, despite multiple shrapnel and gunshot wounds, while under constant mortat and small arms fire, remained on the exposed forward slope of the hill, where he calmly called in artillery fire and directed helicopter strikes on the advancing enemy. When the attack built to the point of overrunning the Marine position, Captain Getlin moved to the most critical position and delivered devastating shotgun fire into the assaulting enemy. The barrel of his weapon split due to the rate of fire. With complete disregard for the danger involved, he reloaded and continued to fire, personally killing at least six enemy soldiers. Realizing that the position was not tenable, Captain Getlin directed his men to move to a better position while he covered their move. At this time three grenades fell within his immediate position. He threw one grenade back at the enemy and was mortally wounded attempting to retrieve the others. As a result of his professional ability, extraordinary courage, and stirring example, the Marines gained the new position and repulsed the enemy attack of over sixty North Vietnamese.

Second Lieutenant John P. Bobo, of the weapons platoon, immediately organized a hasty defense and moved from position to position encouraging the outnumbered Marines despite the murderous enemy fire. Recovering a rocket launcher from among the friendly casualties; he organized a new launcher team and directed its fire into the enemy machine gun positions. When an exploding enemy mortar round severed Lieutenant Bobo's right leg below the knee, he refused to be evacuated and insisted upon being placed in a firing position to cover the movement of the command group to a better location. With a web belt around his leg serving as a tourniquet and with his leg jammed into the dirt to curtail the bleeding, he remained in this position and delivered devasting fire into the ranks of the enemy attempting to overrun the Marines. Lieutenant Bobo was mortally wounded while firing his weapon into the main point of the enemy attack but his valiant spirit inspired his men to heroic efforts, and his tenacious stand enabled the command group to gain a protective position where itrepulsed the enemy onslaught.

In addition to the severed leg, Lieutenant Bobo succumbed to gunshot wounds to the face, neck, back, both arms and the right leg.

In the following witness statement by First Sergeant Raymond G. Rogers he writes: "I personally observed Second Lieutenant John Bobo, Weapons Platoon Commander, kill at least five North Vietnamese soldiers, although he had been seriously wounded previously by an enemy mortar that had severed his leg below the knee. He also killed the NVA soldier who had wounded me in the leg, and was then standing over me. I pleaded with Lieutenant Bobo to crawl to the rear and try to stop the bleeding. He refused, and asked me to tie his leg off and give him his shotgun ammunition and pull him back up on the line. I did and the last time I saw Lieutenant Bobo alive he was in a half sitting position firing his shotgun. By his raw courage, dogged determination, and unselfish devotion to duty, he was a tremendous inspiration to the Marines fighting with him. I have never witnessed a more heroic act in my twenty years of service in the United States Marine Corps."

Lance Corporal John F. Lange, in his witness statement, wrote: "I saw Second Lieutenant Bobo move from position to position organizing the men into a hasty defense. He recovered a 3.5 rocket from a team that had been knocked out and gave it to another team who had their's destroyed by a mortar. After losing his leg below the right knee as a result of a mortar round, he refused to be evacuated and insisted on being moved up to the front line, where he kept up a steady volume of shotgun fireinto the attacking enemy. In this position Lieutenant Bobo drew a terrific volume of small arms fire and grenades. He refused to be evacuated and stayed on the forward slope of the hill covering our withdrawal to a covered position. If it were not for his heroic actions we would not have reached this position.

Rogers saw his company commander's position occupied by the enemy, he single handedly charged through heavy automatic weapons fire and grenades to assist him. Upon reaching the position, he found that Captain Getlin was mortally wounded and engaged the numerically superior enemy force, killing several. In this instant he was severely wounded by rifle fire, but despite his painful injuries, he continued to deliver accurate and effective fire upon the enemy. Seeing the weapons platoon commander, Lieutenant Bobo, go down, he crawled to him, administered first aid and, at Bobo's request, propped him into a firing position. Being the senior Marine present, he assumed command of his remaining forces and crawled through a widely exposed area to radio and reestablish contact with the battalion. While attempting to call in artillery fire on his position, a huey gunship came into the area. He established contact with the gunship and directed fire on the enemy, but the enemy hordes kept coming into his positions. Although seriously wounded, he led six seriously wounded survivors to a covered position and established a hasty defense. As a result of his professionalism, courageous leadership and stirring example, the fanatic enemy assaults were stopped, and he and his forces accounted for sixty two enemy soldiers killed.

67 At the iniation of the action, Corporal John L. Loweranitis moved through the intense fire to the 60mm mortar position, reorganized the crew and delivered effective fire on the machine gun positions that were raking the Marine positions. When the mortar ammunition was expended he again exposed himself to small arms fire and grenades as he moved from position to position evacuating wounded to the reverse slope of the hill. When the North Vietnamese Army attempted to overrun the Marine positions, he moved to the most threatened point and personally accounted for five enemy kills. Although wounded by small arms fire and grenade fragments on two separate occasions, he refused to leave his position and resolutely covered the withdrawal of the command group to a more tenable position until he fell, mortally wounded. His heroic action, with complete disregard for his own life, allowed the Marines to gain the new position and account for numerous enemy casualties.

In his witness statement recommending Loweranitis for the Navy Cross, Rogers wrote: "I observed several acts of heroism performed by our company messenger, Corporal John Loweranitis. At the beginning of the action, Corporal Loweranitis, having been a mortar section leader, prior to becoming the company messenger, aided in setting up our 60mm mortar to return fire on the enemy. As the battle progressed Corporal Loweranitis was everywhere, carrying wounded to the reverse slope and returning small arms fire against the enemy. When the enemy gained a foothold on the forward edge of Hill 70, Corporal Loweranitis voluntarily came with me to see if we could stop the flow of the enemy soldiers from coming on Hill 70. He killed at least five NVA during this action and was himself wounded by enemy grenades. He remained in his position; leaving it only for resupply of ammunition and grenades. Corporal Loweranitis was again wounded by small arms fire, but refused to leave his position. His last words to me were, 'They'll have to kill us, Top, to take this hill.' When I next observed Corporal Loweranitis he was lying face down in the same position that he had defended so well during the fight. He had been struck with several rounds of small arms fire. Corporal Loweranitis was a Marine that could be called on at anytime to do more than what his duties called for. He did not hesitate to lay down his life when the time came. He was an inspiration to all Marines fighting on the hill with him, and his actions were in keeping with the highest traditions of the United States Marine Corps.

Corporal Glen H. Crosby wrote in his witness statement: "I personally observed Corporal Loweranitis, the company messanger, come to the assistance of an inexperienced 60mm mortar gunner. Although under intense small arms fire he laid effective fire into the machine gun positions that were raking our position. When all the mortar ammunition was expended he moved back to the position occupied by the company commander. The NVA troops began assaulting our position. Corporal Loweranitis ran through a hail of fire to our position and picked up an M-79 grenade launcher and ammunition and, despite several painful wounds, returned to the point of our line that was hardest hit. Here he began firing at the advancing enemy. He was killed covering our withdrawal to a covered position."

Lance Corporal James E. Burghardt the radioman for the first platoon under the command of Second Lieutenant John L. Prickett also stood out. In the initial burst of fire, Prickett was seriously wounded and fell in an area exposed to enemy fire. Disregarding his own safety, Burghardt unhesitatingly advanced through the heavy rifle and automatic weapons fire to the side of his wounded platoon commander. Throwing himself on the wounded officer in an attempt to shield him from the heavy volume of enemy fire, Burghardt immediately began administering first aid. As a result of his unselfish act, he was seriously wounded in the area of his neck by small arms fire. His heroic actions, despite grave personal danger, undoubtably saved the life of Prickett.

In a written witness statement recommending Burghardt for Silver Star Medal, Staff Sergeant Roy Knight, Jr. observed that: "Lieutenant Prickett, while maneuvering his units, was seriously wounded and knocked to the ground by small arms fire. Unhesitatingly and with complete disregard for his own personal safety, Lance Corporal Burghardt threw himself on his wounded platoon leader in an attempt to shield him from incoming enemy rounds. Lance Corporal Burghardt immediately began to administer first aid to the wounded officer, and it was at this time that he, too, was seriously wounded by the heavy volume of enemy fire. Burghardt's unselfish actions were undoubtebly instrumental in saving his platoon leader's life and were in keeping with the highest traditions of the Marine Corps."

In another witness statement attesting to the courage of Burghardt, First Lieutenant Richard I. Neal wrote: "The first platoon was moving across an open area in an attempt to reinforce the company command group which was being hit hard by the North Vietnamese Army attack. At this time, the first platoon came under heavy enemy fire from automatic and small arms weapons, which penetrated the immediate front and the right flank. The platoon leader, Second Lieutenant Prickett, was hit immediately, knocked to the ground, and lay seriously wounded by small arms fire. Without hesitation, and without regard for his personal safety, Lance Corporal Burghardt moved to the side of the wounded officer and threw himself upon him in an attempt to shield him from the large volume of fire being delivered by the enemy. He immediately began to administer first aid to the platoon leader, and it was at this time that Lance Corporal Burghardt was hit by small arms fire and seriously wounded."

Suddenly and without warning, the machine gun team of Lance Corporal Thomas E. Butt, became the target of a vicious attack. Exposing himself repeatedly to enemy, he was ordered to man the rocket site. He unhesitatingly dashed to the rocket team's position, picked up the launcher tube and assaulted the enemy. Despite being wounded several times, Butt hept on fighting.

In a hand written witness statement written nineteen years later, Lance Corporal Joe Lempa wrote: "I've been recently contacted by the son of one of the men I served with at that time. This letter is written in honor of that man."

Thomas E. Butt was my team leader on March 30, 1967."

Our company was on an operation in Quang Tri Province on that day. Our platoon was sent out to establish ambush sites. We were attacked and surrounded by a company of North Vietnamese soldiers."

The following is my recollection of that battle. We were on a small hill when we got hit. We set up a small perimeter of defense. Casualties were heavy early in the battle. Several attempts were made to break through our lines. Tom Butt moved our gun position four times to prevent the enemy from overrunning us. We finally settled next to Captain Getlin and First Sergeant Rodgers."

Sam Phillips was machine gunner, Bill Hill was ammo carrier and I was the A gunner."

Tommy directed our fire and gave us covering fire. When the rocket team suffered heavy casualties, Tom Butt went to the aid of Jim Blevins, the rockets team leader. When Blevins was killed, Tom Butt continued firing the rocket until he was wounded by machine gun fire to the arm. He then traded his weapon for Second Lieutenant John Bobo's pistol so he could aid in the defense of the perimeter. He knew every man was needed while the Captain called in helicopter gunships for air support. Lieutenant Bobo died heroically and proud, as did the Captain and many other fine young Marines. We lost our entire rocket team. Two of the men were captured and shot when they refused to be taken prisoner. Luckily they lived to tell about it."

Sam Phillips was wounded. First Sergeant Rodgers was wounded and as Tom Butt attempted to pull him to safety, Tom was wounded again. When we got back to our home base, stories spread that Tommy was to be awarded for his actions that night."

Ninteen years later I'm shocked to hear that it wasn't to be. I can honestly say, 'If I were to face a similar situation today, I can think of no man I'd rather have at my side.'"

I will be forever proud to have fought in that battle with Lt. Bobo, who was awarded this countries highest honor, the Medal of Honor."

I will be forever proud to have fought along side all the other courageous young Marines who gave their lives that night."

I will be forever proud to have known and fought alongside Tom Butt, a true hero in my eyes. Last but not least I'll be forever grateful to his son for reuniting a bunch of old friends after 19 years."

In another carefully, hand printed written statement, which appears without corrections in spelling or grammar, Corporal Roger Turnquist wrote: "I am writing this letter in reference to Sgt. Thomas E. Butt, Ser. # 2172737, USMC Ret., who on March 30, 1967 was badly wounded under fire in a large scale operation at the DMZ. The undersigned and Sgt. Butt were with I Co. 3rd Bn. 9th Marines."

'At the time of the ambush on March 30, 1967, I was a Corporal and Sqd. Leader attached to "I" Co. Our Bn. was on a large scale operation to sweep North Vietnamese Army regulars from the northern most area of Quang Tri Province. We had made some sporadic contact on that fateful day, when later in the afternoon we had set up a perimeter on a hill approximately 1000 meters from the North Vietnam border. My Squad had been assigned a forward patrol to see if we could observe any enemy activity. While we were on patrol a few hundred yards from the base perimeter we heard automatic weapons fire and mortar rounds exploding. The firefight that ensued was intense. My Squad at this point had lost contact on radio with the command post. I took it upon myself to regroup with our platoon. We had received some small arms fire to our rear and crossed a rice paddy dyke under a mortar barrage. When we got back to the main fighting area the scene was chaotic with Marines spread all over digging in. Gunships firing on enemy positions within 100 yards and wounded being evacuated. The casualty total was 16 killed and approximately 27 wounded. Among them was Sergeant Butt, who in the highest tradition of the Marine Corps, had fought valiantly and courageously."

Sergeant Butt on that day, who was a Wpns. Team leader, and was a Lance Corporal at the time, had aided a fellow Marine, Jim Blevins, who was firing rockets at enemy positions. After Blevins had been mortally wounded, Butt took over the rocket launcher and continued to fire on enemy positions. Butt suffered multiple machine gun wounds to his right arm, which rendered it paralyzed. Upon refusing to be evacuated, Butt continued to fight with a .45 caliber hand gun, to ensure the safety of his fellow Marines. At this point the NVA were advancing on the out numbered Marines. Butt desperately tried to pull other Marines from the devastating weapons fire. Butt was wounded again with a bullet which passed through his abdomen and right leg, disabling him. At this time Butt was finally evacuated."

Many brave Marines fought and lost their lives that day, and many more wounded. In the facts at hand, Sgt. Butt had fought with uncommon valor in the highest tradition of the Marine Corps and should so be justly awarded."

In his witness statement, which appears here with no changes in grammatical errors, First Sergeant Raymond G. Rogers, Jr. also wrote: "The mortar barrages continued and increased in intensity as the enemy walked the mortars across the hill. The weapons platoon commander (Lt John Bobo) was seriously wounded during the mortar barrage and his leg was severed, while he was attempting to move our machine gun section to the forward part of the hill. The mortar barrage continued as the enemy ground assault increased. I tried to assist Lt Bobo and ordered the machine gun and rockets sections to the front to catch the enemy in the open rice paddy."

During this time Cpl Thomas Butt brought his machine gun section forward through heavy enemy automatic weapons fire and st up on the forward slope of hill 70 and preceeded to deliver devastating machine gun fire on the advancing enemy. At the time I ordered the rocket section to the front as the enemy had committed more troops to the assault."

Cpl Butt continued to expose himself to the heavy enemy automatic small arms fire by moving from one gun position to another, re-suppling them with ammunition and directing their fire."

When the gunner of the rocket team (PFC Blevins), was shot in the head and killed, I shouted to Cpl Butt to man the rocket launcher. As he was moving from one position to another firing the rocket, he was seriously wounded by enemy's small arms fire. The enemy was laying down intense heavy machine gun fire to cover their assault and we had to change our rocket positions and machine gun positions several times."

About this time, one of the Corpsman informed me that the Compnay Commander (Getlin) had been wounded, the Air Officer (Captain Ralph B. Pappas) killed and the C.P. was in danger of being over run."

I mentioned to Lt Bobo that I was going to the Company Commanders position and had not moved 5 yards before I ran into four NVA soldiers. I opened fire and killed three of them. When the fourth one shot me with an AK-47, hitting me in the right thigh. As I was attempting to get back to my feet, one of the Marines shot and killed the NVA soldier."

I am not certain if it was Lt Bobo or Cpl Butt that shot and killed this NVA soldier, but Cpl Butt did come to my aid and assist me in getting back on my feet and gave me a battle dressing. Cpl Butt and I proceeded to the C.P. and found the Company Commander and both Radio Operators (Private First Class Jerry E. Easter and Private Gregory T. Jochem) seriously wounded. Cpl Butt had armed himself with an M-14 and I was using a captured AK-47."

Several enemy soldiers broke through to our C.P. and threw six or seven grenades into our position. The Company Commander, myself, Cpl Butt and one other Marine returned fire, immediately killing some of the NVA soldiers."

The Company Commander was killed and the Battalion and Company Radio Operators were seriously wounded, again. I lost track of Cpl Butt at this point and did not see him again, until that evening when he led a squad to our position."

While I was waiting to be operated on at the Aid Station, I gave verbal statements to some Marines in our company regarding Cpl Butt and several Marine's actions on the hill that day. I instructed them to give word to our former Company Commander of I/3/9, Captain G.D. Navadel and was under the impression that these Marines had been recognized for their heroic deeds, but apparently these recommended citations were never forwarded from the Battalion."

Cpl Butt had been in the Marine Corps less than two years when this action took place. This stirring example and devotion to duty inspired other Marines who observed him, to continue the fight regardless of the odds."

Hospitalman Third Class Kenneth R. Braun, in close combat conditions and having received serious shrapnel wounds during the initial firefight, quickly began administering medical aid to those around him. Seeing that their current position was untenable and as casualties mounted, the order was given to pull back. Realizing that many wounded Marines were forward of his position, and disregarding his own safety and wounds, Braun moved throughout the fire swept area, treating the wounded and exposing himself to enemy fire. Time and again, he dragged the wounded across open terrain, shielding them from fire with his body while pulling them to safety. Trading his pistol for a rifle, Braun fought his way back down the hill to where Marines were pinned down and suffering heavy casualties; moving from Marine to Marine, with total disregard for his own safety, he fearlessly and skillfully administered medical aid to the wounded while simultaneously engaging the enemy. With enemy forces in position, Braun did not withdraw, but continued to treat the wounded and though armed with a malfunctioning weapon was able to kill two enemy soldiers while protecting the Marines. Maintaining the tenuous position while assisting a seriously wounded officer, and with enemy troops all around him, Braun held his position, continued to treat the officer, and held off the enemy with his weapon until shot three times.

As the enemy attempted to overrun the command post Private First Class William N. Stankowski delivered accurate rocket fire into the advancing enemy force. When he was severely wounded by an incoming mortar round, he continued to advance under fire to treat other wounded Marines while continuing to fight the enemy. Later, while dragging wounded Marines to the landing zone from the surrounded position, Stankowski was critically wounded a second time. He refused evacuation and led a small group of Marines back through the withering fire in an attempt to recover the body of Lieutenant Bobo. Stankowski was wounded a third time while engaged in a heavy fire fight to retake the command post. This time he was evacuated.

When the enemy threatened to envelop the left flank of Ray Gaul's third platoon the fire team of Private First Class Gene Chamberlain was designated to counter the attack.

In a statement made on April 18, 1967 Corporal Elmer R. Blanchard, a squad leader, wrote: "The third platoon had set in on both sides of a tank trail, approximately ten meters into the brush. We were eating chow and making final preparations before moving into our night position. At this time the company command group, first platoon, and second platoon had moved into similar positions ahead of us when they were hit by North Vietnamese Army forces. The main body of the forces was moving in toward our left flank. We immediately moved on line at a 270 degree azimuth sweeping toward the North Vietnamese Army's right flank. At that time, we came into heavy contact with the enemy, engaging in automatic and small arms fire plus a haeavy exchange of hand grenades. We had moved within ten to fifteen meters of the enemy when they began an envelopment on our left flank. I employed Corporal Siemon's fire team to assault the enveloping force, and they broke immediately from our main body and engaged the enemy at point blank range. It was at this point that Corporal Siemon was killed. Immediately Private First Class Chamberlin took charge of the assault and kept it moving. During the Assault Private First Class Chamberlain was knocked off his feet when hit in the upper left arm. He quickly regained his position and continued to assault the enemy. Once more, he received a wound in the upper right leg which, again, knocked him down. He quickly recovered and continued the attack. Again he received wounds from enemy grenade fragments and, for a third time, Chamberlain was knocked from his feet. He continued to fire into enemy positions until the remainder of the fire team had swept through and destroyed the enemy enveloping force."

Blanchard, a short timer and a former football player with the University of Georgia, who had previously served as a body bearer with Marine Barracks, Eighth and I, was wounded by shell fragments in the buttocks during the fight.

Hospitalman Third Class Floyd W. Garrett, Jr., a corpsman serving with Gaul's platoon, made this observation: "When Corporal Siemon was killed in the initial assault Chamberlain immediately took command of the fire team and continued their moving. At that time he took a round in his right arm which knocked him down. He regained his position to assault again when he was hit in the upper right thigh. Knocked down once more, he unbelievingly got to his feet and moved out again. For the third time, he was knocked down by fragments from an enemy grenade which hit his right foot."

When it was all over Chamberlain had suffered gunshot wounds to his right arm, left thigh and hip, buttocks, abdomen and both legs.

Gaul remembers that, "Prior to the fight, Chamberlain complained of problems with the heat, and I thought that he was going to be a shitbird, but during our assault he kept getting knocked down and getting back up. He was magnificent."

Throughout the battle Staff Sergeant Francis X. Muldowney, Gaul's platoon sergeant, performed calmly under fire and by his quick thinking and immediate aggressive action kept the third platoon on line and directed return fire on the enemy. He added to their firepower by firing LAAW's, M-79's and by throwing hand grenades, killing North Vietnamese with all of these weapons and knocking out one machine gun position.

When Lance Corporal Richard G. Weller and his fire team became separated from the platoon during the enemy attack, he competently led his men through the intense small arms fire to the protective cover of a trench and immediately established a hasty defense. Once there he spotted a North Vietnamese soldier moving through the tall grass toward his position, Weller opened fire and killed him. As three more of the enemy charged, he calmly directed his team's fire and killed two of them. As the third one tried to escape, Weller killed him with a grenade. Later, his instanteous reactions and effective fire killed three additional North Vietnamese who had infiltrated the trench under cover of darkness.

As darkness approached two Marine Corps Huey helicopter gunships appeared overhead. They were from Marine Observation Squadron Two (VMO-2) whose squadron's call sign was Deadlock. VMO-2 operated out of Dong Ha in support of Third Marine Division operations in vicinity of the DMZ. On this mission the call sign was Deadlock Playboy with the word Playboy indicating that this was to be a direct fixed wing attack on a target.

One bird, VS-4, was piloted by the section leader Captain Christopher M. Bradley. His copilot was Captain John W. Shoaff. The other aircraft, VS-14, was piloted by First Lieutenant David P. Nazarian and copilot First Lieutenant Ezekial C. Barnard. Their transmissions could not be heard well by Rogers due to the enemy jamming the frequency. But the gunships responded to the directions Rogers was giving and attacked the enemy force concentrated on the top of the hill with a combination of rocket and machine gun fire.

Each helicopter made three passes before running out of ammunition. Their accurate fire, which included that of the door gunners on each craft, wreaked havoc on the NVA assault. Corporal Michael C. Martin, the crew chief, and Corporal Roy A. Reed, a helicopter mechanic, manned the door guns on Bradley's ship. On Nazarians plane CorporalRichard A. Gatchel, the crew chief, and Joseph Zutterman, a nuclear biological and chemical defense specialist, flying as aerial gunner, added to the devastation as they raked the enemy with their guns. On each strafing run the enemy soldiers tried to knock the helicopters down with rifle fire and in doing so gave away their positions. As they did so the Marines of India began to pick them off with well aimed rifle fire of their own. The attack that was coming from enemy soldiers still holding the top of Hill 70 soon abated.

Charles Dockery, India's senior corpsman was combat experienced, and knew what to do. He was all over the battlefield working on the wounded. As Dockery responded to a call for help he was shot in the bottom of his right foot by a round that travelled up his leg. A grenade then exploded next to him and the concussion knocked him about ten feet from his original position. Luckily he did not drop the .45 pistol that he held in his hand and he promptly shot three enemy soldiers at close range. Dockery was soon shot through the right wrist and later took another round that passed through his left foot and lodged in his left elbow. His injuries later resulted in the amputation of both legs.

Rogers found the FAC, Captain Pappas, hunched over his map board, and thought that he was calling in air support. When he got closer, he could see that the Captains face was covered with blood and he had a bullet hole in his forehead. Two of the FAC Team radio operators, Private First Class Donald Krick and Lance Corporal Roman Villamor were dead, and two, Private First Class Paul J. Kain and Private Arthur F. Mercer were missing.

As Rogers passed Pappas' body he found three Marines who were bound with communication slash wire. The wire was tied around their wrists and ankles and then around their necks. The North Vietnamese had them prepared to carry off as prisoners of war. Two of the bound Marines were the missing FAC Team radio operators, Kain and Mercer. Both were badly wounded, Kain with gunshot wounds to both legs and his left foot and Mercer with gunshot wounds through both legs. Neither could walk. Rogers cut them loose with a Case pocket knife. An unwounded Marine freed the other man.

That night, Corporal Jack Madden took charge of the recovery and movement of the wounded and dead back to the security of the 1st Platoon's perimeter.

The artillery FO, Lieutenant Richard Neal had immediately taken charge of the first platoon when Prickett was wounded and now assumed command of India Company. He led throughout the remainder of the engagement and reestablished contact with battalion. Neal also planned night defensive fires and personally directed the medevac lifts. Neal's leadership instilled confidence in all the Marines who witnessed it under fire.

Braun was all over the battlefield. Wounded at least three times and with both shoulders ripped open by bullets; the tough corpsman crawled back into the perimeter under his own power. He had made it back to Rogers and the group of wounded before Neal and the first platoon got there. Once there Braun instructed the troops on how to treat his wounds. The gaping wounds to his back were horrible and so large that the entire spine was exposed. Since the medical supplies were exhausted Braun had them stuff blood soaked bandages just removed from other Marines, who had been killed during the fight, into his back.

Among the wounded were Lance Corporal Pasquale Gigliotti, who had been hit twice and Lance Corporal Bobby J. Jordan with a head wound. Lance Corporal Walter J. Riley had been hit on six separate occasions during the battle by fragments from Chicom grenades and mortar fire. Most of the wounds were to his right arm and right shoulder. Weller had superficial fragmentation wounds to his back and his left hand. Meeks had fragmentation wounds in both legs. Corporal Raymond S. Wolfe had suffered gunshot wounds to the left side of his neck and his right thigh.

Marines from Madden's third squad were involved in the retrieval. Madden assigned Private First Class Carmelo Burgaretta and Lance

Corporal Chedmond R. Browne were assigned to guard the perimeter. Among those bringing in the wounded and the dead were Private's First Class Lebrun A. Abbott, Martin "Frenchy" Landry, Severiano Ovalle and John W. Webster and Lance Corporal David Underwood

Lieutenant Neal saw to it that a landing zone was secured and called for a medevac. The most critical were being moved to the assembly area to be the first lifted out to the triage facility at Delta Field Medical in Dong Ha. There each casualty would be stabilized before further evacuation to the appropriate treatment center.

NOTES:
The Marines who died at Getlin's Corner were as follows:
Anter, Albert G. Pfc. 2289697; I-3-9; 18; Central Falls, Rhode Island
Armenta, Ruben M. Pfc. 2296394; I-3-9; 18; Pico Rivera, California
Blevins, James E. LCpl. 2209756; I-3-9; 19; Empire, Ohio
Bobo, John P. 2dLt. 092986; I-3-9; 24; Niagra Falls, New York
Cannon, Edward E. Pfc. 2300332; I-3-9; 18; Avon Park, Florida
Crumbaker, Larry H. LCpl. 2103772; I-3-9; 21; Salem, Ohio
Getlin, Michael P. Capt. 086661; I-3-9; 27; La Grange, Illinois
Krick, Donald W. Jr. Pfc. 2229201; FAC; 20; Cleveland, Ohio
Loweranitis, John L. Cpl. 2052170; I-3-9; 22; Du Bois, Pennsylvania
Nerad, Walter, J. Jr. Cpl. 2064574; I-3-9; 22; Maple Heights, Ohio
Pappas, Ralph B. Capt. 088493; FAC(VMA-211); 27; Presidio, California
Siemon, David A. Cpl. 2073662; I-3-9; 21; Springdale, Pennsylvania
Thomas, Frank H. Pfc. 2221441; I-3-9; 20; Pompano Beach, Florida
Villamor, Roman R. Jr. LCpl. 2202144; FAC; 19; Warren, Michigan
Williams, Wallace Pfc. 2283032; I-3-9; 19; New York, New York

Captain Getlin had been in Vietnam since April of 1966. On arrival he served as the executive officer of A Company, First Battalion, Twenty Sixth Marines for a period of six months. In October of 1966 he transferred to the Third Battalion of the Ninth Marines and took command of India Company on February 9, 1967.

First Sergeant Raymond Rogers had been in the Marine Corps since 1946. He was a veteran of the Korean War and had been a Drill Instructor and Rifle Marksmanship Instructor on some of his duty assignments. Just before coming to Vietnam he had been the boxing coach at Camp Lejuene.

In The *"Flaming I" At Getlin's Corner,"* Hospital Corpsman Third Class Kenneth R. Braun recalls:

"I was supposed to be on Rest and Recuperation leave when the fight on Hill 70 took place. I was asked by Captain Getlin and Lieutenant Bobo to go with the company on the operation. I felt that I couldn't let them down so I told them that if we were going to see some action that I would go. Captain Getlin guaranteed me that we would definitely get into something.

I had no idea that it would be this bad.

When the mortar attack began I hit the deck and looked for cover. One mortar round hit between me and another Marine but it did not explode. At that time the assault really began to intensify.

The wounded began to call for a corpsman. Charles Dockery and I went out and brought back a couple. Dockery was a second class and was the senior corpsman. He had combat experience and knew what to do. We set up a staging area and he told me to get to the more seriously wounded Marines and bring them back. I was to treat those in less serious condition on the spot so that they could continue to use their weapons against the enemy. I went out but got cut off from Dockery and was unable to return. He got hit several times and was seriously wounded.

They were throwing a lot of grenades and we were picking them up and throwing them back. Captain Getlin was using his shotgun and he killed a bunch. Loweranitis was a fighting machine and was everywhere on the battle field. He would occasionally stand and calmly aim and kill enemy soldiers. I think he finally had enough and just got pissed off and charged the North Vietnamese.

Although I couldn't see what happened to him I could hear the rounds hitting. Loweranitis probably thought that we were finished because we were running out of ammunition and were badly outnumbered. He tried to take as many with him as possible before he was killed.

When Eddie Cannon went down I could not figure out what was wrong with him. He only had a small entry wound in his chest but when I flipped him over there was a large exit wound in the back.

We all had a great amount of respect for Mr. Bobo and he was holding them off when a mortar round hit and took off his leg. I grabbed someone's web belt and used it as a tourniquet around his upper leg. I gave the lieutenant a shot of morphine and marked his fore head with an "M" and tried to move him out of the line of fire. He resisted and asked me to let him die there. I couldn't leave him there so I started to drag him off. He was hit in the back and the rounds came right through his chest. I could tell by the look in his eyes that he was dead.

I was hit badly at the same time and I also thought for sure that I was dead."

The battle, like most in Vietnam, receives very little mention in Corps history but it did produce a disproportionate number of awards for combat heroism. In no other single battle in Marines Corps history have so many of the nation's highest combat decorations been awarded to men whose valor went so far beyond the call of duty

The list includes: Second Lieutenant John P. Bobo the Medal of Honor.

The Navy Cross was earned by Hospitalman Third Class Kenneth R. Braun, Captain Michael P. Getlin, Corporal John L. Loweranitis and First Sergeant Raymond G. Rogers.

The Silver Star was awarded to Lance Corporals Thomas E. Butt and James E. Burghardt, Privates First Class Gene Chamberlain and William N. Stankowski and First Lieutenant Richard I. Neal.

In addition the Bronze Star was earned by Staff Sergeant Francis X. Muldowney and Lance Corporal Richard G. Weller.

The awards to Bobo, Getlin and Loweranitis were posthumous.

Neal would go on to earn an additional Silver Star while serving as an advisor on May 22, 1970. Captain George Navadel would later earn the same medal while serving as the commanding officer of M-3-9 on June 1, 1967.

On September 3, 1966 John Loweranitis was involved in a firefight in which he earned the Silver Star. The actions that led to his receiving this award are described by Captain George Navadel in his statement recommending Loweranitis for the medal. The statement appears in its original form with no correction of spelling or grammatical errors.

The statement: About 1330 on 3 September 1966 in the vicinity of An Hoa, RVN a patrol from Company I, Third Battalion, Ninth Marines composed of two reinforced squads and two ontos engaged a Viet Cong force of two companies. As Company Commander, I personally observed the then PFC John L. Loweranitis, an ammunition man in the 60mm section attached to the patrol, after expending all the mortar ammunition evacuated a wounded ontos operator from a vehicle that was halted in the only available helicopter landing zone. He then stowed the mortar tube aboard the vehicle and drove it out of the area. He then ran across the open ground that was continually under heavy small arms fire and boarded a second ontos with an operational 30 Cal. Machine gun. He assumed command of the vehicle and firing the machine gun, directed the ontos toward the enemy held tree line that was holding up the advance of the rest of the company moving to reinforce the patrol. His action enabled the company to route the enemy. Account for nineteen KIAs and permit the medical evacuation of five wounded Marines. He remained in his exposed position atop the ontos keeping the machine gun in operation and anchoring the company's exposed left flank throughout the three hours of the company's advance.

Stankowski remembers Loweranitis as a field Marine. "He was Gung Ho and knew his shit about being in the field. He never appeared to be scared when we were under fire, at least not that I could tell when I was in his M-60 mortar team. At times he seemed a little cocky and he liked to play practical jokes on people."

On the website, A *Vietnam Veterans Memorial*, Randall Hoffman, a corpsman serving with Mike 3-9 has this recollection of that day:

"I remember the day that Lieutenant Bobo lost his life. I was a corpsman in the third platoon, Mike 3/9. He was the weapons officer for India, Mike's sister company. We were both based out of Camp Carroll. To my knowledge both companies were operating less than a few kilometers apart in the Cam Lo area on the 30th of March, 1967. In the early afternoon, Mike's third platoon ran into a small North Vietnamese unit. Outnumbered, the enemy unit tried to escape. Several did, but I recall 3 or 4 confirmed dead North Vietnamese soldiers. Mike's third platoon suffered only one casualty that being the other corpsman with a gunshot wound. I can only suspect that this enemy unit was part of the battalion lying in wait for India. A few hours later, I remember hearing the fire fight in the distance. It wasn't long after that when I could make out two Huey gunships circling and firing into the ambush site. By late afternoon, Mike's third platoon rejoined the rest of the company and was in route to reinforce India. The fire fight had subsided by this time. Darkness fell and we were arms length from one another slowly and quietly heading to join India. We joined-up with India late in the evening and with no enemy contact. The next day, I saw the terrain and remember a small valley less than a hundred meters wide which had hampered India from regrouping. The image I will never forget is the captured faded green heavy machine gun with large wooden spoke wheels and steel armor plating to protect the gunner. I remember the Marines of India Company talking of Lieutenant Bobo bravery and how he took command of the CP after it had been over-run. Lieutenant Bobo's bravery saved lives that day. He truly is a hero. God bless him."

Captain Ralph B. Pappas was a pilot who loved flying. He guided his Phantoms and Skyhawks with precision and skill. Pappas flew these jet attack aircraft in Vietnam as a member of Marine Attack Squadron Two Hundred Eleven (VMA-211). He was later assigned as a forward air controller with the Third Battalion, Ninth Marine Regiment. On January 26, 1967 Pappas participated in aerial flight as a pilot and flight leader of a jet aircraft section during a night air strike. This strike covered three downed helicopters, their crews and passengers. He skillfully maneuvered the flight with the aid of flares through the target area and carried out daring bomb raids along the defensive perimeter of the downed Marines. His strike helped to protect the beleaguered Marines and contributed to their successful extraction at daylight. For his heroism that night he was awarded the Distinguished Flying Cross.

On the website *The Virtual Wall*, Marine Captain Chris Madsen had this to say about Ralph Pappas: "Barney was one of the most genuine people I ever knew. He was my best man, best friend, and brother in the Corps. I was FACing with 1/9, we pulled back to regroup and resupply on 28 March 1967. Barney's battalion took our place. As he went out I yelled at him to grab his flak jacket. He laughed yelled back, 'You can cover your legs with it while you sleep in my cot'. Two days later his position was overrun.

Before coming to Nam, Barney sold his TR-3 to his girl for a dollar. He would laugh and say '...not in case I don't make it back, but there is pride in ownership. She will take good care of it for me if she owns it.'"

David A. Ballentine, a helicopter pilot with VMO-6 (call sign Klondike) from March of 1966 to April of 1967 made this observation of Forward Air Controllers in his Vietnam memoir, *Gunbird Driver*: FACs were aviators on tours of duty with infantry units for the control of close air support. After the unit call sign. FACs had the numerical designation of "One Four." I was always comforted talking to one of my own, yet felt a hair of pity. Normally the only reason any aviator would go on an FAC assignment was to avoid an unpleasant boss or some other less okay job. Although I don't mind camping out occasionally, lugging all my stuff and "living out" is another matter. I have great and abiding respect for the infantry, but mostly I need a cot when my work is done. A hot meal is a hell of a good idea as well, and fleas, chiggers, mosquitoes, leeches and the like are depressing to even think about. That was my attitude and, if pressed, the attitude of most aviators, maybe all. But some aviators had to be FACs and they performed super. They already knew the "aviator speak" and were used to jabbering on radios. Still, rather than volunteering they were usually assigned.

Stankowski, a career Marine, was medically retired, as a Gunnery Sergeant, after eighteen and a half years of service in 1985 when his old wounds caught up with him. Here are his recollections of that day:

"It was a very hot and humid day and several Marines including myself came down with heat prostration. Top Rogers and Lieutenant Bobo came down to see about me. Top told me to 'Pull yourself together, there would be no medevacs. We need every man that we have on this trip' Rogers took my pack to carry and Mister Bobo took the six LAAW's that I was carrying, and Top told me to get up with the headquarters section."

'We continued to walk until we came upon a stream. After everyone filled their canteens the heat casualties were told to get wet and try to cool down. As soon as this was done we moved out again. The Top and Mister Bobo continued to carry my gear and we humped until just before dusk.'

'As the sun was going down, the sky lit up in a hail of mortar and machine gun fire. Lieutenant Bobo yelled out for the machine guns and rockets to move up. Rogers moved next to me and tossed me my pack and said, 'That means you too Stankowski!' As I jumped to my feet and started to put my pack on Lieutenant Bobo was standing next to me handing me the six LAAW's. He told me to, 'Circle the hill and to concentrate my fire wherever I saw automatic or mortar fire.' Rogers chimed in with, 'And be sure not to shoot from behind our automatic weapons or mortars.' He then added, 'And keep moving.' I had been taught all of these things while going through the Basic Infantry Training School."

'While I circled the hill I had fired off four of the six rounds from the

LAAW's. As I was moving to my next position an enemy mortar round landed close to me, and exploded. Fragments from the explosion ripped open the palm and the back of my right hand. I tore off the bottom of my jungle shirt to use as a bandage. Before wrapping the cloth around my hand I wet it down thoroughly with water from one of my canteens."

"I cared for the wound as best as I could and, then continuing to move, fired off my remaining LAAW's. I came across two Marines who were tied up with communications wire and I cut them free. As I continued to move around the hill I came across Tom Butt and Lieutenant Bobo. The Lieutenant was already wounded and in a firing position. He told me to continue to move around and take care of any wounded that I came across and if I were to find any ammo or weapons to bring them to his position."

'I then saw Top Rogers close to the tree line. Three enemy soldiers popped out. Two were in front of him and one behind. Rogers shot the two in front of him with his Thompson. On the way down one of the soldiers shot him with a three or four round burst which hit his thigh and split it open like a watermelon. I fired at the soldier to his rear and he fell down, so I guess I got him."

'When I got to Top Rogers a corpsman and another Marine were already working on his wound. The corpsman asked for battle dressings and I had two which I immediately gave to him. They got him patched up and we moved him to an area that he had pointed out to me. There he began using the radios and talking to the helicopters in the area and calling in for medevacs."

'I left Rogers to help other wounded get to where he was and to obtain any weapons and ammo that I could for Lieutenant Bobo. It was during this time that I was wounded across my left bicep by either a bullet or a shell fragment. My arm felt like it was on fire but I had to block it out and do my job, just like the Top and Mister Bobo."

'I made it back to Top with a wounded Marine and as I reached him he said, 'We have to help each other get down through the tree line.' The rest of the company was on the other side of the tree line where the medevac choppers were coming in. We all paired up and helped each other move toward the trees. Top Rogers and I were together."

'As we moved through the tree line we could hear the enemy coming through. They were laughing, talking and thrashing about. We all hit the deck and lay very still while they passed by us. They were so busy laughing and talking that they never realized that they were within a foot or so of us."

'Once they were past, we got up and started moving again. As soon as we cleared the tree line we met up with Marines from the first platoon. They helped to get Top Rogers and the other wounded to the landing zone."

'I refused to be evacuated and along with about five other Marines chose to go back for Lieutenant Bobo. We went back through the tree line and started up the hill when we started taking on automatic rifle fire. I was hit in the leg before I could go to ground. From all of the fire we were receiving we knew that the lieutenant's position had been over run. It was better that we return to the landing zone than take on additional casualties."

'Lieutenant Bobo was meek and mild mannered and was well liked by all of the troops. He was like Mister Rogers on television. When he wasn't around we all affectionately called him Mr. Rogers, because their personas were so alike. Even when he chewed you out he didn't really chew you out, but you got the point. He was one of those guys that you didn't want to mistake his kindness for weakness."

Others who were wounded, mentioned in the chapter, or were known participants in the battle are as follows:

Abbott, Lebrun A. Pfc. 2262000; I-3-9; 19 Detroit, Michigan
Arcand, Paul R. Cpl. 2046626; I-3-9; 22; Boston, Massachusetts
Avent, Roger L. Pfc. 2105702; I-3-9; 19; New York, New York
Balkcom, Ronald W. LCpl. 2230917; I-3-9; 19; Columbus, Georgia
Ballentine, David A. 1stLt. 091121; VMO-6; 26; Jacksonville, Florida
Bartley, Michael LCpl. 2192477; I-3-9; 20; Minneapolis, Minnesota
Birch, William L. HM3 7702475; H&S-3-9; 21; St. Petersburg, Florida
Blanchard, Elmer R. Cpl. 2061403; I-3-9; 21; Macon, Georgia
Boden, Edward E. Pfc. 2269361; I-3-9; 20; Bowling Green, Ohio
Borowitz, John M. LCpl. 2226584; I-3-9; 19; Roseville, Michigan
Bradley, Christopher M. Capt. 086094; VMO-2; 25; Bethesda, Maryland
Braun, Kenneth R. HM3 9188839; 18; Eden Prairie, Minnesota
Brees, Marvin F. Pfc. 2231498; I-3-9; 19; Missoula, Montana
Brogan, Bill Sgt. 2115260; I-3-9; 20; Baton Rouge, Louisiana
Brown, Chedmond R. LCpl. 2170256; I-3-9; 20; New York, New York
Burgaretta, Carmelo Pfc. 2244379; I-3-9; 19; New York, New York
Burghardt, James E. LCpl. 2215780; I-3-9 20; Belmont, California
Butt, Thomas E. LCpl. 2172737; 20; Rockville, Maryland
Chamberlain, Gene Pfc. 2217725; I-3-9; 19; Milwaukie, Oregon
Collopy, William LCpl. 2184056; I-3-9; 21; New York, New York
Corey, Robert A. Pfc. 2286161; I-3-9; 19; Milwaukee, Wisconsin
Coriz, Fidel A. Pfc. 2184487; I-3-9; 19; Albuquerque, New Mexico
Crosby, Glen H. Cpl. 2061625; I-3-9; 21; Macon, Georgia
Dockery, Charles M. HM2 3906560; I-3-9; 22; Sumner, Washington
Easter, Jerry E. Pfc. 2258240; I-3-9; 19; Oklahoma City, Oklahoma
Engel, Gilbert L. Pfc. 2251938; I-3-9; 19; Houston, Texas
Errera, John V. Jr. Cpl. 2160351; I-3-9; 20; Boston, Massachusetts
Etheridge, James LCpl. 2188054; I-3-9; 19; Memphis, Tennessee
Findley, Jon L. Cpl. 2152556; I-3-9; 20; Nashville, Tennessee
Frinzi, Joseph D. LCpl. 2201316; I-3-9; 19; Delafield, Wisconsin
Garrett, Floyd W. Jr. HM3 9969938; I-3-9; 20; Harlingen, Texas
Gatchel, Richard A. Cpl. 2074084; VMO-2; 21; Pittsburgh, Pennsylvania
Gaul, Ray H. Jr. 2dLt. 095986; I-3-9; 30; Lancaster, Pennsylvania
Gigliotti, Pasquale LCpl. 2225358; I-3-9; 19; Coral Gables, Florida
Gipson, Billy LCpl. 2234049; I-3-9; 19; Houston, Texas
Gomez, David J. LCpl. 2256016; I-3-9; 19; Omaha, Nebraska
Hill, Billy Joe Pfc. 2340165; I-3-9; 18; Harvey, Louisiana
Hoffman, Randall J. HM3 9995317; M-3-9; 20; Los Angeles, California
Hutson, Charles E. LCpl. 2268186; I-3-9; 19; Indianapolis, Indiana
Jager, Arjen S. HM3 7732593; H&S-3-9; 21; Pomona, California

Jochem, Gregory T. Pvt. 2201429; H&S-3-9; 19; Milwaukee, Wisconsin
Jordan, Bobby J. LCpl. 2227030; I-3-9; 19; Oak Ridge, Tennessee
Kain, Paul J. Pfc. 2271634; H&S-3-9; 19; Albany, New York
Knight, Roy Jr. SSgt. 1587059; I-3-9; 29; Anderson, South Carolina
Landry, Martin Pfc. 2186595; I-3-9; 20; New Haven, Connecticut
Lange, John F. LCpl. 2023608; I-3-9; 22; Wilmington, Delaware
Lechowit, John Pfc. 2333490; I-3-9;19; Albany, New York
Lempa, Joseph S. LCpl. 2250400; I-3-9; 19; Chicago, Illinois
Lewis, Daniel A. Pfc. 2304556; I-3-9; 19; Cincinnati, Ohio
Lloyd, Raymond W. Pfc. 2105442; I-3-9; 20; New York, New York
Lombard, Geoffrey LCpl. 2266858; I-3-9; 19; Manchester, New Hampshire
Luke, Wayne A. Pfc. 2296468; I-3-9; 19; Lancaster, California
Madden, John P. Cpl. 2098416; I-3-9; 21; Deadham, Massachusetts
Madsen, Chris Capt. 088656; H&S-1-9; 25; Akron, Ohio
Martin, Michael C. Cpl. 2058808; VMO-2; 22; Detroit, Michigan
Meade, Ronald D. Pfc. 2230603; I-3-9; 19; Blue Grass, Iowa
Meeks, Oscar A. Sgt. 2012719; I-3-9; 22; Shaker Heights, Ohio
Mercer, Arthur F. Pvt. 2269383; H&S-3-9; 19; Columbus, Ohio
Mitchell, David HM3 7854539; I-3-9; 20; Bainbridge, Maryland
Muldowney, Francis X. SSgt 1599222; I-3-9; 29; Philadelphia, Pennsylvania
Navadel, George D. Capt. 075389; H&S-3-9; 30; Buffalo, New York
Nazarian, David P. 1stLt. 092194; VMO-2; 23; Oswego, New York
Neal, Richard I. 1stLt. 092178; I-3-9; 24; Hull, Massachusetts
O'Connell, Terence S. LCpl. 2165341; I-3-9; 20; Chicago, Illinois
Ovalle, Severiano Jr. Pfc. 2256305; I-3-9; 19; San Antonio, Texas
Phillips, Sammy D. Pfc. 2251764: I-3-9; 19; McKinney, Texas
Plumley, James R. LCpl. 2278404: I-3-9; 19; Bakersfield, California
Prickett, John L. 2ndLt. 092639; I-3-9; 23; Toccoa, Georgia
Pultz, Daniel 2ndLt. 090998; I-3-9; 23; Greensburg, Pennsylvania
Randall, Terry L. Pfc. 2312633; I-3-9; 19; Richmond, Virginia
Reed, Roy A. Cpl. 2177085; VMO-2; 20; Cincinnati, Ohio
Reynolds, William E. Sgt. 1342038; H&S-3-9; 33; Little Rock, Arkansas
Riley, Walter J. LCpl. 2223202; I-3-9; 22; Gadsen, Alabama
Rogers, Raymond G. Jr. 1stSgt. 583291; I-3-9; 39; Cincinnati, Ohio
Rosales, Walter J. LCpl. 2157348; I-3-9; 20; Los Angeles, California
Roth, Gregory J. Pfc. 2259471; I-3-9; 19; Buffalo, New York
Ruester, Michael Pfc. 2305644; I-3-9; 19; St. Louis, Missouri
Ruiz, Joe A. LCpl. 2110525; I-3-9; 20; San Antonio, Texas
Shoaff, John W. Capt. 080453; VMO-2; 29; Pensacola, Florida
Sparrow, George R. Pfc. 2204192; I-3-9; 20; Chesapeake, Virginia
Stankowski, William N. Pfc. 2201257; I-3-9; 18; Milwaukee, Wisconsin
Turnquist, Roger W. Cpl. 2169521; I-3-9; 20; Middle Island, New York
Underwood, David LCpl. 2168425; I-3-9; Jordan, Arkansas
Webster, John W. Pfc. 2168404; I-3-9; 20; Richmond, Virginia
Weller, Richard G. LCpl. 2196588; I-3-9; 20; Rio Vista, California
Wheeler, Larry G. LCpl. 2201250; I-3-9; 19; Portage, Wisconsin
Williams, Joseph R. Jr. 2106044; I-3-9; 20; Bronx, New York
Williamson, Lynn Cpl. 2147206; I-3-9; 20; Tarrant, Alabama
Wilson, James S. LtCol. 050341; CO-3-9; 41; Atlanta, Georgia
Wolfe, Raymond S. Cpl. 1945806; I-3-9; 23; Lubbock, Texas
Zertuche, Adolph Jr. LCpl. 2181565; I-3-9; 20; San Francisco, California
Zutterman, Joseph A. Jr. Sgt. 2067649; VMO-2; 22; Marysville, Kansas

★ ★ ★ ★

On May 3, 1967 the Second Battalion of the Third Marines was commanded by Lieutenant Earl "Pappy" DeLong. Major Wendell O. Beard, who would later command the battalion, was the executive officer. At the helm of Golf Company was Captain James P. Sheehan, Fox Company was led by Captain Merle G. Sorenson and Captain Raymond C. Madonna commanded Hotel Company.

Captain Alfred E. Lyon was the Company Commander of Echo Company. The platoon commanders were Second Lieutenant Frank M. Izenour, Jr., first platoon, Second Lieutenant James R. Cannon, the second, and First Lieutenant John Adinolfi, the third. Second Lieutenant Clifton H. Canter, Jr. had the Weapons Platoon.

The following story is primarily concerned with the participation of Echo Company, Second Battalion, Third Marines in the Hill Fights that took place near Khe Sanh from April 26 through May 11, 1967. It is about the attack on Echo Company that occurred on May 3, 1967.

"Hell isn't the word for it."

Corporal Richard F. Sutter
Mike-3-26

"The NVA were as good as they get. They were the equal of any Marine Outfit in tactics and fighting spirit."

Lance Corporal Harold E. Kepner
Mike-3-3
From *The Hill Fights* by Edward F. Murphy

Grateful acknowledgement is made to Khe Sanh Veterans, Inc. and to Chaplain Ray William Stubbe for permission to reprint previously published material from his fine book *Battalion of Kings. A tribute to our fallen brothers who died because of the battlefield of Khe Sanh, Vietnam.*

Information for this story was also obtained from *The Hill Fights* by Edward F. Murphy, *Operation Buffalo* by Keith William Nolan, and James R. Cannon's unpublished, written account of the attack on Echo Company.

Also used were the awards citations of many of the participants, the rosters and diaries of the participating units, and information obtained from the National Personnel Records Center.

6

Attack on Echo

After a few days of conducting patrols Echo Company attacked Hill 881 on April 30. During the attack Cannon was hit three times by rifle fire. The North Vietnamese had zeroed in on his radio antennae. One round hit him just above the heart and knocked him down him down but it did not penetrate his flak jacket. Another round grazed his left arm and destroyed the C-rations in his pack. The third round creased his hand and destroyed an M-79 grenade launcher that he was carrying. With the help of his radioman, Lance Corporal Carl R. Hovietz, he moved to a position of safety and continued the attack.

Canter was knocked out of the fight on the same day when he was hit by a round from an AK-47 that hit just above his right wrist and tore a large chunk of flesh from his forearm as it exited. A corpsman patched the wound and Canter moved on but was soon hit again. He was spun to the ground by the impact of another round that hit his right hand. As another Corpsman worked on this new wound he was again hit. This time in the right butt cheek by a round that smashed his hip, went down his leg, and lodged in his knee. This one paralyzed his right leg and was extremely painful.

On 1 May as the attack on Hill 881 continued the Marines of Echo Company suffered several more casualties, many as a result of fragmentation from exploding enemy rounds. They included Corporal James D. Blackston who received wounds to the left leg and Private First Class Ronald H. Irvin who was also hit in the face, left thigh and right foot. Private Marvin R. Kraner was hit in the chest and Lance Corporal William P. McManus in the back. Lance Corporal Thomas J. Carstens received a fragmentation wound to the left hand. Corporal Richard D. Backus, Lance Corporals Barry K. Weiner and William Gallegos, and Private First Class Charles C. Davis were all hit in the head by fragments. Weiners skull was fractured by the shrapnel and Gallegos was also hit in the left leg. Lance Corporals Roy Hall and Edward S. Fennell were also wounded, Hall in the lower left arm, and Fennell in the left leg and right elbow. Private First Class Donald LeBlanc joined this number when he was hit in the left elbow. Private First Class Kenneth R. Kostendt was wounded in the back, left knee and right hand. Corporal Leonard P. Wasilewski was hit in the left arm and thumb. Private First Class Garry D. Thurber was wounded in both buttocks and back.

Carstens would later be killed in action, on his birthday, by small arms fire during fighting on 14 August.

Enemy small arms fire claimed several more. Private First Class Curtis L. Miller was hit by rifle fire near the base of the spine and through the left hip. Lance Corporal Juan H. Cruzesquilin was shot through the left leg as was Private First Class Martin D. McGannon who also suffered a fragmentation wound to the right side. Privates First Class Roger L. Morin, Ronald L. Schriner and James H. Taylor were also hit by the intense enemy fire, Morin in the left hip, Schriner in the upper right arm, and Taylor in the right leg.

By the early afternoon of 2 May the second objective of the assault, Hill 881S, had been secured. As soon as this occurred Echo Company moved toward their new objective, Hill 881N. On this day Hospitalman Lloyd L. Heath was killed when shell fragments tore into the right side of his neck and head. As they moved artillery pounded the left side of the hill. The exploding shells caused many fires that provided excellent screening for their approach to the enemy positions.

As Echo Company neared their objective and began deploying it began to rain. About half way up the hill the assault was stopped and they were ordered to return to their previous position which was a small hill about 500 meters south of Hill 881 North. Cannon's second platoon was situated on the northeast section of the perimeter and would bear the brunt of an assault by an estimated company sized unit.

Cannon recalls the unfolding terror: "It became increasingly dark during our movement back to our previous position, and by the time we arrived, it was night. It was still raining. Everyone was tired and miserable and the bone chilling rain didn't help. We searched out the inner perimeter and, after doing so, each platoon took up their old positions. It was still raining hard. We put out our trip flares and claymore mines, put out our security, and established fifty percent security for the night. In each two man fighting hole, one man would be awake while one man slept.

With me in my bunker were my platoon sergeant, Staff Sergeant Robert L. Morningstar, and my radioman, Hovietz. It was 0400 in the morning. The rain had stopped and there was an eerie mist hovering over the hill. I had taken my boots off, wrung the water from my socks, and turned my boots upside down to allow them to drain. Everything was quiet."

Sergeant Billy Joe Like, a squad leader in the second platoon, was walking the lines of the perimeter on the hill checking his squad positions, heard a noise to his front and yelled, "Who's out there?" The answer was a burst of automatic fire that seriously wounded him in the stomach. Though seriously and painfully wounded Like remained standing and warned the others of the attack. This discovery forced the North Vietnamese to prematurely commence their attack and probably saved a number of Marines from being casualties. Despite the wound Like moved among his men and calmly and professionally directed their fire until they quickly gained fire superiority and repulsed the attack

Cannon continues, "With Like's challenge the entire northern half of the hill exploded. We were hit. I shouted to Morningstar, 'Let's go!' and sprang for the bunker entrance. As I did so, one of my Marines was blown in on top of me. He was badly wounded. I told Morningstar to take care of him and told Hovietz to come on. From my right front at about two o'clock, enemy machineguns were raking my section of the perimeter. At the same time enemy mortars were falling over the entire hill. I could hear the machine gun near the first squad chattering in long bursts and I knew this was not a probe."

Hospitalman Second Class Clarence W. Young immediately left the safety of his position and raced about the lines in search of wounded Marines. Disregarding intense rifle fire and the explosion of grenades, he went to the aid of Like and treated him within yards of the attacking enemy soldiers. Young was wounded in both arms as he ministered to the fallen Marine and he continued to do so until he was able to pull Like to a position of relative safety. He then ran into the darkness and helped other Marines until he was in too much pain to continue and had to be evacuated.

Hospitalman Third Class Danny P. Williams, despite wounds to both arms, repeatedly exposed himself to hostile fire as he worked on wounded Marines.

Corporal Larry Welch remembers, "Corporal Dennis Bergenstein, machine gunner, died about four feet from me. Everybody got killed there except one guy. Later, when we recovered his body there were 20, 25 dead North Vietnamese soldiers around his position."

Bergenstein had been hit by shrapnel in the upper left arm, lower right leg and left buttocks.

Private First Class John R. Meuse, the radio operator for the first squad of the second platoon, was seriously wounded early in the fight when he was shot through the left thigh and buttocks and in the right side of his abdomen. He radioed Cannon and told him that he had been hit and needed a corpsman. The sound of his voice gave the impression that he was dying. His squad had been penetrated and their positions had been overrun. The machine gun had been knocked out and the squad leader was about to set off their claymore mines. He was given permission to do so and soon a terrific explosion could be heard.

Although he had been mortally wounded, Meuse refused to abandon his position and somehow managed to hang on. Despite pleas from Cannon he refused to seek the medical help that could have possibly saved his life. He remained on the radio for over an hour and kept Lyon and Cannon supplied with accurate information on the situation in his area. This information helped them make essential decisions necessary for repulsing the enemy attack. As daylight approached his radio went silent.

Sergeant Ronald E. Kolodziej, a machine gun squad leader, was painfully wounded in the left hand by shrapnel in the initial moments of the attack. Despite his injury he quickly deployed his gun section into effective firing positions in front of the perimeter in areas where they would do the most good. When Kolodziej's machine run became inoperable he moved through enemy fire to the position of another infantry squad and remained there throughout the remainder of the fight. Only after the North Vietnamese attack was repulsed, some six hours later, and the company consolidated did he accept evacuation.

Another machine gunner, Lance Corporal James G. Hawley, was down in his fighting hole when he saw that Staff Sergeant David L. Noakes in a nearby hole had been hit in the head and right leg by shrapnel. He moved to Noakes and dragged him to the safety of his own position. Hawley then moved through the fire swept area in an effort to retrieve other wounded Marines. Sustainlng painful fragmentation wounds to his face from an exploding enemy grenade, unarmed, and constantly exposed to enemy fire, he successfully evacuated other casualties to places of relative safety.

Cannon recalls: "Casualties were mounting. The situation was extremely bad. I called the command post and told Captain Lyon that I needed help, that we were under assault by a large force, and that all of my crew served weapons had been knocked out. I also asked for my preplanned artillery concentrations to be fired and called for a flare ship. I was informed that I couldn't have both. I was somewhat pissed. I knew that the flare ships could fly above the maximum ordinance of artillery and I needed both."

Lyon told Izenhour to send a squad to help fill the gaps in the second platoon lines. He selected Sergeant Robert C. Powell's second squad for the job. They were badly hit as they tried to reach Cannon's platoon and Powell was wounded. Hawley went to get him even though he knew that the enemy was waiting. He picked up an operable M-60 and as he moved forward he delivered a heavy volume of fire on the North Vietnamese which helped to repel their assault.

After being treated for his wounds, Powell again moved toward the enemy and was shot and killed along with most of his squad when a deadly accurate burst of machine gun fire cut them down. A projectile had ruptured the top of his skull. They never did reach Cannon.

Another squad was sent, but it too did not arrive. North Vietnamese gunners decimated the squad, killing and wounding most of its Marines.

Lyon sent a group of engineers to take up positions behind Cannon's line. Their steady and accurate fire helped to plug the gap. One of them, Lance Corporal William T. Womble, ran to the overrun positions and retrieved machine gun ammunition and rounds for the 60mm mortars. He moved through extremely heavy automatic weapons fire and redistributed the much needed ammunition to machine gun and mortar positions.

The engineers were still armed with the dependable, heavy hitting

M-14 rifles while the rest of Echo Company carried the new M-16, a rifle which was prone to malfunction. When Womble saw that several Marines were pinned down in the second platoon command bunker, he opened fire with his dependable M-14 and single handedly engaged the North Vietnamese aggressors. His actions drew fire from the enemy and allowed the Marines in the bunker to gain fire superiority. As he poured accurate rifle fire into the enemy ranks he was shot through the neck and instantly killed.

At least four other engineers, Corporal Glenn E. Adams, Private First Class Thomas J. Gandt and Lance Corporals Edward F. Healey and John K. McNeer, were wounded by shell fragments during the vicious fighting. Adams and Gandt were hit in the hand and Healey in the right wrist. McNeer was wounded in the left arm and right ankle.

The fight would be bitter to the end and several other Marines were killed. Staff Sergeant Javier A. Sanchez was shot through the head and killed. Also killed were Privates First Class James Anderson, Charles C. Davis, Michael F. Huwel, Robert H. Kruger, Jr., Daryl L. Johnson, John C. Rice, Robert E. Flannery, Norman R. Martel, Ben Roberts and Robert G. Yagues, and Lance Corporals David B. Koenig, Woodrow Williams, Gary L. Biehl, David E. Holdway and Richard Almeida, and Corporal John S. Almoney, and Sergeant Willis R. Heavin. Anderson had been shot through his left arm, chest and throat. Davis was shot through the head and throat and Huwel in the face and left arm. Kruger had been hit in the right and left sides of the chest and the lower legs by shrapnel. Johnson had been hit in the left side of the chest by a gunshot. Among Rice's many wounds was a gunshot wound to the abdomen. Flannery was killed when hit in the head by a piece of shrapnel. Martel was hit by several gunshots and had a fragmentation wound to the left side of his chest. Roberts was shot through the head and Yagues through the head and chest. Koenig was shot through the chest. Williams died from shrapnel wounds to the head and chest. Biehl was killed by multiple shrapnel wounds, most notably to the neck and throat. Holdway was shot through the spine and right shoulder. Almeida was hit in the left arm and shoulder by a gunshot. Almoney died as a result of gunshot wounds to the head and chest. Heavin was killed by numerous shrapnel wounds to the chest, pelvis and lower extremities.

Cannon made his way to the command post and when he reached Adinolfi's bunker he discovered that he was gone. Cannon feared that he may have been taken captive. He asked for and received a 3.5" rocket launcher team. Lance Corporal Thomas Rice was carrying the launcher and Private First Class Phillip G. Skinner carried the rockets and Rice's M-16. As they prepared to fire the North Vietnamese poured rounds into their area and just as Rice asked for his rifle, Skinner raised up a little to pass it to him he was shot through the left side of the head and killed.

The anguished Rice aimed his M-16 and squeezed the trigger. He got off one round and the rifle jammed. He cleared it and tried again with the same result. He picked up the rocket launcher and ran for the safety of a nearby bunker. He thought, "How can we fight with weapons like this?"

Cannon had set the rocket team up in front of Adinolfi's position and was in the process of directing their fire when an explosion went off right before him. He was left still standing and thinking that the rocket team had been knocked out.

"I was still standing. For some ungodly reason, I thought nothing could kill me." Cannon recalled.

Corporal William K. Downing, a fire team leader with the third platoon, was another Marine who contributed to the success of the Echo Company defense. He moved his team into the sector of the beleaguered second platoon. Downing was quickly wounded in the leg but continued to press forward as he directed the fire and movement of his fire team. As they gained fire superiority and forced the enemy to withdraw he refused to be evacuated and told the corpsman to treat others more badly wounded than he was.

A member of Downing's fire team, Private First Class Ernest S. Alford, III, remembers him shouting, "Don't come down here and help me, take care of the rest of the men first."

When the North Vietnamese counter attacked Downing remained in his position and continued to deliver effective fire into their ranks. He continued to do so until the last wounded Marine was evacuated. Downing's body was found later with six of the enemy dead around his position. He had been killed by a gunshot wound to the chest.

All around the perimeter Marines responded to the calls for help. Private First Class George Sternisha, of the third platoon, crawled forward to help. He recalls, "Sergeant Powell from the first platoon came up behind my hole. I ran some patrols with him back in Mai Loc when we were short Marines. He was already shot in the shoulder. He had a battle dressing on but was still bleeding. I told him he needed to get where the wounded were being gathered for medevac. He refused and said he needed help getting the NVA off of the hill and out of the perimeter as they had over run the company's first and second platoons. I went with him leaving my M-79 with Private First Class Schadler and thinking other Marines from the third squad were coming with me. That was not the case."

"Somewhere, Sergeant Powell and I got separated in the elephant grass. I later found him dead. He had gotten himself killed trying to save other Marines. I was alone now there were no other Marines from third platoon with me."

As Sternisha neared the lines of the second platoon he entered a chamber of horrors. Everywhere he looked he saw dead and wounded Marines. One Marine sat there shaking uncontrollably; the enemy assault had been too much for him. "As I stumbled across the second platoon's position I found Corporal Welch, from machine guns, just sitting there, shaking uncontrollably, in shock. I asked him if he was alright. He couldn't answer me. He was just sitting there shaking. Maybe he was deaf from all the explosions around him. I did notice a trickle of blood coming out of his ears."

A Marine sergeant lay there with the top of his head blown off. Sternisha continues, "I never seen so many dead bodies in my life all in one place like this. The dead Marine may have been Sergeant Heavin from second platoon. I couldn't be sure. Whoever this Marine was his brains were sticking out. I still get flashbacks from what I seen that morning." Although stunned by what he saw around him Sternisha still managed to help some of the wounded back to a collection point for the casualties. He would occasionally fire his pistol or toss grenades at enemy soldiers that he spotted. "I may have killed one or two." Sternisha was later wounded in both hands by shell fragments.

Cannon worked his way to the command post to inform Lyon of his situation. There he found Adinolfi, who would soon take over the company, alive. Cannon recalls, "I informed my company commander that my entire first and second platoons were wiped out, and that there was only one friendly position up there---my bunker. My third squad, although lightly engaged, was still intact and had suffered no casualties."

"Captain Lyon informed me that a platoon from Foxtrot Company was moving up the rear to help block penetration. I passed the word to keep my bunker protected, that it was occupied by friendly troops. Sometime, from somewhere, I had grabbed an M-16 rifle, firing it until it jammed. I threw it down and picked up an M-14. I overheard reports that the North Vietnamese were wearing flak gear and Marine helmets and some were in the trees shouting, 'Marine, you die.' I cannot attest to this. I do know that they were hard to stop. Even point blank firing into their chests would not insure that they would be stopped. Some had been beaten with entrenching tools or whatever one could get his hands on. Consequently, there was much hand to hand combat that morning. We later found that several Marines had died while trying to clear jams from their M-16's."

Before long, the third platoon from Foxtrot Company arrived. It was under the leadership of Second Lieutenant Patrick G. Carroll. He was briefed on the situation and soon after the helicopter gunships came in firing rockets. The Echo Company positions were marked by smoke and the rockets were supposed to strike to the north of them.

One of the first rockets fired hit the command bunker and wounded Lyon. It also killed or wounded several other key personnel including the forward air controller, artillery forward observer, and the senior corpsman, HM2 Kenneth Kleinschmidt. The corpsman was hit in the side of the neck and the point of the chin by shell fragments.

First Sergeant Charles Patrinos was responsible for coordination of the defense of the company command post. Although the command post was under continuous enemy fire, including infiltrating snipers, he maintained its security and made sure that it functioned effectively, by the skillful emplacement of personnel. Throughout this period of combat, he refused evacuation, although he had been wounded three times during the initial attack on 1 May. Patrinos had been hit in the leg, chest, and arm by fragments from exploding enemy mortar rounds.

Lyon, a former enlisted man, was magnificent during the fight. He had twice been wounded during the fighting on May 1 and 2 but when the ambush started he moved quickly and aggressively to reorganize the company, repel the penetration and restore the defensive perimeter. With close combat and hand-to-hand fighting taking place around the command post, Lyon remained calm. When he called for and received the platoon from Fox Company he quickly integrated it with the one infantry squad remaining in the second platoon. He used his supporting arms with devastating effect against an enemy that he had fixed in place with the Echo Company defense. The explosion of the rocket that hit the command bunker wounded Lyon in the arm, chest, and thigh and though he refused to be evacuated the heavy loss of blood caused him to lose consciousness and he had to leave.

With Lyon out of the fight, Adinolfi immediately assumed command of the company. He quickly reorganized his unit which enabled it to continue functioning as an effective fighting force. Adinolfi moved from one position to another encouraging his men, and directing their fire, and then cooridinated a two company assault against the enemy positions. He again stood out on 9 May when he aggressively led a company size patrol to the aid of Foxtrot Company when it was ambushed.

When consolidation of the position began, Patrinos again assisted with casualty evacuation, as he had done on 1 May, and repeatedly exposed himself to the accurate fire of the snipers. Maintaining control of the evacuation of the wounded, he insured the orderly movement of and meticulous accounting for every wounded Marine. Still refusing evacuation, although his wounds were aggravated and infected, Patrinos remained at his position to help Adinolfi reorganize the company after Lyon's evacuation. He displayed uncommon endurance and dedication and remained with Echo Company until 5 May when his evacuation was ordered.

After hours of bitter fighting, the penetration was contained and action was taken to eliminate the remaining North Vietnamese.

Sternisha recalls, "I started helping carry the wounded back. Once there, I helped load all the dead and wounded Marines on helicopters, some were fairly close to me. I recall one chopper lifting off with dead Marines on it and the blood running out like a spigot."

The North Vietnamese attack on Echo this day left the company with 27 killed and 84 wounded.

Included in the wounded that day was Private First Class Elvin H. Cole with shrapnel wounds to the right arm and right leg and Lance Corporal Stanley Ferbos who suffered a fragmentation wound to the forehead. Ferbos would later be killed on July 21, 1967 during Operation Bear Chain.

Cannon writes: "I would cry later and when I did the tears would never completely stop. The memory of these Marines, so young, so brave, would stay with me forever."

NOTES:
The Marines from Echo Company killed in action on May 3, 1967 or listed as follows:
Almeida, Richard H. LCpl. 2209405; 21; Fall River, Massachusetts
Almoney, John S. Cpl. 2195646; 20; York, Pennsylvania
Anderson, James Pfc. 2309318; 19; Philadelphia, Pennsylvania
Bergenstein, Dennis P. Cpl. 2103530; 21; Avon Lake, Ohio
Biehl, Garry L. LCpl. 2225384; 20; Oneco, Florida
Davis, Charles C. Pfc. 2212919; 20; Bethany, Missouri
Downing, William K. Cpl. 2201839; 20; Fort Gibson, Oklahoma
Edie, Kurt C. HM3 9183257; 22, Sunnyvale, California
Flannery, Robert E. Jr. Pfc. 2327756; 19; St. Helena, California
Heath, Lloyd L. HN 7752577; H&S-2-3; 21; Bicknell, Indiana
Heavin, Willis R. Sgt. 1916555; 23; Fort Worth, Texas
Holdway, David K. LCpl. 2130624; 21; Hyattsville, Maryland
Huwel, Michael F. Pfc. 2269144; 20; Cincinnati, Ohio
Johnson, Daryl L. Pfc. 2332905 19; Oklahoma City, Oklahoma
Koenig, David B. LCpl. 2285512; 21; Romeo, Michigan
Kruger, Robert H. Jr. Pfc. 2253209; 18; Clifton, New Jersey
Louis, Robert Y. Jr. LCpl. 2291303; 18; San Antonio, Texas
Martel, Normand R. Pfc. 2181948; 22; Manchester, New Hampshire
Meuse, John R. Pfc. 2249495; 18; Malden, Massachusetts
Moore, William R. Cpl. 2172872; 20; Scotland Neck, North Carolina
Picanso, Leonard Jr. Pfc. 2182098; 19; Acushnet, Massachusetts
Powell, Robert C. Sgt. 2080153; 21; Dallas, Texas
Ramirez, Juan J. LCpl. 2234002; 20; Houston, Texas
Rice, John C. Pfc. 2277415; 20; Seattle, Washington
Roberts, Ben Pfc. 2256836; 19; Savannah, Georgia
Sanchez, Javier A.E. Sgt. 1925849; 24; Laredo, Texas
Skinner, Phillip C. LCpl. 2250322; 21; Chicago, Illinois
Tino, John F. Jr. Pfc. 2311778; E-2-3; 24; Torrington Connecticut
Williams, Woodrow LCpl. 2220265; 20; Philadelphia, Pennsylvania
Womble, William T. Jr. LCpl. 2168383; 18; Norfolk, Virginia
Yagues, Robert G. Pfc. 2275586; 18; Mountain Home, Idaho

During the fighting that occurred on May 3, 1967 many Marines were honored for their heroism. The Navy Cross was earned by Private First Class John R. Meuse and Lance Corporal Frederick G. Monahan. The award to Meuse was posthumous.

The Silver Star was awarded to Second Lieutenant James R. Cannon, Lieutenant Colonel Earl R. Delong, Private First Class Roy W. DeMille, Corporal William K. Downing, Lance Corporals Edwardo J. Gonzales and William T. Womble, Captain Alfred E. Lyon and First Sergeant Charles Patrinos.

The Bronze Star was earned by First Lieutenant John F. Adinolfi, Major Wendell O. Beard, Sergeants Ronald E. Kolodziej and Billy Joe Like, Lance Corporal James G. Hawley, and Hospitalman Second Class Clarence W. Young.

In a letter to the author, George Sternisha wrote: "I got to know Al Lyon pretty good after the war. I talked to him at the reunions quite a bit, often into the wee hours of the morning. He told me about the war from an officer's point of view and leadership. We discussed everything from calling in artillery, air strikes, battalion orders, company operations, chain of command, and screw ups and problems in the battalion."

'Al went to Ohio State and then got his commission from that six month school that Marines give college grads. He told me that it bothered him and he felt responsible for all the dead and wounded Marines from Hill 881 North, all his life. He resigned his commission from the Marine Corps because he felt that we were betrayed by the cover up of the M-16 rifles. He said the Marines knew there was a problem with them and did nothing about it. I last saw Al Lyon and talked with him at the Reno VA hospital in November 2008. My Grenadier, Bill Smith, Oscar Ramos, Jim Cannon and Roy Skelton were all there to be with him and his family on his last days. He died the night we left and was buried at Arlington National Cemetary. Some of the men from Hill 881 North, like Fred Monahan and Don Hinman, attended the funeral. He was well respected by the men of Echo Company."

'In Vietnam, Second Lieutenant John Eller was my platoon commander. We, the third platoon, were called 'The Animals'. Eller was a mustang officer having served with the First Marines as an enlisted man during the Korean War. He was very cool and calm under fire and never seemed to get excited. Eller carried a little duffle bag with him in combat in which he carried his maps and C-rats. He retired from the Marine Corps as a Sergeant Major after making Captain. He said he could retire with a better pension due to time in grade."

'My platoon sergeant was Staff Sergeant Javier Sanchez. He was a very nice guy. I never heard him holler or chew a Marines ass out. Didn't smoke, drink, swear… faithful to his wife… very religious… Never went to the bars with us… dedicated Marine. Even helped us pack the crates with 782 gear when it was time to go back to Vietnam. He joined us on Okinawa in March 1967. He ended up getting killed on Hill 881 North. I have no idea where he was on the hill when he got killed. I know he wasn't with me."

'Sergeant Foreit was our right guide. He was another very nice guy. Didn't smoke, drink, swear… faithful to his wife… He didn't fool around like us younger guys did on Okinawa, going to bars and whore houses. I believe he joined Echo Company in February 1967, just before we rotated back from Okinawa. He came from the 26th Marines with Jim Cannon. He got shot in the arm on May 1 attacking Hill 881 South. Not sure of the date or the hill. He did two tours in Vietnam and ended up with 8 years in the USMC and then 12 years in the Army."

'Corporal Bill Downing was my squad leader for a short time before we rotated back to Okinawa-maybe late January thru February 1967. He was the kind of man that would carry the radio (Prick 25) himself, being the squad leader, than have two of his Marines argue over whose turn it was to carry it. I always said that he was too nice of a guy to be a squad leader… Didn't seem to be mean enough… a little bit too soft… great guy."

'Sergeant Henry Morgan was my squad leader at the time of May 3, 1967 (3rd squad, 3rd Platoon). He was a short man, 5'6" or 5'7" tops. I nicknamed him "Peanut." I think I was the only guy in the squad that could get away with calling him that. I spent a lot of time with as I was his Grenadier. When he got excited in combat, or mad, I would say to him, 'Just calm down Peanut'. He tried real hard to be a good squad leader. He even helped carry the extra gear, like the grappling hook and rope, just to appease the guys in the squad. He ended up getting killed on Operation Auburn in December '67."

'Private First Class Bill Alexander was with me a lot when he first got to Vietnam in November, 1966. We shared fox holes together, listening.. posts… guarded the water… point together at Camp Carroll. He was a quiet guy. He was a golf caddy for "The Four Seasons", back in New Jersey. He got shot in the head on May 1, 1967, attacking Hill 881 South. I believe we were carrying another wounded Marine down the hill to an LZ. I think this is the first time (Battle) that the M-16's started to jam on us. We had quite a few jam."

'Pfc. Bob Kruger was a quiet kid from a large Catholic family of eight kids. We called him "Baby Face", as it looked like he never shaved, yet. Fred Monahan told me that he thinks the NVA tried to capture Kruger after he was wounded dragging him down the hill and Kruger calling for help. He thinks Kruger pulled a hand grenade killing him and 3 NVA as they were all found close together. Nice, clean cut kid."

'Bob Yagues was a big boy. He probably weighted between 250-280 pounds. I use to tell him when he first joined us, back in December '66 at Camp Carroll, to wash the after shave off his face before he went into the bush because the gooks could smell. I think he got shot in the head on Hill 881 North. I remember it took six of us to load him on the chopper when he got killed."

'Kurt Edie was a third platoon corpsman… Gung Ho as hell. Even carried an M-16 after we told him it was against the Geneva Convention for a corpsman to carry a rifle. He took good care of us. I can't remember if he was the corpsman that bandaged my hands before he got killed on May third, or if it was Doc Burke."

'Sergeant Schumacher was supposed to be one of third platoon's new squad leaders. We were supposed to get three. He joined us on Okinawa in March of '67 along with Sergeant Breeding. I am by no means trampling on this man's death in any way, shape or form. When he joined us on Okinawa he tried to push that garrison style, Marine "Shit" on us. Schumacher said that we were undisciplined and out of control. Please keep in mind that most of us had 3-6 months already in Vietnam. We were only going to be on Okinawa a couple of weeks. We tried to tell him that, that garrison shit wouldn't work here in Vietnam and that it would get him killed. He got killed on Operation Bear Chain on the 20th of July. He stood up when we told him to get down. I don't think the guys were real fond of him, especially the older guys that had been in Vietnam awhile. He was the kind of guy that put himself first. An example is that he always picked his C-rats first out of the box so he would get his canned fruit. Not a good leader. He did not lead by example."

'Barry Weiner was in rockets. I met him, too, on Hill 41. He got hit in the head with shrapnel on May 3, 1967, by friendly fire, I believe, from an air strike."

'Private Mark Trout was in ITR with me at San Onofre. We went to Vietnam together on the USS Simon Buckner and joined Echo Company together in Dai Loc, Hill 41. He loved the soul brothers for some reason. He even wanted to be like them by buying himself a mohair suit on Okinawa. Trout was carrying the radio when he was shot in the back approaching Hill 881 South on May first."

Second Lieutenant Jim Cannon was the second platoon commander. He was a gung ho, hard charging, Marine officer who knew his shit… a real Mustang… aggressive as hell. The guys from weapons platoon were afraid to go out with him because he was so aggressive. I talk with Jim, a lot, at the reunions. He still cares about his men from Echo Company, including me. He stays in touch with me and is helping me with my VA claim. He ended up being the executive officer of the Company. He holds all the surviving Marines from Hill 881 North with high regard and deep respect. He is very knowledgeable about Hill 881 North and the war in Vietnam. Cannon retired from the Marine Corps as a Major."

'As for myself, I believe that the story about Hill 881 North can't be told enough. I think it's always been hushed up and silenced because of the M-16's jamming and the number of Marines killed because of it. According to Al Lyon there was some bad leadership, information and intelligence from higher ups."

I joined the Marines in January of 1966. I come from a large Catholic family of 11 kids-7 boys and 4 girls. Five of us boys were in the Marines. I was a Hollywood Marine. I arrived in Vietnam in September of 1966 and was sent to 2nd Battalion, 3rd Marines Headquarters at Dai Loc, south of Danang. I later joined Echo Company on Hill 41. We also went to Hai Van Pass, and later moved to Dong Ha in November, then Camp Carroll, and later Mai Loc. We rotated back to Okinawa in March to get replacements and thosr damn M-16's. I was lucky they gave me an M-79 grenade launcher, which I kept til they made me a squad leader in August. Then I got my M-14 back. It was the only way I would take the squad leader job. It was on Okinawa that 2nd Battalion, 3rd Marines became BLT 2/3. They didn't tell us that the life expectancy in a BLT was not that great compared to being on a fire base or fixed position. You either got killed or wounded. Just look at the Operations BLT 2/3 was on and the casualties they took.

'On the morning of May 3, 1967, on Hill 881 North, there was a light drizzle of rain and fog. Private First Class Schadler, from Ohio, was in the hole with me. He was a great guy who never seemed to sleep…always on the alert. We were the last hole of third squad, third platoon that tied into second platoon. Our side of the hill was a little steeper than the rest of the hill. Behind us the hill was covered with elephant grass. When all hell broke loose we started getting mortared and shot at from behind our backs. I yelled for them to quit shooting at us, not knowing that it was the NVA. I didn't know that it was the NVA. I didn't know that the NVA were already in the company perimeter of the hill. The second time I turned around to yell I saw a dead NVA five feet behind our hole with his RPG aimed at our hole. Another Marine had shot him and saved our lives. I think I found Jesus Christ after that morning, we were throwing hand grenades and I was firing my M-79. We were very lucky on my side of the hill. I took some shrapnel to my hands from the mortar but I was functional. God was looking after me. The main attack was coming through second platoon and first platoons lines where that side of the hill wasn't as steep. Sometime during the initial attack is when I took shrapnel to my hands from a mortar round. I was very lucky compared to a lot of other Marines that morning that got killed or seriously. The NVA would like to mortar us first and then attack, knowing we were crouched down in our holes. I also believe God was looking after me."

Years after the Vietnam War, in a letter to the Veterans Administration requesting assistance for one of his Marines, Captain Alfred E. Lyon wrote: "For young Marines, life in Echo Company of 1967 was a world of chronic mental and physical exhaustion endlessly humping the mountainous area just south of the DMZ or the heavily forested jungles from Con Thien to the coast. Into the hundred yard conscious bubble of their stress and pain were frequent unexpected shelling and sniper rounds that killed and maimed indiscriminately. Often these Marines walked through or rolled up at night in areas that had been defoliated by Agent Orange, drinking from streams making their way through grey jungle. Temperatures were often well over one hundred degrees with rain occurring or just about to begin. The combat load of each Marine was supposed to be eighty pounds, but the reality of moving everything by foot was that most men carried well over one hundred pound loads. These young men had zero control over their destiny and reacted by focusing on the day they could leave. They lived in the one set of clothes, jungle utilities; they had, usually without underclothes because they just caused chaffing. In the wet of Vietnam, boots and clothes wore out fast and it was not unusual to see the Company enroute to the next objective with lots of bare skin showing through. The men had their battle tested M-14 rifles taken away and given an infantry main battle arm M-16 rifle that had a failure rate somewhere around fifty percent in those original issues and which were clearly inferior to their enemy's rifles when working perfectly."

In early April, Echo Company was embarked on the USS Ogden as the assault company of the Second Battalion, Third Marines. We landed south of Dong Ha to interdict North Vietnamese Army coastal supply operations in mid-April. After several days flushing out NVA hidden in trenches and spider holes, the Battalion was hurriedley airlifted onto the base at Khe Sanh. Triggered by the loss of a platoon size reconnaissance, the Third Marines stepped into the largest conventional battles fought to that time in Vietnam. For approximately a month, Echo faced off against elements of three NVA divisions in a few square miles of jungle. Sleep was forgotten, every ravine and tree seemed to sprout NVA soldiers and the stench of death was a constant companion. For the Marines of Echo Company hell did not need any other definition. All of them were shot at repeatedly. It was a very dangerous place."

When Hotel Company was called in to help Echo and moved into the heavily fortified and well constructed enemy bunker system, Private First Class Roy W. DeMille, advanced through the intense North Vietnamese fire and crawled into one of the bunkers. Armed only with a .45 caliber pistol, he shot and killed two enemy soldiers. DeMille then saw that another bunker about twenty feet away was holding up the advance with a heavy volume of automatic weapons fire. He again dangerously exposed himself to the enemy fusillade, moved to the bunker, jumped in, and killed the enemy soldiers with his .45 pistol at point blank range.

Another Hotel Company Marine to distinguish himself this day was Lance Corporal Edwardo J. Gonzales of the first platoon. When several members of his platoon were wounded by enemy fire and pinned down, he crawled to the radioman and discovered that had been killed instantly. Gonzales used the radio to regain communications. He was then able to establish a base of fire by manning an M-60 machine gun. The gun team had been wounded and unable to use the weapon. Gonzales would alternate between operating the radio and placing accurate counter fire on enemy targets. He kept this up until he was shot through the head and killed.

Staff Sergeant Robert Morningstar was a chain smoking Indian who talked abusively to everyone, but actually had a heart of gold. He was shot and killed by a North Vietnamese sniper on July 5, 1967 during Operation Buffalo. During a break in the action he was standing with Captain Robert Bogard and Second Lieutenant James Cannon when he suddenly went down, hit in one side of his face by an enemy round. The projectile then ricocheted internally to blast out his chest.

Cannon had this to say about Staff Sergeant Robert Morningstar. "He was a shotgun loving, beer drinking, Staff Non-Commissioned Officer who wanted everyone to think he was somebody. He was. He was a loyal, dedicated Marine with a heart as big as a barn door that would do just about anything for anybody. He would give you the shirt off his own back, whether you needed it or not. You only had to ask for it. I could and depended on him a lot. He was my platoon sergeant."

The Silver Star earned by Lieutenant Colonel DeLong was the third such award for a Marine fighting in his third war. He first earned the medal on October 10, 1942 in the Santa Cruz Islands. During the war in Korea, DeLong was awarded his second Silver Star on November 2, 1950 during the fighting at the Chosin Reservoir.

Major Wendell O. Beard was a huge man who stood six-foot-four and weighed about 230 pounds. He was known to his men as Moose. Beard had served as an enlisted Marine during World War II and was commissioned a second Lieutenant in 1949. He was a member of the Quantico football team that compiled a splendid record playing against college and service teams during the 1949 season and won the All- Navy Football Championship for the third successive year. Beard then earned his first Bronze Star medal while serving as a platoon commander with the 4.2 Mortar Company of the Fifth Marines during the break out from the Chosin Reservoir. His company provided protective mortar fire for the convoy as it moved over a stretch of road that covered about eight miles. The heavy mortars and ammunition had to be hand carried along that distance as they set up hasty firing positions as they moved. Beard directed and adjusted effective fire on enemy positions, crew served weapons and observation posts as they were, almost always in the open and, on the move.

He was a natural, unflappable, infantry oriented commander. Beard was very outspoken, and would speak his mind, without filtration, no matter who he was speaking to or what the situation was. His attitude, in general, was nonchalant and devil-may-care. He would often rip into his subordinates in a high, whiney voice that left them with feelings of ambivalence about him. They both hated and loved the man.

Captain Richard O. Culver had it covered when he said, "Moose was a competent sonuvabitch, but he was also the world's most abrasive human being. I could have cheerfully killed him a million times, but you had to love him."

Beard survived the war, but his loyal but combative personality led to him being passed over many times for promotion. He retired with the rank of Lieutenant Colonel in 1969.

Charles Patrinos enlisted in the Marine Corps in September of 1942. He was a three war Marine having served in the Pacific with Battery K, Fourth Battalion, Tenth Marines. Patrinos also served in Korea. He retired from the Marine Corps in 1971 with the rank of First Sergeant.

One of the badly wounded Marines, Staff Sergeant David L. Noakes, would recover from his injuries and return to Vietnam for a second tour. This time he was the Company Gunnery Sergeant of Company F, Second Battalion, Fifth Marines. On November 5, 1970 he earned the Silver Star when he led an aggressive assault on North Vietnamese positions on a small island near DaNang. He rushed across fire swept terrain, seized a rifle, and forced the enemy to momentarily fall back. He provided first aid to a wounded Marine lying in an exposed position and then moved the casualty to a covered position. That done, he then moved to an exposed position and guided resupply helicopters to a safe landing.

Others either wounded or mentioned in the chapter are as follows:
Abare, Carl M. Pvt. 2249732; E-2-3; 19; Boston, Massachusetts
Adams, Glenn E. Cpl. 2101248; H&S-2-3; 21; Kansas City, Missouri
Adinolfi, John F. 1stLt. 091509; E-2-3; 23; Brooklyn, New York
Alexander, William F. Pfc. 2170361; E-2-3; 20; New York, New York
Alford, Ernest S. Pfc. 2274024; E-2-3; 19; Jacksonville, Florida
Alvarez, Andrew L. Pfc. 2240290; E-2-3; 19; Kansas City, Missouri
Arant, George F. LCpl. 2224730; E-2-3; 19; Charlotte, North Carolina
Aungst, Terry L. Pfc. 2261940; E-2-3; 19; Detroit, Michigan
Backus, Richard D. Cpl. 2137689; E-2-3; 20; Pittsburgh, Pennsylvania
Barnett, Robert L. LCpl. 2159840; E-2-3; 20; Jacksonville, Florida
Beard, Wendell O. Maj. 050060; XO-2-3; 42; Burilla Carr, Oklahoma
Blackston, James L. Cpl. 2214324; E-2-3; 19; Kansas City, Missouri
Breeding, Fred V. Sgt. 2064493; E-2-3; 21; Cleveland, Ohio
Brown, Glen W. Jr. Cpl. 2185754; E-2-3; 19; Houston, Texas
Brunson, William Pvt. 2181928; E-2-3; 20; Boston, Massachusetts
Cannon, James R. 2dLt. 093569; E-2-3; 31; Fredericksburg, Virginia
Canter, Clifton H. Jr. 2dLt. 092837; E-2-3; 23; Lake Placid, Florida
Carroll, Patrick G. 2dLt. 092848; F-2-3; 24; Elgin, Oregon
Carstens, Thomas J. LCpl. 2201105; E-2-3; 21; Hartland, Wisconsin

Cole, Elvin H. Jr. Pfc. 2230611; E-2-3; 19; Des Moines, Iowa
Crawford, Timothy J. LCpl. 2217652; E-2-3; 19; Buffalo, New York
Cruzesquilin, Juan H. LCpl. 2216228; E-2-3; 19; San Juan, Puerto Rico
DeLong, Earl R. LtCol. 048884; CO-2-3; 47; Jackson, Michigan
DeMille, Roy W. Pfc. 2269614; H-2-3; 19; Phoenix, Arizona
Drakeford, Jackie L. Cpl. 2198818; E-2-3; 19; Charlotte, North Carolina
Dupont, Stephen P. Pfc. 2233884; E-2-3; 19; Los Angeles, California
Eller, Johnnie L. 2dLt. 096610; E-2-3; 36; Raleigh, North Carolina
Fennell, Edward S. LCpl. 2228987; E-2-3; 19; Cleveland, Ohio
Ferbos, Stanley LCpl. 2238687; E-2-3; 20; New Orleans, Louisiana
Fierro, Alejandro C. LCpl. 2222667; E-2-3; 19; El Paso, Texas
Flannery, John V. Jr. Pfc. 2218845; E-2-3; 19; Philadelphia, Pennsylvania
Foreit, Robert L. Sgt. 1508527; E-2-3; 30; Wilkes-Barre, Pennsylvania
Gallegos, William Pfc.2217927; E-2-3; 19; Denver, Colorado
Gandt, Thomas J. Pfc. 2281528: C-2-Eng; 19; Brooklyn, New York
Hall, Roy LCpl. 2233332; E-2-3; 19; Ashland, Kentucky
Harvey, Richard F. 2209941; E-2-3; 19; Pittsburgh, Pennsylvania
Hawley, James G. LCpl. 2220726; E-2-3; 19; Hartford, Connecticut
Healey, Edward F. LCpl. 2160223; H&S-2-3; 20; Boston, Massachusetts
Hinman, Donald E. LCpl. 2099201; E-2-3; 21, Medina, New York
Hovietz, Carl R. LCpl. 2179997; E-2-3; 20; Los Angeles, California
Hunter, Samuel H. LCpl. 2195398; H&S-2-3; 19; Philadelphia, Pennsylvania
Irvin, Ronald H. Pfc. 2268579; E-2-3; 19; Cedar Rapids, Iowa
Izenour, Frank M. Jr. 2dLt. 094281; E-2-3; 24; Washington, D.C.
Johnson, Leon V. Jr. LCpl. 2137723; E-2-3; 20; Pittsburgh, Pennsylvania
Jordan, Willie J. Cpl. 2210080; E-2-3; 19; Indianapolis, Indiana
Kepner, Harold E. LCpl. 2203641; M-3-3; 21; Cleveland, Ohio
Kleinschmidt, Kenneth F. HM2 7761389; H&S-2-3; 21; Minneapolis, Minnesota
Kolodziej, Ronald E. Sgt. 2117474; E-2-3; 21; Hartford, Connecticut
Kopec, Gene R. Sgt. 2079935; E-2-3; 21; Chicago, Illinois
Kostendt, Kenneth R. Pfc. 2246031; E-2-3; 19; Detroit, Michigan
Kraner, Marvin R. Pvt. 2269368; E-2-3; 19; Columbus, Ohio
Kulick, Daniel H. HN 6773448; H&S-2-3; 20; St. Albans, New York
Leblanc, Donald G. Pfc. 2258932; E-2-3; 19; New Orleans, Louisiana
Lee, Gerald A. Pfc. 2288951; E-2-3;19; Detroit, Michigan
Like, Billy J. Sgt. 2076770; E-2-3; 22; Louisville, Kentucky
Loggins, Raymond B. Cpl. 2239782; E-2-3; 19; Wichita, Kansas
Lyon, Alfred E. Capt. 086886; E-2-3; 27; Cincinnati, Ohio
Madonna, Raymond C. Capt. 084233; H-2-3; 28; Annapolis, Maryland
March, Gerald P. HM3 3908833; H&S-2-3; 20; Seattle, Washington
May, Blaine R. Pfc. 2280588; H&S-2-3; 19; Pittsburgh, Pennsylvania
McDaniels, Peter W. LCpl. 2109521; E-2-3; 20; Jackson, Tennessee
McGannon, Martin D. Pfc. 2304733; E-2-3; 19; Chadron, Nebraska
McManus, William P. LCpl. 2209237; E-2-3; 19; Boston, Massachusetts
McNeer, John K. LCpl. 2147345; H&S-2-3; 20; Birmingham, Alabama
McWhorter, Samuel T. Jr. Cpl. 2069069; E-2-3; 21; Birmingham, Alabama
Miller, Curtis L. Pfc. 2257831; E-2-3; 19; Richmond, Virginia
Monahan, Frederick G. LCpl. 2195404; 20; Holland, Pennsylvania
Morgan, Henry L. Sgt. 2034970; E-2-3; 22; Benson, North Carolina
Morin, Roger L. Pfc. 2249856; E-2-3; 19; Boston, Massachusetts
Morningstar, Robert L. SSgt. 1364599; E-2-3; 32; Warfordsburg, Pennsylvania
Nicholas, Warren D. Cpl. 2137037; H&S-2-3; 20; Hartford, Connecticut
Noakes, David L. SSgt. 1932904; E-2-3; 25; Portland, Oregon

Osborne, John E. LCpl. 2200565; E-2-3; 19; South Charleston, West Virginia
O'Shields, Charles A. LCpl. 2236110; E-2-3; 19; Fort Jackson, California
Patrinos, Charles 1stSgt. 410434; E-2-3; 40; Salem, Massachusetts
Ramos, Oscar Pfc. 0000000; E-2-3; 00; Dewey, Arizona
Rice, Thomas E. LCpl. 2342943; E-2-3; 21; Pasadena, California
Richardson, Clarence A. LCpl. 2246613; E-2-3; 19; Richmond, Virginia
Rogers, Gregory HN 9995321; H&S-2-3; 19; Los Angeles
Schirmer, Ronald L. Pfc. 2219130; E-2-3; 19; Philadelphia, Pennsylvania
Schumacher, Ronald K. Sgt. 1981029; E-2-3; 24; Wauconda, Illinois
Sheehan, James P. Capt. 077100; G-2-3; 30; Philadelphia, Pennsylvania
Siemons, Glenn J. Cpl. 2195515; E-2-3; 20; Philadelphia, Pennsylvania
Smith, Clinton Cpl. 2213606; E-2-3; 19; Birmingham, Alabama
Sorenson, Merle G. Capt. 078363; F-2-3; 31; Alden, Minnesota
Spano, Charles D. Pvt. 2220111; H&S-2-3; 19; Philadelphia, Pennsylvania
Sternisha, George Pfc. 2205241; E-2-3; 19; Crest Hill, Illinois
Sutter, Richard F. Cpl. 2208653; M-3-26; 21; Atlanta, Georgia
Taylor, James H. Pfc. 2309657; E-2-3; 19; Philadelphia, Pennsylvania
Thurber, Gary D. Pfc. 2241916; E-2-3; 19; Los Angeles, California
Troutt, Mark A. Pvt. 2224792; E-2-3; 19; St. Louis, Missouri
Turner, Eddie H. Cpl. 2221832; E-2-3; 19; Newark, New Jersey
Wasilewski, Leonard P. Jr. 2221170; E-2-3; 19; Jacksonville, Florida
Weiner, Barry K. LCpl. 2148672; E-2-3; 20; Boston, Massachusetts
Welch, Larry C. Cpl. 2098604; E-2-3; 19; Raleigh, North Carolina
Wells, James L. LCpl. 2237184; E-2-3; 19; Pittsburgh, Pennsylvania
Williams, Danny P. HM3 6859468; E-2-3; 20; Vienna, West Virginia
Williams, Ronald A. Cpl. 2013390; E-2-3; 24; Glenwood, Iowa
Young, Clarence W. HM2 7721185; E-2-3; 21; Montgomery, Alabama
Zebrowski, Anthony J. Pvt. 2137092; E-2-3; 19: Unknown

✯ ✯ ✯ ✯

In 1948 the U.S. Army established the Operations Research Office (ORO) to analytically study a number of problems associated with ground weapons in the nuclear era. One of their early projects was a search for better body armor. During the search it was discovered just how little was known about how individuals were wounded in combat. To find an answer the ORO studied the data from over 3 million hits from World War I, World War II and the war in Korea. They discovered that although aimed fire was significant in the defense it had little impact on casualty figures and that most infantry combat was at close range, usually less than 100 meters.

The ORO came to the conclusion that the United States military needed an infantry weapon with low recoil that fired small rounds so that a soldier could carry more ammunition. In 1957, Winchester and Armalite were asked to design weapons that could penetrate both sides of a standard issue military helmet at 500 meters. This weapon would have a magazine of 20 rounds and would be able to fire on both automatic and semi automatic. The rifle would weigh no more than six pounds. Several weapons were put forward for trials and a number of studies were conducted. The Armalite AR-15 was chosen as the best small caliber weapon and it was adopted as the M-16.

"I believe that the cold, hard facts about the M-16 are clouded over by a fabrication of the truth for political and financial considerations."

First Lieutenant Michael P. Chervenak
Executive Officer
Hotel Company 2/3
from *Con Thien: The Hill of Angels*
By James P. Coan

Grateful acknowledgement is made to Major Richard O. Culver, USMC (Ret.) for permission to reprint previously published material. The following story is a reprint of *The Saga of the M-16 in Vietnam* that has appeared on the internet © 1999 by Richard O. Culver by permission of Richard O. Culver.

7

The Saga of the M-16 in Vietnam

Like most things, the reality of being armed with an ineffective weapon was of little import to those who were not risking their lives on a daily basis. By the time the problem was finally fixed, many friends and comrades had been awarded *"the white cross,"* or in the verbiage of the time, had *"bought the farm."* Many lives *could* have been saved if a few individuals in "decision making billets" had possessed the intestinal fortitude to correct the problem. ....And the problem was "correctable" – all that was necessary was the application of a bit of guts and common sense. Aircraft that are suspected of being flawed are immediately grounded until a problem has been corrected, or a fix has been found. And so it was with the Marines' CH-46 Helicopter during the same time frame. The tail pylons started rather abruptly separating themselves from the bird with catastrophic results. The CH-46 was quite rightly grounded and sent back to Okinawa until the problem was isolated and fixed. For some unexplained reason the same rationale was not applied to a rifle that was costing lives on a daily basis. Perhaps the *"Wingies Union"* was stronger than the "Grunt's Union" – whatever the reason, dead is dead, and the Grunts were not amused! Unfortunately, doing the "right thing" would have cost individuals in positions of authority considerable embarrassment something that no one was willing to risk. The "air types" could blame Boeing, but many of the decisions concerning the M16 were made within the "military industrial complex", making it more difficult to pin Colt to the wall. Individuals within the Military who had given their "yea verily" to the project would have found themselves looking for another job.

Rather than bore you with cold statistics and hard facts to start, I will tell the story as it happened and as I remember it. Making allowances for the dimming of the memory after 32 years, the entire saga still stands in my consciousness as if it happened only yesterday – things like that are hard to forget.

Our outfit, the Second Battalion, Third Marine Regiment, was selected to assume the duty as one of the two Battalions filling the job as the "Special Landing Force." This evolution consisted of a quick trip out of Vietnam to the peacetime home of the 3rd Marine Division Okinawa), for a refurbishment of web gear, worn out equipment, and the fleshing out of a casualty riddled Battalion with fresh replacements. This slight respite from the "free fire zone" afforded new replacements an opportunity to gain experience and training with their new organization. The SLF was in fact a BLT (Battalion Landing Team) with enough attachments to make it into a sort of "bobtailed Regiment." In addition to the standard four line (infantry) companies, and an H&S Company (Headquarters & Service), we also had attached: a Helicopter Squadron, an Artillery Battery, a Recon Platoon, an Engineer Platoon, Amtracs (Amphibious Tractors/Landing Vehicles) and various other supporting elements. At that time, an (unreinforced) Infantry Battalion (before being festooned with the above attachments) consisted of approximately 1100 men. 1/3 (1st Battalion, 3rd Marine Regiment) was to be designated as SLF Alpha, and 2/3 was to make up SLF Bravo.

The SLF's job was to act as a sort of "Super Sparrow-hawk" (cavalry to the rescue stuff) to reinforce any organization actively engaged with the enemy who wound up in a "feces sandwich." when the brass sent in the SLF, someone was already in big trouble! Knowing that you were headed into a "hot LZ" (landing zone) on a rather repeated basis made for a very exciting tour. The normal SLF tour of duty was usually scheduled for a duration of 6 weeks. The outgoing SLF Battalion was then returned to its parent Division (1st or 3rd), and a new Battalion took over the rather thrilling duty as "The I Corps' Fire Brigade." It was an ingenious scheme, as it allowed the Marines to refurbish their battalions occasionally, and allowed time (albeit relatively short), to train new replacements out of the line of fire. Normally, the SLF tour was anticipated by the selected Battalions with some enthusiasm, as it was *supposed* to include one short R&R for the Battalion in Subic Bay, prior to the SLF's reassignment to the RVN. Needless to say, no one in 2/3 ever saw Subic, except as a casualty. *Murphy*, always taking a hand in things, stirred the pot in such a way that the refurbishment and replacement of battalions on the SLF was curtailed after the vicious "Hill Fights" around Khe Sanh in April of '67. 2/3 (and their sister battalion, 1/3) had taken on the best that the NVA could throw at them and whipped them hands down, but it was not without cost. Many a dead or dying Marine was found with a cleaning rod shoved down the bore of the little black rifle...

The constant pressure on I Corps starting that Spring left 2/3 manning the ramparts as one of the two SLFs for a period of nine months (versus the normal 6 weeks)! When the smoke finally settled, 2/3 had taken over 800 casualties and those who survived walked away with a sigh of relief. By August of '67, my company (Hotel, 2/3) had only 5 Marines without at least one Purple Heart, and I was not one of them. Technically, the SLFs were supposed to return to the LPH (and other supporting shipping) after a battle, lick their wounds, get cleaned up, draw more ammunition and standby for the next mission.

By way of explanation to those who have not been in the Corps or associated with the Navy, "LPH" stands for "Landing Platform - Helicopter." The LPH is in fact nothing more than a small aircraft carrier, primarily designed to launch helicopters for a Marine (or perhaps Army) landing force. The supporting shipping usually consisted of an LSD, ("Landing Ship - Dock" designed to launch Amphibian Tractors for a seaborne surface assault), an LST ("Landing Ship - Tank", self explanatory) and an APA (assault transport to house additional troops). All together, they made up the seaborne vehicles for a rather formidable assault force.

Murphy again took a hand, and out of those fateful 9 months, we spent approximately 12 days aboard our assigned shipping. The rest of the time we got "chopped op-con"1 to one of the Infantry Regiments ashore (transferred *to*, and *under their operational control*) - after all, we were those "pogues" who lived aboard ship and had it easy, were they not? Everyone figured that we were well rested and ready to go. The Regiments ashore, of course, took full advantage of such obviously fresh troops, and threw us into the very "choicest" assignments, to allow their units a breather – we were eventually referred to as the *"day on - stay on battalion,"* and brother, they weren't kidding!

It was in the arena outlined above that I got my first introduction to the XM16E1. When 2/3 arrived on Okinawa to refit and train for their duties as SLF Bravo, they were already licking their wounds. The Battalion had been ambushed on a march between two hill masses, losing their Commanding Officer and Sergeant Major, along with numerous other individuals. While they were hardly demoralized, they possessed a particular affection for their CO and Sgt.Maj. and were chomping at the bit to return to the RVN to avenge the Battalion's losses. Shortly after 2/3's arrival on Okinawa, the Battalion learned that it was scheduled to draw a new "experimental rifle." the XM16E1. 2/3 dutifully turned in their M14s to draw a curious little plastic thing that drew lots of snickers and comments from the old timers (we still had a few WWII vets in those days). The Battalion was given an orientation lecture in the Camp Schwab Base Theater by some ordnance folks, sent to the range to fire some sighting in rounds, and pronounced properly prepared for combat... little did they know!

The Battalion was told that they would now be able to carry 400 rounds ashore on each operation, and were now armed with an accurate, hard hitting rifle that would tear a man's arm off if you hit him. The lecture was impressive. The interesting thing is that the Marines WANTED to like the little rifle – it was light, cute, and supposedly extremely effective! Marines are always in favor of a weapon that will dismember their enemy more efficiently and more effectively. The Marines of 2/3 left Okinawa READY to go try this "jack the giant killer" on the NVA or Cong (they didn't care which, as long as it made a good fight!). However, there were several flies in the ointment. First, they only had one cleaning rod per rifle and no replacements – sounds reasonable, but events were to prove this assumption wrong. The second problem was that ordnance had only enough magazines to issue three (3) per rifle, and they were "twenty rounders." The thirty rounders in those days were only being used by the

Special Forces – Robert "Strange" McNamara, (The Secretary of Defense), had decreed that the 20 round magazines were more cost effective than the 30 round magazines (this from the guy who was responsible for marketing the Edsel!)! We were now armed with the latest in weaponry, and able to carry 400 rounds ashore. Our confidence level would probably have been considerably higher if we had been issued more than three 20 round Magazines per gun. We were promised more of course, and as it turned out, it became true, but only because we were able to pick up those left behind by the casualties. The long and the short of this lesson, however, was that they were trying to get the M16 into action well before adequate supplies were available to support the weapon, even if it had been functioning properly. Politics is indeed a strange game!

Ammunition was issued in "white" or "brown" twenty (20) round boxes. Bandoleers with "clipped ammunition" in ten round "strippers" had not yet made their way to South East Asia. While this would have been a handicap under normal circumstances, it turned out to be a "non-problem." A full 50% of the rifles wouldn't shoot semi-automatically! The unfortunate individuals armed with the malfunctioning rifles couldn't shoot enough rounds to need more than the initial three magazines at any rate! Three hundred and forty rounds in 20 round cardboard boxes were stowed in our packs, with the idea that during a firefight, a man who had run dry, could roll over to his buddy and take ammunition out of his pack and his buddy could do the same. As it turned out, this rarely figured into the equation.

The first clue (for 2/3) that something was wrong came during the battle of Hill 881 North… but *all* the Hill Fights at Khe Sanh in April '67 came up the same – dead Marines with cleaning rods stuck down the barrel of their M16s to punch out cartridge cases that refused to extract. At first, we considered that the experiences encountered during the Hill Fights might have constituted an isolated incident, but as experience was to prove, alas, 'twas not so! The regulations of the time required that all such malfunctions were to be documented, and reported to Ordnance Maintenance/Division Ordnance. The 2nd Battalion, 3rd Regiment of Marines must have filled a 6X6 truck with malfunction reports attempting to stay within the administrative guidelines. We submitted the required reports and waited – we wanted the problem *FIXED – NOW*, and we were willing to play the ordnance paperwork game if that was what it took to correct the situation!

Spring stretched into summer, and summer gave way to fall, with reams of paperwork having being sent out to the Ordnance Maintenance Folks on the "Rock" (Okinawa) and to the ordnance folks in Vietnam. We outlined, in great detail, the failure of the much vaunted M16 to perform as advertised… It simply wasn't working! It seemed that if your rifle *would* shoot, it would shoot under almost all conditions (if clean), but if it *wouldn't*, no amount of coaxing would help. *All* of the M16s seemed to be extraordinarily sensitive to carbon build-up, even if the rifle was one that would shoot when freshly cleaned. This meant that in a long and heated firefight, it was possible to have a much larger percentage of rifles "out of action" than the 50% that didn't want to shoot at all. Something was seriously amiss! A rifle that refuses to shoot during a firefight is unsuitable as a combat implement. The NVA was obviously not gonna' allow us a "time out" while we held a cleaning session! My first clue to the solution to the problem came from talking to the Battalion Armorer. He had an M16 that worked under almost all conditions. I asked him what he had done to it, and he replied that he had taken a ¼" drill, attached a couple of sections of cleaning rod to it, and put some "crocus cloth" through the slotted tip (like a patch) and run it into the chamber and turned the drill motor on. He "horsed" the drill a bit and apparently relieved the chamber dimensions just enough to ensure positive functioning. This was a positive functioning. This was a sort of precursor to the "chrome plated chamber fix" that would be applied in days to come.

 FSR (Force Service Regiment – which also acts as a home for the small arms repair folks), sent a trouble shooting team to visit us aboard the LPH shortly after the "Hill Fights" to try and pin down the problem. As soon as the ordnance team arrived, they made it clear that *THEY* were already well informed (meaning they'd already made up their minds) concerning our problem and had decided (without so much as a question to us) that *WE* as a Battalion were responsible for a bad rap being given to a marvelous little rifle! The lads in the rear had decided that WE were simply not keeping our rifles clean, and if we weren't such inattentive and unmotivated "oafs" being led by incompetents, we wouldn't have such a problem. Needless to say, the hackles stood up on the back of our necks. "Them wuz *fightin'* words!" …And we wuz peaceable folks (well sorta' anyway)! To say that they had misread the problem is an understatement!

 Certainly from a personal standpoint, they were full of "un-reprocessed prunes." My background in small arms went back as far as my conscious memory, and when I "screwed up" with a firearm of any kind as a kid, my Daddy left knots on my head and welts on my "stern-sheets"! During this time frame, I had just finished firing on the USMC Rifle Team (in 1965 – this was now 1967) and to say that I had high standards of weapons cleanliness for my rifle company is an *extreme* understatement. If the rifles had been clean enough to eat off of *before* the visit from the FSR clowns, rifle cleanliness moved up a notch to *"autoclaved"* as a result of the insults they were bandying about! We literally fired thousands of test rounds over the fantail (the stern) of the LPH. Each of the issued rifles was fired, cleaned and then fired again! …Same story, about 50% of the rifles were reliable and 50% were "non-shooters." We cleaned the rifles between strings of fire (and *this* test was conducted in the more or less "sterile" conditions encountered in a shipboard environment), with the same results! *NOW* we were getting worried.

The malfunction reports continued to pour into the rear echelon paper mills without any tangible results. On one notable occasion, a stalwart Marine crept around in a flanking movement on an enemy machine gun position. He assumed a quick kneeling position to get a clear shot over the saw grass, and "did for" the hapless NVA gunner! His second shot aimed for the assistant gunner never came, as his rifle jammed and the assistant gunner avenged his dead comrade by splattering the Marine's gray matter all over the stock of the Matty Mattel Special. After the fight, we sent his little black rifle to Division Intelligence with a complete report on the events (without removing the brain matter from the stock). We waited with baited breath for the response to this one, but alas to no avail! Still no action!

Normally aggressive Marines were understandably getting a bit edgy about being assigned to listening posts or outposts. Ambushes were, more likely than not, to result in Marine casualties. We started stealing and or trading the cute little black rifle for M14s. Many rear echelon troops (usually known as *REMFs*) were more than willing to trade their old fashioned M14s for a little lightweight rifle that was easy to carry, (the M16s in those days were reserved for the frontline troops). Supply and demand prevailed, and what we couldn't trade, we appropriated (a polite military term for outright theft!). The Engineer troops assigned to us for support (mine clearing, demolition, and setting up helicopter landing zones) were still armed with the M14 (not being infantry). The Engineers became some of the most popular troops in the Battalion and made up a substantial part of our base of fire. I was always partial to Engineers anyway, and these guys cemented our relations in a big time way – good people those Engineers – and *THEY* were armed with a *REAL* rifle!

It finally became apparent that no one was gonna' come to our rescue! Our reports were falling on deaf ears, and our Battalion Commander was more than a little annoyed. The bayonet had become more popular than before and indeed enjoyed a resurgence of usefulness, until in the throes of hand to hand combat one of the lads gave the enemy a vertical butt stroke that resulted in his holding a "two part" Matty Mattel... Captured AK 47s began to show up in increasing numbers, but they were a double edged sword. The AK 47 had a rather distinctive sound when fired, and would occasionally result in the Marine "wielding" the foreign piece, receiving a bit of "friendly incoming"! This was in addition to the fact that ammo re-supply for the AK was a problem. After a fire fight, the battalion S-4 (supply & logistics) frowned on requests for a couple of thousand rounds of 7.62 X 39...

Things were getting desperate... Our Commandant at the time, General Wallace M. Greene, when queried about the rumors filtering back from the front-line troops, contacted the Marine Corps ordnance people and asked them what the problem was. The Ordnance Brass "bleated" the school solution and told the *"Commandanche"* that the problem stemmed from poor weapons maintenance and a lack of leadership! The Commandant then appeared on TV and announced to the entire world that the only thing wrong with the M16 was there weren't enough of them! How *RIGHT* he was! It took 20 rifles to get off 20 rounds! We were enraged! – And we began to plot! Never let it be said that the average Marine isn't cunning, if not terribly intelligent.

This is probably a good place to describe the actual malfunction that was prevalent with the "mouse gun" – although there were variations the problem was essentially as follows:

1.) The rifle would be loaded normally, i.e., a loaded magazine would be inserted and the bolt would be allowed to go forward, causing a round to be chambered.

2.) The trigger would allow the hammer to fall, with the rifle firing the first round in the expected fashion. Then the problem began…

3.) The bolt would start to the rear, but the cartridge case would remain in the chamber. There were two variations to this one, one in which the extractor would "jump" the rim, and one where the extractor would "tear through" the rim. Either version left the case in the chamber.

4.) The bolt would start forward stripping the next round from the top of the magazine.

5.) Since the chamber was already occupied by the cartridge that had just been fired, the newly fed round would shove the bullet tip firmly into the stuck case effectively jamming the rifle.

This "jam" could be cleared by:

a.) Removing the magazine from the rifle, pulling the bolt to the rear, and locking it in this position by depressing the bolt catch.

b.) If the newly fed live round did not automatically fall free (it often did), you had to shake the rifle to allow the round to fall free of the magazine well.

c.) A cleaning rod was then inserted in the muzzle and the "stuck case" was driven out of the chamber.

d.) The magazine was then reinserted and locked into the magazine well, and the bolt allowed to go forward by depressing the bolt catch. The bolt would again strip a round from the magazine and reload the chamber.

e.) This round could then be fired and the entire cycle started all over again.

Essentially we had been reduced to a *"magazine fed, air cooled, single shot, muzzle ejecting shoulder weapon"* shooting an inferior cartridge. How lucky can you get?

Mike Chervenak, my XO (executive officer) was a man of rare moral fiber. Not only was Mike one hell of a good Marine, but he cared for and about our Marines… and the M16 was continuing to get them killed. On one of the very few days we spent aboard the LPH preparing for our next thrilling adventure, Mike came to see me in my quarters.

"Skipper" said Mike, "what the heck are we gonna' do about this miserable little rifle?"

"Well Mike," I replied, "I guess we're doing about all that can be done – I'm about out of options! All we can hope for is that ordnance'll find a fix!"

Mike being smarter than the average bear, drug his toe in the dirt and asked, "Skipper, do I have your permission to write a letter to my congressman?"

"Well Mike," I said, "I can't tell you *NOT* to write such a letter, it's a free country!"

"Well Skipper," said Mike, "what would *YOU* do?"

Uh oh – now I'm trapped! "Well," I told him, "I'd probably write a letter to the Commandant!"

"But Skipper," Mike says, "you *KNOW* he won't ever get to see it!"

"Wrong," sez I, "all you have to do is put 'copy to: Senator Zhlotz' (or whoever) at the bottom of the letter, and military paranoia will kick in! The staff will be afraid *NOT* to show it to him, lest he get a call from an outraged Congressman!"

"Yeah," said Mike, "but I'll bet that nothing will be done about it even if he *DOES* see it!"

"Well, you're probably right," I tell him, "but it might be worth a try!?"

Mike, somewhat discouraged at this point, allows as how it'll probably be more effective to send one to his Representative. I agree without overtly suggesting that he do so. He turns to go, but just as he reaches the Water Tight Door (WTD), he turns around with a slight grin and says "Skipper, would *YOU* help me write it?"

Hummm... the rest is history. Mercifully we did a workmanlike job on the letter, and simply explained the problem (much as above) and made note that it took precious seconds to clear a jammed rifle that an Infantryman doesn't have in a firefight. We were also careful not to call names or point fingers, and that's all that saved us in the light of things to come! I'm not too sure who Mike sent the letter to, but a copy of it WAS published in that "Communist Rag," *The Washington Post!*

Mike was on R&R when the thunder came rolling in. He received a "person to person" phone call from "Wally" (Wallace M. Greene, the Commandant, who hangs his hat in Washington, D.C.) in Vietnam! Alas, Mike was not there to take the call! The brass came to me of course, asking where Mike had gone when he left on R&R. Since Mike had earned his R&R in spades, and I didn't want to screw it up for him (knowing the problem would still be there when he returned). I did the only honorable thing I could and lied! Hee, hee, hee... Mike finished his R&R in good order and without harassment.

When they discovered that I had aided and abetted Mike in his endeavors, the feces struck the ventilation! That letter kicked off *FIVE* simultaneous investigations; one from the Third Marine Division, one from the 9th Amphibious Brigade, one from the 3rd Marine Regiment, one from the 2nd Battalion, 3rd Marine Regiment (us) and last but not least a Congressional Investigation led by a Congressman from Louisiana named "Speedy O. Long" (yes, that was really his name!). During the investigations, the Battalion hid me so far back in the "ding toolies" that it was necessary to pipe in air and sunlight. Mike and I had become the "pariahs" in the Marine Corps in general, and the 3rd Marine Division in particular. However...

At long last people started doing something overt for a change. We were pulling an operation down in the 1st Marine Division AO, south of Da Nang (AO stands for "area of operations") – the SLF was essentially a "hired gun" and went wherever there was hate and discontent). The Corps "flew-in" a C-130 with 400 brand new XM16E1 rifles along with a Marine Warrant Officer considered to be an expert in the small arms ordnance field. The ordnance Warrant was an old friend of mine who had been the Marine Representative to Cadillac Gauge when they were building the "Stoner 63" System. He had been a Staff Sergeant at the time and we used to sit on my living room floor and disassemble the Stoner System over an occasional beer (well, maybe several beers) when the Marine Corps was running its Stoner tests in at Camp Lejeune. Now, I tell myself, we'll get some results; Bob is a pretty savvy guy! ...Wrong again *gopher breath!"* – Bob Baker (the Marine Warrant Officer), had suddenly and inexplicably switched to (what we thought of as) the enemy camp!

In a private and rather heated conversation with Bob, he allowed as how the problem was that we weren't keeping them clean enough!

"BS." I said, "Bob, you know me better than that!"

"Nope," he said, "the M16s will work if they're clean!"

Seeing that I had reached a dead end, it was time to try a different approach. Another Captain/Company Commander and I (he having just as much a case of the "$%#^" over the "16" as the rest of us) watched as WO Baker utilized his $800 ultra powerful chamber scope to examine the M16 rifle chambers of a line of troops brought in out of the lines for evaluation of the condition of their rifles.

This marvelous chamber scope was supposedly powerful enough to make any imperfections in the chamber look like the surface of the moon. The first man stepped up to the front of the line and handed over his rifle. Bob sticks the chamber scope in the chamber, shakes his head and throws the old rifle in a pile that was to grow materially in the next couple of hours. The Marine was then issued one of the new rifles brought in on the C-130. Watching the lad with his " brand new" rifle stride off. Bob Bogard (the other Company Commander) and I chased him down (out of sight of course). We talked him out of his rifle, threw it into the dirt, kicked a little over it, picked it up and dusted it carefully off (to make it look like a "used" rifle). We then waited awhile until a number of folks had gone through the line and "number 1's" face had faded from WO Baker's recent memory. We put the trooper back in line and hid and waited. When (Warrant Officer) Bob stuck his chamber scope into the new rifle, he again shook his head and threw the new rifle on the pile of discards! Gotcha! When we pointed out to Bob what we'd done, he went orbital (not a word to come into general use until '69 of course)! He accused us of not taking his efforts seriously, and trying to make him look bad – not hard to do at this point! While we had outraged the brass, a seed of doubt had been planted, and it grew!

Back at the Command Post, a rather short civilian gentleman of Asian extraction wearing a Colt Detective Special on his belt strode over to see me. I recognized him as a Mr. Ito, the Colt Representative that had flown in with the 400 rifles.

"Howdy," he sez, "my name is Ito!"

"I know," I said, "and my name is Culver."

"Yes, I know," sez Ito, and at that point, I figured that my fanny was truly gonna' be grass.

My instincts in this case were wrong.

Mr. Ito turned out to be a heck of a nice gentleman and told me all sorts of revealing stories. Among other things, he told me that Colt had offered to chrome plate the bores and chambers of the M16s for the sum of $1.25 each, but that Robert "S" McNamara had vetoed it as being non cost effective. Mr. Ito sent me a "care package" when I got home, guess what it contained? A double handful of Colt M16 tie tacks2. Grrrrr...

Ultimately, Colt wound up chrome plating the chambers (and later the bores) of the M16s, thus reducing the coefficient of friction between the cartridge case (not necessarily a good thing, incidentally) and the chamber. The bolt then began battering the frame from the excessive velocity in its rearward movement, and they again gave the *"patient"* with a brain tumor an aspirin tablet as a "fix" – they simply made the buffer group heavier! But the real story had yet to be told. The story eventually leaked in bits and pieces but was never made public in the headlines it deserved. The rifle was eventually fixed, but at what a price... Much like the guy unjustly accused in print - when the real culprit is found, the headlines don't shout out his innocence, a retraction is usually printed in extra small type on the last page. The guys who died for this folly can never be brought back, and the people responsible who fought the problem by placing the blame where it wouldn't get *their* fingers dirty came away clean.

Somewhat later, a new Battalion Commander, who hadn't fought with us in the old days when the rifle was at its worst, inherited 2/3 in time to preside over the ensuing hate and discontent. He called me in during the ongoing investigations, and chided me about my stance on the rifle.

When I stood firm, he asked me, "Culver, just what would be *YOUR* solution?"

"Easy." I said, "it's only been 9 months since we turned in our M14s, all that's necessary is for us to draw the 14s again until ordnance can work the bugs out of this little piece of #$@&!"

"Unfortunately," said the Colonel, "it's not as simple as that!"

"Unfortunately," sez I, "it's *EXACTLY* that simple! What you mean is that it's not *'politically'* that simple!"

I was dismissed without another word.

The aftermath? The rifle was eventually fixed of course, but at great cost in life and suffering. Unfortunately, "fixing" the M16 left us saddled with a service rifle that shoots a cartridge not powerful enough to be used on anything larger than groundhogs according to most state hunting laws. The latest version is almost as heavy as the M14 without any of the 14's redeeming features. In retrospect, the cost of saving reputations and enhancing corporate well being was high... too high. Mike and I both spent an extra year in grade and Mike decided not to stay on in the Marine Corps even though he was a regular officer, and a damned fine one at that! That was one of the larger tragedies, as Mike was one of the truly good guys. Men of principle are rarer than the Hope Diamond in real life, and he was one of those. After the decorations had settled on the scene in SE Asia, they decided to keep me around and I was too stubborn to quit. The Corps, with what can only be described as a rather macabre sense of humor, sent me to Naval Post Graduate School in Monterey, California and made an Ordnance Engineer out of my somewhat "frayed" fanny. Life is often rife with seemingly contradictory incidents. Most of these give truth to the statement of George Burns in the movie, *"Oh God"* where he describes God as a comedian playing to an audience that's afraid to laugh! Amen...

NOTES:

Lieutenant Colonel Victor Ohanesian was the Commanding Officer of the Second Battalion, Third Marines. He was fatally wounded in a North Vietnamese ambush on February 28, 1967 at the beginning of the Hill Fights. Also killed was the battalion Sergeant Major, Wayne N. Hayes.

The Marines were trying to rescue a reconnaissance team north of Camp Carroll. They were on a trail between two hill masses when the North Vietnamese sprung the ambush. Several Marines were hit by automatic weapons fire and everyone hit the deck. As Ohanesian tried to gain control of the situation the enemy started to drop mortar rounds, with exceptional accuracy, on the trail. One hit close to Ohanesian, mortally wounding him, and four others nearby. Fragments from the exploding round torn into his right side and he was bleeding badly. Both of Sergeant Major Hayes hands were blown off by the blast.

The M-16 rifle was a drastic change from the dependable M-14 the Marines had carried since it replaced the M-1 in 1963. The venerable Garand was used effectively by our troops during World War II and Korea. John J. Culberson, a Marine sniper with 2/5 in 1966 and 1967, in his book *13 Cent Killers*, called it the finest battle rifle ever fielded by the United States.

Instead of wood, the M-16 had a plastic stock which significantly dropped the weapon's weight. The M-14 weighed eleven pounds while the M-16 weighed less than seven. The new weapon did not look like any weapon that any army had used before. It was black, very light to carry and had an unusual appearance. The little rifle even had a carrying handle on top. The M-16 was ridiculed by the Marines and called many derisive names. The "Black Stick" being one of the more positive ones.

The M-16 used a smaller cartridge than the M-14. Both the M-14 and the M-60 machine gun used the standard NATO 7.62mm round. The 5.56 round used by the new rifle was about the size of our familiar .22 caliber. It was, and still is, nothing more than a varmint round.

With the smaller and lighter rounds a Marine could carry more ammunition. Each man could now take 400 rounds with him into combat. Since the M-16 fired at a faster rate than the M-16 it would use rounds at a much faster rate than the M-14. The new rifle was designed to spray out rounds in a wide, random pattern while the heavier M-16 was designed for well aimed fire.

The little rifle was about to get its first real test during the hill fights that took place near the Khe Sanh Combat Base during late April an early May of 1967.

In Edward F. Murphy's book, *The Hill Fights,* he wrote: "On Okinawa the Marines of 2/3 turned in their trusty M-14s for M-16s. The weapons training consisted of a brief lecture followed by the firing of a few magazines at the rifle range. Most of the troops took an instant dislike to the new weapon. Lance Corporal Frederick G. Monahan, who had been with Echo Company for seven months, was one of them. "Most of the guys thought it was a 'sissy' weapon because of the lightweight plastic stock," he recalled. "But worse than that it jammed all the time. Plus, the magazines had to be individually loaded; we had no 'speed clips.' And then, we could only put seventeen or eighteen rounds in each magazine, not twenty like it was designed for, or else you'd overload the spring. We were only issued three magazines each and one cleaning rod for every four guys. The men in the rear were giving their magazines and cleaning kits to the grunts before we headed south so we'd have enough."

Private First Class Gary D. Murtha, a rifleman with Fox /2/7 at the time of the rifle exchange wrote: "We were told that turning in the M-14 would be voluntary. The M-16's looked like toys and soon our leaders had a minor revolt on their hands. After threats of Office Hours we reluctantly turned in the M-14's. In time our reluctance would be proven justified."

Lance Corporal Ira G. Johnson, a rifleman and radio operator with Mike 3/9 recalls, in *The Hill Fights* that, "I was given one magazine to practice with. After just one round the weapon jammed. I was furious. I couldn't believe they wanted us to go into combat with these things."

Johnson determined that a faulty bolt was the problem with his rifle and he decided to obtain a good one. It took a heated argument with the company armorer before he got one but once he did he had no further trouble with the rifle.

During an attack on Hill 881 on April 29, 1967 many Marines from Johnson's company were wounded by North Vietnamese Army automatic weapons fire. He aggressively moved to the side of a badly wounded Marine who was lying in an area exposed to enemy fire, and administered first aid before carrying the man to safety. Johnson then helped a corpsman move another casualty to safety and then calmly resumed his duties as the platoon radio operator.

Another Marine from Mike 3/9, Private First Class Ernest M. Murray, during the fighting for Hill 881S, had an M-16 that worked. Asked to help carry a wounded Marine to the rear he handed his weapon to another man. After carrying the wounded Marine to a position of relative safety he went looking for his rifle. "I found the guy who had taken my rifle, but he seemed in a daze and didn't remember what he'd done with it. I finally found a pile of discarded M-16s. The first one I picked up was jammed. So were the next three I looked at. Finally I found one that worked."

Second Lieutenant Frank Teague, a platoon commander with Mike Company, Third Battalion, First Marines, in Otto Lehrack's book, *Road of 10,000 Pains*, remembers, "We were issued the M16 rifle. It was a nightmare. Mine jammed all the time. Everybody had been trained on the M14, which was like your right arm, and we got this Tinker Toy that didn't work. I remember thinking that I'm a college grad, and I'm having trouble working this friggin' thing. Some of my men just don't get this shit. This thing doesn't work."

Second Lieutenant Jim Cannon of Echo 2/3 recalls that on May 3 1967, "We were attacked by an estimated battalion of North Vietnamese Army. We were hit so hard and with such force that our defenses were penetrated and we found ourselves fighting inside our perimeter. Our weapons (The M-16 rifles) began to jam …..rifle after rifle jammed. We had to fight with what we could get our hands on…rocks, sticks, knives, bayonets or entrenching tools. Many of our young Marines perished while trying to unjam their rifles with cleaning rods."

Second Lieutenant Billy D. Crews was the leader of the first platoon of Mike 3/3 as they assaulted Hill 881S on 30 April. During the fighting his radioman, Lance Corporal Francis Palma, was shot in the eye and instantly killed. Within minutes Crews spotted several North Vietnamese soldiers running right through his lines. He grabbed Palma's M-16 and fired. It jammed after firing just two rounds and Crews disgustedly threw the rifle away.

He could hear Marines, in the area, yelling that their rifles had also jammed. Sometimes these men with now useless weapons were killed. The doubts that the Marines had harbored about their new rifle were being realized.

Private First Class Philip G. Curtis of the third platoon of Mike 3/3 was one of the few Marines on Hill 881S armed with an M-14. When the new rifle was being issued he chose to keep the older more trusted one. Somehow he had gotten away with it.

Lance Corporal Richard A. Russell of Fox Company, Second Battalion, Ninth Marines, in his memoir *Hell in a Helmet: Memoirs of a Marine Infantryman in Vietnam*, writes, "The change over from our familiar and reliable M-14's to the unfamiliar and unreliable M-16's was the direct cause of many casualties (especially on May 8, 1967: see *Honor the Warrior* page 129), regardless of what many politicians, arms' dealers and generals said. How many of their asses were on the line with malfunctioning weapons?"

During Operation Swift on September 10, 1967 nine Marines from Hotel Company, Second Battalion, Fifth Marines were killed during an assault, across some rice paddies, just in front of the enemy fortifications. Captain Gene W. Bowers their commanding officer recalls, "They were found with their M-16 rifles broken down in an attempt to remove cartridges jammed in the chambers. They had powder burned bullet holes in their heads."

The Marines of the Second Battalion, Fourth Marine Regiment were involved in terrible fighting during Operation Kingfisher in which they incurred serious losses on September 21, 1967. Bob Bliss of the first platoon of Golf Company wrote, "We were fighting with a defective weapon. The M-16, at this time in 1967, would often jam after firing only a few rounds. We were told by our officers we were not keeping our rifles clean. Keeping the M-16 clean in Vietnam was difficult at best. This was due to the environment and the weather. But this was not the problem. There were real problems with the M-16, and many Marines were killed because of the weapons malfunctions. I personally went through three

M-16s during the September 21 fight, serving with Golf Company until I was wounded. It was like fighting with one arm tied behind your backs. We were outnumbered, using a weapon that jammed, but we still fought like Marines have always fought. And we gave as good as we got."

Barry Smith kept a diary of the daily activities of Fox Company, Second Battalion, Seventh Marines during the year 1967 in Vietnam. It is entitled the *History of Fox 2/7*. This diary includes a statement made by 2/7 Headquarters in its Command Chronology for March of 1967 that reads: The battalion received the M-16 on 30 March 1967. Issue of these M-16's and retrograde of the M-14 began without the benefit of prior planning or coordination. However, no major problems have arisen and this issue and retrograde is progressing well. 449 weapons were exchanged as well as a basic allowance of 5.56 ammunition for each weapon.

Barry Smith wrote: "Foxtrot issued the new M-16 rifle. No training or maintenance classes given for this new weapon. We gave up the old M-14's with great reluctance."

In June of 2000 Terrance R. Gulch, the armorer for Fox Company 2/7 wrote a letter to Barry Smith. The letter is included here, as it was written, with the written permission of Terry Gulch. It is intended to correct the battalion statement in Smith's diary with this concise explanation:

"In March 30, you have M-16's issued. This is correct. As I remember some of the line companies were still out on operation, and were brought in and issued the M-16 and returned to the field. The M-16 was made by Colt, and all we could give was two magazines and one bandoleer of ammunition. We did give everyone a cleaning kit, but no bayonets were available.

It really did bother me to take in an M-14 with sometimes as many as twenty magazines, and only give out two. As we received more magazines they were handed out to the line companies first. This caused some dissension with the people in the rear and supply. I felt the guys on the line everyday needed them more than some of the people in Headquarters Company.

In about one week we took in and reissued almost a thousand weapons. No time was given us in the armory to give classes on maintenance of the weapon. It seemed the only concern was to take in the M-14, check the serial number, and reissue the M-16. Orders.

The Colt weapons had a polished bore and receiver. This would later be the cause of many young Marines getting wounded or killed. On July second, you state that an M-16 division contact team arrived to give schooling on the M-16. We went out to all the line companies as we could schedule the time needed to conduct classes. You may have forgotten that before we went out to give classes, all M-16's had to be turned in to have the barrel and receiver group exchanged with a new chrome plated group.

The reason the weapons needed to be turned in was the serial number was stamped on the outside of the barrel and receiver. It was found that the polished units after firing rounds through them that the polished metal would pit. Thus the brass would expand from the heat into the pits this, in turn, was causing the extractor to pull the lip of the cartridge off, leaving the cartridge inside the chamber. That was the reason the weapon had to be broken down and the cleaning rod shoved down the barrel from muzzle to chamber. Not a real pleasant task to do during a fire fight. I recovered many an M-16 from a medevac unit that was broken down, and many with spent cartridges still in the chambers.

Also the bolts were replaced with chrome plated units. The problem with this was that the bolts were shiny. This was overcome by coloring them with markers.

Contrary to the report of Congressman Richard H. Ichord, the cyclic rate of fire was never changed in the M-16's. There also never was a problem with the power or lubricant. The only lubricant used was to lubricate the bullet or projectile. This again was not causing the problem. The problem was Colt sending weapons that were unserviceable to a climate of monsoons, mud, dust and very high temperatures, with little or no time to properly clean the weapons. I have always felt that Colt knew what they had done, and I'm sure that they were paid twice for the weapons.

I, as the battalion armorer, always carried an M-14 and an M-1911A1/45 when I went out in the field. I was allowed to keep a low number of M-14's in the armory. Some of the M-14's were attached with the old style infrared sniper scopes. Then as we received them, the starlight night sniper scopes were attached. I, personally, used both and the difference was unbelievable. Neither scope would mount on the M-16 for they were too small. I hope this will help you Barry. Anything else I can help you with, let me know.

Richard Culver also wrote, "During my latter days with 2/3 I served as the Battalion Intelligence Officer. One of the reports that came in was an intercepted message from the Viet Cong. The VC were not as well equipped as the North Vietnamese Army who operated primarily in the North just south of the DMZ.'

"We, as Marines, usually policed all, or as much as we could find, of our equipment left on the battle field by our wounded. Most, of course, was sent to the rear with the wounded man, but there would occasionally be items on the battle field by the departing units. The VC used this recovered equipment to equip their own rather meager supplies as would any good guerilla force. The intercepted document reinforced this practice, exhorting the VC troops to police the battle fields for usable equipment. This document however, had one telling exception to the rule. It stated that all equipment was to be picked up with the exception of "the little black rifle" which is useless to our cause!"

In his fine book, *What It is Like to Go to War*, Karl Marlantes wrote, "The M-14 was the standard rifle in use before the adoption of the M-16. It was much heavier than the M-16, with a heavier bullet, so had greater range and accuracy, neither of which was needed in jungle warfare, but which was greatly desired in open space. At that time the M-16 was also considered unreliable. It had improved (by 1969), but only after needless deaths. A small act of tiredness during the design phase? A small favor granted to a defense contractor? A small burecratic slip about power composition? Small things, magnified terribly by war."

Others mentioned in the chapter are as follows:
Baker, Richard L. CWO 098341; H&S-2-3; 43; Houston, Texas
Bliss, Robert E. LCpl. 2207297; G-2-4; 20; Hastings on Hudson, New York
Bogard, Robert N. Capt. 078262; E-2-3; 29; Little Rock, Arkansas
Bowers, Gene W. Capt. 079956; H-2-5; 29; New Orleans, Louisiana
Chervenak, Michael P. 1stLt. 092836; H-2-3; 22; Spangler, Pennsylvania
Crews, Billy D. 2dLt. 095920; M-3-3; 28; Bryceville, Florida
Culver, Richard O. Capt. 075696; H-2-3; 31; Hopewell, Virginia
Curtis, Philip G. Pfc. 2272019; M-3-3; 18; Madisonville, Kentucky
Gulch, Terrance R. LCpl. 2261963; H&S-2-7; 19; Toledo, Ohio
Hays, Wayne N. SgtMaj. 533657; H&S-2-3; 39; Oceanside, California
Johnson, Ira G. R. II LCpl. 2257167; M-3-9; 19; Oklahoma City, Oklahoma
Monahan, Frederick G. LCpl. 2195404; E-2-3; 20 Holland, Pennsylvania
Murray, Ernest M. Pfc. 2127602; M-3-9; 20; Martinez, California
O'Dell, Harry J. Pfc. 2246058; K-3-9; 26; Three Rivers, Michigan
Ohanesian, Victor LtCol. 050708; C0-2-3; 40; New York, New York
Palma, Francis M. LCpl. 2220373; M-3-3; 25; Philadelphia, Pennsylvania
Russell, Richard A. LCpl, 2258527; F-2-9; 18; Milford, Michigan
Smith, Barry J. Cpl. 2300594; F-2-7; 21; Orlando, Florida

✷ ✷ ✷ ✷

In January 1968 the North Vietnamese started the most important military campaign of the Vietnam War. The surprise Communist offensive was launched on, Tet, Vietnam's most important holiday. A key objective of the Tet Offensive was the ancient capital city of Hue, once considered the jewel of Indochina's cities. When the enemy launched their massive invasion of the city, they expected a general civilian uprising and hoped for an easy victory. Instead they were faced with a counter attack spearheaded by the United States Marines with a little help from the South Vietnamese. What followed was a devastating battle of street by street and house to house fighting that the Marines had not experienced since the liberation of Seoul during the Korean War in 1950. The Marines suffered many casualties, but those of the enemy were horrendous with losses estimated to be around 5,000. In the end, the battle for Hue was a decisive military victory for the United States.

By February 24 the Marines had retaken almost all of Hue. The North Vietnamese, however, received an unexpected political victory when on February 27, Walter Cronkite, dealt a devastating blow to American troops when in a thirty minute television special, asked the question: "Who won and who lost in the great Tet Offensive? He then, for unknown reasons, answered his own question by saying that, at best, the American military in Vietnam was "mired in a stalemate," when, in fact, we were winning and winning convincingly. Cronkite even predicted that "Khe Sanh could fall with a terrible loss of American lives, prestige and morale." Of course, that didn't happen either, we kicked their butts there too, but the damage had been done. Many of the American people were now against the war and most soon would be.

Unfortunately Cronkite's decision to make the news, as opposed to simply reporting it, has produced the likes of Dan Rather, who used to claim that he was a former Marine, until it was discovered that he did not complete Marine Corps Recruit Training.

"Until Delta could finally take and hold the tower, 1/5 was stuck."

First Lieutenant R. Nicholas Warr
Charlie 1/5
from *Phase Line Green*

Referances used in compling the following story are the following books; Fire in the Street: *The Battle for Hue, Tet 1968* by Eric Hammel, *Battle for Hue: Tet 1968* by Keith William Nolan, *Semper Fi Vietnam* by Edward F. Murphy and *Unheralded Victory* by Mark W. Woodruff.

Also used were the awards citations of many of the participants, the rosters and diaries of the participating units, and information obtained from the National Personnel Records Center.

8

Attack on the Dong Ba Tower

On February 10, 1968, the Marine Corps chose to commit the First Battalion, Fifth Marines to the Citadel in Hue. Then on February 14, 1968, Delta Company was assigned the job of leading a planned dawn assault on the Dong Ba tower. Most of the company was temporarily stranded on the wrong side of the river. Captain Myron C. Harrington, his company command group, and one rifle squad were the only ones to cross, at this point, but everyone was certain that the rest of the company would be across in time. They were.

The selection of Delta Company to lead the February 15 attack on the Citadel was a well thought out choice. At this time the company roster showed that only about 100 Marines were in Hue. One platoon, consisting of twenty men, had been detached for convoy security duty for trucks running supplies between Phu Bai and Hue. Delta, also, had not been involved in the February 13 battles that occurred within the walls of the Citadel, as had the other three companies of the battalion. They were also experienced in street fighting having participated in a two day sweep along the Perfume River, west of the Phu Cam Canal. At that time they were attached to the Second Battalion, Fifth Marines.

Harrington had no combat experience. He had spent his first six months in Vietnam with a supply unit back at Da Nang and had just taken over Delta Company on January 23.

In the morning before the break of day artillery fire pounded all objectives along the former front of the First Battalion, Fifth Marines. Mortars and the shells from the five and six inch guns of a U.S. Navy cruiser added to the bombardment of the tower. It had begun to crumble into a huge pile of rubble as parts of it collapsed. All that surrounded it had also been destroyed.

As rain drizzled from a gray sky the wind blew cold. Charlie and Bravo Companies closed up to the right and Delta Company moved in between Bravo's left flank and the northeast wall of the Citadel. Charlie and Bravo moved easily through the rubble and regained all ground that they had given up the previous day. For Harrington's Delta Company it was much tougher going as they approached the Dong Ba tower.

Harrington and his officers had been unable to do any reconnaissance of their objective and had no knowledge of the ground or the location of any enemy positions. They were essentially attacking blindly. Second Lieutenant William C. Conrad, Jr., who led the second platoon, moved to a second story balcony that faced the tower so that he could better control the fight. The North Vietnamese fired a B-40 rocket that exploded among the command group and Conrad, with shrapnel wounds to both legs, the platoon sergeant, Staff Sergeant Edward M. McNamara, who suffered gunshot wounds to the left hip, left hand, and right leg, the radio operator, and at least one of the squad leaders were seriously wounded. The all important radio was also destroyed and communication became difficult. The platoon guide took over temporary command of the platoon.

Verbal communication was almost impossible because of the noise made by the sound of gunfire reached a deafening roar as it bounced off of the walls that surrounded the restricted battlefield. The intense close quarter's battle that Delta Company now faced made their previous two day experience with street fighting seem like a walk in the park.

Bravo and Charlie companies achieved their objectives as the North Vietnamese conceded the ground. The Marines of Delta Company stalled short of Mai Thuc Loan Street and the Dong Ba tower as they were subjected to a withering fire from its tenacious defenders. They made it clear that they would not give up the high ground easily.

Two weapons that helped the Marines were the World War II vintage 3.5 inch rocket launcher and the company's 60mm mortars. These were used to fire point blank at the enemy.

Sergeant Frank Jandik, Jr., serving as a platoon commander, with Delta Company requested tank support and then fearlessly moved about the fire swept terrain locating enemy positions for the tank gunners to shoot at. This succeeded in suppressing the North Vietnamese fire.

At 1400, Harrington sent one infantry squad from the first platoon, led by Second Lieutenant Jack S. Imlah, to climb up on top of the northeast wall about 150 meters from the Dong Ba tower with orders to sweep the high ground. They were to come in behind the tower. Their progress was a helped by the deadly blasts from the 90mm cannon mounted on the M-48 tank. The wall was about seventy five meters thick and as heavily defended as the streets below. The Marines advancing along the wall were exposed to intense enemy fire. As they made small gains they began to ask for hand grenades from those more lightly engaged. These were proving to be the most effective weapon used on the top of the wall.

That afternoon Jandik's platoon began receiving fire from two enemy snipers and as a result sustained several casualties. The intrepid sergeant ran forward to a more advantageous firing position and directed a heavy volume of fire upon the snipers, killing both.

Private Willie M. Smith had been slightly wounded earlier in the fight. It was not a serious wound but it was his third purple heart which meant that he would automatically be sent back to the states. Knowing that the wound was not serious and not knowing what to do with Smith, Harrington had decided to keep him with the command group. When the Marines on the wall started to call for grenades, the idle Smith went to work. He collected all that they needed and made numerous trips to the wall carrying grenades, small arms ammunition, batteries, and whatever else was needed to supply the squad on top of the wall. Smith helped bring wounded Marines down from the wall and led replacements up. The peripatetic Marine was everywhere doing whatever had to be done and at great risk to his life.

Lewis C. Lawhorn, a Private First Class, with the first platoon moved forward with his squad. They moved forward slowly and, as he moved from his covered position, the fire from the North Vietnamese intensified. Lawhorn quickly dropped down again as other Marines moved in rushes behind the cover of crumbling brick walls and destroyed houses. They fired at the tower with quick bursts from their rifles as they moved. Lawhorn got back to his feet and ran and stumbled to the cover of the next brick wall. Overcome with fear, he lay there struggling to breath and sweating profusely. It was mass chaos as bullets and shrapnel hit everywhere. The Marines were firing furiously and trying to shout above the din. It was mad hysteria and Lawhorn felt alone and confused, He had no idea of what lay ahead of him or who was behind him. He didn't know what Marines had been hit or if any of his squad mates were still with him. Lawhorn raised his head and looked through the debris at the tower. He could see enemy soldiers moving in the rubble there, occasionally popping up and firing their rifles at the attacking Marines. The tough, little bastards would fire and then drop down to reload. This was definitely a new experience for him. Although only about a hundred yards away, the tower appeared to be a hundred miles in the distance. Mustering all of the courage that he could, Lawhorn, brought up his rifle and started to fire madly at the North Vietnamese.

Among the many wounded during the initial moments of the attack was Private First Class Horace Smith. As the Marines neared their objective, Smith saw that the platoon radio operator was wounded and lay exposed to the heavy fire from the North Vietnamese defenders. Although painfully wounded himself he moved to the fallen Marine, skillfully administered first aid, and then moved him to a position of relative safety. Realizing the need to maintain communications, Smith then went back for the radio and brought it to his platoon commander. On two additional occasions he ran back into the exposed area and pulled fallen Marines from the line of fire.

During the attack, Lance Corporal Raymond D. Sexton, a squad leader, repeatedly exposed himself to the intense North Vietnamese fire as he skillfully placed his men in positions where they could bring a heavy volume of accurate fire on the enemy emplacements. They then moved aggressively through the hostile fire and his squad seized their objectives.

Realizing the necessity of maintaining the momentum of the assault, Lance Corporal Herbert D. Hammons, a fire team leader, bravely ran out in front of Delta Company firing his rifle at the enemy and shouting words of encouragement to his fellow Marines. Despite the heavy fire coming from the tower he charged up the only entrance ramp to the tower and placed suppressive fire on the enemy positions. He continued to direct the movements of his companions until he was hit by automatic weapons fire and killed.

Staff Sergeant James G. Davidson recalls, "The assault was made under heavy fire from the front and flanks. Hammons led the assault and charged forward up the only entrance ramp to the tower. He was hit and killed but his efforts were not in vain for the attack succeeded. If not for him I don't know if we would have made it."

Also killed were Privates First Class Ronald L. McElroy and Daniel M. Stone.

Davidson also stood out during the fight as he skillfully set up the 60mm mortars to be used against the enemy on the wall. At great risk to his life he moved to a vantage point from which he directed a devastating volume of fire on the North Vietnamese defenders.

Corporal Maurice P. Whitmer, a squad leader, disregarded his own safety and skillfully maneuvered his squad through the intense enemy fire to the point of heaviest contact. He ignored the enemy fire and led an assault against the North Vietnamese positions and enabled his company to seize its objective.

Throughout the attack, the Marines moved through a haze of gun powder and dust from the pulverized brick. There were dead North Vietnamese soldiers scattered in the rubble. Some of them had been killed a day or two before and they were stiff and bloated and their rotting bodies put out a terrible stench.

Jandik boldly led his men in the assault on the Citadel wall. Unfazed by the hostile rounds impacting around him, he continuously moved about firing his weapon, shouting words of encouragement to his men and directing their fire. Jandik was soon seriously wounded but continued to control his platoon and throw hand grenades into the North Vietnamese positions until he was forced to be medically evacuated.

Lance Corporals Douglas M. Fracker, a fire team leader, and Roy F. Swed were killed during the assault. Swed had already been wounded in action on January 4 when he suffered a fragmentation wound to the left shoulder. Privates First Class Harry G. Bloomfield and Marion F. Ferguson also perished. Among the wounded were Privates First Class William

Boney and Stanley L. Skiles. Boney suffered a gunshot wound to the lower abdomen. It was the third time that Skiles had been wounded. He had suffered fragment wounds to the chest and left shoulder in early January and then had been hit again, by shrapnel, in the chest, lower back and spine about a week earlier. On this day he was hit in the head and right hand, again, by shell fragments.

Private First Class Arturo H. Campos was hit in the left side of the neck by a piece of shrapnel that penetrated the jugular vein. Due to quick and efficient work by the corpsmen he would survive and be evacuated.

Harrington, sensing that the attack was losing momentum and using whatever troops he had available, ordered another quick assault on the tower. It was about 1600 when he did so. At the last moment a squad from Delta's third platoon, under the leadership of Staff Sergeant Robert L. Thoms, arrived to help in the attack. Thoms was the commander of the third platoon and, as soon as he found out that the company needed help, had on his own volition brought in the badly needed reinforcement's. It was his first day in Hue and his first time in combat. The squad was sent to the wall to help Imlah's first platoon.

The Marines on top of the wall, reinforced many times by Marines replacing others that had been wounded, closed in on the tower. It had been reduced to piles of rubble which they went over and around as they advanced. Below it other Delta Company Marines pressed forward too and finally at 1630 following a blistering 60mm mortar barrage, Imlah's first platoon, began the attack from positions south of the tower.

Thoms led his men in the attack on the enemy redoubt from the northwest. Although the enemy fire was intense he repeatedly exposed himself to the maelstrom of hand grenade and heavy automatic weapons fire and moved his men to within thirty meters of the tower. When the attack momentarily stalled due to an increasing number of casualties, and an increase in the volume of fire coming from the enemy held tower, Thoms, moved to the point of heaviest contact and aggressively led the attack against the objective. Although wounded by fragments from an exploding hand grenade he refused treatment and continued to lead the attack. Following his lead, the Marines in his squad routed the North Vietnamese and seized the Dong Ba Tower.

This done, Thoms then established a hasty defense, redistributed the ammunition and supervised the evacuation of the wounded. Despite intense North Vietnamese mortar and rocket fire throughout the night, he maintained his position until directed to withdraw. Wounded a second time by fragments from an exploding mortar round, he steadfastly refused medical evacuation and remained with his platoon to lead it in an aggressive counterattack the following morning, completely routing the North Vietnamese.

Imlah's Marines pushed through the enemy at the tower archway at street level. The grunts on the wall captured the tower. During the assault on the tower, seven Marines were killed and another thirty three were wounded and evacuated. Twenty four North Vietnamese and Viet Cong bodies were pulled from the rubble.

With the objective secured, Harrington moved his men off of the massive rubble heap and they set up for the night in the houses below. He realized that the tower was a good observation point and placed a five man team on top. This group was led by Private First Class Dennis S. Michael, a member of the first platoon. After darkness set in they picked their way to the top, and finding a crater among the heaps of blasted brick and concrete, set in for the evening.

During the morning hours, North Vietnamese infiltrators suddenly tossed four Chicom grenades into the pit occupied by the Marines. One hit Michael in the back and another grenade rolled under his leg. Chicom grenades were notoriously unreliable and sometimes would not explode. All Michael could do was stare at them and say a quick silent prayer, asking that they not explode. They didn't. One of the other Marines grabbed the other two and tossed them from the hole. Another grenade came in and this one exploded, wounding two of the Marines. All scrambled to the rim and opened fire on the attackers.

Wounded in this attack was Private First Class Russell G. Chapman with superficial fragmentation wounds to the face, both hands, and both lower legs. He had previously been wounded on September 4, 1968 with a gunshot wound to the left leg in Quang Ngai Province.

Private First Class Larry T. Hays was seriously wounded with gunshot wounds to the left thigh, and left and right knee. Private First Class James P. Walsh, Jr. suffered gunshot wounds to the knee and thigh.

The North Vietnamese counterattacked the tower at 0430 on February 16, 1968. Using a deadly effective mortar barrage the enemy moved in close and, by using a combination of grenades and B-40 rockets and forced the Marines from the tower.

Harrington did not hesitate and firing his .45 caliber pistol at point blank range led the command group in a counter assault of their own.

Although wounded in this attack, with fragmentation wounds in both legs and a gunshot wound to the right hand, Corporal Whitmer again rose to the occasion and ignoring his painful injuries led an aggressive counter attack against the North Vietnamese which disrupted their assault an enabled his unit to regain its objective.

The enemy poured fire on Michael's group and down on Harrington's men. Green and red tracers lit up the night and bounced off of the rubble at weird angles. The Marines managed to pin the attackers in between the two groups but the North Vietnamese continued to fight. Michael and his men were forced to the bottom of their pit by an exploding round from a recoilless rifle. A figure suddenly stood at the rim of the crater and dropped two more grenades on them. These also did not explode and the Marines scrambled back to the rim of the hole and continued to fire at their attackers.

Using all of the small arms weapons at their disposal the Marines took it to the enemy and by first light the tower was again in the hands of Delta Company. The bodies of fifteen North Vietnamese soldiers lay sprawled in the rubble between Michael's small group and Harrington's position.

There was little time to rejoice as the artillery started up again and concentrated their fire further down the wall. The Marines of Delta Company settled down around the tower, finding spots in the piles of rubble, and sitting on their helmets and flak jackets, smoked cigarettes and ate tasteless meals of C rations.

NOTES:
The following Marines were killed during the assault on the Dong Ba Tower:
Bloomfield, Harry G. Pfc. 2366690; D-1-5; 19; Jacksonville, Florida
Ferguson, Marion F. Pfc. 2403921; D-1-5; 19; Romulus, Michigan
Fracker, Douglas M. LCpl. 2365308; D-1-5; 18; Jackson, Michigan
Hammons, Herbert D. LCpl. 2332783; D-1-5; 20; Pocasset, Oklahoma
McElroy, Ronald L. Pfc. 2336079; D-1-5; 24; Duncanville, Texas
Stone, Daniel M. Pfc. 2391443; D-1-5; 20; Ilion, New York
Swed, Roy F. LCpl. 2243664; D-1-5; 21; Shrub Oak, New York

On this day the Navy Cross was earned by Captain Myron C. Harrington and the Silver Star medal was awarded to Lance Corporals Herbert D. Hammons and Raymond D. Sexton, Sergeant Frank Jandik, Jr., Staff Sergeant Robert L. Thoms and Corporal Maurice P. Whitmer. The award to Hammons was posthumous.

The Bronze Star medal was earned by First Lieutenant Jack S. Imlah and Private First Class Horace Smith.

Other Marines wounded that day were Privates First Class Cecil E. Bullard with a gunshot wound to the neck, and Damian Rodriguez with fragmentation wounds to both legs, and his right wrist. Lance Corporal David P. Schultz suffered a gunshot wound to the left leg and a fragmentation wound to the right ring finger.

First Lieutenant Jack S. Imlah, returned to Delta Company after recovering from his wounds and was killed in action on August 29, 1968.

Private Gilbert Apodaca remembers him, "Lieutenant Imlah was a very great man. He served his country well. He was my Lieutenant and a good Lieutenant."

Private First Class Dennis S. Michael was killed in action on February 20, 1968 when he was shot through the throat by a round fired from an AK-47 rifle by the North Vietnamese as the Marines of Delta Company pressed their attack along the wall of the Citadel.

Staff Sergeant Edward M. McNamara was again wounded in action on May 30, 1968, while serving as a platoon sergeant with Delta 1/5 during Operation Houston II. He was hit by fragments from an exploding device in the right shoulder, left forearm, left side, right calf, left side of the neck and left side of the cheek and mouth. McNamara died the next day from his injuries. He had been a Marine since March of 1958.

Corporal Maurice P. Whitmer was serving his second tour in Vietnam. He had previously been with Charlie Company, First Battalion, First Marines from January 1966 until he was wounded with a gunshot wound to the right leg on November 17, 1966 during Operation Arcadia in Quang Nam Province. Whitmer was hospitalized and returned to the United States. He returned to Vietnam in August of 1967 as a member of Alpha Company, for a short while, until transferred to Delta Company in September. Whitmer was wounded again by fragments from an exploding enemy device during Operation Essex on 18 November.
After being wounded during the attack on the Dong Ba Tower he rejoined his unit in March and was wounded again on 19 April. This time it was a missile wound to the chest.

Other Marines who participated in the attack on the tower distinguished themselves either before or after that action, while serving with Delta Company. Private First Class Harvey E. Thompson, a radioman, was returning from a road clearing mission on January 5, 1968, when the truck in which he was a passenger came under an enemy attack as it slowed to cross a bridge near the Hai Van Pass. In the attack the Viet Cong used hand grenades and intense automatic weapons fire. The driver, Lance Corporal Robert W. Doyle, was killed by a burst of automatic weapons fire as he attempted to drive through the ambush. The truck, which carried eighteen passengers, careened over a 150 foot embankment. Thompson sustained only minor injuries when he was thrown from the vehicle. He immediately regained his feet and attempted to contact his unit by radio. After organizing the wounded Marines, he led them up the embankment to the road in order to fight off a second attack by the enemy soldiers. As they reached the top of the embankment they saw nine enemy soldiers approaching their position, Thompson did not hesitate and aggressively attacked them, firing his pistol as he moved and wounding two of the Viet Cong. Following his lead his fellow Marines routed the enemy and drove them from the area. When enemy soldiers began to fire on them Thompson directed counter fire as he contacted his command by radio. He then requested and adjusted effective artillery fire on the enemy emplacements. This done, Thompson then established a well organized defense and administered first aid to the wounded. He then directed their movement to a landing zone and stayed in contact with his unit until reinforcements arrived and a medical evacuation took place.

On February 22, 1968, Corporal Richard A. Morris, was serving as a platoon sergeant as Delta Company continued its attack on the wall surrounding the Citadel. The Marines were working against a well entrenched unit of the North Vietnamese Army when they suddenly came under a heavy volume of small arms, automatic weapons and antitank rocket fire and sustained several casualties, including the platoon commander. Morris assumed command and moved forward to reorganize the platoon and direct fire upon the enemy positions. As he boldly moved his men toward the enemy emplacements he was painfully wounded by the fragments of an exploding enemy grenade. Morris ignored his injuries and led a determined assault into the North Vietnamese positions. His men overran them, killed four enemy soldiers, captured one and seized two machine guns.

Corporal Jim A. Henderson, Jr. was a machine gun squad leader with Delta Company on June 2, 1967 during Operation Union II. The company was advancing across open terrain in the vicinity of Tam Ky in Quang Nam Province when it suddenly came under intense automatic weapons and mortar fire from a much larger enemy force. When two of the company's platoons were pinned down Henderson's was deployed on the right flank to launch an attack against the enemy. He led his team in the attack and directed highly effective fire against the enemy emplacements. When two gunners from Henderson's squad were wounded, he moved across a rice paddy to recover one of the unmanned guns. Although two enemy rounds grazed the top of his helmet he managed to cross over about 150 meters of rice paddy until he reached the gun. Once there, he quickly placed the gun in position and placed a heavy volume of fire into the hostile positions, neutralizing them, and allowing the Marines from his unit to advance to more advantageous positions. He was soon seriously wounded by a heavy burst of enemy fire.

On the evening of May 24, 1968, Sergeant Thomas E. Mitchell, Jr., was a platoon sergeant with Delta Company. He had been a Corporal squad leader during the battle for Hue City but had since received a meritorious promotion to Sergeant. His platoon had established an ambush on a suspected enemy infiltration route in Thua Thien Province and became heavily engaged with a much larger enemy unit using small arms and B-40 rockets. The platoon commander was seriously wounded during the initial moments of the attack and Mitchell took charge of the platoon. He led it in an aggressive attack on the enemy and forced them into a panicked and confused withdrawal. Under Mitchell's leadership the platoon killed ten North Vietnamese soldiers.

Others mentioned in the chapter are as follows:
Alvarado, Julio A. Pfc. 2291974; D-1-5; 19; Brooklyn, New York
Apodaca, Gilbert Pvt. 2394711; D-1-5; 18; Portland, Oregon
Boney, William Pfc. 2336074; D-1-5; 18; Dallas, Texas
Bullard, Cecil E. Pfc. 2338904; D-1-5; 19; Raleigh, North Carolina
Campos, Arturo H. Pfc. 2324164; D-1-5; 19; Chicago, Illinois
Chapman, Russell G. Pfc. 2377384; D-1-5; 19; Detroit, Michigan
Conrad, William C. Jr. 2dLt. 0103416; D-1-5; 21; Fairmont, West Virginia
Davidson, James G. SSgt. 1876899: D-1-5; 26; San Francisco, California
Harrington, Myron C. Capt. 081869; D-1-5; 27; Augusta, Georgia
Hays, Larry T. Pfc. 2369242; D-1-5; 18; Cincinnati, Ohio
Henderson, Jim A. Jr. Cpl. 1984516; D-1-5; 24; Albuquerque, New Mexico
Imlah, Jack S. 1stLt. 0113384; D-1-5; 24; Corvallis, Oregon
Jandik, Frank Jr. Sgt. 2132892; D-1-5; 21; Miami, Florida
Johnson, Carrol J. Pfc. 2344827; D-1-5; 19; Houston, Texas
Lawhorn, Lewis C. Pfc. 2337665; D-1-5; 20; Vandergrift, Pennsylvania
McNamara, Edward M. SSgt. 1814405; D-1-5; 26; Chicago, Illinois
Michael, Dennis S. Pfc. 2318585; D-1-5; 20; Vacaville, California
Morris, Richard A. Cpl. 2172310; D-1-5; 22; Charleston, South Carolina
Rodriguez, Damian Pfc. 2398049; D-1-5; 19; Los Angeles, California
Roland, Charles Pfc. 2258452; D-1-5; 18; Oklahoma City, Oklahoma
Schultz, David P. 2268354; D-1-5; 29; Indianapolis, Indiana
Sexton, Raymond D. LCpl. 2305797; D-1-5; 18; Ullin, Illinois
Skiles, Stanley L. Pfc. 2303455; D-1-5; 20; Abilene, Texas
Smith, Horace LCpl. 2352079; D-1-5; 20; Detroit, Michigan
Smith, Sterling M. III Pfc. 2365008; D-1-5; 19; Memphis, Tennessee
Smith Willie M. Pvt. 2335703; D-1-5; 19; Dallas, Texas
Taitt, Selwyn H. Pfc. 2281748; D-1-5; 20; Brooklyn, New York
Thompson, Harvey E. LCpl. 2077393; D-1-5; 23; Albany, New York
Thoms, Robert L. SSgt. 1951269; D-1-5; 24; Baton Rouge, Louisiana
Walsh, James P. Jr. Pfc. 2297240; D-1-5; 19; Los Angeles, California
Warr, R. Nicholas 1stLt. 0102370; C-1-5; 22; Coos Bay, Oregon
Whitmer, Maurice P. Cpl. 2157170; D-1-5; 20; Spring Valley, California
Widener, Richard H. Pfc. 2370707; D-1-5; 18; Nashville, Tennessee

✯ ✯ ✯ ✯

Second Lieutenant Donald Jacques was the youngest Marine Corps Lieutenant to complete The Basic School in eighteen years. On February 25, 1968 he commanded the third platoon of Bravo Company, First Battalion, Twenty Sixth Marines at the Khe Sanh Combat Base.

"The Marine tradition of always retrieving our wounded and dead from the battlefield has been shaken. Our dead are still out there."

Private First Class John A. Corbett
81mm Mortar Platoon 1/26
From *West Dickens Avenue*

Grateful acknowledgement is made to Khe Sanh Veterans, Inc. and to Chaplain Ray William Stubbe for permission to reprint previously published material from his fine book *Battalion of Kings. A tribute to our fallen brothers who died because of the battlefield of Khe Sanh, Vietnam.*

Also used were the awards citations of many of the participants, the rosters and diaries of the participating units, and information obtained from the National Personnel Records Center.

9

The Jacques Patrol

For Private First Class Ronald L. Ridgeway the morning of February 25, 1968 started when his squad leader Lance Corporal Kenneth Claire briefed his men that they were going on a sweep that was not to exceed 1000 meters from the combat base perimeter. It was basically to be a perimeter wire check to determine if the wire had been cut or sabotaged, or to see if there were any North Vietnamese forward observers. They were told not to go out of sight of the base perimeter.

"Claire said we were looking for any trench positions that were being dug toward our perimeter and for Bangalore torpedoes. He also said that there was a possibility of receiving enemy mortars because it was open terrain."

This patrol was to be far from a normal patrol for Ridgeway. It was to be the beginning of a five year nightmare as prisoner of war.

For Corporal Gilbert Wall, the 81mm Forward Observer, the day began with the Second Platoon Commander, First Lieutenant John W. Dillon, telling him that he was to accompany the third platoon on today's patrol. Wall already had two purple hearts and was due to be rotated out of Khe Sanh, but he pleaded with Lieutenant Colonel James B. Wilkinson the Commanding Officer of the First Battalion, Twenty Sixth Marines, to stay at Khe Sanh. Only a few days earlier, during a night in which the base received heavy mortar fire the first platoon corpsman, Richard Blanchfield, was about to medevac Wall, noting that Wall's eardrums had been blown. Wall refused.

Wall had been on a patrol along Route 9 on 21 July that collided with a North Vietnamese force that was preparing ambush positions. He was a part of the 81mm mortar section that accompanied First Lieutenant Benjamin L. Long's Second Platoon of Bravo Company. On this patrol Sergeant Jose Castillo was shot twice in the chest and died instantly. Corporal Bruce Jones was shot through the throat and Wall was hit in the shoulder by a round that then went through his lung and broke two of his ribs. The bullet lodged in his back. He also was hit by a ricochet in the face and another shot under the eye. This was his second Purple Heart and he received orders to report to the Third Marine Division Headquarters as an Military Policeman. Wall wanted to remain with his unit so he worked his way up the chain of command to Wilkinson who allowed him to stay.

On this day Wall did not have time to clean his M-16, a fact that would save his life. The previous night he estimated that he had fired almost 700 rounds through the rifle and the carbon had built up so much in it that it would jam after firing three rounds.

Wall had been on numerous patrols with Jacques and they knew each other well. The Lieutenant invited him to breakfast and coffee in his bunker. There Wall was introduced to the new platoon sergeant, Staff Sergeant George McClelland. Most of the Marines who were to go on the patrol were new. They stood out for their faces were white and yet to be dyed by the red clay pigment of Khe Sanh. Everybody was in a surprisingly good mood. All were happy, loose and relaxed. It was a nice day and just to move out of the base for awhile was a good feeling.

The patrol consisted of 41 men including the third platoon led by Jacques with McClelland second in command. The squad leaders were Corporal Robert E. Matzka, second squad, and Corporal Kenneth Claire, third squad. Jacques accompanied the second squad which was first in the line of march. The third squad was second in the order of march. Attachments included Wall, the 81 Forward Observer, one S-2 Scout, and one Kit Carson Scout, two machine guns and one rocket team.

Not all members of the third platoon went on the patrol. Although he pleaded to go, Lance Corporal Edward Prendergast, a 60mm mortar man, was kept behind to better provide protective fire for the patrol. Lance Corporal William H. Woolman II, his section leader wanted him to stay on the guns.

Lance Corporal William D. Holliman had been wounded the day before and medevaced when a round split his helmet and creased his skull. Private First Class Jeffrey E. Culpepper who was superficially wounded earlier, but not medevaced, was on Okinawa attending Infantry Weapons School. Private First Class James Hebron was with him.

The mission of the patrol was to locate a North Vietnamese 82mm mortar position, believed to be about 1200 to 1300 meters from the base perimeter, and learn whatever they could. They were to check for Bangalore torpedoes, damaged wire, spot any positions and not to make contact unless it was with an enemy Forward Observer. The patrol was to sweep around the platoon area and not proceed beyond 1,000 meters of the wire, and remain within sight of the perimeter. By now the North Vietnamese had crept close to the base perimeter as part of their overall plan.

The diamond shaped patrol departed at 0806 and moved through the trash pit. They immediately noticed a trench built by the North Vietnamese that half encircled the trash pit and went down the face of a small hill, across a river, and up the face of another small hill.

Wall remembers that the trench was about 2-3 feet deep and about a foot wide, just wide enough for an American Marine to stand sideways up to his waist, but through which the North Vietnamese would be able to run.

They proceeded down this trench for about 30 meters, then decided to move out of the trench and moved down the slight side of a small hill into and across a small creek and across an open slope which led to a heavy tree line. When they reached the tree line they were in a single column and spread out very well. There they stopped, all got down on one knee, and a couple of fire teams were moved up to the front.

Private First Class Calvin E. Bright was walking point and in his words, "was scared to death!" Private First Class Clayton Theyerl told Bright, "This is your first patrol and I think you and I should switch." Bright recalls, "Probably 20 minutes later he was laying on the ground dead and I was standing. I think that if it wasn't for him, I'd be in his position and he'd be in mine right now."

Bright would soon distinguish himself. When one of the Marines was shot down, he ignored the intense enemy fire and crawled to within a few feet of a North Vietnamese machine gun emplacement. Bright dragged the Marine to a covered position, only to discover that he was dead. When the beleaguered patrol was ordered to break contact he provided covering fire.

As they advanced to the tree line, three North Vietnamese soldiers dressed in camouflaged utilities and wearing helmets were spotted at

0845 running along the road. Jacques moved up along with Hospitalman Third Class Frank Calzia, and the enemy soldiers turned off the road and disappeared after the Marines shot at them. Jacques requested permission to attempt capture of a live prisoner. Captured North Vietnamese soldiers were prized by intelligence because they almost always willingly supplied a wealth of vital military information.

Captain Kenneth W. Pipes, the commanding officer of Bravo Company, thinking that they were on the road that led to the trash dump and not the parallel road further to the west, granted permission but told them to be extremely careful not to get sucked into an ambush. The North Vietnamese were using a tactic as old as warfare itself.

The patrol was much farther out than they were supposed to be. Pipes assumed that they were near the defensive perimeter of the base and on the road closest to it. They were actually 600 to 700 meters from the wire.

For some unknown reason—disorientation and confusion by Jacques, the shrouding fog, negligence of some radio operator, mis-interpretation of check points, being too aggressive, not being able to restrain further examination of the trench system which both fascinated and attracted, the desire to capture a live prisoner—the patrol was not where they were assumed to be by the command bunker at Khe Sanh.

The Kit Carson Scout warned the patrol that it was a trap. Hospitalman Third Class John A. Cicala, knowing it was a trap said, "Are You crazy?" Cautiously, Jacques moved his patrol across the access road and stepped into the brush.

From his position in the formation, Wall noticed that the Marines to the front looked confused. But Jacques ended this confusion by ordering them to charge by shouting and waving his pistol in the air. As the Marines moved after them the North Vietnamese opened up with heavy automatic weapons and machine gun fire. This fire was coming from fortified positions in trenches located in two trench lines. One was running from north to south and the other east to west forming a classic L-shaped ambush.

It was 0915 and the patrol was in heavy contact with the enemy. The unit's leaders were among the first hit. Private Alexander Tretiakoff, a rifleman, suddenly found himself face to face with an enemy soldier. "I wanted to shoot him but my rifle wouldn't shoot. So I.. took the magazine out and threw it away and put another one in. To my right, Lance Corporal McKenzie was shot. The Lieutenant crawled up to him and took his weapon. The enemy was firing from my right and from my left."

When Sergeant Patrick J. Fitch, a radio operator with Bravo Company, heard that they were in a fire fight he said, "I left the command bunker to see where, precisely, they were, and they were almost out of sight. They were far beyond where they were supposed to be."

Jacques immediately dispatched Claire's third squad accompanied by the Platoon Sergeant, McClelland, and one of the machine guns, to move to the right and hook around the rear of the North Vietnamese position and deter the flanking action. This maneuver might have succeeded but Claire did not hook sufficiently far enough to the west before turning back in on the North Vietnamese positions. Instead of hitting the tender flank the squad walked into more blistering frontal fire.

Initially, the Marines had fire superiority. Calzia noticed from his position with the second squad that the Corpsman, Cicala, with the third squad had been wounded and was not on the road. "When I went down to get him I noticed that Claire's squad, from all indications, had been wiped out. There was no one left. He was the only person left and he was wounded."

Cicala had accompanied the third squad as it charged into a mouth of a crescent of bunkers and were immediately cut down by fire from three sides. He managed to treat one dying Marine, Lance Corporal Jerry Dodson. 'I ran over to him and I took care of him. He had caught a round through his left eye. It came out the side of his head, but he was still conscious. There was nothing that I could do. I put a dressing on him and he told me---it totally freaked me out---but the last thing he said to me was, 'Doc, make sure I got my weapon.' I gave him his piece back and I laid a couple of clips by him."

As Cicala dashed to another cry for help a North Vietnamese soldier fired and hit him. The round struck his taped neck chain, which included his dog tags; a large St. Christopher medal and a GI can opener, and pushed it into his windpipe. Another round ripped into his flak jacket and tore through his chest and right lung. Cicala fell backwards and, to his horror, realized that he had suffered one of those dreaded, sucking chest wounds. Still trying to kill him the enemy threw a grenade which landed between his legs. When it exploded it blew away his kneecap, shredded his legs and tore into his groin. Despite the severity of his injuries, Cicala lived, and managed to cover the sucking chest wound with cellophane that he tore from a cigarette pack. Once this was done he began to crawl toward the Khe Sanh perimeter.

Ridgeway recalls what he saw of the action in vivid detail:

As the Marines continued to pursue the two North Vietnamese soldiers, they ran through a tree line, across a Montagnard village (Clong Sareh) and then into an adjacent field that had been ploughed by the Montagnards. The field was bare and was approximately 50 meters by 50 meters. There they came upon a newly dug trench that was approximately 6 feet long and 4 feet deep. The Corpsman, Jacques, McClelland and the radioman moved into it for protection. McClelland, back in the tree line, told Claire to get his squad up and moving. While one squad moved into the trench and set up a base of fire, Claire stood in the middle of the field and directed his squad to move in fire team rushes toward a series of interconnecting trenches to their front. When Private First Class Edward C. Rayburn didn't move,

Claire turned and asked, "Rayburn, what's the matter with you? Get up and Move! Are You Hit?"

To which Rayburn replied, "No. I'm just laying here for my health."

"We began our fire team maneuvers across the field, receiving two or three rounds here and there. Off to our front we could see a trench running the full length of the field. The trench was approximately 50 to 60 meters long curving off to our left, back around the end of the field, against waist high shrubbery. We couldn't see anyone as we started across. We gradually picked up fire until we had just a full hail of fire coming in on us. We were caught in the open about half way out, but we kept our assault up. The first man hit was to my left. The Corpsman came and got him when he started screaming. Another man was wounded off further to my left."

"When we assaulted, Claire was standing up in the middle of this field telling us to move, hollering and ducking bullets. He told us, to just get up and run, forget maneuvers that we were to try to get to the trench. So we got up and ran, three of us."

The three were Ridgeway, Private First Class James R. Bruder and Lance Corporal Charles G. Geller. Another three man fire team also made it to the trench.

Once they got into the trench, Ridgeway, Bruder and Geller moved through it under the cover of the machine gun fire of the other fire team, until they arrived at another small trench with a small bend. As they advanced, a grenade exploded and they returned to their initial trench, engaging in an exchange of tossing grenades. Geller stood up in an attempt to locate the rest of the squad when he was hit by a round that creased his forehead and knocked him back into the trench.

"Geller rolled back into the trench and said, 'Everybody's dead. Everybody behind us is dead. There's nobody left alive. What are we going to do?' We had been down below ground level for awhile and the North Vietnamese had taken it for granted that we were dead."

Ridgeway and Geller stood up and spotted four enemy soldiers approaching them. Ridgeway fired a burst from his M-16 killing two and forcing the others to cover while Geller killed one with his .45 pistol. They dropped back down in the trench as Bruder stood up. When he did a round came through the dirt in the side of the trench and hit him in the chest killing him instantly.

Geller, the fire team leader, told Ridgeway that they had to move.

"We had to try to get back inside the perimeter because there was nobody left but us. He got out and took off running back across the field. At first I got stuck in the trench because it was so small with my pack on. I pulled loose, got out, and started back out across the field."

Ridgeway continued, "Geller came upon a black Marine, Private First Class Willie Ruff, who had a broken arm and was lying on his back. Geller knelt beside Ruff and was hit on the right side of his face. The round took out the whole left side. He was spitting teeth all over the place. Just as I ducked another round came from the back and I got hit in the shoulder. I dropped down to the ground and just laid there."

Ridgeway checked Geller and found him to be alive but in shock. The three of them decided to wait for a reactionary force and play dead until dark and then try to make it to safety.

"We were surrounded by NVA, so we decided to stay where we were. We laid there talking." Ridgeway added.

While this was happening, Jacques and Wall moved to the left of a little access road where they were pinned down by enemy fire. Wall moved away from Jacques and managed to reach a tree but could move no further.

"We were right in front of their trenches and bunkers." Wall said. "The air was filled with screaming, shouting, open fire and blood."

It was fortunate that his rifle jammed. Wall explains, 'My jammed rifle probably saved my life, because it would jam about every three times that I fired it, and I would get back and they would just chew up the tree."

As Wall took out his map to call in fire missions from the 81's, he realized that trees obstructed his sight of the other Marine squad and he feared calling in a mission that was too close. While he was able to hear the initial round exploding, he was unable to see where it struck. Wall couldn't move, due to the intense enemy fire, until another Marine gave him covering fire. As he moved he noticed wounded screaming for help everywhere—ours as well as North Vietnamese.

"By this time I was terrified and could see clearly that we were in deep trouble. We didn't have enough men to match their fire power." Wall explained.

Meanwhile, the second squad under the leadership of Matzka was on the left flank, further down the road. Matzka was seen exposing himself to enemy fire with no regard for his own safety as shrapnel and heavy machine gun and small arms fire showered the area. He was going from man to man to direct fire and encourage them.

Observing one of his comrades fall mortally wounded while attacking a North Vietnamese machine gun, Bright, ignored the intense enemy fire and crawled to within a few meters of the hostile position to retrieve his fallen comrade and move him to a covered position.

As Calzia, and the other members of the second squad, moved to the area where the third squad was slaughtered they assumed that all were killed but one of those in Claire's squad was still alive. It was Rayburn, the first one in the squad to be hit, who had his jaw shot off. Claire had placed him in a trench from where he observed the entire squad being killed in minute detail. Writing from a hospital shortly after the action he told Captain Pipes that, "I saw and heard them die for three hours."

Private First Class Thomas A. Detrick was a member of the machine gun crew that had gone out with the third squad. When the gunner was hit he picked up the weapon and was firing it when he too was hit. He describes what happens next, "I was on the verge of blacking out, and then I yelled 'machine gunner out of action, replacement up!' and crawled into a crater and passed out. When I came to, a Marine had joined me and he had his lower jaw blown off. This Marine and I spent the next several hours in agony and hell. We came to realize that either the patrol had been wiped out or had been pulled back."

As Wall was making his way back to the perimeter of the base he heard a wounded Marine by a tree. "He was screaming at me, hollering, 'Marine, Marine, help me!' So I ran back to get him when two other Marines came running and took him. They gave me his rifle. Mine was jammed and I had thrown it away. I had already taken a rifle from a dead Marine. I turned around and was using both rifles, just shooting them both. Marines in front of us were feeling a little relief because we were getting closer to the base. Then they ran into another ambush closer to the tree line. We were under their fire again. Bullets were flying all over the place. I thought of calling in a fire mission and lay on my stomach to get the map and realized that I had lost it, maybe when I was carrying the wounded Marine. Anyway, the fire mission was out of the question. Bullets were cutting up the big banana leaves above my head to pieces. Everyone was lying there dead or wounded, and there was no time to be scared of death."

Next to Wall, a Marine who was shot in the chest was moaning, obviously in deep pain. Wall placed a bandage on his chest but he was bleeding profusely. The man was dying so Wall moved on. Then he came upon Jacques. When he attempted to put a field dressing on the wound, "All the blood went through the bandage and ran down my fingers. I couldn't do anything to stop the bleeding. All this time he was trying to talk to me but I could not understand him."

At this time, Wall was hit by North Vietnamese fire. "One round hit the shoulder of my flak jacket and knocked me on my ass and another one went between my arm and my ribs and left a hole in my sleeve."

Doc Cicala was with Jacques when he was hit. "When I was wounded, the Lieutenant came running by me. I was just laying there. The grenade had just gone off and I was stunned. He came by me and said, 'We got to get out of here! Get out the best way you can. We're getting wiped out.' I just started crawling. I left my rifle and just had my .45 with me. As I was going back I heard him. I looked over and he stood up for some ungodly reason and that's when he got it. He got hit and I crawled over to him. He caught it right across the femoral arteries and he was dead within minutes. There was nothing that could be done. It looked like he got hit with machine gun fire and it caught him just below the groin area. It severed both arteries."

When Jacques' patrol made contact with the North Vietnamese Captain Pipes realized that, for whatever reason, the Lieutenant and his patrol had veered several hundred meters from its assigned patrol route.

Pipes requested permission to deploy the first platoon of Bravo Company as a reaction force. They were ready and when the request was granted they were moving out of the wire within minutes. The platoon commander was Second Lieutenant Peter Weiss. As they reached the access road to Route 9 they were subjected to extremely heavy small arms fire. They were unable to advance but were able to hold their position. Pipes then requested assistance from the tanks, which was approved, and they provided excellent fire power.

Walking point for the first platoon was Private First Class Donald E. White. As soon as they were pinned down Lance Corporal George W. Jayne realized the need for a radio went looking for the radioman. He found him, the squad leader and the corpsman. All had been killed.

Jayne explains, "We were separated from Lieutenant Weiss and the rest of the platoon and I just felt like we needed to have a radio. I figured he must have been wounded, and started to crawl back, and ran into part of Delta Company which came out also. I led them back to where we had been. By the time we got there, there was nobody there. People had started to pull back, and so we started to pull back."

The Corpsman, Hospitalman Lloyd W. Moore had fearlessly moved through the heavy barrage of fire to provide medical aid to the seriously wounded and saved the lives of at least three Marines. As he was moving to a fourth wounded Marine an incoming mortar round took his life.

One of those he rescued, Lance Corporal Johnny F. Bellina, had been next to a buddy who was shot through the forearm---"just like a paper punch." When Bellina rolled to his right side with his left leg sticking up in the air a North Vietnamese soldier fired and the round hit ripped across his leg.

"I saw the muzzle flashes keep shooting and I put a couple of magazines in that position about 100 yards off. I stopped him. He didn't shoot anymore. I rolled down in a ditch and I was bleeding so bad I thought I was going to pass out. I thought that they'd run up and bayonet me or something. I just laid and a corpsman got out to me, gave me 4 or 5 morphine syrettes in the leg. He threw me across his shoulders and carried me back." Bellina added.

Pipes then requested the deployment of another line company so that they would be able to break the North Vietnamese ambush and blocking positions. This done, they would be able to recover the wounded and dead. This request was denied. "I was, at first, extremely bitter at this decision, however reflections indicated that it was a wise one."Pipes said.

By noon the battle broke. First Lieutenant Thomas F. O'Toole, Jr., flying as an Artillery Observer, had sad things to report. He saw 15 to 20 bodies in Marine helmets and flak jackets lying near a trench line. Three of the bodies were lying together as if they were a machine gun crew. One was bandaged on the chest and another lay close by, as if a corpsman, giving aid. On a second pass O'Toole received fire from nearby and from one figure that stood up from among the bodies. It was probably a North Vietnamese soldier who was searching the bodies. He then dropped to the ground in an attempt to avoid detection. The bodies were in the center of a trench complex, in open area.

The survivors of the ambush trickled back to the base throughout the afternoon and night through an open area that afforded no cover. Those who returned did so singly and included Cicala, Rayburn, Bellina and Detrick. All could say little.

HM3 Thomas E. Casey had just arrived at Khe Sanh and, along with a few others, ventured outside the perimeter to assist any returning wounded. They came upon Rayburn, not far from the perimeter.

"He walked to us as we moved to his aid. I was surprised to see that his lower jaw was completely gone! I examined him quickly and saw that he was not bleeding and did not appear to be in shock. One of the Marines wanted him to lie down but he didn't want to do that. The Marine said to me, 'Shouldn't he be laying down, Doc?' I said, 'Well he made it all the way back here, so if he wants to walk, let him.' I don't recall putting any dressing on him. At that time I had no serum albumin which may have been a good start. He did not want to leave and kept looking back from where he came. I asked him then if he was trying to tell us something and he nodded his head 'Yes.' I asked him if there were others still alive out there and he seemed relieved and very much shook his head 'Yes!' Once he communicated to us he was ready to be helped back to the base, but only then! We found no others that day and darkness was coming. We looked for awhile and returned. That wounded Marine did all he could possibly do for his buddies that day!"

Among the other wounded were Ricky Brunk, William Bryant, Michael Estep, James Jewett, James Melanson, Ronald Pressley, Donald Santner, Richard Schlup, Stanley Smith, Doss Thrasher, Kenneth Totten, Donald E. White, Daniel Wolff and Gary Woodell.

In the center of the trench complex and still alive were Ridgeway, Geller and Ruff. Wounded and surrounded by North Vietnamese they moved in and out of consciousness. Ridgeway remembers a birddog aircraft flying over them. "My left arm was broken. Geller's right arm was broken and he was on his back. He reached down to the man that was wounded (Ruff) and pulled off a smoke grenade. We used this to notify the birddog that someone was alive. I held it and he pulled the pin and I threw it to the right, looked up and all I could see was a jet coming straight down. We ducked and the napalm hit back behind the North Vietnamese trench lines. We got small fragments of the napalm on us but were able to knock it off as we laid there. I came to in the afternoon around 3 or 4 o'clock. Both men were still alive. The man with the broken arm and I talked just to keep each other's morale up because we didn't know what was going to happen. The other man was delirious. All he wanted was a cigarette. He kept moving around. We told him to be still. He was afraid we were going to leave him."

Geller came up on his knees and as he did, grenades were thrown, when they exploded the fragments from one killed him. When the throwing of the grenades ceased, Ridgeway saw small arms fire from the right front traversing the field from right to left, firing into the bodies. A round hit between Ruff and Ridgeway causing Ridgeway to flinch. Another burst was fired in his direction. A round ricocheted off his helmet and hit him in the left buttock. Ruff was hit again, this time in the leg, breaking the bone. Ridgeway lost consciousness.

When Ridgeway awoke it was dark. He decided that they should attempt to reach the trench. Ruff was unable to crawl because of the wounds that had broken his arm and leg. He pleaded with Ridgeway not to leave him behind. Ridgeway told him that he would not be left behind and tried to calm him down and bolster his morale. Both Marines curled up under their flak jackets and helmets and decided to wait until morning in hopes that a reaction force would find them.

Incoming artillery continued to fall about them and Ruff was hit in the head by shrapnel and was bleeding badly. Unable to help him, Ridgeway instructed him to place his hand over the wound to stop the bleeding. As the shelling continued Ridgeway again lost consciousness and did not awaken until early morning.

At first light he became aware that someone was tugging at his wristwatch. Ridgeway opened his eyes and a North Vietnamese soldier was standing over him. The soldier must have thought that he was dead for Ridgeway's movement startled him. The enemy soldier jumped back and pointed his weapon at him a move that made Ridgeway believe that he was about to be shot. The North Vietnamese motioned for him to get to his feet but Ridgeway was too injured to comply. More soldiers appeared, stripped him of his flak jacket and pack, and lifted him to his feet. As Ridgeway moved toward the trench he saw Ruff lying with his head in a pool of blood. One of the soldiers kicked Ruff's head to make sure that he was dead. He also saw the bodies of Baptiste, Geller and the demolitions man, a black Marine nick named 'Stoney'. There were other bodies that he was unable to identify.

"When they captured me and moved me I could see the trench line at the base and see Americans walking around inside," said Ridgeway. From there, He walked through Khe Sanh village and then to a large rear area of the North Vietnamese Army in Laos.

In early March, Sergeant First Class Pierce N. Durham, Army Special Forces, led a ten man patrol to the area of the ambush. They reached the bodies of the Marines but when suddenly showered with grenades and taken under fire they decided to return to the base. "I reported that the bodies were there. They were in the best condition that they could possibly be. The only thing missing were flak jackets, helmets and gas masks. We got in contact and had to get out of there. Colonel Lownds wanted me and my ten men to carry the 25 bodies back!" Durham, a former Marine, explained. "It was obvious that we couldn't do that."

On April 6, 1968, Delta Company 1/26 moved to the North Vietnamese trench/bunker complex parallel to the Route 9 access road and secured it. Bravo Company moved in trace of the Delta Marines and together they policed the ambush site. The area was strewn with the detritus of battle including the head sets used by the North Vietnamese to call in mortar rounds. These had Chinese and Russian marking's on them.

Private First Class Theodore H. Golab, artillery Forward Observer with Alpha-1-13, remembered it well, "I walked over to a boot and picked up the boot and there's the rest of the bone in the leg. At least on the boot was a dog tag. A lot of the guys at that time used to put the dog tags on the boot. And there were a couple of guys scared. They didn't want to touch them because these guys were laying there a month. I said, 'Put them in the bag; let's get out of here. I don't like this anymore than you do but these guys are Marines."

The bodies of Seventeen Marines previously declared Missing in Action on February 25 were recovered. The recovery team put the bodies and body parts into Willy Peter bags, sand bags, ponchos and anything else that would carry them. All of the remains were in bits and pieces but it was officially determined that this patrol recovered all bodies of the missing Marines.

When the remains reached Graves Registration at Khe Sanh the task of identifying the almost totally decomposed bodies fell to the company commander Captain Pipes. "The shock of losing 25 men, who were missing in action for a month and a half, then going to graves registration several weeks later to help identify the decomposed remains is an indescribable feeling."

All but nine of the bodies were positively identified. The remaining pieces of bodies were so badly mutilated that identification was impossible. They were buried in a group grave at Jefferson Barracks, Missouri on September 10, 1968.

The nine unidentified bodies included Lance Corporal Frederick A. Billingham, Jr.; Lance Corporal Michael J. Brellenthin; Lance Corporal Bruce E. Jones; Private First Class David C. Scarborough; Private First Class John A. Lassiter; Baptiste, Bruder, Dodson and Ridgeway. The latter was, of course, not dead but a prisoner of war.

NOTES:
The following members of Bravo-1-26 were killed in action on February 25, 1968.
Akins, Ronald P. LCpl. 2225422; 20; Akron, Ohio
Baptiste, Michael B Pfc. 2366757; 19; Tampa, Florida

Billingham, Frederick A. Jr. LCpl. 2279502; 18; Trenton, New Jersey
Brellenthin, Michael J. Pfc. 2355263; 20; North Bergen, New Jersey
Bruder, James R. Pfc. 2374658; 18; Allentown, Pennsylvania
Claire, Kenneth W. Cpl. 2327737; 21; Redwood City, California
Clay, Doyle G. Pfc. 2380920; 20; Chicago, Illinois
Dodson, Jerry L. LCpl. 2235054; 20; Collinsville, Illinois
Geller, Charles G. LCpl. 2225009; 20; East St. Louis, Illinois
Hayes, Phillip III LCpl. 2340296; 19; New Orleans, Louisiana
Jacques, Donald 2dLt. 0102013; 20; Rochester, New York
Jones, Bruce E. LCpl. 2249963; 20; Rockland, Massachusetts
Laderoute, Michael J. Pfc. 2394093; 19; Boston, Massachusetts
Lassiter, John A. Pfc. 2363909; 19; Slidell, Louisiana
McClelland, George SSgt. 1883252; 25; Passaic, New Jersey
McDonald, Henry III Pfc. 2374642; 19; Philadelphia, Pennsylvania
McKenzie, Richard W. LCpl. 2297953; 19; Oxnard, California
Meads, Kim E. Pfc 2359332; 18; Chicago, Illinois
Moore, Lloyd W. HN B-1-26; 19; Wilmington, North Carolina
Rivera, Arnold J. Pfc. 2390939; 19; El Paso, Texas
Ruff, Willie J. Pfc. 2308049; 20; Columbia, South Carolina
Scarborough, David C. Pfc. 2291001; 20; Marietta, Ohio
Skinner, Walter F. Pfc. 2278383; 19; Soledad, California
Smith, Douglas W. Pfc. 2246477; 18; Ft. Worth, Texas
Theyerl, Clayton J. Pfc. 2351478; 18; Racine, Wisconsin

Donald Jacques joined the Marine Corps while attending college. His parents had little money and he didn't want to ask them for more to be able to continue in school. When he arrived in Vietnam in early October, 1967 he wrote home: 'Tonight I had a 19 year old come to me for help and advice. He is married to an 18 year old and he was having problems. If he knew that I was only 20 I wonder if he would have come."

On the evening of February 24 he wrote to his parents, wishing them a happy anniversary and adding: "The days go by quite quickly but the nights are long around here. I haven't been getting much sleep. I stay awake all night sometimes and others not. We get our sleep in the day a couple of hours at a time. That is if the work load isn't too great. It is a constant digging deeper and the work really never stops. It can't."

In what was to be his last day on earth he also wrote to his sister, Jeanne, and her family: "We're just sitting here waiting for all hell to break loose with each day that goes by the incoming increases and the third platoon digs deeper. We are already for them in my opinion but no matter how well prepared you think you are there is always something more you can do. We have been lucky so far and I hope that the good fortune holds for us. Although my platoon has had 30 wounded since January 21. None have been too serious and the majority of the men are still on duty and all healing. One or two have made it home early but that is all to the good. For they have done their part. They seem to think that Khe Sanh will be the turning point to the negotiation tables. I hope so but if not we will go on until this is over…I just hope we can do our part well enough so that your

Billy and Mark and Matt don't have to come... I'm fine and still kicking, I'll write again soon. Love brother, Don."

Second Lieutenant Donald Jacques was as "Green" as the piece of camouflaged parachute he always wore around his neck.

Ronald Ridgeway turned 18 in boot camp and went straight over to Vietnam, arriving in November of 1967. In early February of 1968 he was knocked unconscious during a mortar barrage. "I did a foolish thing. I jumped into the trench line to get away from the mortars and I broke my nose." He woke up in a helicopter in DaNang. "I got medevaced, I guess, because I was out, unconscious, for so long.

Kenneth Claire was a tall and rugged squad leader. Before joining the Marine Corps he had briefly played football at the College of San Mateo at 6-2 and 245 pounds, which was a big man in 1968. His father had been a both a professional football and baseball player.

He was well known for his Tarzan yell as well as other atavistic demeanor and actions. During a battalion size sweep in late December Claire led his squad which included Hebron, Ridgeway, Rayburn, Culpepper and others, together with another squad and a machine gunner up a creek bed covered with about six inches of mud. Suddenly they spotted four North Vietnamese soldiers carrying a wounded man. When the enemy saw the Marines they dropped their wounded officer and ran. Claire and Hebron fired 29 rounds from a grenade launcher and killed the wounded officer. Ridgeway and Culpepper threw the dead officers body into a poncho.

Culpepper explains, "He was bleeding like crazy and the blood was sloshing around in the poncho and got all over us. When we got back we finished our chow and went on."

Upon returning, Jacques took the officer's pistol. "It should have been Claire's," but the issue became academic since both men were soon to die. "Jacques asked Claire, "Is this a fresh kill?" Claire responded, "Of course," and took the blood on his finger and licked it."

Claire had a reputation of being hard but fair. He took care of his men. Claire was very strict about falling asleep on watch and making certain that there was sufficient separation between the men as they walked on patrol. If someone was walking with his head down, looking at the ground, he'd come by and let that person know it---physically as well as vocally! Culpepper continues, "But then in quiet times when you'd be standing watch or something he'd come around checking lines. He'd apologize to you and say, 'I'm not trying to give you a hard time; I'm just trying to instill in you the importance of what we're doing. He did care for us."

James Hebron notes that Rayburn had been a very handsome guy, but after having his jaw shot off, his girl friend left him. In the hospital his jaw was fastened to his shoulder and he used his two front teeth to mash French fried potatoes. He became addicted to drugs and committed suicide in 1979 while in a VA hospital.

Tom Winston, who had been badly burned during the Khe Sanh siege on April 9, 1968, spent a year in the U.S. Naval Hospital, Balboa, along with Ed Rayburn. According to Winston, the doctors took a rib out of Rayburn in order to construct a jaw and also made a graft on his chest to look like a chin. The Doctors failed, however to clean the jaw bone where it was shot off, and after gangrene developed, they had to remove it, repeat the procedure, and Rayburn became very depressed. Winston recalls that he was a fantastic person. "He used to blow smoke rings out of his trach!"

Laderoute was a Canadian citizen and in his last letter home wrote "We haven't had any mail in four or five days. The place is fogged in and no planes landed. A few today. For here is hoping for mail. Not much else to look forward to except mail and chow and hoping for contact. As soon as the North Vietnamese attack and we beat them off the wire we will go on a sweep through the mountains and jungles to Laos and Cambodia after the remainder of their forces."

Culpepper recalls Rivera as, "A real good looking Mexican American, kind of quiet, easy going, soft spoken and real friendly." Skinner was, "A kind of hippie type, free spirit, kinda cocky Marine from San Francisco."

McClelland enlisted in the Marine Corps at age 17. He was married and had raised a family prior to his first Vietnam tour. On January 21 the base was attacked and the ammunition dump exploded. The explosion scattered and covered the area occupied by Bravo Company with many unexploded artillery rounds. Corporal John N. Tracey remembers that, "Even though the shrapnel from the incoming rounds was still heavy, Staff Sergeant McClelland did not hesitate to clear them away until the entire area was secure from dud rounds."

Decades later Ed Prendergast remembered Billingham. "I've had friends since then but never a best friend that Fred was. Now double his age I still haven't lived life the way he did! Sometimes I think he knew he'd only be here a short time."

Brellenthin had been married for only three weeks when he departed for Vietnam. In letters written to his wife he tried to reassure her. "I know that everything is going to work out right… There is nothing to fear. I believe now more than ever that a person who thinks enough can shape his own fate… An intelligent man who uses his training can easily survive.

Don't worry about me Ruth… I've decided that I'm going to be an architect when I get out of here."

After the siege started his letters were optimistic. "Things started popping on the 21st of January. It all should be over pretty soon. The way I understand it we could end the war here… Don't worry about me Ruth. I have too much waiting for me at home to let anything happen to me here… I got your package today. The one with the rum in it. I'm saving it until all this blows over… I love you beaucoup."

The Marines of Bravo Company had to wait until March 30 to exact their revenge and attempt to recover the bodies of their fallen brothers from the ambush of February 25. The horror of what had happened that day continued to gnaw at those still alive at Khe Sanh, especially within the command staff of 1-26. They had almost immediately started to make plans to return to the ambush site and retrieve the MIA.

Shortly after the action Captain Pipes was to say, "I believe all of the young Marines that we had in the company had been waiting several weeks for this opportunity…Once the kids started, there was nothing, it seemed, that could stop them. From the point of initial contact until we reached our final objective, some 500 to 600 meters, perhaps more, they covered the ground in 20 to 25 minutes. A later body count confirmed that approximately 115 North Vietnamese were killed…by kids using small arms fire, grenades, bayonets, M79s, all weapons organic to an infantry company."

A section of flame throwers also participated in the attack and Pipes felt that these were a complete surprise to the enemy. "When we first used them there was a notable slack in their fire. I think the shock effect at that particular moment of the NVA seeing the flame throwers, knowing that we have them with us…put a bit of fear in them and enabled us to continue the initial momentum of the attack clear through to the final objective."

The company quickly secured the first objective then paused momentarily to re-organize and then moved out on a line in a platoon wedge/squad wedge with the third platoon in a V formation. They advanced about 25 meters and then came under extremely heavy fire from a combination of weapons which included RPG, 82mm and 60mm mortars, heavy automatic weapons, small arms and grenades from mutually supporting and very well built bunkers. These fortifications were stronger than anticipated. Despite this the momentum of the attack never slackened.

Private First Class Ted D. Britt, a fire team leader with the third squad of the first platoon, maneuvered his fire team by individual rushes until his squad secured the outermost defensive position. There they were pinned down by heavy enemy mortar and automatic weapons fire. Britt left the cover of the trench and assaulted the enemy position from which the largest volume of fire was coming. He was able to destroy the position, the weapon and killed four North Vietnamese soldiers. Britt was fatally wounded as he attacked yet another position. His efforts were greatly responsible for reducing the number of friendly casualties.

Lance Corporal Jerome C. Foster, another fire team leader, unhesitatingly ran across the fire swept terrain and boldly assaulted the nearest enemy position and in doing so forced the North Vietnamese soldiers into a nearby trench where they were killed by his fire team. Although wounded he destroyed the emplacement with hand grenades and then attacked a second hostile position, killing three enemy soldiers and destroying the bunker. Inspired by his actions his buddies quickly destroyed a third redoubt. When the company was ordered to withdraw Foster disregarded his wounds and organized his men to provide covering fire for the withdrawal of the wounded and dead.

Hospitalman Third Class Thomas E. Casey continually moved from casualty to casualty despite the heavy enemy fire around him. As he jumped into a crater to render medical aid, three mortar rounds exploded in and around the crater, killing two Marines. Casey disregarded the danger around him and continued to treat the wounded.

Private First Class Melvin D. Frazier assaulted and silenced two enemy machine gun emplacements. Although seriously wounded he continued to attack and silenced the fire from several additional North Vietnamese positions. Frazier then moved to an injured Marine and assisted him to a covered position.

Private First Class Gene E. Edgar moved about the battle field and accurately threw hand grenades into the apertures of hostile bunkers, killing several enemy soldiers. He determinably assisted several wounded Marines from danger and when seriously wounded himself launched another aggressive assault against the North Vietnamese redoubt and silenced several more positions. When a withdrawal was ordered he assisted in evacuating the wounded.

Corporal Samuel Boone, Jr., a squad leader, rapidly deployed his men and without hesitation launched an assault against the enemy positions. He moved to the front of his unit and directed a heavy volume of suppressive fire. Boone quickly destroyed two bunkers and then maneuvered his squad from one hostile emplacement to another, effectively silencing the enemy fire. When the withdrawal was ordered he continued to provide covering fire while the casualties were evacuated.

'Boone led this assault himself," said Lance Corporal Lonnie W. Morrison. "He assaulted position after position, and destroyed and killed the enemy within. Marines who had been hesitant to join the assault became inspired, and assaulted the remaining positions with great enthusiasm. The enemy hit the company with a heavy mortar barrage, and we received orders to break contact. Corporal Boone provided covering fire so that the casualties could be taken from the field."

Morrison, a rifleman from the point squad of the third platoon, was one of the Marines that Boone inspired and he rushed across the fire swept terrain and fired into the apertures of enemy bunkers and threw numerous grenades to knock out several enemy positions enabling the rest of his platoon to advance. Morrison also provided suppressive fire as the wounded were evacuated.

Golab watched the entire action: "Everything looked like it should be. They all had fixed bayonets. They all had flak jackets on. They got helmets on. They looked good. They all had their formations down pat and they were moving out. You could see the enemy walking in their trenches, just nonchalantly. One guy had a cup in his hand. I called artillery on Khe Sanh ville because I knew something was going to come out of there, to cut off the road. And then there was this mortar round, it lands and doesn't go off. It must of landed 6 yards from me, went into the mud and its fins are sticking out."

For some unexplained reason our artillery barrage stopped as the 'Cease Fire' command was given. Using imitative deception the North Vietnamese had managed to give the order to cease fire. During this lull in our barrage the enemy managed to lob several mortar rounds into the middle of Pipes command group.

The mortar barrage was deadly. First Lieutenant Henry Norman, the Artillery Forward Observer, was killed and both Norman and Pipes radio operators were wounded. Norman's radio operator was hit in the face. Lance Corporal Thomas N. Quigley, the radio man for Pipes, witnessed the entire incident. When Doc Blanchfield got hit Lieutenant Norman said, "Doc's hit," and he started getting up. "I turned back to look at him, and that's when the rounds hit, and I saw him get it. He didn't know what hit him. It picked him up and slammed him right back against the trenchline. It took the whole side of him off. I screamed… It scared me, because Lieutenant Norman was one of the best officers besides Captain Pipes and Ben Long that I ever had the privilege of serving with, just a straight guy, and when I saw him, I screamed. My hand busted out blood. The shrapnel had broke my radio."

Lieutenant's Dillon and Long led a group of Marines to Captain Pipes' position and found that he had already put battle dressings on one of his radio operators and had already evacuated four others. Pipes, with the help of Dillon and Long, rapidly regained communications with his maneuver elements, the battalion, the artillery and air support.

Lance Corporal Michael E. O'Hara made these observations, "Pipes stood his ground to call for air and artillery and just wouldn't quit until all of his men were back safe. He set an example for my whole life. He was standing up, bleeding like a stuck pig. He had two handles (radio headset), one on each ear, controlling. Demaagd couldn't get him to hold still. He was trying to keep him bandaged up, and the blood would explode out— 'pfssh'—I mean I thought the man was going to die. He was tore up in his armpit real bad, real bad. I was simply in awe of this guy. Here most of his good friends were lying dead or wounded right at his feet, and he was still in control. I bent over and picked up Lieutenant Norman's body. When I saw who it was, I was devastated. Lieutenant Norman was a slightly built man, but I remembered he seemed so heavy. My legs were beginning to ache with pain and I was very nearly ready to drop him. I laid him on the back of an APC, fell back on my butt, and began to cry. I began to wonder, 'How much can a man take?'"

Prendergast also managed to reach Captain Pipes and saw him "standing in an exposed position under the continuing heavy enemy fire, calling in artillery fire on the forward observer's radio as well as directing the operations of the three platoons of the company… If Captain Pipes had not acted in the manner that he had, even though severely wounded, and had not maintained the communications with the fire support center, the enemy would have been able to launch a counterattack to our advancement into their fortifications."

The Marines of the second platoon continued the momentum of the assault into the North Vietnamese positions. O'Hara captures the moment: "When I reached their trenches and jumped in, it was unreal. The trenches were filled to the brim with freshly killed enemy soldiers. My foot never hit the ground while I was in their trenches. The rats were everywhere. Those bastards fear nothing and I would later reflect that we had become like rats. We were like sharks on a feeding frenzy. We were passing by wounded Marines on our quest for the kill. We went clear over the top and began charging down the backside before we realized we had inflicted total destruction on an entire North Vietnamese battalion. For just a moment we all cheered like a come from behind football team who had just won a pennant or something. We began to pull back and started to realize there was a very high price for our exhilaration. Dead and wounded Marines were everywhere you looked and the danger had certainly not passed."

Lance Corporal Author C Smith, the third squad leader with the first platoon, left the already secured trench line and rushed across open terrain to silence a North Vietnamese position. This done he then returned to the secure trench to consolidate his squad which was now under intense enemy fire. Once again he leaped from the trench and assaulted another enemy position that was putting out heavy automatic weapons fire. Smith silenced that position and as he continued his assault on a North Vietnamese spider hole he was killed.

Corporal Donald A. Warren, the Right Guide for the first platoon, volunteered to be one of four men who were to recover the bodies of the MIA's. He carried the body bags to be used for this endeavor. After the initial burst of North Vietnamese fire, Warren led his fire team to the front of the platoon so that he could direct the assault against the enemy positions. After eliminating two North Vietnamese strong points and while directing accurate small arms fire against the enemy, Warren was knocked down by an enemy round.

Corporal Wood W.J. Lathrope saw Warren go down. "I just left him. I was the squad leader and was leading a charge into the enemy trench. I really had no choice, but all these years that has bothered me, that here was my best friend. I had seen him get hit. I had seen him fall. And I don't go to him."

During the assault five Marines, from the weapons platoon (60mm mortars), were located directly behind Pipes command group about 25 meters from the two assault platoons. Led by Corporal Kenneth A. Korkow, they were without the base plates and sights for their mortars, having left them at the point of contact. After they were pinned down by enemy mortar fire, Korkow left his position of safety and holding the mortar by hand fired it at an angle that placed him within the bursting radius of his own rounds. His accurate fire destroyed four enemy bunkers and one enemy mortar. Korkow was later critically wounded as he helped another wounded Marine from the field.

When the Marines of his squad were pinned down the grenadier of the point squad of the third platoon, Private First Class Albert Moguel, assaulted two North Vietnamese bunkers. The range was so close that he was unable to use his grenade launcher but instead used his .45 caliber pistol and proceeded to kill four of the enemy. Moguel was later seriously wounded during an enemy mortar barrage but he ignored his injuries and provided effective covering fire while the other wounded were being evacuated.

Private First Class Julio M. Miranda, another grenadier from the third platoon, rose up from his position and made a one man assault on a North Vietnamese redoubt with his M-79, giving cover for the rest of the squad to advance. He leaped from the trench and rushed a fortified hostile position and killed two enemy soldiers with his pistol. Despite sustaining a serious wound he continued to employ his M-79 against the North Vietnamese enabling his company to evacuate the wounded and dead.

Private First Class Donald R. Rash, also with the point squad of the third platoon, overcame three enemy positions with grenades and small arms fire. When the company was ordered to break contact he remained behind to provide effective suppressive fire which enabled the evacuation of the dead and wounded. Rash was killed when struck by mortar fragments from one of the enemy rounds.

The following members of Bravo-1-26 were killed in action on March 30, 1968:
Aldrich, David A. Cpl. 2288904; 20; Gibsonburg, Ohio
Anderson, David B. Pfc. 2376560; 18; Avoca, Iowa
Britt, Ted D. Pfc. 2378809; 19; Decatur, Georgia

Jones, Jimmie L. Pfc. 2353406; 20; Cordova, Alabama
Moore, Wayne P. LCpl. 2289947; 21; Plymouth, Massachusetts
Norman, Marion H. 1stLt. 0100563; A-1-13; 27; Houston, Texas
Rash, Donald R. Pfc. 2230103; 19; Pocahontas, Virginia
Ruiz, Jose Jr. Pfc. 229191619; New York, New York
Sanford, Albert R. Cpl.; 20; Russellville, Kentucky
Smith, Author C LCpl. 2329263; 20; Glen Allen, Alabama
Totten, Kenneth R. Jr. Pfc. 2320773; 18; Brewster, New York
Warren, Donald A. Cpl 2151640; 22; San Diego, California

Private First Class Kenneth R. Totten who was killed this day was also wounded during the ambush of February 25. Lance Corporal Doss B. Thrasher was wounded on both occasions.

For heroism on March 30, 1968 the Navy Cross was earned by Corporal Kenneth A. Korkow and Private First Class Donald R. Rash. The award to Rash was posthumous.

The Silver Star Medal was awarded to Corporal Samuel Boone, Jr., Private's First Class Ted D. Britt, Gene E. Edgar, Melvin D. Frazier, Albert Moguel, and Lonnie W. Morrison, Lance Corporal Jerome C. Foster, Captain Kenneth W. Pipes and First Lieutenant Peter W. Weiss. The award to Britt was posthumous.

The Bronze Star Medal was earned by Private's First Class David B. Anderson, Calvin E. Bright (2/25/68), Julio M. Miranda and Donald E. White, Hospitalman Third Class Thomas E. Casey, Navy Lieutenant's Dr. Harvey J. Demaagd and Ray W. Stubbe, First Lieutenant's John W. Dillon and Benjamin L. Long, Lance Corporal's Wayne P. Moore, Author C Smith, and Corporal Donald A. Warren. The awards to Anderson, Moore, Smith and Warren were posthumous. Demaagd would later earn a second Bronze Star.

Others who were wounded or were mentioned in the chapter are as follows:
Bellina, Johnny F. Cpl. 2228507; B-1-26; 20; Jacksonville, Florida
Blanchfield, Richard HM3 B812166; B-1-26; 27; San Francisco, California
Boone, Samuel Jr. Cpl. 2294933; B-1-26; 20; Baltimore, Maryland
Bright, Calvin E. Pfc. 2377388; B-1-26; 19; Detroit, Michigan
Brunk, Ricky M. Pfc. 2366105; B-1-26; 19; Milwaukee, Wisconsin
Bryant, William P. Pfc. 2363483; B-1-26; 19; Richmond, Virginia
Calzia, Frank V. HN B817950; B-1-26; 21; El Segundo, California
Castillo, Jose Sgt. H&S-1-26; 20; Fresno, California
Cicala, John A. Jr. HN B503201; B-1-26; 20; Detroit, Michigan
Corbett, John A. Pfc. 2321157; H&S-1-26; 19; Nyack, New York
Culpepper, Jeffrey E. Pfc. 2367117; B-1-26; 19; Cleveland, Ohio
Demaagd, Harvey J. Lt. 677479; H&S-1-26; 27; Grand Rapids, Michigan
Detrick, Thomas A. Pfc. 2365672; B-1-26; 19; Baltimore, Maryland
Dillon, John W. 1stLt. 0101863; B-1-26; 24; Oneida, New York
Durham, Pierce N. SFC Unknown; ArmySF; 31; Kingsport, Tennessee

Edgar, Gene E. Pfc. 2337757; B-1-26; 19; Pittsburgh, Pennsylvania
Estep, Michael H. LCpl. 2272405; B-1-26; 19; Knoxville, Tennessee
Foster, Jerome C. Pfc. 2145219; B-1-26; 21; Philadelphia, Pennsylvania
Frazier, Melvin D. Pfc. 2366536; B-1-26; 19; Jacksonville, Florida
Golab, Theodore H. Pfc. 2323366; A-1-13; 18; Chicago, Illinois
Hebron, James Pfc. 2283087; B-1-26; 20; Brooklyn, New York
Holliman, William D. Pfc. 2353663; B-1-26; 18; Montgomery, Alabama
Jayne, George W. LCpl. 2333426; B-1-26; 20; Albany, New York
Jewitt, James F. Pfc. 2367544; B-1-26; 19; Cleveland, Ohio
Korkow, Kenneth A. Cpl. 2258125; B-1-26; 21; Blunt, South Dakota
Lathrope, Wood W.J. Cpl 2303879; B-1-26; 20; New Orleans, Louisiana
Long, Benjamin L. 1stLt. 0100437; B-1-26; 29; Monmouth Illinois
Lownds, David E. Col. 015530; CO-KSCB; 47; Westerly, Rhode Island
Matzka, Robert E. Cpl. 2327516; B-1-26; 20; Detroit, Michigan
Melanson, James W. LCpl. 2290185; B-1-26; 20; Manchester, N.H.
Miranda, Julio M. Pfc. 2359901; B-1-26; 19; Chicago, Illinois
Moguel, Albert Pfc. 2384561; B-1-26; 19; Houston, Texas
Morrison, Lonnie W. Pfc. 2338568; B-1-26; 19; Raleigh, North Carolina
O'Hara, Michael E. LCpl. 2353070; B-1-26; 19; Indianapolis, Indiana
O'Toole, Thomas F. Jr. 1stLt. 093732; AO; 28; Boston, Massachusetts
Pipes, Kenneth W. Capt. 081285; B-1-26; 30; Fresno, California
Prendergast, Edward I. LCpl. 2347421; B-1-26; 19; Philadelphia, Pennsylvania
Pressley, Ronald L. Pfc. 2354193; B-1-26; 19; Charlotte, North Carolina
Quigley, Thomas N. Cpl. 2266063; B-1-26; 20; St. Louis, Missourik
Rayburn, Edward C. LCpl. 2315616; B-1-26; 20; Phoenix, Arizona
Santner, Donald F. Pfc. 2247543; B-1-26; 20; Milwaukee, Wisconsin
Schulp; Richard A. Pfc. 2276102; B-1-26; 20; Hammond, Indiana
Smith, Stanley R. Pfc. 2392125; B-1-26; 19; Albany, New York
Spencer, Ernest E. 1stLt. 092282; D-1-26; 25; Honolulu, Hawaii
Stubbe, Ray W. Lt. 667678; H&S-1-26; 30; Wauwatosa, Wisconsin
Thrasher, Doss B. LCpl. 2332814; B-1-26; 20; Oklahoma City, Oklahoma
Tracey, John E. Cpl. 2121590; B-1-26; 21; Philadelphia, Pennsylvania
Tretiakoff, Alexander Pfc. 2379763; B-1-26; 19; Oakland, Califironia
Weiss, Peter W. 2dLt. 0103881; B-1-26; 24; Bronx, New York
White, Donald E. Pfc. 2393851; B-1-26; 19; Boston, Massachusetts
Wolff, Daniel P. Pfc. 2361216; B-1-26; 19; Kansas City, Missouri
Wooddell, Gary B. Pfc. 2342613; B-1-26; 19; Columbus, Ohio
Woolman, William H. II LCpl. 2309729; B-1-26; 19; Philadelphia, Pennsylvania

★ ★ ★ ★

On April 16, 1968 First Battalion, Ninth Marines occupied the high ground of Hill 689, in the area of Khe Sanh, with three of its companies, Alpha, Charlie and Delta. On that date Alpha Company was ordered to conduct a search and clear operation to the southeast of Hill 689 in an area where they had made contact with the enemy a few days earlier. It was assumed that there were little or no enemy soldiers in the area. The mission was to secure the high ground and check it out for forward observer teams.

The initial prep fire came from the 105mm guns from a battery at the Khe Sanh air base. A short round nearly killed some of the men so a cease fire was called for after only a few rounds.

At 0900, Alpha Company, under a new skipper, Captain Henry D. Banks, the replacement for Captain Henry J.M. Radcliffe, moved out of the battalion perimeter with two reinforced platoons. The leader of the first platoon was First Lieutenant Francis B. Lovely, Jr. The second platoon was led by First Lieutenant Michael P. Hayden. The patrol consisted of 85 men and included two 60mm mortars.

The movement of the patrol was held up for a short while because of thick fog in the area. A few rounds, probably less than ten, of 81mm mortars were dropped on the area in advance of the patrol. Using the terrain as concealment they moved out. It was considered to be a normal, routine patrol.

"We didn't move fifteen feet when Lieutenant Hayden got hit and was killed. Suddenly they opened up and everybody hit the ground. We stayed that way for at least an hour."

Sergeant Paul J. Cogley
Alpha Company 1/9

"He's a preacher. He's over here fighting along with us."

Sgt. Thomas E. Dubroy
Charlie Company 1/9

Grateful acknowledgement is made to Khe Sanh Veterans, Inc. and to Chaplain Ray William Stubbe for permission to reprint previously published material from his fine book *Battalion of Kings. A tribute to our fallen brothers who died because of the battlefield of Khe Sanh, Vietnam.*

Also used were the awards citations of many of the participants, the rosters and diaries of the participating units, and information obtained from the National Personnel Records Center.

10

PREACHER

Two squads from the first platoon were sent to secure the ridgeline with one squad on either side. When the squad moving up the forward face of the slope reached the top a fire team was sent to check out the low ground on the western side of the ridgeline. This squad took sniper fire and the Platoon Sergeant Robert Rice was shot through the head and instantly killed. Two other Marines were wounded.

Private First Class Dennis A. Sykes was a radiomen assigned to Captain Banks that day. He makes these observations: "I was supposed to be with Lieutenant Hayden that day but the new company commander asked for a radio operator to accompany him. Having been in a fire fight a few days earlier with Banks, I wanted to be with Hayden, but to my chagrin I was assigned to Banks. When the ambush was triggered by the North Vietnamese and Captain Banks and I heard the fire fight begin, from the bottom of the hill, Banks dropped to one knee and banged his fist into his hand and with great euphoria said 'Contact!' At that moment I felt that he didn't understand or connect that contact meant death. He was so new that he was still clueless to the terror of close combat. In that moment Captain Banks did not begin to comprehend the trap we had fallen into, let alone the lethal consequences and mental anguish and trauma that his Marines were facing in the NVA's close combat trap. Banks was oblivious to the fact that the North Vietnamese now had us, we did not have them. Amazingly the Marine leadership had not learned from previous hill fights around Khe Sanh, as long as a year before, that the North Vietnamese knew exactly how to run their traps.

Lovely came down and asked Banks to come up to the top of the hill and see for himself what was going on. He told the Captain, 'You have men up there dying.' Banks never went up the hill that day. During previous engagements I saw many Marine officers demonstrate total disregard for their own safety. They would face lethal danger, at, or near the actual combat. The man that preceded Banks, Captain Radcliffe, was one of those Marines. If he had been there that day I am certain that the outcome would have been different."

Lovely deployed his two remaining squads to help the beleaguered group and then all came under heavy fire from an eight bunker complex that was set up in a horseshoe shape. They had walked into a classic ambush prepared by the North Vietnamese. Three or four Marines were killed instantly with head shots. Corporal Robert H. Littlefield appeared to have been hit by a round that went through his lower torso just above his left leg and almost hit Corporal David R. Ford. He walked from the hill with the assistance of a corpsman but died while waiting for a helicopter.

Sykes found Littlefield midway down the hill and he drank water from Sykes canteen. Sykes recalls, "When I came across Littlefield he did not appear to be in a lot of pain and was fully conscious and able to lift his head and drink from my canteen with some assistance from the corpsman. The blood on his jungle pants was drying already and there did not appear to be any fresh blood. I held his hand and told him he looked good and he was lucky that he was going to make it out of here. Less than an hour later someone reported that Littlefield was dead. This would have been at least three hours after he was wounded and by then, which was early afternoon, the medivac helicopters could no longer land due to the mortar fire and the ferocity of small arms fire from the North Vietnamese."

Banks then sent Hayden's second platoon to sweep the ridge area from which the first platoon was receiving fire. The North Vietnamese held their fire and when Hayden's men assaulted they were cut down by small arms fire. Hayden was shot and killed and Sergeant Homer D. Compton, standing about 15 feet behind him was also hit.

With the loss of Hayden and Rice, Lovely had to take over both platoons. Banks wanted to withdraw at this point but he was informed by Lovely that some of the wounded Marines were down within ten meters of the North Vietnamese bunkers. Alpha Company estimated that they had ten men killed and another twenty wounded. Among the wounded were Privates First Class Thomas A. Martinez, Thelbert R. Banks, Thomas E. Davis, Jr., with shell fragment wounds to the scrotum, right arm, face and left leg, and Charles T. Haase, who had the first and second fingers of his left hand taken off by a piece of shrapnel. Corporal John W. Slaughter, a radio man, was hit by shell fragments in the left leg and hip. Another radioman, Corporal Peter W. Johnson, suffered fragmentation wounds to his right shoulder and along his right flank.

Banks asked to break contact and requested assistance in extracting casualties. Alpha established a landing zone that could not be observed by the enemy to be used for removal of the casualties.

Private Raymond R. Dias, A machine gunner, moved to the aid of his fallen comrades and was wounded by mortar fragments during one of his rescue attempts. He refused medical aid and crawled into a bomb crater and pulled two injured Marines to safety. Dias returned to his weapon and provided accurate suppressive fire until it was destroyed by an enemy hand grenade. He then obtained an M-79 grenade launcher and engaged several enemy bunkers. On another occasion, Dias crawled dangerously close to an enemy sniper position and tried to knock it out with hand grenades. When he realized that Marine casualties were mounting he manned a radio and assisted in the recovery of the dead and wounded.

The corpsman for the second platoon, HN Richard R. Itczak, started to help Lance Corporal James A. Hunt who had been wounded in the abdomen. He recalls: "I had to put a dressing on his stomach before I took him back because his intestines would fall out if we didn't block them. After that I was informed of many WIA and KIA. At the time, I wasn't aware of the number because everybody was sufficiently excited. Nobody was really sure of themselves, then. There were three that were dead for sure, Hayden, the first squads M-79 man, Private First Class Haberman, and our platoon's right guide, Lance Corporal Herve Moise. Lance Corporal Terry J. Rampulla was also dead. This will give you some idea of the condition the platoon was in. I couldn't get to the dead and wounded who had fallen right by the machine gun site. I was going to make a try and the Marines that were left said they were dead so don't worry about them."

The badly wounded, Hunt had also been hit in the face, right arm and right hip by shell fragments. He would survive the battle and evacuation to the naval hospital at Yokosuka. From there he was flown to the Hospital at Camp Pendleton. In September he was transferred to the Veteran's Administration Hospital at Long Beach, California where on October 1, 1968 he died.

Throughout the ensuing ten hour engagement, Itczak worked tirelessly, administering first aid to the wounded and assisting in moving them down the hill to the landing zone for evacuation. Although wounded by fragments from an exploding mortar round and no longer able to walk, he remained at his forward position under fire and directed the treatment of the casualties. When he was eventually moved to the landing zone for his own evacuation he continued to comfort the wounded and assist in the administration of first aid. By his actions, Itczak probably saved the lives of several Marines.

Sykes remembers that Rampulla had only a few weeks left on his tour when he was killed. When 1/9 was over run at Con Thien he was wounded and played dead while a North Vietnamese soldier took his watch from his wrist. "I thought that he was indestructible. Somehow Rampulla was not going to die in Vietnam. Either he was protected by the divine or some other force that kept him safe. During the battle when it was reported that Ramps was killed, it just drained the last vestiges of hope for our own survival. It seemed that no one in Alpha 1/9 would live to finish out their tours of duty."

The Commanding Officer of the First Battalion, Ninth Marines Lieutenant Colonel John J.H. Cahill asked Banks if he needed assistance. Banks requested no fire because his men were intermingled with and pinned down by the North Vietnamese within a U-shaped bunker complex.

Cahill ordered Delta Company, under the command of Captain John W. Cargile, to assist Alpha. Second Lieutenant David W. Tuckwiller first platoon and First Lieutenant Kenneth J. Wilkinson's second platoon moved out and were in position to assault at 1430.

Charlie Company was assigned the mission to sweep the bunker complex from the rear and eliminate enemy resistance. Led by Captain Lawrence Himmer the company left Hill 689 with two platoons. Second Lieutenant David O. Carter led the first platoon and the second was led by Staff Sergeant Seth L. West.

At about 1600 the first platoon got on line and moved through a wooded portion of the hill until they were about twenty meters from the top. The second platoon also moved on line behind them. Carter was operating with only two squads and they ran into another bunker complex from which the enemy opened fire on his left flank. They immediately took four casualties and among them were Carter whose arm was shattered by an enemy round. He was also hit in the jaw by shrapnel, but continued to lead the assault.

Private First Class George Cimmerman and Carter's radioman Lance Corporal Michael J. Federowicz, were also hit. Corporal Roy D. Hurlburt and Privates First Class Larry G. Moore and Richard W. Johnston were killed.

Carter was to recall, "The only choice for Charlie Company was to get on line in classic Marine fashion with bayonets and whatever we had and come up the hill at them. When we got to the top of the hill they were everywhere. They were dug in bunkers. I was hit three times. I went up with 32 men and only eight of them didn't get hit. We did one helluva job up there. I think that the only thing that saved us was basic aggressiveness at this time."

Carter received a third wound, this time in the hip, when he jumped into the bunker complex. His life had been saved by Private First Class Eddie R. Pritchett who knocked Carter down, to keep him from being hit again, and placed him in a protected place. Although wounded himself, Pritchett continued the assault and jumped into a North Vietnamese emplacement and engaged them.

Fedorowicz, the radioman, already wounded in the hand, continued to the summit and was wounded several more times. He remained at Carter's side and relayed details of their situation through the action and later guided helicopters into the landing zone. Once the crest of the hill was taken Fedorowicz repeatedly ran across open terrain to rescue fallen Marines and carry them back to be evacuated. He refused to be evacuated until the twelve hours of fierce fighting subsided.

Carter's right flank squad, under the leadership of Corporal Jose Ruiz, had advanced about 75 meters and taken four casualties. Ruiz was killed as was Private First Class Wendell Guillory and, a new man, Lance Corporal Smith was wounded. When the platoon leaders were wounded, Ruiz took charge and moved through enemy fire to render first aid to the wounded.

Corporal John S. Kocsis later wrote, "Ruiz got down and gave each and every one of them the best medical care he could. After this he and his men continued on the assault. There he gave his life by jumping in front of a wounded Marine to save his life."

Lance Corporal William T. Parr moved across the fire swept terrain and administered first aid to the casualties and while completely exposed to the accurate enemy fire assisted several badly wounded Marines to the helicopter landing zone.

Himmer then directed West to come forward and pass with his platoon through Carter's group and continue the assault. This done, they cleared the crest and took some fire from the front but the left flank came under serious assault. They received a heavy volume of enemy fire and Private First Class Robert D. Cicio was killed. Corporal Nathaniel E. Jackson and Private First Class George Panykaninec were seriously wounded. Jackson would later die from his injuries.

West ran to a wounded Marine who was lying in an exposed area and as he was administering first aid he was hit and mortally wounded. Sergeant Leslie P. Hagara, the platoon sergeant for the second platoon was also killed. He had moved forward in an attempt to locate the North Vietnamese and took a round in the face.

When West was killed, Corporal Kocsis became the platoon commander. From this squad, Private First Class Aubrey McClelland was killed and Corporals Earl R. Best and Jose A. Kosaka, and Lance Corporal Parris E. Joyce and Private First Class Lawrence E. Hutson were wounded. At first, Corporal David G. Redenius was missing, but it was later determined that he had been killed. Reacting instantly, Corporal Melvin R. Curry took charge of the platoon and deployed the Marines into a defensive perimeter and directed the delivery of a heavy volume of fire into the enemy redoubt. He later supervised the evacuation of his platoon's casualties.

The attack by Charlie Company allowed Alpha Company to extract all of their wounded and most of their dead and bring them to a collection point at the bottom of a steep hill protected from mortar fire. They started the evacuation at about 1600 and worked until around 2400 getting them all out. Lance Corporal Gerald W. Post, a FAC with Alpha, used a cigarette lighter to guide the helicopters after darkness set in. When Banks was wounded about 1600, Lovely assumed command of the company.

As Charlie Company pressed the attack, the third squad of its second platoon made the best progress. Advancing on the right flank they moved about 100 meters across the top of the hill. This good advance would later prove tragic. The squad leader, Corporal Hubert H. Hunnicutt, was wounded as was Private First Class Darwin G. Rice. Killed were Privates First Class Mauro Martinez and Nathaniel M. Williams. Martinez was killed when he used his body as a shield to keep Corporal William P.

Holland, already suffering from a fragmentation wound to the head, from being hit again.

Captain Himmer was also hit and knocked down. Kocsis recalls that, "We tried to take care of him and bandage him up. But we were pinned down by a sniper that I would say was about six feet in front of us. I told Captain Himmer not to get up but he did, and started walking right on top of the sniper who shot him. He fell right beside a bomb crater. The Captain sat right straight back up and looked over in my direction and I saw that the whole right side of his face was torn up. I tried to get him at this point, but I couldn't because of the sniper. This was about 1800. Then me and two more Marines, with a stretcher, went up there and tried to get him. The sniper popped up and killed one of ours and put some rounds in the back of Captain Himmer."

At about 1630, Delta Company moved over the crest of the hill and tried to locate Charlie Company without success. Their movement in any forward direction was limited by accurate enemy rifle fire and mortars. The senior corpsman of Delta Company, HM3 William G. Gessner, helped Lance Corporal James C. Lee who had been hit in the hand, and then helped HM3 Larry D. Houtz in the treatment of badly wounded Private First Class Floyd R. Reno, who had been hit in the right side of his chest by a gunshot. The erstwhile corpsman, Gessner, would later be wounded himself.

Cargile remembers, "Already I had five KIA of my own and I could see that I would have my hands full getting my own casualties out, let alone A and C. We were receiving the snipers fire from two to three directions. Still we could not pinpoint the source. The sniper or snipers were very good. Four of my five KIA's had been shot in the head. To continue to charge an unseen enemy would be foolish. So, with darkness approaching, I consolidated and attempted to retrieve all the dead."

The Marines were able, by slowly crawling, to pull back the bodies of three of those killed but had to leave the body of a fourth, Corporal Robert F. Owens. The bodies retrieved were those of Corporals Alfred L. LeBlanc and William C. Averitte and Lance Corporal Hugh E. Shavelin. All died from gunshot wounds to the head. One of the wounded, Private First Class William J. Dolan, would die the next day, also from a gunshot wound to the head. Private First Class Marvin G. Rush, last seen helping Reno reach the landing zone was missing and later reported killed.

At 1830, First Lieutenant William C. Connelly was told to extract all wounded and dead and pull back. By 1915 this was accomplished with the exception of Hunnicutt who was trapped in a crater in front of the lines.

Lance Corporal Ross M. Kasminoff and Private First Class Antonio Borjas were in another bomb crater setting up a base of fire when they heard Hunnicutt yelling to them from about 125 meters away. With him was Martinez, two dead Marines and another two who were wounded, and unable to move. Hunnicutt did not want to abandon the wounded.

At 1940, Captain Charles B. Hartzell told Connelly to withdraw and bring the eleven wounded Marines with him. This left Hunnicutt and those with him alone. One of the wounded, Redenius, who was shot in the head was bleeding profusely and was going into convulsions.

Hunnicutt remembers, "He must have lost two quarts of blood in about ten seconds. I got him down in the bottom of the crater and put his head up a little bit, sort of stop the bleeding, but then he started throwing up dark blood. I didn't know what would be the best thing to do in order to get these men out of here."

Hunnicutt waited until dark and told Martinez to go. "If I had been physically able we could have dragged both men down and gone down to the gulch and gone back up to the machine gun position behind us, however, I couldn't pull with my arm. I was weak. Martinez left and then, later, I realized that the North Vietnamese must have got him. With the two wounded now almost gone I decided to crawl out of the crater and came upon Captain Himmer who was laying on his side. He had been shot in the mouth, arm and neck and apparently had lost a lot of blood because he could barely speak. He called to me and said, 'Marine, help me.' I said, 'Skipper, we got to get out of this place, but we can't move right now. We'll just wait about another five or ten minutes.' He said, 'OK.' About that time I moved away from him a little and an AK opened up on me and hit me in the leg, hand and elbow. I managed to pull a grenade and toss it, but by then I was too weak to do much more. I played dead for about five hours. Some NVA passed by and kicked my canteen. I managed to get closer to Captain Himmer."

At first light on April 17 Hunnicutt yelled for Kasminoff but went unanswered. At 0800 he yelled as loud as he could and someone answered and said that help was on the way. Hunnicutt and Himmer then slid downhill until they reached a ledge that dropped off sharply into a valley.

Hunnicutt recalls, "Captain Himmer's face was covered with blood and his legs were very weak. He was much, much heavier than me and I was unable to carry him. I heard a Marine yell, 'Preacher, is that you' and I replied, 'Can you come and help?' They answered, 'Yeah, we're coming.' But no one came that day. That night an air strike dropped ordnance almost on top of us."

On the morning of 18 April, Hunnicutt managed to move the company commander to a covered position and then tried to seek assistance but, overcome by weakness from loss of blood, fell into a gully where he lay for several hours.

About 1200, an Army helicopter spotted Hunnicutt and dropped a red smoke on him to mark his position. He remembers, "It scared me to death. I thought that they were going to blow me away. They set down and snatched me away. When we reached the medical facilities they asked me if there were any more alive and then asked if Captain Himmer had been killed. All I could say was that, 'He couldn't be dead. I had him with me for two days.'"

In the very late hours of 16 April all of the surviving members of the three companies returned to the 1/9 perimeter on Hill 689. The company commanders immediately began to assess their casualties. Companies A and D were able to provide a fairly accurate report, but Charlie Company, who had been hit the hardest, had a difficult time accounting for their personnel. They did not make it back to the perimeter until about 0300 of April 17. What happened on 25 February to Bravo 1/26 had happened again.

Connelly remembers, "We were forced to do something we had never done before---just leave bodies out, to report people missing. I don't feel that this was through carelessness of the individuals. It was simply that there weren't enough walking bodies left to carry the limp bodies out."

Dennis Sykes painfully recalls, "The dense fog began to descend again. I began to hear the calls for help from those left behind. In the fog you could hear sound clearly from a distance. It was like a microphone. I remember clearly the overlap of voices. I realized that there had to be more than one Marine left behind and that one of the voices might belong to Littlefield. As I crouched down in my fighting hole I could clearly hear the pleading of the wounded. 'Mama, help me, Mama!' and the words, 'Please help me!' over and over. Also repeated was the phrase, 'Please don't leave me out here to die!' In that moment I knew that this would be the most agonizing and tormenting feeling that I would ever experience. I crouched in my hole and pulled my poncho over my head in an attempt to muffle the pleading voices of those left behind. A rescue attempt would be mass suicide. Sometime After midnight the voices went silent."

Alpha Company listed seven Marines missing, Delta two, and Charlie twenty five. Plans were made to recover the bodies on April 18, but higher headquarters did not allow the battalion to do so.

About 1200 on 17 April one of the artillery observers flying over the area of contact spotted a Marine in a crater that appeared to be alive. A rescue team, consisting of volunteers, was flown in and landed on top of the enemy bunker complex and immediately killed four North Vietnamese soldiers trying to escape.

Captain Hartzell remembers what they found, "They went around the left side of the helicopter and checked the bomb crater. There were five Marines in there. Two of them had been partially decapitated, one of them disemboweled and the other two, they checked out, were dead. After checking these five, they got back on the helicopter, and the Marine that was alive was off to the right of the chopper."

An Army helicopter from the First Air Cavalry flew in and rescued the Marine thinking that it was Hunnicutt, but it was not. The man rescued was, the badly wounded, Panykaninec.

The Second Battalion, Third Marines was given the task of recovering the missing bodies. They took Hill 689 at a terrible cost but by mid afternoon of 22 April they had recovered the bodies of 38 of the missing Marines.

NOTES:
The names of those killed on April 16, 1968 are as follows:
Averitte, William C. Cpl. 2124858; D-1-9; 23; Dallas, Texas
Boynton, Charles B. Jr. HM3 B207688; H&S-1-9; 20; Baltimore, Maryland
Brown, David D. Jr. Pfc. 2389168; C-1-9; 18 Wrangell, Arkansas
Christian, Daniel K. LCpl. 2137269; A-1-9; 21; Wadsworth, Ohio
Cicio, Robert D. Pfc. 2360699; C-1-9; 20; Farmingdale, New York
Craig, Bruce K. Pfc. A-1-9; 20; Escanaba, Michigan
Craun, Gale E. Pfc. 2153597; C-1-9; 20; Portland, Oregon
Dolan, William J. Pfc. 2346026; D-1-9; 19; Hartford, Connecticut
Guillory, Wendell Pfc. 2364082; C-1-9; 20, Church Point, Louisiana
Haberman, David Pfc. 2367366; A-1-9; 20; Cleveland, Ohio
Hagara, Leslie P. Sgt. 2133344; A-1-9; 21; Saltsburg, Pennsylvania
Hayden, Michael P. 1stLt. 0101969; A-1-9; 23; Detroit, Michigan
Himmer, Lawrence Capt. 083022; C-1-9; 29; Chula Vista, California
Hinkle, Jack L. Pfc. 2404238; D-1-9; 22; Lyons, Ohio
Hurlbert, Roy D. Cpl. 2373656; C-1-9; 22; Wheatridge, Colorado
Jackson, Nathaniel E.L. Cpl. 2206955; C-1-9; 21; Georgetown, South Carolina
Johnston, Dennis N. HN B401600; H&S-1-9; 20; McDonald, Ohio
Johnston, Richard W. Pfc. 2354664; C-1-9; 20; Lock Haven, Pennsylvania
Kilgore, Danny R. Pfc. 2394610; C-1-9; 19; Myrtle Point, Oregon
LeBlanc, Alfred L. 2303764; D-1-9; 20; Ponchatoula, Louisiana
Littlefield, Robert H. Cpl. 2223201; A-1-9; 19; Birmingham, Alabama
Martinez, Mauro Pfc. 2373680; C-1-9; 20; Hudson, Colorado
McClelland, Aubrey D. Pfc. 2335671; C-1-9; 20; Dallas, Texas
Medeiros, William C. LCpl. 2378126; A-1-9; 22; New Bedford, Mass
Moise, Herve J. Pfc. 2410468; A-1-9; 20; Los Angeles, California
Moore, Larry G. Pfc. 2326450; C-1-9; 19; Owensboro, Kentucky
Owens, Robert F. Cpl. 2210453; D-1-9; 20; Wheatland, Indiana
Owens, Timothy E. Pfc. 2377785; C-1-9; 19; Kansas City, Kansas
Rampulla, Terry J. LCpl. 2283825; A-1-9; 21; Easton, Pennsylvania
Redenius, David G. Cpl. 2265922; C-1-9; 20; Plymouth, Illinois
Rice, Robert Sgt. 1885428; A-1-9; 26; Flushing, New York
Ruiz, Jose Cpl. 2282540; C-1-9; 24; New York, New York
Rush, Marvin G. Pfc. 2212097; D-1-9; 21; Memphis, Tennessee
Schavelin, Hugh E. LCpl. 2309307; D-1-9; 20; Norma, New Jersey
Sweet, Jerry A. Pfc. 2391151; C-1-9; 20; Lebanon Springs, New York
Wells, Robert J. Jr. Pfc. 2391171; A-1-9; 18; Schenectady, New York
West, Seth L. Jr. SSgt. 2003650; C-1-9; 24; Kinston, North Carolina
Williams, Nathaniel M. Pfc. 2241057; C-1-9; 19; Genevia, Arkansas
Wilson, Wilmer D. HN B703161; H&S-1-9; 20; Sweetwater, Texas
Wright, Robert Cpl. 2320317; C-1-9; 20; New York, New York

Wymer, William W. Pfc. 2367520; A-1-9; 19; Ravenna, Ohio

Of the Marines killed in action that day, three, Privates First Class Danny A. Kilgore and David D. Brown and Private Robert H. Collins, had arrived at Khe Sanh the day before, April 15, 1968.

For heroism on April 16, 1968 the Navy Cross was earned by Corporal Hubert H. Hunnicutt.

The Silver Star was awarded to First Lieutenant William C. Connelly, Private Raymond R. Dias, Lance Corporals Michael J. Federowicz and William T. Parr, and Corporal Jose Ruiz.

Second Lieutenants David O. Carter and Kenneth J. Wilkinson, Corporals Melvin R. Curry, David R. Ford (2/8/68) and John S. Kocsis, First Lieutenants Michael P. Hayden and Francis B. Lovely, Privates First Class Mauro Martinez and Eddie R. Pritchett, Lance Corporal Gerald W. Post, and Staff Sergeant Seth L. West, Jr. earned the Bronze Star.

Among the wounded that day were corpsmen HN Joseph R. Cloutier, HN William R. Harper and HN Paul D. Kelly.

Sergeant Leslie P. Hagara had extended his tour in Vietnam so that his son could have an operation. He had only a little amount of time left to do, but the surgery his son needed was expensive and if he extended, the military would take care of it.

Captain Henry D. Banks first enlisted in the Marine Corps in 1957 and from the very beginning planned to make the Marine Corps his career. He served as an enlisted man in the reserves as a member of the Ninth Engineer Company of Phoenix, Arizona until 1961 advancing to the rank of Corporal. Banks then applied for and was accepted into Officer Candidate School and then the Basic School at Quantico. He was commissioned in December of 1961.

Banks earned the Navy Commendation Medal in Vietnam while serving as the executive officer and eventually the commanding officer of Company C, Third Motor Transport Battalion from May 7 to August 21, 1965.

He was serving a second tour in Vietnam when he was wounded during the action that occurred on April 16, 1968. Banks was evacuated and eventually ended up at Tripler Military Hospital in Hawaii. Upon recovery he returned to Vietnam and on July 12, 1968 assumed command of Company A, First Battalion, Fourth Marines.

Banks was killed in action on July 14, 1968 as he led his company's assault on a fortified hill top.

David Sipperly wrote: "Captain Banks gave his life doing his duty as Commanding Officer, Company A, First Battalion, Fourth Marines. I was with Captain Banks when he died. He was a strong leader and a good "Skipper." He was helping direct the company's assault on a fortified NVA hill top, not from the rear, but up front with the lead platoon when he died."

Sykes had a totally different recollection of Banks: "He was not an effective leader or decision maker. I feel that Banks was incapable of exercising common sense combat decisions. I do not contend lightly that his incompetence unnecessarily killed Marines. He did not go anywhere near the fight when Alpha Company took Hill 471 0n 4 April. Once again he was well away from the close fighting. I remember him asking on the radio what was going on up there while Hayden and Lovely were in the fight at the top with their Marines. The North Vietnamese had nowhere to hide because 471 was bombed heavily in the days prior to the assault, unlike Hill 689 on 16 April. So when Alpha Company got on line and swept up to the top the line assault tactic was successful."

Radcliffe was named for his grandfather, Henry James MacDonald, and was known by his nickname, "Mac." He was born in 1939 when his father was over fifty years old. His father, William Martin Radcliffe had served in World War I, where like his son was to do in Vietnam, twice earned the Silver Star for heroism in action. In the Globe and Laurel pub at Quantico there is a corner that honors him. Born in 1887, he first enlisted in the Marine Corps in 1904 as a seventeen year old, and retired as a Captain in 1935 with thirty years in the Marine Corps, and before Mac was born.

Radcliffe had commanded Alpha Company four or five months by the time they arrived at Khe Sanh. On 8 February he chose Hayden's platoon to send as a relief force for the embattled Marines on Hill 64. Radcliffe accompanied the force when it moved out and they encountered heavy fire from the North Vietnamese. He aggressively led the platoon, knocking out two enemy positions with hand grenades. When one of the enemy grenades landed between him and another Marine, Radcliffe picked it up and tossed it back killing a North Vietnamese soldier.

On Lieutenant Colonel John F. Mitchell's Bronze Star Medal recommendation for Radcliffe, Major General Rathvon M. Tompkins wrote: "In my opinion deserves the Silver Star. I am intimately familiar with this action and the terrain on which it took place. Had it not been for Radcliffe we never would have thrown the enemy back out of that position."

In a letter written to Radcliffe on April 6, 2005 Sykes wrote: "Without hesitation or reservation I can attest that you are the best officer I have ever seen. I was close to you several times in close combat situations. You always did the right thing for your men even when the risks were potentially, and likely, lethal to you personally. You demonstrated enormous personal courage by exposing yourself to almost certain death or injury in order to instruct your men under fire. I am delighted that you survived. Many officers that followed you simply couldn't fill your boots. I often think how April 16, 1968 might have been different if you had directed that engagement. In any case, I think that you are the best that comes out of the Marine Corps and our country. And you sir are the best warrior that I have ever seen. I am very proud to have served, in combat, under your command."

Dennis Sykes wrote, "I knew Michael Hayden about as well as you could allow yourself to know another Marine when you are engaged in sustained combat. I served in his platoon as a rifleman and later as his radio operator. I enlisted in the Marine Corps at nineteen. When our Battalion deployed to Khe Sanh in January 1968 I was a PFC rifleman with six months in the Corps. Two months into the Khe Sanh siege Lieutenant Hayden reassigned me from my rifleman position to platoon radio operator and as platoon radio operator I had frequent contact with him. In the bunkers that we dug at Khe Sanh, when I was not on line duty or a listening post outside the wire, I would be invited to the platoon command bunker with the Lieutenant and Gunny Lee.

Michael Hayden was a recent graduate of the University of Chicago and had just married a girl from England. He was a proud Irishmen who did not want his men to know his middle name was Pym. Hayden was like the rest of us except he had absolute authority over life and death decisions. He would occasionally talk about how much he missed his bride and life back in Chicago. Underneath we were all the same in our desire to be safe and to live to return home. Michael knew well the lethal danger we all faced but he took that awesome responsibility of giving life and death orders with utter seriousness and as much compassion as war permits. Michael knew that combat meant some would die and he was willing to accept the same risks as his Marines.

I was a rifleman in Hayden's second platoon of Alpha Company on 8 February when our platoon recaptured Hill 64 from the North Vietnamese. Hill 64 had been almost completely overrun the night before and most of Alpha Company's first platoon including Second Lieutenant Terence Roach was killed. The battle scene the next morning revealed the ferociousness of the bunker to bunker fighting. The wounded enemy soldiers left behind prayed throughout their capture. It was exceedingly gruesome with the bodies of dead American and North Vietnamese soldiers scattered across the hill top. Body parts were all around and in some cases the only way you knew a lone leg was not a Marine's was it had a sneaker on the foot instead of a jungle boot.

I began serving as Lieutenant Hayden's radio operator on 24 March when two squads of our platoon tried to take out the 50 caliber North Vietnamese gun emplacement near the end of the Khe Sanh airstrip. We were ambushed by the enemy before the objective could be taken. We were able to get the wounded out and some of our dead but had to leave some of those killed behind because if the intensity of rifle fire from the enemy. On 4 April I again served as his radio operator when Alpha Company took Hill 441 from the North Vietnamese. Hill 441 was located several kilometers outside the Khe Sanh perimeter. I was with him again on a patrol on 13 April when we made contact with a small contingent of the enemy near the site of a major battle to follow a few days later on 16

April. Each of these close in fire fights involved the death and wounding of Marines under Lieutenant Hayden's command. I witnessed firsthand his actions and bravery in close combat situations. In every fire fight I was in with Hayden, he would move to where he could see and know what the ground conditions were so that he could command it. Not all officers did that. Some would attempt to direct a fire fight from a safer distance. Not Michael.

On 16 April when elements of the second platoon walked into the carefully orchestrated NVA ambush Second Lieutenant Michael Pym Hayden was one of the first killed by a bullet from an AK-47. Those who were with him said he died quickly "leading" his Marines. He was an exemplary Marine Officer. He was an authentic American hero."

Sykes remembers Lovely as an Ivy Leaguer, a graduate from Dartmouth that was fresh out of OCS, and the Basic School. "He was about 22 or 23 with red hair and a very freckled face and looked to be about eighteen. Lovely like Hayden and Radcliffe, was prepared to do his duty even if it meant certain death." He was especially good friends with Hayden. At Khe Sanh two of his officer peers, Hayden and Roach, were killed.

It was Marine Corps policy during the Vietnam War that officers served six months in a combat platoon and were then rotated to a rear area for the rest of their tour. So the officer corps because of deaths, injury and rotation almost always consisted of new personnel with no combat experience.

Major General Rathvon McClure Tompkins joined the Marine Corps in 1939 and served for 32 years retiring in 1971. He was highly decorated for his heroism in three wars. During World War II he earned the Navy Cross on June 17, 1944 as the Commanding Officer of the First Battalion, Twenty Ninth Marines on the island of Saipan. He also earned the Silver Star during the battle of Tarawa and the Bronze Star on Guadalcanal. He passed away on September 17, 1999.

Private First Class Danny A. Maguire wrote a tribute to Hunnicutt, at the time thinking that he had been killed. He observes, "This letter certainly won't win any Pulitzer Prize but I think you'll see how much affection I had for the man. Guys like him helped the rest of us make it through. For the life of me I can't remember when I wrote it. I think it was while I was in the hospital. I also don't have a clue as to how I would have known that we occupied Hill 471. I couldn't tell you who the others were that I was evacuated with. I remember that two men beside me were killed and that two others were hit, one very bad and one very minor. The letter follows:

Corporal Hubert Hunnicutt, USMC, known to the men around him only as "Preacher." From a small town somewhere in Georgia came this warm and wonderful human being. His plans included becoming a minister and living in the great southern city of Atlanta and, of course rooting for the Atlanta Braves baseball team.

I first met Preacher on a lonely hill in Vietnam about a week after I got there. We were introduced and he greeted me with a warm handshake and a friendly smile. I outweighed him by nearly forty pounds and stood a head taller than he, but there was no doubt in my mind that I had met a man with deep inner strength.

He told me that he held lay services occasionally and asked me if I'd attend. He also asked my religion and the very appropriate question, "Are you a Christian?"

"I try to be." I said.

I remember one particular service where he spoke about our believing in God.

"Gentlemen, we could very easily die here and I'd hate to think any of us might spend eternity in Hell."

He was a God fearing man, one you might not think would fit in properly in war.

One night a group of us were listening to country and western music, his favorite, when someone asked, "Why'd you join the Corps, Preach?"

From somewhere in the dark he answered, "To fight for what I believe in." He was that type of a man.

Late January found us at Khe Sanh for the start of the famous seventy two day siege. During those anxious and trying times he brought me peace and faith. His small services now and then, and his joyous personality, and undying love of God helped me keep my faith when it faltered as it often did. Many are the times we laughed and joked over coffee and cocoa, and I will never forget his good humor and especially the wild tales he told about his high school football days.

Then in early April we moved out to take the offensive.

"I hope the hills are empty, Preach. Maybe the bombs got them all."

But the bombs didn't get them all and as Charlie Company moved up from the rear to occupy Hill 471 I saw him there comforting a badly wounded Marine. This man's greatest gift was compassion.

There were more hills to take and for awhile we had hopes the enemy had withdrawn. Our hopes grew as we took the next two hills unopposed, but after we had set in, our hopes were dashed when the enemy mortared us and we sustained heavy casualties. Men were killed and wounded trying to get their buddies onto the Medevac choppers. Yet through the hell of it all preacher was there, comforting and carrying the wounded.

I was wounded early on a hot April morning and he was there to help carry me out. Four of us had to be evacuated. Two of my friends were hurt much worse than I and one was in great pain.

He'd ask where the choppers were and Preacher would comfort him.

"They're coming now, chap, I can see them. Just hang on; you're going to be alright now."

The first chopper came and while it was on the ground a round came in somewhere.

"Where'd that hit?" I asked him. "Way up on the hill, but they'll try to adjust." He answered.

The second chopper came in and I buried my face in the poncho liner I was being carried on. I felt myself being lifted into the big bird and being drug forward. Above the roar of the whirling blades I heard him shout.

"Ya'll take care now, Moose."

Grinning, I turned my head and yelled. "Get the hell out of here, Preach!"

Then he was gone and we hit the air. Eventually I was to end up in an Air Force hospital at Cam Ranh Bay.

That was the last time I saw Preacher. I was told at the hospital that he had been killed somewhere in the lonely hills of the DMZ. But for those of us who mourn his death, I can only offer this small story.

I remember early in our days at Khe Sanh when four men were killed about twenty yard's from me. One of them was Preacher's best friend. Sometime later he said, "He was my best friend, but I don't worry about him now. He was a Christian."

And after my initial pain and sorrow, I no longer worry about Hubert Hunnicutt. I'll always remember him when I hear country and western music or see the Atlanta Braves play. Yet I'll know that the world is a better place because he was a Christian.

After his rescue Hunnicutt was recuperating from his wounds when he was told that he would receive the Medal of Honor. He had been transferred to a hospital in Virginia and issued a set of dress blues to be worn at the presentation, but it was not to be. It was downgraded to the Navy Cross by a Navy Admiral.

Corporal Robert H. Littlefield older brothers were already in the Marine Corps when he joined. He had been married for only three months when he received orders to Vietnam. His older brother Earl received his orders on the same day. His middle brother, Ernie, had only been back from a tour in Vietnam for about two weeks when Robert left. Littlefield only had 45 days to go when he was killed.

Lieutenant Colonel John J.H. Cahill, was a "Mustang", and was known as "Blacky" throughout the Marine Corps. He was an enlisted Marine during World War II and after the war was commissioned through a meritorious promotion program as an officer in 1947. Cahill served in Korea as a platoon leader and executive officer of Golf Company, Third Battalion, Fifth Marines. As a Second Lieutenant he earned the Silver Star in early August, 1950, during the fighting at the Pusan Perimeter.

Cahill was a staff officer until requesting duty in Vietnam and performed similar duties until assigned as the commanding officer of the First Battalion, Ninth Marines. He suffered fragmentation wounds to the left buttock and right arm on April 16.

Daniel P. Baker, a corpsman assigned to Charlie 1/9, writes about Second Lieutenant David O. Carter: "He was helping evacuation procedures on his wounded men when he was finally ordered to go to the landing zone himself. He could not find the landing zone, so he returned to Hill 689 on foot, alone. When he reached Hill 689 he was in shock from loss of blood and the exertion of the trip. He would not succumb to medical treatment until he had reported to his superior officer about the condition of his men."

Staff Sergeant Seth L. West was married and had two sons. His remains were not recovered until 22 April. They were already badly decomposed and the mortuary at DaNang was unable to determine the specific cause of death due to the poor condition of the remains.

Others who were either wounded or mentioned in the chapter are as follows:
Baker, Daniel P. HM3 B808960; C-1-9; 20; Olympia, Washington
Banks, Henry D. Capt. 082380; A-1-9; 29; Prescott, Arizona
Banks, Thelbert R. Pfc. 2330623; A-1-9; 19; Dallas, Texas
Best, Earl R. Cpl. 2231699; C-1-9; 20; Butte, Montana
Borjas, Antonio Pfc. 2379116; C-1-9; 19; St. Louis, Missouri
Cahill, John J.H. LtCol. 049898; CO-1-9; 43; Wallaston, Massachusetts
Cargile, John W. Capt. 083154; D-1-9; 28; Tulsa, Oklahoma
Carter, David O. 2dLt. 0103397; C-1-9; 22; Los Angeles, California
Chippero, James A. Pfc. 2399408; A-1-9; 20; Unknown
Cimmerman, George Pfc. 2329941; C-1-9; 20; Unknown
Cloutier, Joseph R. HN B511274; H&S-1-9; 20; Detroit, Michigan
Cogley, Paul J. Sgt. 2250390; A-1-9; 20; Chicago, Illinois
Compton, Homer D. Sgt. 2215065; A-1-9; 21; Raleigh, North Carolina
Connelly, William C. 1stLt. 0101824; C-1-9; 25; Cleveland, Ohio
Curry, Melvin R. Cpl. 2137702; C-1-9; 20; Pittsburgh, Pennsylvania
Davis, Thomas E. Jr. Pfc. 2356079; A-1-9; 19; Little Rock, Arkansas
Dias, Raymond R. III Pvt. 2190084; A-1-9; 20; Honolulu, Hawaii
Doney, Basil C. Pfc. 2427250; A-1-9; 18; Kansas City, Missouri
Dubroy, Thomas E. Sgt. 2137266; C-1-9; 21; Cleveland, Ohio
Federowicz, Michael J. LCpl. 2347563; C-1-9; 20; Philadelphia, Pennsylvania
Ford, David R. Cpl. 2309699; A-1-9; 20; Philadelphia, Pennsylvania
Gaspard, Robert J. HM2 1386211; C-1-9; 20; Abbeville, Louisiana
Gessner, William G. HM3 9199490; D-1-9; 21; Kent, Rhode Island
Haase, Charles T. Pfc. 2379208; A-1-9; 19; St. Louis, Missouri
Harper, William R. HN B202168; H&S-1-9; 20; Hopkinsville, Kentucky
Hartzell, Charles B. Capt. 081176; H&S-1-9; 28; Overland, Missouri

Holland, William P. Cpl. 2182234; C-1-9; 21; Boston, Massachusetts
Houtz, Larry D. B207341; HM3; D-1-9; 20; Washington, DC
Hunnicutt, Hubert H. III Cpl 2277992; C-1-9; 20; Suwanee, Georgia
Hunt, James A. LCpl. 2235251; A-1-9; 20; El Monte, California
Huskey, James H. Pfc. 2363681; C-1-9; 22; Albany, New York
Hutson, Lawrence A. Pfc. 2399598; C-1-9; 19; Philadelphia, Pennsylvania
Itczak, Richard R. HN B109061; A-1-9; 20; Hartford, Connecticut
Johnson, Peter W. Cpl. 2259923; A-1-9; 20; Minneapolis, Minnesota
Joyce, Parris E. LCpl. 2369218; C-1-9; 19; Cincinnati, Ohio
Kasminoff, Ross M. LCpl. 2320567; C-1-9; 20; New York, New York
Kelly, Paul D. HN B323853; C-1-9; 20; Moore, North Carolina
Kocsis, John S. Cpl. 2312572; C-1-9; 20; Richmond, Virginia
Kosaka, Jose A. Cpl. 2190115; C-1-9; 20; San Diego, California
Lee, James C. LCpl. 2257891; D-1-9; 20; Richmond, Virginia
Logan, Monty R. LCpl. 2272413; D-1-9; 19; Knoxville, Tennessee
Lovely, Francis B. Jr. 1stLt. 0103109; A-1-9; 22; Boston, Massachusetts
Maguire, Danny A. Pfc. 2264415; C-1-9; 19; New Cumberland, Pennsylvania
Martinez, Thomas A. Pfc. 2409595; A-1-9; 19; Denver, Colorado
McClain, Larry L. Pfc. 2369488; C-1-9; 19; Cincinnati, Ohio
Mitchell, John F. LtCol. 044003; CO-1-9; 39; Portland, Oregon
Nation, Roy G. Pfc. 2378554; C-1-9; 19; Atlanta, Georgia
Panykaninec, George Pfc. 2291634; C-1-9; 19; Brooklyn, New York
Parr, William T. LCpl. 2358150; C-1-9; 20; Syracuse, New York
Porter, John E. Jr. GySgt. 643477; A-1-9; 39; Columbia, South Carolina
Post, Gerald W. LCpl. 2312154; A-1-9; 20; Omaha, Nebraska
Pritchett, Eddie R. Pfc. 2367818; C-1-9; 19; Cleveland, Ohio
Reno, Floyd R. Pfc. 2385874; D-1-9; 19; Detroit, Michigan
Rice, Darwin G. Pfc. 2275591; C-1-9; 20; Boise, Idaho
Sipperly, David W. 1stLt. 0104776; A-1-4; 22; New York, New York
Slaughter, John W. Cpl. 2267244; A-1-9; 20; Indianapolis, Indiana
Sykes, Dennis Pfc. 2367407; A-1-9; 19; Cleveland, Ohio
Tompkins, Rathvon M. MajGen. 05269; CG-3-MarDiv; 55; Boulder, Colorado
Tuckwiller, David W. 2dLt. 0103852; D-1-9; 22; Morgantown, West Virginia

✯ ✯ ✯ ✯

Con Thien is located in the Gio Linh District of South Vietnam's most northern province, Quang Tri. It is about two miles south of the DMZ and about twelve miles from the coast. Con Thien is a small hill that rises to a height of about 160 meters in elevation that is surrounded by lowland. When the French occupied Vietnam they valued it as an observation post. Its Vietnamese name, Nui Con Thien, translates into English as "the hill of angels."

On June 6, 1968 James N. Kaylor was a 19 year old Private First Class assigned as a grenadier with the Second squad. Second platoon, Company E. Second Battalion, Twenty Sixth Marines. The reinforced second platoon was assigned to perform a security sweep to the southeast of Con Thien. These were sweeps that were done daily by each rifle company assigned to the base to prelude North Vietnamese forces from encroaching on their position. Each platoon rotated this responsibility within the company Operating within this system the second platoon had the security sweep every third day over familiar terrain. The men of Echo Company had seen very little action around Con Thien since they had replaced First Battalion, Forth Marines in late May. They fully expected this patrol to be just hot, humid and dusky walk in the sun. The following is a story about an engagement that occurred between those forces on that day.

"I did not pick these men.
They were delivered by fate and the US Marine Corps.
But I know them in a way I know no other men.
I have never given anyone such trust."

Sergeant Michael Norman
Golf 2-9
from *These Good Men: Friendships Forged from War*

Grateful acknowledgement is made to James N. Kaylor for permission to use information from a story that was originally written in April of 1990 and first appeared in its original form on the *Khe Sanh Veterans* website.

I took the liberty to rewrite much of the story but it is essentially James Kaylor's account as originally written. None of the facts have been changed.

James P. Coan's outstanding book, *Con Thien: The Hill of Angels*, was used for essential background information.

Also used were the awards citations of many of the participants, the rosters and diaries of the participating units, and information obtained from the National Personnel Records Center.

11

FIX BAYONETS!

On June 6, 1968 the platoon leader of the second platoon of Company E was Staff Sergeant Allen J. Baker. The former platoon commander Second Lieutenant Alec J. Bodenweiser had become the battalion executive officer and because of a shortage of officers at the time Baker took over the platoon. He was a career enlisted Marine who first enlisted in 1956 and was a very good athlete. His wife owned a dry cleaning business in Oceanside, California.

The day started out routinely with each man carrying at least five hundred rounds of 5.56m ammunition and two hundred rounds of belted 7.62mm machine gun ammunition. Each Marine also had to carry four fragmentation grenades and one or more M-72 anti-tank rockets. The lower ranking members of the platoon carried additional 60mm mortar rounds and 3.5 inch rockets.

We moved out in a southeasterly direction along a dirt road that intersected the main road to Con Thien. This road was about one kilometer south of the main base. The area was relatively flat and open. This land had been used as rice paddies and appeared to be hedge rowed into parcels of about one acre blocks. The only significant land marks were three old, bombed out, Catholic Church structures located next to a dirt road.

We continued southeast with first and third squad on line. The second squad was in a staggered column to the rear where the lead squads intersected. Our crew served weapons, the machine guns and the mortar, were with the second squad.

At about 1100 we were about 500 meters east of the third church when our right flank erupted in a heavy volume of enemy automatic weapons fire that was interspersed with rocket propelled grenade fire. Our entire right flank was under heavy North Vietnamese fire.

The squad on our right flank had walked into the midst of a hidden enemy bunker complex. Our second squad returned the enemy's fire as we moved to our left flank. We found cover in a sunken trail that extended parallel to the North Vietnamese positions. The first squad was on our left flank. They moved and joined us in the sunken trail as we engaged the enemy.

Our battle lines were roughly parallel to each other, approximately 20 meters apart. The second platoon was on the north and we mostly occupied the sunken trail which was lined with hedgerows. The third squad, down to the south, was pinned down around the enemy's hidden positions.

The exchange of fire was intense and very confusing especially for the newer Marines. I had to tell a couple of them to fire at the enemy. Being pinned down in the sunken trail we were unaware of the plight of our right flank. We were unable to contact them by radio and the enemy fire was too intense for us to be able to move closer to them.

We continued to engage the enemy positions that were south, east and west of us. Our 60mm mortar was deployed and supporting our effort along with our 3.5 inch rocket launcher. The mortar was firing at such close range that it was almost as dangerous to us as it was to the North Vietnamese. Lance Corporal Leldon D. Barnett detached the mortar from its bipod and used it in the lever fire mode as he aimed it with his hands. One of the rounds he fired struck a fleeing enemy soldier directly in the back at a distance of less than 50 meters. Barnett fired the mortar in much the same manner as I was firing my M-79 grenade launcher until he exhausted his supply of mortar rounds.

I was moving up and down the sunken trail firing at targets being pointed out by other Marines. I had fired quite a few high explosive rounds when I was given the rounds of another grenadier. He had been wounded and his M-79 had been destroyed. I was now loaded down with twice the amount of grenades. The minimum load of rounds per grenadier was between 100 and 110 and was very heavy and difficult to carry.

At one point during the action Barnett acted as my spotter, adjusting my fire onto an enemy position. About the third round I fired, while my projectile was in flight, an enemy soldier exposed himself by sticking his head above ground. After the round hit Barnett moved over to my position. He excitedly and half laughing exclaimed, "You got that guy right on the steel pot." My last round had apparently hit the North Vietnamese soldier directly on his head as he peered out from his position.

As the engagement continued in various stages of intensity we learned that Staff Sergeant Baker had been killed at the beginning of the fight. We were temporarily without leadership until the second squad leader Corporal Kenneth Schauble assumed command of the second platoon.

Schauble wanted to call in artillery and gun ship support but our artillery forward observer Lance Corporal Steve McDonald felt that we were too close to the enemy for support. Schauble and I discussed firing artillery to the enemy's rear to prevent their withdrawal. I believe that McDonald took care of this request.

Schauble wanted to send a runner to the right flank to determine the situation there. I volunteered to go but he wanted me to remain with him because I was armed with the M-79. There were a number of volunteers and Private First Class Mikal J. Sullivan was selected.

The battle had become a stalemate when the enemy decided not to advance or disengage. We were unable to receive helicopter support for they were all engaged elsewhere.

Schauble sent a fire team north to determine an avenue for a possible flanking movement. They were immediately taken under heavy small arms fire and driven back to the safety of the sunken trail. At this time we realized that we were completely surrounded.

This sunken trail was about two to three feet deep and six to eight feet wide. The sparse vegetation provided little shade and the day was very hot and humid. As the midday sun rose overhead many Marines nearly succumbed to heat exhaustion.

The battle remained a stalemated exchange of fire. As I lay on my back in the trail I made mental aiming stakes in my head. I calculated where the enemy positions were in my mind and using these mental aiming stakes I fired fairly accurately on the enemy positions. I must have been getting their attention for they concentrated their fire on my location. Exploding rocket propelled grenades and small arms fire forced me to change positions several times.

At one point we had a LAAW that we wanted to use on an enemy position located in a large clump of bamboo about 50 feet to our front. We were all reluctant to fire the weapon because in order to do so it required that the Marine doing the firing had to be almost completely exposed to the North Vietnamese. After a brief discussion with those around me I came up with a plan.

I extended the LAAW, removed the safety clip, and pointed the launcher in the general direction of the target. Squatting on my haunches I told the other Marines to give me suppressive fire and I would spring to my feet like a "Jack-in-the-Box." This they did as I quickly popped up, discharged the weapon and fell on my back. I heard the rifle bullets crack by where I had just stood to fire the launcher. We all kind of laughed as the rocket hit the center of the base of the clump of bamboo.

During one of my many moves while stuck in the sunken trail I passed behind a 3.5 inch rocket launcher as it was fired. I was struck by flying debris and the concussion knocked me to the deck. It was an incredible blast and it felt like my lower jaw was uncoupled from my head. We had been warned that anyone exposed to the direct back blast from any recoilless weapon would be killed. Luckily for me the back blast had hit me at a glance.

As time passed Schauble became concerned with the fate of the rest of the platoon located on the right flank. It also concerned him that we were low on both water and ammunition and the fact that we were without radio communication. We all understood that Schauble was the leader of the second platoon and no one questioned his authority.

The North Vietnamese had engaged us with ferocity and determination and in what was an unusual tactic for them, during daylight and for an extended period of time. We had been at it for four hours and they did not appear to be in any hurry to break off contact.

Another concern was the lack of reinforcements from Con Thien. We were in close proximity to the base and help was yet to arrive. We knew that the North Vietnamese force that we faced was larger than us but we did not know its exact size. Schauble was concerned that any rescue force sent from ConThien may have been ambushed and would not be able to help us. Our situation was becoming desperate.

We continued to exchange sporadic fire with the enemy as early afternoon soon became late. Some of the newer Marines were affected by the uncertainty of our predicament. The heat and lack of water did not help either. We had started this patrol with a canine and his handler. The dog immediately became a problem as it soon overheated. We had to take turns carrying the animal. Dogs had proven to be very ineffective in extreme heat over long distances and this day was no exception.

By late afternoon the dog handler was concerned with the welfare of the dog. The animal was hyperventilating and in need of water. The handler asked me and other Marines to share our water with the dog. We all declined and added various disparaging remarks about dogs in general and this one in particular since he did not alert us to the ambush.

By this time we had used up most of our small arms ammunition and much of the water. Our mortar, rocket and machine gun ammunition had been used up long ago.

Since our situation had become desperate Schauble decided to aggressively challenge the enemy and attempt to consolidate with our right flank. We still did not know of their circumstances.

Corporal Schauble deployed the first squad to our right flank to provide cover for the second squad who would press the attack. They would do so, 'on line,' in a skirmish formation. The first squad successfully moved into position and supported us with grazing rifle fire.

Schauble then ordered most of the remaining Marines to join with the second squad in the assault. All eyes were on him as he placed himself in the middle of the skirmish line. I was to the right side of the line and next to me on my left was Private First Class Richard Eernisse. As we waited for the command to attack we were surprised when Schauble yelled "Fix Bayonets!" Eernisse and I just looked at each other in amazement and I said, 'I don't think they make them for this!" I saw a couple of Marines attach bayonets to their rifles as we waited for Schauble to signal attack.

Schauble raised his right hand and then dropped it forward and verbally gave the command to attack. We leaped to our feet as one, yelling and firing our weapons as we ran toward the North Vietnamese. I had what remained of the platoon's remaining M-79 rounds. There were only five or six and I fired them sparingly as we ran forward.

Our first objective was a hedgerow about 20 meters to our front. It contained many enemy positions and we easily took it with only a few casualties. During the attack on this hedgerow my M-79 ceased to function.

It became inoperable because of excessive wear to its interior mechanism. After I had fired a couple of rounds I could not close the breech. I threw the useless weapon behind me as I continued to run forward. I pulled my .45 caliber pistol from its holster just as we reached the first hedgerow.

The first squad maintained a steady flow of grazing fire from our right flank in their effort to suppress enemy fire. The momentum of the attack carried us through the first hedgerow into another field that was approximately fifty meters across and bordered by another hedgerow. Eernisse was still to my left and was throwing hand grenades when he was hit by enemy gun fire. The grenade that he was trying to throw fell from his hand and exploded about ten feet directly in front of me, and surprisingly did not injure me.

We continued about half way across this field and approached Corporal Jerril Callaway's position. My attention was drawn to my left front along the edge of a very large bomb crater. Located there was a very extensive defensive position that was not well concealed. The North Vietnamese had dug trench lines connecting their positions and I could see several of the enemy moving from position to position. An extensive amount of fire was coming from these positions. Two enemy soldiers were operating a light machine gun and several were firing AK47s. They were pouring fire into the Marines on my left.

Private First Class William A. Hayes was a few feet to my right. We looked at each other and without exchanging a word we simultaneously knew that we were caught out in the open and would probably be killed. We were too far from the hedgerow and if we tried to run back to it we would certainly be shot in the back. Dropping to the prone position would have left us completely in the open and exposed to the murderous North Vietnamese fire.

Knowing that we faced certain death we changed directions and sprinted directly toward the enemy machine gun. Hayes was helmetless and bare-chested. He held a hand grenade in each hand as he sprinted toward the enemy position. The North Vietnamese must have been concentrating their fire to the front and did not notice us at first. We got to within twenty to twenty five feet of the gun before we became the target of their fire. They were aiming at Hayes who was the closest. I was firing my .45 pistol while running as fast as I could directly at the enemy. We were right on top of them very quickly.

Hayes tossed the grenade that he held in his left hand, underhanded, toward the machine gun. He raised his right hand over his head and attempted to throw the grenade straight down into the enemy position. At that moment he was hit and mortally wounded by machine gun fire and the second grenade sailed over the bunker. Both grenades exploded almost simultaneously. The North Vietnamese in the bunker were frantic. As he died, Hayes momentum carried his body headlong into the position. They pushed his body to their left as the gunner tried to bring the gun to bear on me. I looked down at their faces as I fired my pistol. The enemy gunner continued to fire but could not get the gun to turn on me quick enough. All rounds were fired off to my right. I quickly adjusted to my left and dropped to my right knee in an attempt to avoid being shot. I was muzzle to muzzle with the machine gun.

I continued to fire my pistol at the enemy faces although I could not tell what effect my rounds were having on them. The machine gun was blasting close to my face so I turned my head to the left in an attempt to avoid being shot. I felt the heat and the blast as the bullet passed my face but when I realized that I was still alive I looked forward and saw the muzzle of the machine gun pointing straight up in the air. The gun and the four North Vietnamese disappeared below the edge of the bunker and the firing ceased. With only three rounds left in my pistol and knowing that there were at least four enemy soldiers in the position I decided that to continue my assault would not be wise. All of this action had taken place in the span of about four to five seconds. With a loaded rifle and a bayonet I could have continued my attack.

Taking advantage of this lull in the enemy fire I quickly retreated to a crater that was about twenty feet behind me. As I jumped into the crater an enemy grenade exploded directly behind me.

The machine gunners were defending their position by throwing hand grenades. A right arm, incased in a long sleeved, olive drab uniform shirt, rolled up to the elbow, appeared from the enemy position. Its owner cocked the arm and threw a chicom, stick grenade in my direction. I could tell that it was going to fall short and it did, exploding about three feet in front of my crater.

I took an M-26 grenade from my pocket, pulled the pin, and threw it at the enemy bunker. There was another brief lull in the fighting and I took time to survey my predicament. I first determined that I was not wounded. I then checked my .45 pistol and reaffirmed the fact that I had three live rounds left from the three magazine that I had fired during the attack. I always counted my rounds and had done so this day. The one grenade that I had thrown had missed its mark. I started to tremble and I calmed myself by planning my next course of action.

I did not know how many, if any, casualties the North Vietnamese had sustained. I visually surveyed my surroundings but was unable to detect any sign of life.

When I called out the names of several of my fellow squad members I received no reply. Looking to my left I saw the olive clad bodies of two dead Marines. I yelled several more times hoping to find someone that was alive. It then crossed my mind that this would be my last day to live.

Then I heard a familiar voice from way behind me. It was Private First Class James M. Terry. He and Private Brien A. Hayes, who I think, was the only private in the second platoon, were also pinned down about 30 meters to my rear. Other than me, they were the only two Marines involved in the charge that were not wounded.

Since I was within throwing distance of the enemy position I hollered to them to throw me some hand grenades. They responded that neither one of them had any grenades. I told them to move around and salvage the ammunition and grenades from our casualties.

After they had accomplished this task Terry threw about four grenades to me but none landed in my crater. They landed to the right and left and somewhat to the rear of my position. It was frustrating and I was angry that I could not gain procession of the grenades. I would have had to expose myself to the enemy if I tried to retrieve them and I am certain that I would have been killed.

Since they had thrown all of their fragmentation grenades near me I then asked them to throw a smoke grenade if they had one. One of them asked, "What for?" and I hollered back, "Just do it!" When they threw the grenade it landed in the crater with me.

Judging the direction of the wind, I threw the smoke grenade so that the North Vietnamese would not be able to see us. When I thought that they could not see me I jumped from the crater and ran to retrieve the grenades that Terry and Hayes had thrown to me.

As I ran for the grenades the enemy fired a few bursts of gunfire blindly through the smoke but to no effect. I returned to my crater unscathed and the smoke cleared shortly thereafter.

While placing the grenades in my pockets I became distracted and I must have raised my head for I heard Terry yell a warning, "Don't stick your head up one of them is looking out." He repeated it once more and then I heard the crack of an M-16 bullet as it passed directly over my head. It had come from Terry's position.

I heard Terry yell, "I got him!" He was one of the best marksmen in the company and had used this deadly skill to help save my life. I made up my mind not to expose my head again.

I used the lip of the crater to rest my .45 caliber pistol on while aiming it with my left hand. I was determined not to take my eyes or my pistol off of the enemy position.

I pulled the grenades, one at a time, from my pockets. While maintaining my concentration on the enemy position, I straightened the bent ends of the grenade safety pins with my fingers. With this done I was easily able to remove the pin of each grenade with my mouth. Using this procedure I was able to aim my pistol and arm the grenades at the same time.

The first grenade I threw went long. I was too pumped up so for the next attempt I tried to calm down and throw more accurately. After removing the pins from the next couple of grenades I held them in my right hand with my thumb holding the spoon down. I then released the spoon which armed the grenade and I held it for a couple of seconds before throwing it.

This was a technique for grenade throwing that I had learned during the siege at Khe Sanh. Once the grenade lever is released it will explode in about four to five seconds. The North Vietnamese knew that once they heard the metallic 'clang' sound made as the spoon flew from the grenade that they would usually have time to pick it up and throw it back before it exploded. If the thrower released the lever and held it for two seconds it would usually detonate on or over the enemy position when thrown. They would not have time to pick up and throw it back.

These two grenades were thrown more accurately with the correct distance. One detonated on the left end of their position and the other on the right

My last throw took one bounce and went right into the enemy bunker and exploded immediately. There was no time for anyone to throw it back. Parts of enemy uniforms and equipment were blown skyward. It was my last grenade and I now had only the pistol with which to defend myself.

The North Vietnamese had thrown William Hayes' body out of their bunker and I could see it lying there.

In the meantime James Terry and Brien Hayes continued to find and strip our casualties of their munitions. They found two more grenades which they threw to me. One of these was a white phosphorus grenade which we called a 'Willy Peter.' It was particularly lethal and was designed to start fires and burned with such an intense heat that it could melt most metals. If it got on your skin it burned through and could not be contained with water. I knew that I was well within its bursting range but I threw it anyway. The grenade exploded in the North Vietnamese position and white hot burning particles rained down all around me. I had to move around quickly to avoid them and fortunately I was able to.

Nothing took place for a long time and as I sat in the crater rain started to fall. It rained for a while and when it stopped I saw a spotter plane circling overhead. I asked James Terry and Brien Hayes to try and find another smoke grenade and they did. One of them tossed it to me and I, in turn, pulled the pin and tossed it in front of the North Vietnamese position.

As the smoke masked my position I ran back to their position. Once there I learned and could see that our casualties included all but the three of us. We also determined that between the three of us we had less than twenty five rounds of ammunition for our rifles. I went to the body of Private First Class William R. Campbell and retrieved his rifle which he was still clutching it his hands. I checked the magazine but it contained only two rounds.

We decided not to give up the ground we had gained so I suggested that Terry and Hayes remain where they were while I tried to find more ammunition. We all agreed and I took an empty claymore mine bag and I ran back to the sunken trail. As I ran I prayed that I would not get hit.

It had been about forty five minutes since the second squad had started the attack.

When I reached the sunken trail I found McDonald there. He wanted to know what the situation was and I told him that everyone was dead except me, Hayes and Terry. The look on his face was one of desperation and real concern. One of the mortar men there heard me and I could also see the look of alarm on his face.

McDonald told me that of the six radios we had started out with only one was in operable condition. I told him that the enemy was much stronger than we had anticipated and that a second line of defense had decimated us. I also informed them that Callaway was the only Marine on our right flank that I saw alive and that he was pinned down among the North Vietnamese positions.

I told them of our intention to continue the attack until we linked up with our comrades on the right flank. I gathered up about eight loaded magazines and deposited them in the claymore bag. I told Barnett to maintain his defensive position until I signaled to them that we had been successful and found the third squad.

Exiting the trail I ran back across the first field, through the hedgerow and out to Terry and Hayes. They were now using the bodies of our dead buddies for protection. Both were in the prone position and were using this cover to rest their rifles on.

I distributed the ammunition evenly which left us with about three magazines apiece. We loaded our rifles with full magazines, got on line about five meters apart and resumed the attack on the bunkers. As we walked we directed slow, deliberate fire at our target. The rifle, which I had taken off of the body of Campbell, jammed so I removed the magazine and discarded the weapon. I quickly picked up the rifle of another dead Marine, loaded it and continued to advance.

When we were within twenty feet of the enemy position and near the crater that I had used earlier, we stopped. We had not received any fire during this attack but directly behind the bunker we could see a large body of troops approaching. This group, of at least fifty men, was about two to three hundred meters away and was led by a man waving a pistol.

Since this force was coming from the southeast we initially thought that it was a counter attack and, if so, we were in serious trouble.

Terry assumed the prone position with the intention of shooting the leader, the man waving the pistol. We were in a helpless position and I was demoralized. As I waited to be cut down in a hail of gunfire I made no effort to hide but as we watched their advance we realized that we were not being fired upon.

Terry was about to squeeze off a round when I recognized that they were American Marines and pushed the barrel of his rifle down. We waited for them with a great sense of relief as we realized that the fighting was over for the day. The North Vietnamese had fled from the battlefield leaving their dead for us.

Our rescuers were from Hotel Company, Second Battalion, Twenty Sixth Marines led by First Lieutenant Tyrus F. Rudd. His troops performed the task of cleaning up after the fight.

I walked along our line of attack and found many of our casualties. I heard groaning coming from where I had emerged from the first hedgerow during the attack. There I found my buddy, Eernisse, gravely wounded. At first, I could not see the wound, but when I tried to roll him over; my fingers went into his head. A bullet had entered the top of his helmet striking him at the forward scalp line. It split his skull all the way to the back of his head. The wound was about two inches wide and brain matter was exposed. He was evacuated immediately.

We found Lance Corporal Ronald Kalbhenn at the extreme left flank of the second squads attack. He had been driven to cover at the first hedgerow and was forced to remain there until the battle ended.

Kalbhenn and I came across a North Vietnamese soldier hiding in a spider hole. He was face down and in the fetal position. We could not see his hands but the launch tube of a rocket propelled grenade protruded over his right shoulder. We could see his back heaving up and down from his breathing. We wanted to take him prisoner but felt that it was too dangerous. We did not know if he had anything in his hands, if the RPG was loaded or if he was booby trapped. Both of us shot him through the head with our rifles.

Kalbhenn reached down, grabbed the corpse and carefully pulled it from the hole. The RPG was loaded and he had a grenade ready to go beneath him.

The body of Schauble was found near the first hedgerow. An enemy bullet had entered his open mouth and exited near the back of his head.

Lying in the field in front of the enemy bunker complex were the bodies of Private's First Class Dennis Lobbezoo and Walter Seawright. They were killed during the attack on the bunkers by the second squad.

Only one Marine survived the fight on the right flank. Most of the others died among the enemies hidden positions. Staff Sergeant Baker, our platoon leader, and his radioman Lance Corporal Daniel Prock were found with this group of dead. Sullivan, the runner sent to locate the right flank, was also found dead. That would have been me had Schauble allowed me to go.

The only survivor found in this area was Callaway who had been pinned down from the very beginning of the action. He was now the highest ranking Marine left in the second platoon.

Bodies from both sides were being collected and taken to the middle of a field. Marine helicopters were flying in and taking out the wounded first and then bringing in supplies.

I carried Sullivan's body to the landing zone. Rigor mortis had set in so it was like carrying a mannequin. The men from Hotel Company were putting our dead into green, plastic body bags and they put Sullivan's body into one for me. This done they then lined the bags in front of one of the helicopters that was sitting on the ground with its rotors turning.

Retching, I walked away from the line of bodies, doubled up and started throwing up uncontrollably. Shortly, I noticed that one of the pilots was taking home movies of me throwing up.

Six to eight North Vietnamese bodies were recovered from the bunker complex that Hayes and I had attacked.

During the course of the day I had fired more than two hundred grenades through my M-79 launcher. It was most of our platoon's supply of the projectiles. The day after the battle my shoulder was so sore from firing this weapon that I could barely move my arm.

Of the number of grenades that I fired about ten of them had been duds. One of these duds had landed at Callaway's feet while he was pinned down in a crater. A few months later after we had returned to the states and were stationed at Camp Pendleton he reminded me of this.

One of our two machine guns had been captured and used against us. It was sent to support our right flank and after the battle we found it there, unattended, sitting on its bipod and pointing back at us. All of the belted ammunition that had gone with it was lying, expended, next to the gun. There were no bodies found near the weapon. The North Vietnamese simply abandoned it when the ammunition ran out.

Rudd and some of his men were going through the enemy bunkers checking the dead North Vietnamese for items that might be useful to intelligence. Other Marines were picking up the equipment and weapons of both sides and loading it on the helicopters.

After all of the Marine wounded and dead were removed from the battle field, the North Vietnamese dead were collected, counted and thrown into a heap at the bottom of a large crater. An attempt was made to burn the bodies with phosphorous and thermite grenades. I did not look to see if they were successful.

After the dead and debris was cleared away there was a casualty count. The numbers that I recall were about fourteen Marines killed in action and a lesser number, about eleven or twelve were wounded from small arms fire and taken to hospitals for treatment.

I cannot remember the exact number of North Vietnamese bodies recovered but the number was similar to our own.

After replenishing our ammunition supply the survivors of the second platoon fell in with the Hotel Company Marines for the march back to Con Thien. When I walked through the line all I had was my helmet, flak jacket, pistol belt with pistol and two M-16 rifles with one slung over each shoulder.

Our company commander, First Lieutenant Joseph R. Meeks, met us, grim faced, at the perimeter. I do not think that he knew of the extremely heavy casualty count until Hotel Company arrived on the battlefield.

That evening Staff Sergeant Melvin R. Proctor tried to interview us regarding the second platoon's fight that day. I believe that he was with Headquarters and Service Company of 2/26. I think that he was with the Scout Sniper platoon and was trying to find out what happened for battalion headquarters. Proctor's interrogation was not well received by the survivors who felt that he was insinuating that the blame for this disaster rested somewhere with the second platoon or its previous command staff, all who were now deceased. There was some 'finger pointing' about who did what and with that the inquiry was concluded. The survivors all felt that we should have been reinforced in a timelier manner. This would have eliminated the necessity for the charge and we would have suffered far fewer casualties. We had all lost good friends and were reluctant to discuss this action.

NOTES:
The following members of Echo-2-26 were killed in action on 6/8/68.
Baker, Allen J. SSgt. 1561325; 29; Houston, Texas
Campbell, William R. Pfc. 2337411; 19; Butler, Pennsylvania
Flyte, Forrest J. Cpl. 2096863; 23; Pen Argyl Pennsylvania
Hale, William E. Pfc. 2382474; 19; Columbus, Indiana
Hayes, William A. Pfc. 2359775; 19; Chicago, Illinois
Lobbezoo, Dennis L. Pfc. 2365524; 19; Grand Rapids, Michigan
Picciano, Terrance A. LCpl. 2366238; 19; Baraga, Michigan
Prock, Daniel L. LCpl. 2356648; 20 Gridley, California
Schauble, Kenneth W. Cpl. 2355122; 20; Closter, New Jersey
Seawright, Walter L. Pfc. 2375101; 18; Philadelphia, Pennsylvania
Shepherd, Peter M. Pfc. 2352611; 20; Portland, Oregon
Sheppard, Lonnie Jr. Cpl. 2253618; 19; Newark, New Jersey
Sullivan, Mikal J. LCpl. 2351481; 19; West De Pere, Wisconsin

For heroism this day the Silver Star was earned by Private's First Class William A. Hayes, James N. Kaylor and Daniel A. Staggs and by Lance Corporal Terrance A. Picciano. Hayes and Picciano's were posthumous.

Other Marines wounded that day were Private's First Class James W. Adams, Russell B. Brockman, Thomas W. Evans, Dennis S. Flannery and Cameron Grey and Lance Corporal Hendrik G. Vande Velke.

Brockman received a gunshot wound to the right arm and a fragmentation wound under the right eye. Evans was wounded in the neck by a gunshot. Vande Velde suffered gunshot wounds to the mandible, left hand and back. It was his second Purple Heart for he was wounded in the right arm by fragments from an explosion that occurred on April 17, 1968

Staggs was serving as a Scout with Company E when the ambush occurred and he handled himself very well during this action. First, He helped another Marine carry a mortally wounded man to a bomb crater and then he picked up a machine gun an accompanied by another Marine rushed across the fire swept area to rescue another one of the wounded. When his companion was killed Staggs picked up the radio carried by the Marine and established communications with an aerial observer. Although he knew very little about medical evacuation procedures, he quickly organized a helicopter landing zone and supervised the movement of wounded Marines to a central location. Staggs continued his determined efforts by skillfully guiding helicopters into the zone, thereby providing lifesaving evacuation for numerous casualties.

Picciano volunteered to help move the wounded to a covered area and after he helped carry one Marine to the apparent safety of a bomb crater a grenade landed in their midst. Without hesitation he seized the lethal object and hurled it back toward the North Vietnamese. This courageous act undoubtedly prevented serious injury or death to him and his buddy. When Picciano realized that intense fire from an enemy emplacement would prevent further rescue attempts he and another Marine boldly rushed the position in an attempt to silence the hostile fire. As he approached it he was mortally wounded. This heroic and timely act enabled his unit to evacuate the wounded from the hazardous area.

Second Lieutenant Alec Bodenweiser earned the Silver Star medal for heroism in action on February 5, 1968. Echo Company was occupying a defensive position on Hill 861 near Khe Sanh when they were subjected to a fierce coordinated attack by the North Vietnamese Army that was supported by heavy artillery and mortar fire. The enemy soldiers moved right in under their own mortar fire. Bodenweiser, then the company executive officer, was in charge of coordinating the defense of the western, southern and eastern portions of the perimeter. He had at his disposal the second and third platoons. During the attack North Vietnamese sappers were able to breach the wire. He redeployed his Marines, halted the attack and personally led a counter attack against the enemy inside the perimeter. This attack involved the use of hand grenades at close range, bayonets and vicious hand to hand combat. At great risk to his own personal safety, Bodenweiser's leadership helped inspire his Marines to completely disrupt the enemy attack and decisively defeat them. The courageous actions of he and his men forced the North Vietnamese to abandon their attempt to envelope Echo Company.

First Lieutenant Joseph R. Meeks, the company commander, first enlisted in the Marine Corps in 1954 and served as an enlisted Marine until temporarily promoted to second lieutenant in 1966. He later reverted to enlisted status and retired as a Master Sergeant in 1974 with twenty years of active service. Meeks was a veteran of several military operations in Vietnam and on May 12, 1969 was wounded, by shrapnel in the right arm, while participating in Operation Lancaster II.

Other Marines in the chapter are as follows:
Adams, James W. Pfc. 2393879; E-2-26; 20; Unknown
Bodenweiser, Alec J. 2dLt. 0103347; E-2-26; 22; Salem, Oregon
Barnett, Leldon D. LCpl. 2334688; E-2-26; 19; Amarillo, Texas
Brockman, Russell B. Pfc. 2415273; E-2-26; 19; Kansas City, Missouri
Callaway, Jerril L. Cpl. 2256416; E-2-26; 20; San Antonio, Texas
Coan, James P. Capt. 0100944; A-3 Tanks; 24; Unknown
Eernisse, Richard F. Pfc. 2373879; E-2-26; 19; Denver, Colorado
Evans, Thomas W. Pfc. 2386534; E-2-26; 19; Salt Lake City, Utah
Flannery, Dennis S. Pfc. 2410911; E-2-26; 19; Los Angeles, California
Grey, Cameron Pfc. 2360937; E-2-26; 19; Phoenix, Arizona
Hayes, Brien A. Pvt. 2368789; E-2-26; 19; Hartford, Connecticut
Kalbhenn, Ronald F. LCpl. 2207114; E-2-26; 20; Newark, New Jersey
Meeks, Joseph R. 1stLt. 097408; E-2-26; 32; Ashland, Wisconsin
Proctor, Melvin R. SSgt. 1815206; E-2-26; 28; Los Angeles, California
Rudd, Tyrus F. Jr. 1stLt. 0103774; H-2-26; 27; Dallas, Texas
Staggs, Daniel A. Pfc. 2371706; E-2-26; 21; Jackson, Michigan
Vande Velke, Hendrik G. LCpl. 2387614; E-2-26; 19; Oakland, California

★ ★ ★ ★

Prior to the launching of Operation Dewey Canyon, United States Marine infantry units in the northern I Corps region had been tied to their combat bases along the South Vietnam border as part of the McNamara Line. This "line" was a combination of infantry units and ground sensors devised to stop North Vietnamese infiltration into South Vietnam along the Ho Chi Minh Trail. When Lieutenant General Raymond G. Davis took command of the Third Marine Division, he ordered Marine units to move out of their combat bases and engage the enemy. He had noted that the manning of the bases and the defensive posture they developed was contrary to the aggressive style of fighting that Marines favor. In early 1969, intelligence reports indicated that there had been a large NVA build-up in the A Shau Valley. The A Shau was just 6 miles (10 km) east of the Laotian border and some 21 miles long. Based on this intelligence, Colonel Robert H. Barrow's Ninth Marine Regiment was ordered to depart Vandegrift Combat Base some 50 miles to the east and sweep west to deny use of the valley to the enemy.

Operation Dewey Canyon was divided into three parts: 1) the movement and positioning of air assets, 2) the movement of the 9th Marines south out of their combat base, and 3) the sweep of the A Shau valley. As the 9th Marines moved towards the A Shau valley, they established numerous firebases along the way which would provide them their artillery support once they entered the valley and guard their main supply route. All of these bases needed to be resupplied by helicopter because of their distance from the main combat bases and because resupply via ground was very difficult during monsoon season.

The Marines encountered stiff resistance throughout the conduct of the operation, most of which was fought under triple canopy jungle and within range of NVA artillery based in Laos. Marine casualties included 130 killed in action and 932 wounded.

In return, the USMC reported 1,617 killed enemies, the discovery of 500 tons of arms and munitions, and denial of the valley as an NVA staging area for the duration of the operation. They claimed the operation as an overall success.

"A winning team requires rehearsal and practice
of the plays, even while involved in combat."

First Lieutenant Wesley L. Fox
Alpha Company 1/9
Medal of Honor

References used to compile this story were the books, *Don't Tell America* by Michael R. Conroy, and *Marine Rifleman* by Colonel Wesley L. Fox, USMC (Ret.). Additional information was gleaned from the Website of Fox Company, Second Battalion, Ninth Marines.

Also used were the awards citations of many of the participants, the rosters and diaries of the participating units, and information obtained from the National Personnel Records Center.

12

"Follow Me!"

By 20 February the Marines had moved mall the way to the Laotian border. The North Vietnamese continued to assault the Marines and then retreat to their sanctuary across the border into Laos. Colonel Barrow had seen too many of his men succumb to this unfair advantage and choose to do something about it. He sent Hotel Company, Second Battalion, Ninth Marines across the border and set up ambush positions inside Laos.

With a company of the Ninth Marines now inside Laos, the other battalions moved out to take up positions along the border. The First Battalion moved to ridge overlooking Laos on 22 February. The Marines of 1/9 were called the Walking Dead. For the Marines of Alpha Company this name would be all too real.

First Lieutenant Wesley L. Fox was in command of Alpha Company. As dawn broke a heavy mist and fog covered the area. The company's mission was to seek out and destroy a suspected enemy force operating in the area. The third platoon had made contact with them the previous day, and now the company was looking to finish the fight. The First Battalion was also low on water and a detail was sent to obtain a supply from a stream below. Fox and his men were to lead the way and provide security as they searched for the North Vietnamese.

The company moved due east with the goal of investigating the area of contact near the stream bed. There was heavy foliage to the east and the terrain dropped down steeply to the stream. If they did not make contact, Fox would send for the water detail by radio.

The first platoon under Second Lieutenant George M. Malone was in the lead. They were followed by Fox and the CP group, Second Lieutenant James H. Davis's second platoon, and third platoon, led by Second Lieutenant William J. Christman, III. They moved down the west side of the ridge on the way to checkpoint one. At 1100 one of Fox's flankers found a bunker that looked a though it had been recently used. It was abandoned and contained no supplies or materials. Private First Class James Decker found some increments from an enemy mortar. He showed Christman what he had found and he warned Lieutenant Fox of the obvious presence of the North Vietnamese. Decker recalls, "Lieutenant Christman felt that this was one of the positions from which Charlie Company was mortared earlier. After such an attack it was not unusual for a patrol to locate increments, rounds, and even the mortar tubes in firing positions. Concerned with Marine rapid counter battery fire, the North Vietnamese would fire a round or two, note what sight changes were needed to place their rounds on target, desert the area, wait out the counter battery fire, and creep back hours later to place accurate fire on their targets."

Fox, a mustang, had been in the Marine Corps for nineteen years and this was his second war. He had served a previous tour in Vietnam and had recently extended this one. The men had benefited from this experience and he had trained them well. Fox had a great deal of confidence in them and used the simple leadership principle of exercising care and concern in the handling of Marines. He explained, "Do this and they will always answer the call. If the commander does right by them, they will charge right up the enemy's gun barrel for him and never ask why."

Christman was up front with the point element although his platoon was at the rear of the column. Around 1200, he guided the company toward their first checkpoint and took the wrong turn off the trail. The main body of the company continued on the right path and Christman was forced to backtrack and fall in with his own platoon at the rear of the formation.

As they approached the creek bed the point fire team came under intensive fire from a squad of North Vietnamese soldiers that were entrenched in two fortified bunkers that defended a hospital complex. Private First Class Fred Butler, III, walking point, was shot through the head and killed. The Marine walking behind him was also seriously wounded.

Without hesitation, Davis's point squad advanced on the redoubt and destroyed the bunkers and killed seven North Vietnamese soldiers in the process. No other Marines were wounded during the fight. They searched the complex and discovered many beds and abandoned medical supplies. There was much blood indicating that the hospital had been used extensively. The engineers demolished the complex with explosives.

After conferring with his platoon leaders they determined that that the NVA staging area was located to their rear and the enemy must have chosen to abandon it.

Fox formed a large perimeter around the watering point by sending the lead platoon across the creek to provide security from that side and positioned the other platoons on the near side. He explains, "At this point it looked like that was all the resistance we had. Everything was quiet so I radioed up to battalion to send the water details down to the creek. We were in bad need of water. The helicopters could not get in due to the weather, and the battalion was low."

While they waited, the Marines from Alpha Company rested and ate C-rations. The water from the mountain stream was pure, clear, and cool and they drank their fill. Charlie and Headquarters & Service Companies sent a water detail, of about twenty Marines volunteers under Staff Sergeant Donald F. Myers which arrived in about thirty minutes. They brought their canteens with them strung on ropes, and began filling them with water from the creek.

Private First Class Oscar F. Borboa, a machine gunner with the third platoon of Alpha Company, was assigned the task of refilling the canteens of his squad. As soon as he started his job a mortar tube could be heard firing off to the company's right rear. This was the area where the third platoon had made contact the day before. The enemy rounds began to explode in the thick canopy of the tall trees over their position. Because of the triple canopy, the tree bursts caused no casualties but did draw the attention of all hands.

Borboa recalls, "I must have had about fifteen canteens and as I was at the creek filling them the incoming mortar fire started getting heavier. As the NVA were 'walking' the rounds toward the creek, Lieutenant Davis ordered me to 'get the hell out' and to try to get back to my squad. I had only filled about four canteens and I felt obligated to fill more so I stayed where I was and continued filling canteens. Lieutenant Davis forced me to leave the area and I remember being upset over leaving empty Canteens behind. I turned to retrieve them but Lieutenant Davis again forced me out of the area."

As soon as the duty bound, Borboa was forced to leave the creek mortar rounds landed in the exact spot where he had been filling the canteens. The actions of Davis probably saved the lives of seven Marines.

Fox reported that, "Mortar rounds started popping all around us. Machine gun fire and movement to my right rear positioned the NVA location for me. It was the triple canopy overhead which made it difficult to maintain contact."

Some of the Marines had not been able to leave the creek. Many were wounded and screamed in pain as while unit leaders tried to bring order to the chaos. Calls for corpsmen were being heard repeatedly. A Charlie Company radio operator, Private First Class Robyn W. Fish moved toward the rocks that formed the sides of the riverbank to escape the enemy fire. Myers was hit in the face and the right knee by shell fragments and seriously wounded as was Sergeant Levoid White.

Private First Class Paul G. Runick from H&S Company, who was on the water detail, ignored the enemy fire and twice moved to the side of wounded Marines to administer first aid and help them up the hill to relatively safe positions. On his third trip into the fire swept area, he was wounded by hostile fire and lost the use of both hands. Undaunted, Runick remained in his precarious position and verbally encouraged the wounded and helped direct movement of the last casualties up the hill.

Corporal Randall B. Parnell and the second squad from Davis platoon escorted the water detail and helped to carry the dead and wounded to safety. Lance Corporal William H. Daily and he spent the rest of the day carrying dead and wounded Marines up the Hill. "If it wasn't for Captain Kelly I don't believe a lot of us would have made it back", said Daily.

A rear guard was set up downstream by Sergeant David A. Beyerlein and his squad. When a lot of senior people were wounded he was called forward and command of the squad was passed to Lance Corporal Leonard Cosner. He recalls, "I took over the squad. We killed two North Vietnamese soldiers then pulled back to join the company. It was a shocking sight. There were wounded men everywhere. Staff Sergeant Joseph Guy was hit and out of action. Sergeant White was hurt bad. I think he lost an eye."

White, a platoon guide in Alpha Company, was on his second tour in Vietnam, having served as a rifleman with Bravo Company, First Battalion, Seventh Marines from May of 1965 until June of

1966. He then went to the Drill Instructor School at Parris Island and was on the drill field from December of 1967 until April of 1968 when he ran into a little trouble, as many do, was reduced in rank, and returned to Vietnam in June of 1968.

Cosner continues, "We were too shocked to be doing very much. A machine gun was firing at us from the ridge behind us. Tracer bullets could be seen in mad, crazy flights up to the treetops as they ricocheted off rocks and jungle growth. By using tracer rounds the enemy was making sure that they could be located. They were evidently spoiling for a fight. That's when Lieutenant Fox yelled, 'what are you going to do, Marines, lay here and die? Remember Belleau Wood. Charge!" I swear, that's what he said. That got us moving on line and in the attack toward the sound of enemy guns."

Fox directed the two lead platoons to deploy on line and attack. Control and contact were maintained in the thick jungle by an initial formation of company and platoons on line while squads moved in columns, coming on line as contact was made.

The lead element of the company came under intense fire from a heavily fortified bunker complex that was supported by at least six machine gun teams. Alpha Company was pinned down by this well camouflaged North Vietnamese redoubt.

The fire team leader of the forward element was Lance Corporal Michael P. Hester. A member of his fire team, Private First Class Larry J. Boehm, was shot down during the opening volley. Hester ran into the kill zone in an attempt to help the fallen Marine but when he reached him discovered that he was dead. Hester discovered the source of the North Vietnamese fire and called for covering fire from his fellow Marines. With this he moved forward to within range of the enemy emplacement and was able to hurl grenades through the bunkers aperture. The resulting explosions killed several soldiers and allowed Hester to seize the automatic weapon that had wreaked havoc on the assaulting Marines.

But, Hester wasn't finished yet. He was determined to silence the vicious North Vietnamese fire and he aggressively moved toward its source, firing his rifle as he advanced. Hester was seriously wounded by the enemy crossfire and went down, unable to continue the assault, but remained alert and was able to direct the rifle fire of his team for the rest of the battle. The dauntless Marine was later medically evacuated.

Lance Corporal Darrell H. Chapman, a fire team leader in the forward squad, dropped to the ground at the first sound of the enemy gunfire and began to crawl toward their positions. He was able to throw several grenades as he directed his team in the assault. Chapman continued to encourage them with the words, "keep moving, just keep moving," as he and his men inched forward on their bellies and dug the North Vietnamese out of their bunkers. Two Marines manning a nearby machine gun were hit and no longer moving. Chapman moved through the intense hostile fire and pulled both to a position of relative safety and the treatment of a corpsman. Both men, Privates First Class Norman P. Chittester and Anthony L. Johnson, soon died.

Chapman returned to the abandoned machine gun position and picked up the weapon and ammunition. He then moved to the point of heaviest contact and took an enemy machine gun emplacement under fire. Chapman placed a heavy volume of accurate, suppressive fire on the position drawing its fire and allowing his squad to overrun the bunker with minimum casualties.

Another brave Marine, Lance Corporal David A. Chacon, led an assault by members of his platoon against one of the North Vietnamese redoubts. The enemy was momentarily stunned by the momentum of this attack but soon recovered and returned to directing intense close range machine gun fire at the Marines. Chacon got to his feet and charged the hostile position, silencing it with deadly accurate rifle fire. He overran the position and continued into a small clearing where he was gunned down by interlocking enemy machine gun fire.

Cosner saw Chacon go down. "I didn't know how bad he was hurt and I sent Private First Class Larry Stigall out after him but Stigall was shot the instant he hit the clearing by a sniper on the right flank. In the meantime, Private First Class John Frost, a big guy, was hit by a round from the main force in front of us. After being wounded, Frost stood up and took his flak jacket off. I don't know why. I was amazed that he did it without being hit again. I ran out to get Stigall, a good looking black guy, and drug him back to our squad's position. I could see that Chacon was dead. I believe Lance Corporal Frank Stoppiello got the sniper as I don't remember being fired upon from that direction. Stoppiello was a small, lean Italian. He was a funny guy but good in a firefight."

Stigall had been magnificent since the action began as he delivered suppressive fire upon the sniper infested trees. After being seriously wounded he continued to deliver accurate fire upon the enemy positions.

Corporal Willie Winsley and his squad were on the left side of the hill when the North Vietnamese opened up on them with a withering fire. Winsley went down with his whole biceps muscle shot off and the rest of his squad were either killed or seriously wounded.

Reacting instantly, George Malone, moved among his men and directed them out of the fire swept area. He was soon seriously wounded by the fragments of an exploding enemy mortar round but he chose to ignore his injuries and continued to direct the fire of his machine gunners and grenadiers against the North Vietnamese. When his radio operator was hit and knocked out of the fight, Malone took the radio and used it to direct a coordinated assault against the enemy. They were halted by an intense enemy cross-fire of a combination of enemy weapons, including machine guns, mortar, rocket-propelled grenade, small arms, and automatic weapons fire from bunkers and sniper positions in trees.

Malone moved to supervise the removal of casualties to protected areas and reorganized his squads. He then led an attack by four Marines against an enemy redoubt. Malone fired a light anti-tank assault weapon and threw hand grenades on a daring charge against the North Vietnamese emplacement. He was again wounded, as were his four companions, but he continued the attack and destroyed the machine gun position killing six enemy soldiers.

Fox was finding it difficult to maintain a position near the center of his attacking forces as they moved forward. His Marines were swallowed up by the thick jungle foliage. The Alpha Company Marines were veterans of fighting in the bush, and they went to ground and used the foliage for cover and concealment as they advanced on the enemy. The resourceful Fox was forced to use the sound of their weapons as a guide.

Corporal Richard P. Hodges, a squad leader, took the fight to the enemy. He overran a machine gun position and silenced the North Vietnamese fire from that redoubt. When Hodges spotted several of the enemy soldiers approaching his squads left flank he moved across this hotly contested area and tried to stop this threat to Alpha Company. He was mortally wounded but managed to throw a hand grenade that killed three enemy soldiers and allowed his squad to protect the threatened flank.

Hospitalman Thomas E. Penney crawled under the deadly fire of the enemy machine guns to care for his wounded buddies. He repeatedly exposed himself to hostile fire and took care of the downed Marines. On one occasion, he ran across an open area through enemy machine gun fire, to treat two casualties, one who would not have lived without the first aid that Penney provided. When the corpsman from an adjacent platoon was killed he moved to the aid of the injured Marines from that unit. Penney said, "I guess I was just too busy to think about firing back."

An RPG exploded against the foliage a few feet behind Fox. He was hit and painfully wounded by the hot fragments from the bursting grenade, in the left shoulder and leg, and momentarily dazed. He made a rapid recovery and was able to connect with Lieutenant Colonel George W. Smith, the battalion commander over the radio. Smith directed him to attempt retraction from the area so that it could be pounded by artillery. Fox believed that if a withdrawal was attempted at this time it would result in even greater casualties. They agreed that Delta Company should be sent to assist Alpha Company.

The enemy force was stronger than expected and Alpha Company was low in numbers. The mortar section had been left on the ridge for they would be unable to fire in the thick jungle. The company was now down to ninety-nine Marines without Parnell's squad and two earlier casualties. Delta would not be able to arrive in time to help with the fighting but they would be help in carrying out casualties and providing security.

The Company was facing a battalion of North Vietnamese soldiers and due to the foul weather would be without support from the air. Fox recalls, "Artillery support was out of the question." Charlie Company was up on the nearby hill with two 122mm guns that had been captured the day before by a platoon led by First Lieutenant Archie J. Biggers. They were too close to chance using artillery. "This was definitely a rifleman's fight." Fox concluded.

As the assault stalled, because of the intense enemy fire, Fox moved forward and a sniper fired and missed him by inches. "I saw him as his rifle fired," He recalled, "and returned fire with an M-16 rifle that I had just picked up from a fallen Marine. Though the sniper was close, I did not take a chance with my strange battle sights and fired a half magazine. He slumped into the fork of a tree where his blood flowed down the trunk in a small stream."

Fox decided that he could choose one of two options. He could withdraw or attack. He felt that they might not be able to recover and carry out all of the wounded or dead or defend against a counterattack after the battle. Fox believed that he would lose more men if he attempted to withdraw. That left only one other option and that was to attack.

Gunnery Sergeant Richard Duerr gathered all of the wounded Marines who were not ambulatory, but could still use a weapon. He put them in fixed positions and organized them into a rear defense.

Fox met with the command group and selected the commander of the second platoon, Davis, to lead the attack. He explained to him that it was an all or nothing effort and if Davis's men didn't get it done there would be no other source of help. As Davis started to leave a mortar round landed within the command group and seriously wounded him. Shell fragments also hit many of those gathered there including Fox and both of his radio men.

Cosner remembered that although Davis had been wounded in the back he continued to fight. "When my rifle jammed I took his. He wasn't too happy about that but he couldn't move forward in the attack and he did have his pistol for personal protection."

First Lieutenant Lee R. Herron, the executive officer, was there and uninjured so he was ordered to lead the second platoon in its assault. He quickly organized the platoon into an assault force and took Stoppiello along as his radioman.

As they moved forward to help out the lead elements that had been temporarily pinned down, Lance Corporal William C. Northington, a machine gunner, located and began to pour fire on it until his weapon malfunctioned. He took a rifle from a fallen Marine and continued his assault on the position, encouraging others to join him. Following Northington's lead the platoon soon achieved fire superiority. He moved from position to position and assisted in administering first aid to the wounded.

Northington efforts helped his company regain its momentum. Herron skillfully deployed his men and they aggressively attacked the North Vietnamese until pinned down by a heavy crossfire. They had entered a well designed enemy bunker complex that consisted of machine gun bunkers, mortars, rocket propelled grenades, small arms and hand grenades. They had also placed snipers in the jungle canopy.

Through it all Herron was undaunted and repeatedly exposed himself to the enemy fire. With his encouragement his men inched forward and by doing so were able to deliver more effective fire on the North Vietnamese. They were being held up by fire from two mutually supporting automatic weapons positions and were unable to evacuate their wounded. Herron directed the firing of a LAAW that scored a direct hit on one of the bunkers.

He jumped to his feet and charged across the heavily defended ground, running through heavy North Vietnamese fire. As he reached his objective Herron hurled grenades and fired his rifle into the redoubt and single handedly killed nine enemy soldiers within the emplacement. With the intention of attacking the second bunker he stood up and yelled, "Follow Me!" over his shoulder to his men. At that moment Herron was shot down, and killed by the interlocking fire that also wounded his platoon sergeant. The platoon guide, Sergeant David Beyerlein, took over and continued the attack.

Beyerlein called in and reported that Herron had been killed. Cosner was an eye witness when Herron was shot through the head. Stoppiello, was also hit by the same machine gun fire. He went to him and recalls, "Frank was hit bad. I talked to him and he said that he couldn't feel anything from the waist down."

When Beyerlein assumed control of the platoon Cosner took over as squad leader and quickly organized his men to launch an assault. He led his squad forward through intense enemy fire and attacked a hostile machine gun bunker. They were successful in silencing the enemy weapon and in destroying the emplacement.

The platoon was again pinned down but Beyerlein knew the location of the other gun emplacement and stood in full view of the enemy soldiers and repeatedly fired LAAW's until the North Vietnamese position was destroyed. The second platoon was then able to continue their assault. But Beyerlein was not finished. He repeatedly exposed himself and then killed an enemy sniper with an accurate shot from another LAAW. After he almost single handedly routed the North Vietnamese, Beyerlein supervised the efficient evacuation of the wounded and set up a defensive position.

Lance Corporal John R. Baird, Jr., armed with an M-79, started to put out rounds at a rapid rate as he blasted the enemy positions with grenades. Without the mortar section his weapon was badly needed. His fire enabled his fellow Marines to assault the remaining bunkers. One of the bunkers was protected from Baird's grenade fire by thick foliage. He moved to a far more dangerous position that allowed him to take it under fire. Now exposed, Baird was soon seriously wounded but continued his efforts to knock out the redoubt. He eventually lost this duel with the enemy gunners and was killed.

Christman realized that the lead element was pinned down so he moved his platoon to the right flank and led them in a coordinated attack until they were also pinned down by the murderous fire from hostile positions which included snipers in the trees.

With enemy rounds impacting all around him, Christman fired

a LAAW at a bunker, and then charged across an open area, through heavy enemy fire and tossed grenades into another emplacement. In the process, he killed seven North Vietnamese and destroyed their machine gun. His aggressive and courageous actions allowed his platoon to move through the enemy positions. While attempting to fire a LAAW against another redoubt, Christman was wounded.

Borboa was close to Christman when the Lieutenant received his first wound. "He continued to direct the troops and he was fighting with an M-16 himself. He requested two additional LAAW's. He fired both of them and threw hand grenades. Lieutenant Christman was severely wounded but continued fighting until he was wounded again. He was hit at least four times during a period of about twenty minutes. He rose to fire the last LAAW and got caught kneeling by a volley of rounds which hit him in the pelvic and abdominal area and he was unable to get up."

Propping himself up on one arm and summing what strength remained, Christman directed his men in outflanking and destroying the enemy emplacement. Meanwhile, Malone boldly moved across the fire swept left flank and directed the transfer of casualties to covered positions and reorganized his squads. He then fired a LAAW at an enemy bunker and hurled hand grenades as he led four of his Marines in an assault on a North Vietnamese position. Malone was wounded once more as were his four companions but he continued his assault and destroyed the enemy redoubt and killed six soldiers within.

Malone was now down and now out of the fight but Staff Sergeant Robert R. Jensen took over the platoon and let another assault on a position held by an estimated battalion sized unit. As they moved through the contested area, Jensen led by example and pointed out important targets to his anti-tank assault men. He remained on his feet during the fight and continuously fired at the enemy snipers.

The seasoned veteran, Duerr made this observation. "All the time the Reds were firing at us, Lieutenant Fox was directing our return fire and even after he was hit by shrapnel, he stayed on the radio and gave orders to the platoon commanders."

It was a dark and dreary day with low clouds and rain moved constantly through the thick jungle growth. Fox soon learned that every other officer in Alpha Company was down. He recalls, "Sergeant Beyerlein called to say that Lieutenant Herron had been killed by machine gun fire. Staff Sergeant Michael L. Lane soon added that Christman was seriously wounded with machine gun bullets in his chest and George Malone was down with a second wound. But in Marine tradition, the platoon sergeants knew what had to be done, and Marines were doing it."

Lane unhesitantly assumed command of his platoon and continued the assault against the North Vietnamese positions. He moved about the fire swept area and provided his men with encouragement to move forward by setting an aggressive example. Lane skillfully directed the fire of the members of his platoon and coordinated the attack with Lieutenant Fox and other elements of the company. At the same time he coordinated with fixed wing aircraft. Lane provided calm and heroic leadership when it was desperately needed.

Lance Corporal Thomas D. Horn, new to the company and considered to be a foul-up by many of his fellow Marines rose to the task near the end of the battle. When most of his buddies were running out of ammunition he ran from one downed Marine to the next and retrieved the ammunition from those to who could no longer use it. He then distributed these munitions to those still able to carry the fight to the enemy.

The machine gun that killed Herron was still in operation and holding up the advance of the Marines. It was located in a commanding position and had an excellent field of fire. This gun was the key to the North Vietnamese defense and it appeared that others would be killed or wounded if an attempt were made to knock it out.

The thick mist that still covered the jungle had prevented any assistance from the air. But all of a sudden there was a break in the clouds and the area was bathed in sunshine. Two OV-10 Marine Broncos flying overhead spotted the opening and called in and asked if Alpha Company needed any help in the assault. They answered in the affirmative.

As he led the advance through the heavy North Vietnamese fire, Fox knocked out one emplacement himself and calmly directed an assault on the others. The Marines aggressively drove the enemy from these positions with rifle fire and hand grenades. As the North Vietnamese soldiers retreated Fox, coordinated the rocket and cannon runs of the Broncos with the advances of his platoon. He had his men mark their forward positions with smoke to guide the aircraft during their attacks. Fire from the deadly machine gun was finally halted.

In the almost continuous action Alpha Company had used up all of its LAAW's and most of its small arms ammunition. Fox received word that Delta Company was on its way to assist them. With these factors added to the fact that there were no sounds coming from the North Vietnamese bunkers he decided not to continue the attack.

During the final attack Fox had been wounded again in the leg and hips by fragments from exploding RPG rounds, but he refused evacuation. By nightfall he had reorganized his severely depleted platoons and with the help of Delta Company led the evacuation of the wounded.

When Delta Company arrived the North Vietnamese were gone and there was nothing left to do but gather up the dead and wounded and carry them up the ridgeline. Twelve Marines had been killed and sixty one had been wounded.

Many of the wounded from Alpha-1-9 were hit by shell fragments, Privates First Class Daniel R. Barker in the vicinity of the right arm and hip, George V. Black in the left forearm, Robert A. Kluesner in the left hand, head and neck, Jack D. Moore above the right eye, Dale S. Sperry to the right arm and hand and the left side of his back. Lance Corporals Albert H. Canton to the right side, Harold S. Dickson to the forehead and Louie J. Shepheard in the left arm.

Gunshots wounds were suffered by Lance Corporals Patrick D. Ashby to the left hand, Bobby R. Thomas to the left shoulder and chest, Privates First Class Robert S. Cave to the left hand, Julius Drummond to the left hand, Thomas R. Ryan in the left shoulder and chest and Juan A. Tamez to the left side.

Also wounded were Lance Corporals Daniel E. Fry, Michael P. Hester, William S. Logan and Richard C. Schrader, Jr., and Privates First Class Ira E. Hutto, Lester T. Jackson, Isaac Johnson and William B. McHughs. Lance Corporal Isieah Releford, Jr. was wounded in the right side, thigh and back by shrapnel on this day and died from his wounds on March 13, 1969.

Several Marines from the Headquarters and Service Company were also wounded. These included Corporal Fred M. Walters, Jr. and Privates First Class Simpson Conner, Jr., Richard T. Converse, Nicholas R. Peretta and Salvador Vargas, and Lance Corporals Harold A. Cathie and Billy Jordan.

Of the original strength of five officers and 148 enlisted men, only Fox and 65 enlisted men remained available for field duty following the day's violent and decisive combat. The North Vietnamese had left the bodies of 105 soldiers littering the battlefield.

NOTES:
The following Marines were killed on February 22, 1969:
Baird, John R. Jr. LCpl. 2423144; A-1-9; 19; Oak Lawn, Illinois
Boehm, Larry J. Pfc. 2469072; A-1-9; 19; San Felipe, Texas
Butler, Fred III Pfc. 2491612; A-1-9; 20; Miami, Florida
Chacon, David A. LCpl. 2409589; A-1-9; 20; Gilcrest, Colorado
Chittester, Norman P. Pfc. 2445870; A-1-9; 19; Falls Creek, Pennsylvania
Christman, William J. III 2dLt. 0106531; A-1-9; 23; Gaithersburg, Maryland
Dedek, John F. Pfc. 2303149; A-1-9; 22; Oak Hill, New York
Herron, Lee R. 1stLt. 0102874; A-1-9; 23; Lubbock, Texas
Hodges, Richard P. LCpl. 2172138; A-1-9; 20; Fulton, Georgia
Johnson, Anthony L. Pfc. 2486087; A-1-9; 20; Buena Vista, Virginia
Joyce, Walter A. LCpl. 2321491; D-1-9; 19; Scarsdale, New York
Parker, Richard E. LCpl. 2321643; A-1-9; 20; New York, New York
Pollard, Richard Pfc. 2450770; A-1-9; 19; New Sarpy, Louisiana
Releford, Isieah Jr. LCpl. 2400900; A-1-9; 21; Yatesville, Georgia (DFW 3/13/69)
Thomas, Allen Pfc. 2479137; A-1-9; 20; Youngstown, Ohio

There were many awards for heroism on this day. The Medal of Honor was earned by First Lieutenant Wesley L. Fox.

The Navy Cross was awarded to Second Lieutenants William J. Christman, III, and George M. Malone, Jr., and First Lieutenant Lee R. Herron. The awards to Christman and Herron were posthumous.

The Silver Star was earned by Lieutenant Colonel George W. Smith, First Lieutenant Archie J. Biggers, Lance Corporals John R. Baird, Jr., David A, Chacon, Darrell H. Chapman, Michael P. Hester, Richard P. Hodges and William C. Northington (KIA 5/4/69), and Sergeant David Beyerlein and Staff Sergeants Robert R. Jensen and Donald F. Myers. The awards to Baird, Chacon and Hodges were posthumous.

The Bronze Star was awarded to Staff Sergeants Joseph L. Guy and Michael L. Lane, Privates First Class Paul G. Runick, Jr., and Larry S. Stigall and Hospitalman Thomas E. Penney.

Lance Corporal Thomas A. Mehl, an artillery forward observer, with Battery D, 2nd Battalion, 12th Marines, attached to Delta Company considered the story of Alpha Company's ambush as one of his most vivid memories of Vietnam.

"We were digging in for the night when word came for us to saddle up. We moved past the perimeter of Charlie Company and started down the steep side of a hill. It was dark and the ground was slippery. Stop and go, stop and go, then we stopped for what seemed like an eternity. Word was passed back looking for IV's. Hell, I didn't even know what an IV was. Then we began to move again and, then, there was a sight like I have never seen in my life. At the foot of the hill lay Alpha Company—the foliage, mostly banana trees and elephant grass—was completely cut down for an area of about fifty square feet and within that area lay bodies strewn everywhere.

Dead enemy snipers were hanging out of trees with their blood and guts dripping onto the banana trees. I thought to myself, 'My God, this isn't real. This is a scene from some strange war movie!' But the moaning and groaning of the wounded and dying men caused me to realize that it was for real."

Second Lieutenant William G. Brown of Delta Company also vividly recalls the evacuation. "It was what I considered one of the most loyal and courageous moves of the whole operation. It was a brilliant display of not only physical courage and mental endurance but of the fidelity that these Marines had for their dead and their wounded who they believed it ultimately necessary to get up to the landing zone so they could be choppered out the next morning." Brown distinguished himself with his heroic actions on 19 February when he led his mortar section up a steep incline to the top of a hill. Once there, Brown realized that the vegetation was too thick for the mortars to be fired so he had his men clear several positions and commenced delivering devastating 60mm fire upon the enemy.

Hospitalman Third Class Paul A. Perzia remembers, "Alpha Company's wounded were given basic life sustaining aid which fundamentally included the basics: start the breathing, stop the bleeding, and treat for shock.

All evening and throughout the morning hours we treated wounded Marines who had been intact and healthy earlier in the day and who were now mangled, torn, and dying. Some of the Marines had penetrating gunshot wounds, the most serious being to the chest. Other Marines had both blunt and perforating shell fragment wounds, while still others had partially blown off legs and arms.

Usually the wounded men were quiet. They just lay there wondering, waiting their turn to be treated. Most of Alpha Company's casualties were low ranking enlisted men who looked like kids. They weren't angry or bitter. I suppose that would come later. Right now, they were satisfied just to be alive. They weren't especially frightened, at least not nearly as scared as they probably had been during the battle from which they had escaped. Their pain and demeanor was highly impressive."

The book, *Don't Tell America* by Michael R. Conroy is now out of print. It was published in limited numbers and it is now difficult to find. I tried to obtain one for James L. Johnson, Jr., who earned the Navy Cross on Operation Dewey Canyon, and whose exploits are described in the book. I located one in fair condition that was selling for $166.00. The book is chock full of information and includes the names of many Marines who participated in this operation. It was an invaluable source of information in writing this story.

Others that were either wounded or mentioned in the chapter or as follows:
Ashby, Patrick D. LCpl. 2353026; A-1-9; 20; Beschutes, Oregon
Barker, Daniel R. Pfc. 2446137; A-1-9; 19; Pittsburgh, Pennsylvania
Biggers, Archie J. 1stLt. 0107246; C-1-9; 25; San Diego, California
Black, George V. Pfc. 2339760; A-1-9; 19; Raleigh, North Carolina
Brown, William G. 2dLt. 0105880; D-1-9; 22; Graceville, Minnesota
Borboa, Oscar F. Pfc. 2423905; A-1-9; 19; Phoenix, Arizona
Canton, Albert H. LCpl. 2423561; A-1-9; 19; Chicago, Illinois
Cathie, Harold A. Cpl. 2395656; H&S-1-9; 20; Pittsburgh, Pennsylvania
Cave, Robert S. Pfc. 2477364; A-1-9; 18; Hartford, Connecticut
Chapman, Darrell H. LCpl. 2459791; A-1-9; 20; Claremont, New Hampshire
Connelly, Edmond J. III 1stLt. 094238; H&S-1-9; 23; Los Angeles, California
Conner, Simpson Jr. Pfc. 2401244; H&S-1-9; 19; Atlanta, Georgia
Converse, Richard T. Pfc. 2267007; H&S-1-9; 21; Dallas, Texas
Cosner, Leonard LCpl. 2350489; A-1-9; 19; Baltimore, Maryland
Daily, William H. LCpl. 2439806; C-1-9; 19; Chicago, Illinois
Davis, James H. 2dLt. 0106541; A-1-9; 22; Unknown
Decker, James H. Pfc. 2414018; A-1-9; Long Beach, California
Decker, Joseph T. LCpl. 2342597; A-1-9; 20; Cincinnati, Ohio
Dickson, Harold S. LCpl. 2403967; A-1-9; 20; Detroit, Michigan
Drummond, Julius III Pfc. 2454532; A-1-9; 19; Shreveport, Louisiana
Duerr, Richard G. GySgt. 1361278; A-1-9; 34; Chicago, Illinois
Fish, Robyn W. Pfc. 2470216; C-1-9; 18; Bloomington, Indiana
Fox, Wesley L. 1stLt. 096702; A-1-9; 37; Round Hill, Virginia
Frost, John H. Pfc. 2491685; A-1-9; 19; Hollywood, Florida
Fry, Daniel E. LCpl. 2434285; H&S-1-9; 18; Des Moines, Iowa
Foulkes, Gilmore I. Pfc. 2467472; A-1-9; 19; Brooklyn, New York
Guy, Joseph L. SSgt. 1866259; A-1-9; 28; Pittsburgh, Pennsylvania
Hester, Michael P. LCpl. 2434020; A-1-9; 20; Dallas, Georgia
Hickman, David W. Cpl. 2407987; A-1-9; 19; Blooming Valley, Pennsylvania
Horn, Thomas D. LCpl. 2404217; A-1-9; 19; Detroit, Michigan
Hudson, Charles H. HM2 B717638; H&S-1-9; 21; Houston, Texas
Hutto, Ira E. Pfc. 2429223; A-1-9; 18; Jacksonville, Florida
Jackson, Lester T. III Pfc. 2428032; A-1-9; 19; Houston, Texas
Jensen, Robert R. SSgt. 1641814; A-1-9; 29; Janesville, Wisconsin
Johnson, Isaac Pfc. 2475165; A-1-9; 18; Salt Lake City, Utah
Johnson, James L. Jr. Cpl. 2288355; E-2-9; 20; Plymouth, Michigan
Jordan, Billy LCpl. 2415706; H&S-1-9 24; Chicago, Illinois
Kelly, John A. Capt. 088887; C-1-9; 20; New York, New York
Kluesner, Robert A. Pfc. 2457842; A-1-9; 18; Des Moines, Iowa
Kuhl, Kenneth S. LCpl. 2374282; H&S-1-9; 19; Ashland, Kentucky.
Lane, Michael L. SSgt. 1920154; A-1-9; 26; Pine Bluff, Arkansas
Lemay, George C. Pfc. 2434255; A-1-9; 18; Des Moines, Iowa
Logan, William S. II LCpl. 2405250; A-1-9; 19; Lexington, Kentucky
Malone, George M. Jr. 2dLt. 0105740; A-1-9; 22; Portland, Oregon
Martin, Robert L. LCpl. 2391123; A-1-9; 20; Jacksonville, Florida
McHughs, William B. Pfc. 2458824; A-1-9; 19; St. Louis, Missouri
Mehl, Thomas A. LCpl. 2417071; D-1-9; 22; Baltimore, Maryland
Moore, Jack D. Pfc. 2455903; A-1-9; 19; Oklahoma City, Oklahoma
Myers, Donald F. SSgt. 1277825; C-1-9; 34; Indianapolis, Indiana
Northington, William C. LCpl. 2407330; A-1-9; 20; Prattville, Alabama

Parnell, Randall B. Cpl. 2353634; A-1-9; 20, Montgomery, Alabama
Penney, Thomas E. HN B321541; A-1-9; 22; Fort Lauderdale, Florida
Peretta, Nicholas R Pfc. 2467198; H&S-1-9; 19; Brooklyn, New York
Perzia, Paul A. HM3 7756224; C-1-9; 22; Tampa, Florida
Pritchett, Eddie R. Pfc. 2367818; H&S-1-9; 20; Cleveland, Ohio
Releford, Isiah Jr. LCpl. 2400900; A-1-9; 19; Yatesville, Georgia
Runick, Paul G. Jr. Pfc. 2485954; H&S-1-9; 19; Chicago, Illinois
Ryan, Thomas R. Pfc. 2458502; A-1-9; 19; Minneapolis, Minnesota
Schrader, Richard C. Jr. LCpl. 2438075; A-1-9; 19; Philadelphia, Pennsylvanai
Shaw, Franklyn W. LCpl. 2255153; H&S-1-9; 19; New York, New York
Shepheard, Louie J. LCpl. 2384762; A-1-9; 19; New Orleans, Louisiana
Smith, George W. LtCol. 050104; CO-1-9; 43; Camp Hill, Pennsylvania
Sperry, Dale S. Pfc. 2351384; A-1-9; 19; Milwaukee, Wisconsin
Stigall, Larry Pfc. 2447110; A-1-9; 19; Louisville, Kentucky
Stoppiello, Frank LCpl. 2256921; A-1-9; 19; Newark, New Jersey
Tamez, Juan A. Pfc. 2224745; A-1-9; 19; Salt Lake City, Utah
Thomas, Bobby R. LCpl. 2339409; A-1-9; 19; Raleigh, North Carolina
Vargas, Salvadore Pfc. 2426190; H&S-1-9; 19; Oakland, California
Walters, Fred M. Jr.LCpl. 2364132; H&S-1-9; 20; New Orleans, Louisiana
White, Levoid Sgt. 2073108; A-1-9; 23; Raleigh, North Carolina
Winsley, Willie J. Cpl. 2415819; A-1-9; 18; Chicago, Illinois
Winter, Gary M. Pfc. 2444533; H&S-1-9; 20; Unknown

✶ ✶ ✶ ✶

On February 23, 1969 the Fourth Marines, under Colonel William F. Coggins, initiated Operation Purple Martin in northwest Quang Tri Province. It was a multi-battalion search and clear operation in the area west of Khe Sanh. The 246th NVA Regiment was believed to be moving on a broad front through this area.

When the Marines of Charlie Company, First Battalion, Fourth Marines attempted to reoccupy LZ Mack on the morning of 2 March they clashed with the 246th. LZ Mack was located on a hilltop north of the Elliott Combat Base (the Rockpile). The North Vietnamese pounded Charlie Company with mortars and they had to be reinforced by Lima Company, Third Battalion, Fourth Marines by that afternoon. Dense fog and a steady drizzle made air support impossible and the two companies were forced to withdraw under heavy pressure. They consolidated their defensive positions and waited for the weather to clear before resuming the attack.

The rain did not stop the enemy and for the next three days the North Vietnamese kept up an almost continuous mortar barrage. In addition they used sniper fire and nightly ground attacks against the Marines. During these assaults fifteen Marines were killed. On 5 March, in the afternoon, the weather cleared and the Marines placed an extensive air and artillery bombardment on the enemy positions and the attack on LZ Mack was resumed.

> "People far removed from the reality always seem to possess such certainty, such righteousness."
>
> First Lieutenant Ernest Spencer
> Delta 1-26
> From *Welcome to Vietnam, Macho Man*

Information used to write this story was obtained by interviews with the participants, the book Matterhorn by Karl A. Marlantes and the award citations of many of the participants.

Also used were the rosters and diaries of the participating units and information obtained from the National Personnel Records Center.

13

Hill 484

On February 28, 1969 Charlie Company, First Battalion Fourth Marines was flown into action as the Bald Eagle reaction force as part of Operation Purple Martin. Second Lieutenant Richard D. Porrello's platoon was assigned the mission of securing an enemy held hill near Khe Sanh and north of the Rockpile.. During the assault the Marines came under intense automatic weapon and machine gun fire from several fortified and mutually supporting bunkers. When the platoon was pinned down and unable to advance, Porrello calmly rallied his men and directed an orderly withdrawal so that he could call in fixed wing air strikes against the North Vietnamese positions. As they pulled back he discovered that two wounded Marines were missing. Porrello immediately made his way back up the hill and located the two casualties lying fifteen meters from an enemy machine gun position. He quickly dragged the two Marines out of the line of fire. After the aircraft strafed the area, Porrello again led his platoon up the hill in an aggressive assault, and despite heavy resistance, the Marines were able to reach the top. He destroyed an enemy position and directed his men as they destroyed and captured the remaining North Vietnamese positions.

One of Porrello's squad leaders, Corporal Clifton L. Broyer, fearlessly charged one of the enemy redoubts whose fire had wounded several Marines. He threw hand grenades and fired his weapon as he moved toward the emplacement and in doing so was shot and killed by the small arms fire of the North Vietnamese. Inspired by Broyer's aggressive leadership, his men routed the hostile soldiers and secured their objective. Also killed in the assault was Lance Corporal James L. Vroom.

Among the wounded were Corporal David L. Goodnight, and Privates First Class John M. Kennedy and Octavio Soto, Jr., who was wounded in both legs.

On March 1, 1969, Second Lieutenant Thomas E. Noel was directed to seize the hill against a well fortified bunker complex. He quickly led his Marines up a slope, and as they neared the top the men came under enemy fire from small arms and automatic weapons. They were soon pinned down in a dangerously exposed position by mortar fire. Noel then pinpointed the major sources of the North Vietnamese fire, took two grenades pulled the pins, and boldly raced toward the enemy bunkers. As he charged the emplacements he was repeatedly knocked down by the concussion of the exploding mortar rounds and wounded by shell fragments. Despite a desperate struggle he reached the bunker complex and was able to toss the grenades through the apertures of two of the bunkers, killing all of the occupants. Noel then led his Marines in an aggressive attack on the remaining emplacements. During the fierce engagement he personally carried casualties to places of safety. Noel would not accept evacuation until the position had been consolidated and all of his wounded had been evacuated.

On this day, Porrello again stood out. When he realized that the second platoon was pinned down he moved his Marines around the flank of the enemy positions and approached them from the rear. When the North Vietnamese detected the movement they delivered accurate mortar fire on Porrello's men. He was severely wounded in the neck and chest but refused to quit. Porrello's wounds determined that he required an emergency evacuation but he refused it and directed the fire of his platoon until the enemy fire was silenced. He refused to leave until all of the wounded had been evacuated from the field.

When the first platoon came under heavy mortar fire, Gunnery Sergeant Edwin R. Ring rushed across the fire swept terrain and kept his men on line in their proper assault positions. When the platoon commander, Porrello, was seriously wounded he assumed command of the platoon and successfully completed the assault of the hill. Once this was done, Ring returned to the wounded Porrello and skillfully treated his wounds, while the corpsmen worked on others who were also seriously wounded. He then carried Porrello toward the landing zone to await medical evacuation. When the area was again subjected to another mortar barrage, Ring, undaunted, continued with his leader to the landing zone. As he moved, he too received a serious fragmentation wound and also had to be evacuated.

When Ring went down, Corporal Cleveland King, Jr., a radio operator, assumed command of the platoon and directed the men to the hilltop objective. This area was urgently needed as a landing zone to extract Marine casualties. After the hilltop was secured and the helicopters were called in, King rushed out into the open and used a smoke grenade to mark the landing zone. During this action he was knocked down by a mortar round that exploded near him, wounding him in the hand. King quickly got to his feet and using his radio directed the helicopter to a safe landing. After the first helicopter departed with a load of wounded Marines, he again exposed himself and directed another helicopter into the zone. King then helped to load the aircraft with more wounded Marines.

Corporal George V. Jmaeff, a Canadian citizen, was a huge man who was extremely popular in the unit and a natural leader. When he realized that a frontal attack would produce excessive Marine casualties, he directed three men to provide covering fire, and armed with his own custom made M-60 machine gun, initiated a lone assault on the enemy redoubt. Although seriously wounded by the fragments of an exploding enemy hand grenade, Jmaeff was determined to obtain his objective. He ignored his painful injuries, destroyed the first enemy position, and killed its occupants. Jmaeff was ordered to have his wounds tended too, but during his treatment he saw that several Marines were in trouble. He ripped the intravenous fluid tube from his arm, and leaving the relative safety of his position rushed to their aid. As he struggled forward in a splendid display of valor he was killed by shell fragments from an exploding mortar round.

While the Marines were pinned down, Navy Petty Officer Stanley M. Bell, III, rushed to the point of heaviest contact. Although wounded, he commenced to treat the injured Marines. Bell remained in this hazardous area, working on the wounded until all of the casualties could be evacuated.

During the heaviest of fighting, the platoon guide, Corporal Edward D. Bull, left his covered position and ran from one fighting position to another, redistributing ammunition. He provided first aid to wounded Marines lying in the open, and then gathered the ammunition no longer needed by these casualties and redistributed it to men who were still able to fight. As Bull continued his effort to keep his fellow Marines supplied with needed ammunition he was seriously wounded by fragments from an exploding mortar round.

Other Marines killed that day were Sergeant Joe G. Rodrigues, Jr., Lance Corporals Raymond R. Delgado and David A. Parker, and Privates First Class Robert P. Menninger and Stanley E. Ur.

Among the wounded were Navy corpsmen Stephen A. Sapp and D.V. Yankey.

The weather refused to clear and the rain continued. For the next three days the Marines of Charlie Company could neither be reinforced nor resupplied from the air. They were under siege conditions imposed by the uncooperative weather and the North Vietnamese soldiers that surrounded them. On 3 March they were attacked and several Marines were wounded and some were killed. Among those killed were Privates First Class Michael E. Angel, Robert E. Gooding, Ray T. Shaum, Ronald J. Shier and Philip J. Taylor.

On March 5, 1969 the weather cleared a little and the Marines of Charlie Company commenced their final assault. First Lieutenant Karl A. Marlantes skillfully combined and reorganized the remaining members of two platoons and led an aggressive assault up Hill 484. Under his leadership the attack surprised the North Vietnamese and gained momentum up the slope and through several enemy emplacements before they were able to recover and muster a determined resistance. When the enemy began to deliver a heavy volume of fire the Marine attack temporarily stalled.

Marlantes, jumped up and ran through the enemy fire until he reached an enemy bunker. He threw a grenade into it and knocked it out. He was now flanking the bunkers and under less direct fire and proceeded to destroy three more enemy bunkers in succession. He was not alone in the assault for within seconds, he had a whole lot of help as, other Marines acknowledged his initiative and joined their leader. One of the first to follow Marlantes was Lance Corporal Jimmy H. Harding, a temporary squad leader with the first platoon, who performed magnificently.

Once the first line of bunkers was captured they then began an assault on a second line of enemy emplacements where Marlantes was hit in the face by fragments from an exploding grenade. He refused to be evacuated until the objective was secured, a perimeter defense established, and all other casualties evacuated.

During the assault the third platoon was pinned down by a heavy volume of machine gun fire. Corporal Yale G. Allen, a fire team leader, with the platoon left a relatively secure position on the opposite side of the hill and worked his way up the steep slope. Although the enemy concentrated their fire on him he used whatever cover he could find until he reached an enemy bunker. As he was throwing a hand grenade into it he was knocked down and seriously wounded by shell fragments from an exploding mortar round. Allen quickly regained his footing and climbed on top of the emplacement and tossed another grenade inside. He then ran to the front of the redoubt and sprayed the North Vietnamese soldiers inside with rifle fire killing all of them and destroying the position. Allen was painfully wounded but continued to lead his men until the objective was attained.

When Corporal Michael C. Biehl, a squad leader, and his men came under fire from a North Vietnamese bunker, he placed elements of his squad on either side of the emplacement and with his men providing covering fire, assaulted the position. When he reached the bunker he threw two grenades into the aperture killing the occupants. Biehl entered the position and captured the weapon within. During the course of the battle he was painfully wounded but continued to lead his men and participated in the destruction of other fortifications. Biehl continued his courageous efforts until the hill was secured.

Serving as a Forward Observer with Charlie Company was First Lieutenant John C. Kegel, who skillfully directed mortar and artillery fire upon the enemy on top of the hill. He was instrumental in the destruction of three North Vietnamese bunkers, but one bunker, manned by three enemy soldiers using hand grenades and automatic weapons fire remained to hinder the attacking Marines. When Kegel realized that he was in the most advantageous position to destroy the enemy emplacement he picked his way across the dangerous terrain toward the redoubt. Armed only with a rifle and two hand grenades he closed to within three meters of the position. Kegel tossed the grenades through the aperture of the fortification.

The explosion killed its three occupants and destroyed the emplacement. His actions allowed his fellow Marines to complete the assault.

NOTES:
The following Marines were killed in action on Hill 484:
Angel, Michael E. Pfc. 2482778; C-1-4; 19; Hayward, California
Brake, Boyd L. LCpl. 2371205; C-1-4; 20; Adairville, Kentucky
Broyer, Clifton L. LCpl. 2249149; C-1-4; 23; Cataumet, Massachusetts
Delgado, Raymond R. LCpl. 2425228; C-1-4; 19; Watsonville, California
Gooding, Robert E. Pfc. 2446796; C-1-4; 20; Ashland, Maine
Jmaeff, George V. Cpl. 2436055; C-1-4; 23; Oliver, BC, Canada
Menninger, Robert P. Pfc. 2480574; C-1-4; 19; Gladstone, Missouri
Parker, David A. LCpl. 2354912; C-1-4; 19; Clarks Summit, Pennsylvania
Rodrigues, Joe G. Jr. Sgt. 2058221; L-3-4; 22; Dallas, Texas
Shaum, Ray T. Pfc. 2499355; C-1-4; 20; Mansfield, Ohio
Shier, Ronald J. Pfc. 2484748; C-1-4; 19; Cadillac, Michigan
Taylor, Philip J. Pfc. 2466389; C-1-4; 19; Holden, Massachusetts
Thomas, Charles W. Cpl. 2280328; C-1-4; 20; Gary, Indiana
Ur, Stanley E. Pfc. 2401227; C-1-4; 20; Austell, Georgia
Vroom, James L. LCpl. 2382343; C-1-4; 18; Fort Wayne, Indiana

During the battle for Hill 484 the Navy Cross was earned by First Lieutenant Karl A. Marlantes, Second Lieutenant Thomas E. Noel and Corporals Yale Allen and George V. Jmaeff. The award to Jmaeff was posthumous.

The Silver Star was awarded to Corporals Michael C. Biehl, Edward D. Bull, Clifton L. Broyer and Cleveland King, Jr., First Lieutenant John C. Kegel, Second Lieutenant Richard D. Porrello, and Gunnery Sergeant Edwin R. Ring. The award to Broyer was posthumous.

The Bronze Star was earned by HM3 Stanley M. Bell, III and First Lieutenants James M. Herron and Richard A. Tilghman, Jr.

Karl Marlantes wrote: "On the first day of the battle, February 28, 1969 we had to take a smaller hill just east of 484. Both were on the same ridgeline (an extension of Mutter's Ridge or Mutter's Ridge itself, I think the actual definition of where that geographical formation starts and ends is somewhat flexible). Porello's platoon got the assignment. We had another platoon enveloping the hill from the south, moving up a north-south finger. We had a third platoon in reserve, set up on a small knob just east of the hill we were going to assault.

"I'd just been moved up from platoon commander of First Platoon. We were involved in three assaults from February 28 through March 6, against fortified positions and a hammer and anvil fight that started it all off which we'd been launched into as the Bald Eagle reaction force on February 28, 1969 and two defensive fights once we had taken those positions, so a lot more ground was covered than in what was cited in my medal.

The weather was bad. Low clouds. The tops of the hills were in and out of the clouds early in the morning, just around dawn. We heard the drone of an 01-Charlie, a two-seater observation aircraft, pilot in front, air observer/naval gunfire/artillery spotter behind him. The 01-C was about the width of a canoe, overhead wings. It was up above the cloud cover, so we couldn't see it. I got on the radio and got in touch with it. I described the situation and asked if they could take a look at the hill we had been ordered to take. This was asking a lot, as it required the plane to descend through clouds to hopefully come out from beneath them and still be above the ground. In that sort of mountainous terrain, low clouds, it was a very dangerous thing to do.

As I was talking to the AO (air observer - the back seat) I got this odd feeling that I recognized the voice. My college roommate from Yale, Richard A. Tilghman, had also joined the Marines and had been a platoon commander with the 26th Marines, down south. He was transferred up north to the 3rd Marine Division when the 26th Marines were pulled out and I ran into him when I first arrived in country. He'd told me then that he was considering volunteering to be an AO. So I said, "Is this by any chance Character Tango?" We never used names over the radio. There was this pause. "Is this character Mike?" I said it was. It was one of those coincidences that if you wrote it up as fiction the reader would have said it was impossible odds and really dumb. Anyway, we talked a little more about the pending assault and Rick and the pilot said they'd go in to take a look. They dove into the soup, came out (luckily) from under it just above the ridgeline and went right over the top of the hill. I think they took fire. They pulled up and disappeared back into the soup and Rick radioed me. I think I remember his exact words, but time of course could make the memory faulty.

They were, "I wouldn't go up there if I were you. There's a machine gun pointing right down the ridgeline." He also reported seeing that there were bunkers and possibly a second machine gun, and that there were quite a few NVA waiting and dug in. He was damned sure of the machine gun pointing right down the obvious avenue of approach, the top of the ridgeline. I told Lt. Herron, the skipper, all of this. He asked Rick if we couldn't get some air support, and the answer was a negative. Bad enough to fly through the soup with a slow spotter plane. Any jet wouldn't be able to find the target and the risk of coming out under the cloud cover and hitting rocks was just way too high.

We'd asked for artillery prep earlier. I don't remember getting any, but we may have. We approached the ridgeline in the dark all that morning, pushing off from a sort of temporary position down in the valley to the south of the ridge around one in the morning. We did radio back the information that the enemy was obviously well positioned for defense and with one and probably two machine guns trained on the main avenue of approach.

At the time we were op-conned to second battalion, so it isn't likely that the skipper talked with George Sargent, but probably the commander of second or third battalion. The reason for this is because a company from second or third battalion, I cannot now remember which company, was supposed to help us in the assault. They too were to approach the ridgeline in the dark, join forces with us to make the assault. They ran into mines. I heard them go off as they got triggered. It spooked the hell out of me as we could hear people screaming who'd triggered the mines. Anyway, their skipper decided to stop moving up toward the ridgeline. I can understand his reluctance, but this sort of left us hanging out there by ourselves and I was not too happy with this guy. A platoon commander from that company actually took his platoon and left the company to come and join us I think a couple or three nights later. I'm not sure if he did this with his skipper's permission or not, I have the feeling he did it without permission, but whether it was or wasn't, I do know he was damned embarrassed about his company leaving us in the lurch. I'd known him from training, but didn't know him well. I wish I remembered his name. Brave guy and a good guy (as was his entire platoon who certainly didn't have to join us.)

It is true that we came up with "the flying wedge breakout" and luckily we didn't have to execute. After two full scale assaults (the first hill, which we called Helicopter Hill because there was a wrecked chopper on it, and the first assault on 484) and then I think two night defensive battles, and no supply, I remember having the machine gunners assemble up on the top of the hill and even up ammo. We had less than a minute of machine gun ammo left and I do believe we were down to fewer than ten bullets per rifleman. If we didn't break through that night, or didn't get resupplied, it would be down to knives and e-tools. We had quite a few wounded.

We hadn't had water for several days, and although we kept redistributing all the water from our dead, we were now getting real thirsty and starting to have people going ineffective because of dehydration. (All of which seemed very ironic given that we were continually surrounded by monsoon clouds. Water, water everywhere but not a drop to drink.) Since we were going to break through the lines that night, Herron felt it was best to get what water we had into the people who would have to do the fighting to get us out, so we distributed the IV fluid.

We'd had one chopper get in, I think on the second day. It got into a very small niche on the east side of the hill, sheltered from 484 where the NVA were dug in and could hit it with machine gun or rifle fire. We took mortar rounds, but I think the place the chopper hovered over on the east side was out of sight from their observers so they were just firing knowing it was there but couldn't see it. Anyway, that chopper had brought in the IV fluid that we were now going to drink. As Pete said, the clouds lifted in the afternoon and Lieutenant Timothy J. Rabbitt, former XO of the company and now with the battalion staff, arrived with the palest-faced most overloaded bunch of replacements you could ever want to see. He had them near buckling with just two things, ammunition and water. Tim told me that he'd talked two pilots into making the run. They still had the same problem as Rick (the AO) did, clouds with rocks in them, but he said he basically appealed to their sense of honor, that the company was desperate and waiting for the "right conditions" might mean losing most of the company. Rumor had it after the whole battle that he had pulled a .45 on them, but Tim told me several years ago when I asked him about it that it was just a rumor and false. He was, however, responsible for getting them to fly that afternoon - and they were brave enough to do it.

With the clouds lifted we got air support and artillery support that we could direct since we could now see where the shells were landing. With the enemy as close to as as they were, adjusting by sound alone was silly and pretty dangerous. That night I led a patrol to see if the enemy was still in place. Pretty scary patrol. I reached their line of holes and found them empty. We checked both ways on the line of holes and made sure they had indeed abandoned the position. The combination of air and artillery, now being able to be directed, made them pretty vulnerable. They were not idiots. So they abandoned their circle around us and probably moved to the west, heading toward Laos or maybe up through the DMZ over into North Vietnam, leaving a unit on 484 to make sure we couldn't pursue them quickly and probably to make us pay if we decided to move to the higher ground of 484.

No one knows for sure how many NVA were directly involved. I was later told that a regiment was moving to the east, along Mutter's Ridge, and that we stumbled right into the middle of that movement. Obviously, we never faced an entire regiment. Even the Marines would have lost that battle. When we kicked the small unit off of Helicopter Hill and then kicked another small unit off of 484, my guess is that what probably happened was that they dropped off sufficient men to isolate us so we couldn't cause any more trouble and would have us fixed in place to do as much damage as they could with mortars. (Which they did, but once we'd dug in, our casualties from the mortar fire were minimal. Where we got creamed was with the mortars when we took 484 the first time and were above ground without being dug in. That's when Pete Porello got hit, and also Canada.)

Also, to be totally honest with you, I did charge four bunkers "by myself", but within seconds, I had a lot of help, as you can well imagine. I always feel a little chagrined, although I had nothing to do with it, that this isn't mentioned in the citation.

Our second (and last) assault, against Hill 484, (the third assault had been on March 1 against a lower adjacent hill, also fortified) had been pinned down by two machine guns from the bunkers (along with a lot of AK and SKS fire) and the whole assault was flat-ass down on the ground and stopped cold with healthy and wounded alike laying out on the open ground under the fire.

The NVA had been plastering us with mortars for days, so I knew within a very short time the mortars would be falling on us as well, strung out as we were on this bare hillside without cover. We had already begun to take RPGs from a couple of points just behind their line of bunkers. However to stand up and go forward, there were the machine guns and NVA firing from bunkers with AK's and SKS's. Decidedly bad ju-ju. To retreat was not only dishonorable, but would have achieved the same effect as standing up and going forward,

I remembered then a red-headed Major at the Basic School telling a bunch of us young butterbars after a field problem that most of the time, quite frankly, some sergeant could do our job just fine or better, or the Marines under them could just do it themselves, but there would come a time, our time, when we would earn all of our pay all at once, and we would know when that time was.

So, this was my time, and I knew it. I stood up and went forward up the hill by myself. It was a long lonely run that felt like hours, but I know that it was only a very brief time. I sensed some movement behind me. I rolled to fire (what was I thinking? That the NVA would get out of their bunkers to attack me? So much for cool level-headed thinking) and saw it was (Harding? I think was his name), one of First Platoon's fire team leaders when I had First Platoon, now temporarily a squad leader because of the losses we'd taken. Coming up behind him, spread down below me across the hillside, was the whole line of Marines. My heart soared. I get choked up just writing about it now. I kept running up the hill, threw a grenade into the first of the bunkers and then just started after the others, because I was now flanking them and under less direct fire. By this time, as I said, I had a whole lot of help. Everyone was coming up behind me and we were all over them.

The NVA retreated to a second line of open fighting holes that they had prepared behind that first line of bunkers and we had to clean them out of those as well. It was during the fight for the second line of holes where I got wounded by a Chicom that knocked me out for a while and I lost the sight in one eye. (It's now OK. They took the shrapnel out some days later on the Repose and the surgeon told me afterwards that a micron one way or the other and I'd have probably lost it. I was very lucky).

Porrello took over my platoon when I was moved up to XO. King was a squad leader in my platoon when I had it and I remember watching him standing in the middle of a bunch of incoming mortar rounds on the first assault on 484, guiding in medevac birds with a radio on his back while shells were falling all over the LZ. I assume that was what the Silver Star was for. Allen was also in my platoon and he was the first man over the top of the hill on 484 on the second assault, having worked his way up the right side. I watched him blow up a bunker, go up in the air with the explosion, and I thought he was dead. But up he popped a few minutes later. Jmaeff died on the first assault on 484 and was a terrific loss to the company as he was one of the natural leaders. He was a Canadian. We had two in the company. He was in a shell hole with a corpsman and I am told that he tore the IV tube from his arm and went after a machine gun that had pinned down that part of the assault, took it out, but died doing it."

Marlantes describes the first time he saw Jmaeff in Vietnam in his fine novel *Matterhorn* when he describes the Marine called Vancouver:

A single Marine had jumped off the back of an incoming helicopter and was walking slowly across the landing zone toward the dirt road that led to the regiment's rear area. The Marine stood six-four or six-five, but his size wasn't nearly as interesting as the sawed-off M-60 machine gun dangling from two web belts hung over his shoulders. An M-60 usually took two men to operate. The book assigned a crew of three. A crude handle had been welded onto the barrel so the Marine could control the kick without resting it on a bipod. Two cans of machinegun belts also hung from his shoulders. In addition to all this weight he carried the usual full pack of sleeping gear, food, extra controls, hand grenades, books, letters, magazines, ponchos for shelter from the rain, shovel, claymore mines, bars of C-4 explosive, trip flares, hand-made stove, pictures of girlfriends, toilet articles, insect repellant, cigarettes, jars of freeze-dried coffee, "long rats" of freeze-dried trail food for special occasions, rifle-cleaning gear, WD-40, and even the occasional can of steak. On his head was an Australian bush hat, one brim folded up at the side. Matted blond hair, discolored with grime, showed beneath it. His uniform was a tattered mass of holes and filth. One trouser leg had been torn off just below the knee, revealing paste-white flesh covered with infected leech bites and jungle rot. His hands face, and arms were also covered with jungle rot and open sores. You could smell him as he walked by. But he walked by as if the LZ belonged to him, seemingly unaware of the hundred or more pounds that he carried. He was a bush Marine and Mellas wanted fervently to be someone just like him.

Karl Marlantes was a graduate of Yale University and Rhodes Scholar at Oxford University before joining the Marine Corps.

In addition to two Silver Stars, Second Lieutenant Richard D. Porrello earned the Bronze Star medal for heroism on January 4, 1969. Charlie Company was conducting a search and destroy operation on the Suoi Tien Hien River Valley near Khe Sanh when the point man was mortally wounded by the explosion of a command detonated mine. Porrello was painfully wounded and temporarily stunned by the same explosion but quickly recovered. When the company came under a heavy attack by a

North Vietnamese Army force he directed the effective fire of his men, and then skillfully adjusted supporting tactical air strikes upon the attacking enemy force. Porrello remained in an exposed position until the North Vietnamese fire subsided. Under his aggressive leadership his platoon killed four enemy soldiers and suffered no more casualties of their own.

Porrello wrote, "Ed Ring was my platoon sergeant. I don't believe there is a finer Marine on this earth than Ed. I have the utmost respect for him and I will never forget him."

"Cleveland King was my radioman and once again I cannot say enough about the character and professionalism he exhibited when we were together. I have not seen or heard from Cleve since I got hit but I certainly would like to."

"I knew George Jmaeff as a member of Charlie Company. He was an extremely gregarious guy who carried a sword and was tragically cut down on 484 while engaged in extraordinary heroism. I believe that George posthumously received the Navy Cross and deservedly so."

"Karl Marlantes was my XO and another whose character and professionalism I greatly respect. I know he also received the Navy Cross."

Tobby Baca writes about George Jmaeff: "He was from Canada and everyone called him that. Canada was the biggest guy in our unit and he was strong as an ox. He must have been six-four or six-five. Nice Guy! I was a machine gunner in Vietnam and that was something that he wanted to be. I think he wanted me to give him my gun, however most gunners would never do that. Jmaeff would ask me what it was like to have all that firepower in a fire fight and he always wanted to know how many of the enemy I had killed. He wanted to kill enemy soldiers. Canada took a machine gun off of a tank and showed it to me. He told me that he was going to have a handle welded on it. My question to him was, "who was going to be carrying his ammunition?" Canada told me that, "if you can carry 2,000 rounds than so can I." He did and he walked point most of the time."

"There was another side of Canada that most guys did not know. I was raised in foster homes and boys homes most of my life. He seemed to have an interest in that. "Why did your parents leave you?" was one of his questions to me. It was not until forty years later when I had the opportunity to talk to his sister that I understood why he asked. The relationship between Canada and his father had not been a pleasant one. This was a secret that he kept to himself."

"One day at LZ Stud, Jmaeff was sitting by himself, writing a letter. I walked up to him and joked about him always writing to his girlfriends. Canada looked up at me with tears in his eyes and said that he was writing to his mother. He asked me to sit down and told me that he was not going home after the next operation. This was so out of character for him. Not about him not coming back, hell, everyone knew that they would get hit.

He was there for one reason and one reason only and that was to kill the NVA.'

"A few days later I was hit bad and sent to the hospital in Guam for a few months to recover. Later, one of the guys from our unit was hit on 484 and also sent to Guam. That was when I got the word about all of the guys that were killed. I was told that Canada was up for the Medal of Honor. Of course that never did happen.'

"Here is an interesting story. One day after pushing the gooks back we found a cave that was full of guns, ammo, medical supplies and rice. We had to blow a landing zone to get that stuff out. There were two things that I remember about Canada that day. After blowing the LZ we had to move the trees that had been knocked down. I was helping to move one of them and looked up. There was Canada on one end and three of us on the other. He sure was a strong man.'

"After we finished clearing the zone Canada went around collecting everyone's sugar and cream. He added it to some of the rice we had found and mixed it in his helmet. I'd have to say that it was the best rice pudding that I ever had.'

"There was another two day period that I remember. Canada went out on patrol and came back. My squad was sent out after his came in and we were hit. He wanted to go back out and join in the fight. The very next day my squad went out and we were hit again. Here comes Canada running down the trail with his gun and funny bush hat but by the time he got there it was all over. I have to laugh because at that time he had yet to see action. By the time he was killed I had been wounded six times. Canada told me that he wanted to be next to me at all times so that he could see more action.'

"There was no better Marine or friend than George "Canada" Jmaeff. I was blessed to have known him.'

"Raymond Delgado was a happy go lucky type of guy. Nice guy and a good Marine. He always had a good story to tell. I have a vest with a patch that states, "In memory of George 'Canada' Jmaeff and Raymond Delgado."

"I remember when Robert Menninger first came to Vietnam. We stayed up all night. He was so scared and worried that he would not be able to keep up in the bush. He was on my gun team when I was wounded. I heard that he got a medal but I am not sure what it was.'

"We simply called Michael Angel, "Angel." His squad leader on Hill 484 told me twice in the last year that he felt terrible about sending Angel out on the Listening Post for two nights in a row. He was killed on the second night. The squad leader could not remember his name but I could not bring myself to tell him that it was Angel. I have been in contact with Angel's sister and brother who found me less than a year ago. They wanted to know what had happened to Angel and it was a difficult question to answer. I never told them about his being on the Listening Post two nights in a row.'

"I'm not certain but I think that Leo Mangold and Ron Shier were on my gun team also."

About Jmaeff, Bob Hatcher wrote: "George Jmaeff was known as "Canada." When I arrived in country and first met George I thought he was the CO. He seemed to know everyone and also took the lead at point. Later after asking around I was told that he was a Lance Corporal. He fashioned an M-60 machine gun into a gun toting sling with some modifications to the stock and barrel.'

"Cliff Broyer was like a big brother to me when I got in country in 1968. I was a machine gunner and Cliff was my squad leader. I will always remember his big smile and the time he spent preparing his Marines for each operation. He left us in 1969 on a hill far from home. He saved many Marines that day with his actions. For that he will always remain my Hero."

Lance Corporal Donald D. Groves was a grenadier with Charlie Company during the assault on Hill 484. He had already distinguished himself during an earlier engagement on January 14, 1969. While on a combat patrol near Khe Sanh the Marines came under heavy small arms and automatic weapons fire from a North Vietnamese Army unit occupying well concealed mountain caves. The point man was seriously wounded and had fallen in a dangerously exposed area. Groves ran across the dangerous area and attempted to reach the wounded Marine. He was driven back four times by the intense enemy fire but would not give up. In a maneuver skillfully coordinated with the fire from helicopter gunships, and despite being wounded during a previous attempt, Groves was able to reach the casualty and bring him back.

Second Lieutenant Robert A. Linn, II did not accompany the other Marines of Charlie Company when they made the final assault on Hill 484. He was assigned to lead a squad that was left at the foot of the hill to guard the gear left behind. In Vietnam he came to be known as "Scar."

Linn remembers, "The name was given to me by my platoon sergeant, Sergeant Raymond Monroe. After my second Purple Heart, he told me that if I continued to get hit, I would look like a big scar when I got home. The name just stuck to me and everyone called me that."

"Charlie 1-4 was in the field almost all of the time and usually came to the rear to resupply or update shots. One patrol lasted around 58 days."

In January of 1969 while on a search and clear operation near Fire Support Base Neville one of Linn's men was wounded by the detonation of an antipersonnel mine. At the same time the squad was subjected to a heavy volume of enemy small arms fire. Linn immediately established a defensive perimeter and directed his men in delivering a heavy volume of return fire on the enemy. He then requested a medical evacuation helicopter and moved his men to a landing zone where the wounded Marine could be evacuated. When this was accomplished, Linn led his men in a search of the area that resulted in the discovery of a large enemy supply cache.

On March 23, 1969 while Charlie Company was assaulting North Vietnamese emplacements near Fire Support Base Argonne another outfit was ambushed in a bunker complex. They pulled out leaving their dead and wounded there. Linn volunteered his platoon to go in and get the dead and wounded out of the enemy bunker complex. It was a mission that they accomplished without anyone getting killed.

On this day, Linn was magnificent as he fearlessly led his platoon in an effort to rescue one of the wounded that lay dangerously close to the enemy redoubt. When he spotted an enemy soldier aiming at the point man, Linn charged the fortification and, firing his rifle from the hip and tossing grenades killed the North Vietnamese soldier. As he supervised the search for documents that might be on the dead soldier's body, two grenades, thrown by another of the enemy, located within the bunker, landed near a Kit Carson scout. Linn shoved the scout away and continued his assault on the emplacement and killed the other North Vietnamese soldier.

But Linn was not done yet, as he directed the evacuation of the casualty and the continued search for intelligence information, a hostile soldier arose from a concealed position and threw a satchel charge at the Marines. He then commenced firing his rifle at them. Linn again, acting alone, assaulted and killed the attacker. For his actions this day he earned the Silver Star.

On April 13, 1969, Linn's platoon was conducting another search and clear operation near Fire Support Base Argonne when the Marines ran into a squad of North Vietnamese soldiers. As the firefight heated up an enemy hand grenade landed beside his radio operator. Linn moved to him and knocked the Marine to the ground, which resulted in the man being only slightly wounded. The intrepid Marine then continued to direct a heavy volume of fire at the enemy causing them to break contact and leave behind two of their dead.

Others mentioned in the chapter are as follows:
Allen, Yale G. Cpl. 2406509; C-1-4; 20; Cincinnati, Ohio
Baca, Tobby P. Cpl. 2410973; C-1-4; 18; Los Angeles, California
Bell, Stanley M. III HM3 B213007; C-1-4; Roanoke, Virginia
Biehl, Michael C. Cpl. 2353385; C-1-4; 22 Newark, New Jersey
Bull, Edward D. Cpl. 2306688; C-1-4; 21; Memphis, Tennessee
Garza, Franco Pfc. 2469061; C-1-4; 18; San Antonio, Texas
Goodnight, David L. Cpl. 2410106; C-1-4; 20; Los Angeles, California
Groves, Donald D. LCpl. 2409669; C-1-4; 21; Denver, Colorado
Harding, Jimmy H. Cpl. 2291090; C-1-4; 20; Beckley, West Virginia
Herron, James M. 1stLt. 013563; C-1-4; 24; Port Hueneme, California
Kegel, John C. 1stLt. 0104321; C-1-4; 26; Pittsburgh, Pennsylvania
Kennedy, John M. Pfc. 2252389; C-1-4; 21; Baltimore, Maryland

Linn, Robert A, II 2dLt. 0106604; C-1-4; 23; Bloomsburg, Pennsylvania
Marlantes, Karl A. 1stLt. 0103269; C-1-4; 24; Seaside, Oregon
Martinez, Gilbert L. Cpl. 2424883; C-1-4; 20; San Francisco, California
Monroe, Raymond Sgt. 1925138; C-1-4; 26; New York, New York
Noon, Patrick J. Jr. Sgt. 2016255; C-1-4; 22; Cheverly, Maryland
Porrello, Richard D. 2dLt. 0105432; C-1-4; 22; East Lynne, Connecticut
Rabbitt, Timothy J. 1stLt. 0104749; AO-1-4; 23; Holyoke, Massachusetts
Ring, Edward R. GySgt. 1201472; C-1-4; 35; Hartford, Connecticut
Sargent, George T. Jr. LtCol. 051686; CO-1-4; 39; Auburn, Alabama
Sapp, Steven A. HM3 B607118; C-1-4; 21; Minneapolis, Minnesota
Soto, Octavio Jr. Pfc. 2448593; C-1-4; 19; Fort Myers, Florida
Spencer, Ernest Capt. 010000; D-1-26; 23; Honolulu, Hawaii
Tilghman, Richard A. Jr. 1stLt. 0103014; AO-H&S-3MarDiv; 24; Philadelphia, PA.
Yankey, D.V. HM3 B5233981; C-1-4;

★ ★ ★ ★

Robert Dalton Buster was a 21 year old fire team leader with India Company, Third Battalion, First Marines in early April of 1969. His company was operating in an area located about twenty kilometers south of Da Nang called Dodge City by the Marines.

It consisted of forty square kilometers of hostile terrain. Long a hotbed of VC activity it was called Dodge City because of its shoot-'em-up atmosphere. This area is lowland dotted with rice paddies and swamps and honeycombed with camouflaged caves and tunnels. Eight villages were spaced across the area. It was estimated that elements of three NVA regiments called Dodge City home and that the VC operated a tight infrastructure there.

"Maybe I've died, gone to Hell and ain't been told."

LCpl. Robert Dalton Buster
I-3-1
From *The Walking Dead*

"War is a time of exaggerated emotions.
Love and hate were as close to us as life and death."

First Lieutenant Ernest Spencer
D-1-26
From *Welcome to Vienam Macho Man*

Grateful acknowledgement is made to Nancy Jane Buster for permission to reprint previously published material. The following excerpt is taken from *The Walking Dead*© 1987 by Robert Dalton Buster by permission of Nancy Jane Buster.

Buster writes in a unique and sometime confusing style that seems very appropriate for the mess that was combat in Vietnam. No corrections have been made in spelling or grammatical form.

The notes and information following this story was gleaned from the award citations of many of the participants.
Also used were the rosters and diaries of the participating units and information obtained from the National Personnel Records Center.

14

Dodge City

134 and a wakeup, April 1, '69
Dear Ma,

It is hot and tomorrow at Dodge City is going to be bad. There's no telling how long we'll be gone, so don't worry if you don't hear from me for awhile... Hope I'm wrong but I think we'll lose a few men in the next couple days... The new men in my squad are OK-they're all pretty young 18-19 and we don't have that much in common--plus Viet Nam is still new to them and they simply are boot and boots do some pretty stupid things.

'THE WORD' a one day gig.

Revved/up Major hustlin' around, says no flak jackets. Lt Coarser and Sarge put an end to that shit. Told newer troops, roll shirt tight and weave into web belt and take some bug fuck. Yeah, seen one day things stretch-on--out. Issued one meal, 4 frags and guns' ammo. 'the word' large NVA unit--go get 'em.

Took fire comin' off the birds--no one hit. Sweepin' is the name of the game. Got on line and started doin' it. About half an hour and someone tripped a big one-7 hit, 1 or 2 probably won't make it. It was down range, 'the word' "hold your position'. Moseyed over to Cherry, Hoa Bro, what you know?" Got fisherman's pack off helmet, flippo/Zippo, "You smoke too much."

Medivac came and went. Started back up--almost immediately BOOM!! Schrapnel cuttin' brush and bamboo--3 hit. This one was closer-could hear'n see the pain-the decent news, no one dead or gonna be--soon, the Morphine takes holt and it is quiet. Bird came'n went.

Man, this area is 'for sure' trapped--back in gear, I'm LOOKIN' for those monofilament tripwires. BOOM!! and RFC (real fuckin'close) HIT THE DECK, someone is screaming-UP and moving to the sound. Man, it's Blair--both legs gone, he grabs my right hand, I'm lyin' my ass Off, Its OK, Man; everythings gonna be OK." "Button Vermillion! Button Vermillion! This IZ, an emergency medivac!" Tourniquets on stumps, Morphine for the pain and red M greased on forehead. The bird came--me and Cee loaded him--the bird left.

After more sweepin', we setup for the night-had enuff bug fuck--THE ONLY decent news. Dead quiet-ate our rats.

Resupply came in next morning and sweep went ok til around noon, when napalm setoff brushfire(s). It was up Lima's way--we ran to help out. And son of a, son of a, son of a bitch WALL OF FLAME!! Comin' right at us! newer troop tried to run, grabbed his flak jacket--fired magazine straight/up, pafted my helmet /w right hand and pointed- deep breath and ran straight INTO IT-5 yards and 10 and 15 and 20 and 25-I'm on fuckin'fire but am past the wall. Squad had followed--time to go help out.

It waz NUTS CITY-cats down and goin'down everywhere. We started draggin'to a center point--they're throwin'/in'/up--smells of burnt hair'n flesh. A guy in new gear staggered up. Tryin' to help, we got tangled and he went down hard, "OOMPH." After 90 seconds or so, he acted like he was gonna try'n sit/up-put hand on his chest, "Stay down!"

At least 60 medivaced and more that should've been. It 'for sure"most definitely' ain't no lie, we don't need an enemy; we got the Corps. Heat casuality business took the afternoon--we setup. "Tell you what, Cee, they keep this shit up and there ain't gonna be nobody left." 'Tell ME about it."

Come dawn 'the word' was, we're headed' in'. Man, no mail in 3 days. Told my people, "Keep your minds on fuckin' business!" Long ago Tom told me and experience has reinforced it-this is where people get sloppy-- thinkin' about Hot Chow, showers and The Mail...

QUICK 'N HOT

"Never saw a bomb bounce before." I said flatly and from the mutterings, neither had anyone else.

Thought I'd seen it ALL but this bomb at treetop level, doin' 400 plus mph was a first. If it bounced right, or wrong, there's a ville and an old, abandoned catholic church down that-a-way.

A piece of Hot Metal froze everything--comin' quick n' hard. No one had a chance to move. Instinctively/turned, following the sound--no more than 15 feet up/slope stood a headless grunt!! My mirror image, rifle in right hand-arterial blood gysering straight up! His head rolled towards me til I stopped it with my hand--face all pull/twitch 'n sneer. Got up and wedged it under his right arm. Believe it took me awhile.

Turned on squad. (Mercifully, platoons ended and began with me-- we didn't know him.) "Once a head, always a head." (People who smoke dope are called 'heads', short for dopeheads.) <Right, it ain't funny--tell you what, 'at ease', 'stand down'--100% guaranteed, there ain't no more jokes.> Bein' on one/knee had saved my ass--standin', it'da hit me dead in the chest.

Biggest piece of shrapnel anyone had ever seen, call it 15 inches square- -part of a napalm cylinder. Down to 7 smokes-fuck it--I'm chainin"em. Cut his goddamn head off!

R&R is 8 and a wakeup. Been jokin 'for weeks about bustin' left arm the last nite.

NO, we ain't goin'"in'-bantom, bastard of a major has choppered out and is gonna send us back across this river. It's a set/up-it IZ TFS (too fuckin' stupid)-bein' sucked 'in'. Actually heard 'em tell gathered lieutenants, "Let's go see what we got." Yeah, LET'S.

What it iz--a couple zips fired at our mosquito (small, observation chopper) and then at spotter plane marking the area. Phantoms worked out with bombs and napalm.

What we should do, is split up, go a couple hundred meters in both directions and pinch/oblique in. What we're bein' ordered to do, is go from here to there, A-to-B. Yup, dudley was a genius compared to this clown.

You keep readin'; guess you wanna know--women and kids have taken refuge in derelict church.

"LOW SLOW AND CAREFUL"

We were the last crossin'-callit ll3O—told my troops, "Low, slow and careful." It was takin' 'em awhile to get people in position--got to rappin' with a Splib--doin' some stich maybe--just can't remember--that headless guy had been DIRECTLY BEHIND ME.

"Move out." "Bustaahh, common." says Watson, motioning. "Yeah, yeah, I'm comin'." Fuck it, wasn't exactly hustlin'--would catch up soon enuff.

Had closed distance to 10 or 12 meters-DOWN HARD!!-rounds hittin' all around but mostly to the right--rollin' left, something hits me on right cheek and helmet iz snapped away/silmultaneous-and rollin'!! Ended up behind a small hump of ground with Balance. (That bastard major has waltzed us into the hard teeth of NVA L-shaped ambush.) Bein' back had kept me out of the crossfire. Don't know where people are--can see 4 cats who look dead, but that's what you're supposed to do when hit, play dead.

<If nothing else, haven't and won't blow no smoke up your skirts--there is no linear time-remember what I do--no memory of Coarser or Belcher past here.>

Thousands-of-rounds--roarin ------ to starboard gunners are melting their first barrels--ground shakin'-dirt clods, dust'n ricochets-can there be anything worse than not being able to get to your wounded? Gun ships had to be comin'-where, the fuck, are they? I can't put out no rounds w/o probably hitting my own people. I am bleeding--fell terribly exposed w/o helmet--dig at cheek, something is stuck in there--spitting' up bile and a little blood.

F-i-n-a--I--I--y Hugeys crankin'with rockets and miniguns-we jumped out and got two each. More gunships, kept pullin' towards the river. <In spite of ALL THAT ROAR, heard 3 wet tearing sounds, right between my ears-tore left inner thigh where it meets trunk, diaphram on right side and

above right angle bone--still got'em.>

We stopped, both of Balance's still breathing, one of mine dead. One/knee whip out bayonet, cut his belt loose--need it for suckin' chest wound, took plastic from around his wallet-positioned and cinched it up fight but not too tight and roll onto bad side, "Sorry, Partner." That's what you gotta do, seal puncture and position so good lung can work. Behind, Phantoms dropping bombs and napiam. <6 or 8 gunships would come over the river, one right behind the other--last one pulling/up and jets comin' in at 90 degrees; parallel to (river--6 or 8 in-a-row-this goes on and on and on til dark.>

Grabbed his helmet and took if forward-have to retrieve my rifle and then, hard to port-came up on Tex, on his knees with head on the ground--got there just in time to hear him drown in his own blood (call me an asshole, good enuff for 'em)--kept movin' low, another wounded--the intensity has slacked 'some', can hear guys callin' out to God, for their Mamas and to finish 'em off. This cat, shot twice-thigh bleedin' bad--cut loops/get belt-cinch 'em up-cut his bandoleres

'n snap guns' ammo-get' em on my shoulders and dee dee moi. Sure didn't want to set him down hard but did--hit' em with Morphine and worked on his shoulder.

Jesus, had seen that major on far side with a radioman--where's the bird? Me and Balance moving forward-a Lt shows, seemingly from nowhere. To Balance, "You come with me." To me, "give 'em you ammo." points, "Stay back there and take care of the wounded." Another being dragged towards me--grab flak jacket collar and get'em on back, Jack.

Workin' on wounded boucoo//lookin' for Morphine--I live thru this motherfucker and ever swingin' dick will have his own. A guy walked up, new boots/new meat, eyes dartin'around chafterin'/slobber--stood, one step, grabbed shoulders/headbutt (like warming up before a football game), shook til he focused. Gave 'em bayonet, pointed at the dead, "Tshirts and web belts!"

What to do? What to do? What, the fuck, to do? Every time I turn around, there's more wounded-got a 2nd helper, who keeps askin' me 'what to do'. Hard decisions--work on a cat hit bad and probably lose others in better shape.--leave'em to die. Goin' Dude-to-Dude-to-Dude with empty syringe, lyin' my ass off-"Tell you what Partner, you'll feel better in a minute." Gotta make 'em belive it.

Sea Knights try to come in--fire too intense and they veer off. Here comes one--he ain't fuckin' 'round, draggin' his tail--run out, wavin' arms like a maniac. 'Common, we'll get these people out! 'and BOOM! Hit by RPG/comin'down sideways! Guess I should have hit the deck but turned and ran--wrenching sound as props break--pieces whiz'n whine past and over me. Had to get crew out--not burning but awash in fuel--SHA-IT!! Piece of prop finished off one of the wounded--move to dead pile--I did pull his smokes.

Fifteen plus dead--triple that wounded. They don't get a bird in here, they'll all be dead and then it won't fuckin' matter. Covered in blood 'n flies--wash down some with canteen. I ain't got no battle dressings, ain't got no plasma, ain't got no Morphine-lit one, watched Phantoms, one-two--three--but fourth don't pull/up, left wing catches treetop, it careens out of sight--a couple seconds, an explosion and black smoke.

Back to the wounded--lies, lies, lies and more lies--"It's in the air, hold on." Fight rages on-overwhelming air power IS NOT getting it done. <Charlie's battle plan was always something they described as 'cling to their belts'---all the rockets and miniguns were long and Phantom's bombs and napalm were wide and to the right (North)--No one could stand up to the incredible amout of ordinance raining down, IF it was hittin'.> More wounded coming back--more lies, No sweat Gee Eye, million dollar wound." (I'm losin' it--knock til your knuckles bleed, Buster don't live here no mo'--don't think he ever did.) "Common, Partner, stay with me-they're on the way." Don't you die on me you son of a bitch --- BUT THEY DO.

Call it 1400, a bird got in--and then a second--and then a third--door gunners lighting it up as we loaded.

OK, there's ammo in downed bird, took the two over with me, got all I could find and fed it out. We belt it up--8 or 9 across my chest, double/fold one in each hand. Up and to starboard--call it 2 o'clock--runnin' hard, see my spot--GOTTA get there--roll on my back and rodeo it to 'em. And how can all this be happening in slow motion? 8 or 9 more steps-got tangled-trip/stumble 'n fall--land hard--'bout then, Charlie lit up my spot--dirt clods and richochets--Close, Partner. If I hadn't 'tripped'--i'da been a dead scumbag marine.

Sea Knights came and went--their floors slippery with blood. There's more people around. 'the word', no one is to go forward, Mike Co. is comin'. Had got a pack of Marlboros off a crew chief-one/knee lookin' at the river-fuck that fight--watchin' ain't gonna help--put left fore finger in mouth, push right cheek/out and DIG w/ right forefinger 'n thumb--finally get small piece of metal. Crazy-I'm lookin' at it-how can something that small have hurt so much? A Lt. walked up, handed me a can of rat pears--I ate 'em.

<Someone could/should have figured I needed medical and dropped some--not TFS but criminally so. Charlies 'doin' what he's gotta do-we lost people, cause command has no brains or don't care, and/or both.>

<40 or so old ladies and kids now lie under rubble of collapsed church. Command can't get unfucked to help us, so for them, slow-/-crushing death. If a few manage to survive and make it to DaNang-the sisters will take them in. Yeah, these brides of Christ are good at having their pictures taken surrounded by orphans and even better at makin' sure they never run out.> The afternoon waz so fuckin' UGLY-you'd think I would remember

more-I stop the bleeding I lie my ass off--was told/ordered to deal with the wounded--hard corps/MARINE CORPS; discipline under fire.

W-A-I-T-I-N-G; CEE, WHERE ARE YA?

It iz what it iz--all afternoon, noticing color and size--not big 'n Black, it ain't Cherry. Where the hell is he? And Balance, Watson, Woody, Murf?? So much for the'ol theory of 'if we could ever make Charlie stand and fight'.

Sun sets, so cold, put on ute shirt and leave flak jacket on the ground-Hard Fact, Jack--ain't got what it takes to hump it no mo'. Where, the fuck, are ya, Cee?

Mike came across-guns' ammo Xed across their chest-bristling with LAWS-pockets bulging with frags. So loaded, we have to help'em out the water. They got 'on line' and moved forward.

Top of what would be a full moon/risin'--what was left of India and Lima started pullin' back. I'm starting to get messages--am torn up--some fingernails ripped-hurts to breathe-hurts turnin' my head—something 'wrong' with left knee 'n thigh.

Moved forward-one butt lights another-weapons' fire picks/up and finally, here comes Murf--we do a Roman, forearm-to-forearm grasp, "Good to see ya."

Dudes stragglin' back--here come a couple pullin' a wounded but he ain't, "He's dead, Man, get back to the river!"

Moved forward again-and Man, here come Amigo and Waterbull, Texmex guns' team. Hugged and then, hugged again. They went back. One/knee, here comes another Cat—it's Woody. Shook hands good 'n hard, "Good to see ya." Back to one/knee, "You seen Cherry?" He hadn't.

Woody stayed--and we wait. F-i-n-a-1-1-y--a dark shape--up'n runnin', threw down my rifle and bear huggin"em. He smelled of gunpowder and exhaustion...

QUICKSILVER MADNESS

<'the word'--India 'n Lima started lean, traps got some and Heat casualities boucoo--128 crossed the river--30 dead and 70 wounded.>

Watson and Balance came in and we went back. Mike Sgt said Lima had already left, for us to go 'in'. It seemed like he waz talkin' to me. He may have said other stuff--fight had intensified-couldn't hear much.

Started towards crossing point-tripped and went down--AND MAN, IT'S WILLIE LEE. Seems like I laid awhile--grabbed nearest guy and we dragged him to shore.

Jesus, 15 troops, including me-2 walkin' wounded, 2 dead--thats whats left. Full moon well above the trees--ricochets slashin'/ --some skipping across the water. Best get across and maybe lose no more people.

Fellow Kentuckian, Willies' mine. He was put on my back--Xed his arms across my chest. Cherry got in front, grabbed my shirt and started pullin'. The water was terrible cold and he weighed as only the dead can. Current was strong-Dudes got on each side, got belt loops and helped. Willie is big for a grunt, gotta weigh 200 w/o gear. Wanted to quit, to die, to scream 'n go flat, fuckin' crazy--half/drowin', strugglin' blindly forward--trustin' the Big Cee.

"Cee--Cee--Cee." Water belly deep, I stop, let Willie slide, turn and hold him. We got'em under armpits and to shore.

Still had his helmet on w/ Big Cross-moonlight mirrors in his eyes-Man, he don't look dead. Cannot look away-peace, love, brotherhood lookin' dead at me. Zoned/out--hung in his eyes. . someone has my upper arm, is shakin', then, 3 short--one long and later, "Common, we gotta go."

Unbelievably, it's one click--all this came down 3/4's of a mile from the bat wire. OK, got ponchos, but all around is brush, no bamboo, so no poles.

What it iz-one Cat out front, 6 humpin' the other dead--us and other 2 in back with the walkin' wounded. Started out-soon, torn fingernails/stretched sinew. Maybe 50 meters, Willie's arms are trippin' us. Tuck'em in and back to it--happens again. Third time, "Hold it." Squat over him, pull up T-shirt and there's the wound, has to be bullets but looks like he's been stabbed-left side, just under the ribs. Something made me/again zoned out--reached and touched it.

Got belt loose, tucked hands in his britches and cinched them down. Stood, "LET'S DO IT!"

Carries gettin' shorter 'n shorter--his arms loose again. This time, get belt and CINCH forearms and I check around. Jesus, we ain't got 300 rounds. Biggest, brightest moon imaginable. (Should I split the troops; half guarding the bodies, half goin' for help?)

Headed for large, dead tree where trail splits which is half/way. And finally, we can see it. Humps down to 35 meters of less-one long pull and we can turn. Over the years, it'd been a check point, there's consertina boucoo. Lowe and his poncho get tangled AND UNIT SNAPPED. Went insane in the razor-wire; pullin'/tearin'/cussin/screamin' and IT was feeding on itself. Gettin'cut/up; blood, sweat, corded muscle--quicksilver madness--I'm howlin' at the moon... "HOLD IT!; just, fuckin', hold it!"

My people standing, bleeding, breathin' ragged and hard (God, I ain't never asked for nothin', let me get 'em in). "Look, we gotta get Willie and then the poncho." We did it. Ask walkin' wounded, "Can you make it?" They say yes--that's a big maybe.

"LET'S DO IT!"--made one long pull startin' off-then short, shorter and shorter, still. All-the-time pullin', nothin' for it but to be lookin' at Willie Lee.

F--I--N--A--L---Y in the distance, some lights and bat wire.

Another 100 meters or so, Marines(?) met us. Fuck 'em, WE took our dead on in-made the long pull—laid 'em at the side of the road. Wounded were stretchered away and we all just stood there. Seemed like there should have been something more to do.

Stood there a long, long time, "So long, Willie."

<Always 'n forever, voices are far away and aware of the 'other'

to my left.> 6by was waiting-- Willie and other dead were put in body bags and gently loaded--a terrible green rubber/plastic sheen.

Cold, so so cold; shivering, "Buster?" "Buster?" "Buster?" someone is talkin' and tappin' on my shoulder. It was Kowalski; handed me a lit cigarette, reached and took my rifle, "You gotta get out of those clothes." Has dry with him.

Changed right there and he led me to Chow Hall. Had his tray- I sat, he put pack of Winston's on table and left. Back with chow--didn't want to eat but had to.

... and Ski is talkin' like he's plugged into a wall socket ... been listenin' on radio all day, no names, just numbers-(he's anticatholic as only those who've been FORCED can be) been sayin' Hail Marys and Novenas round-the-clock.

Mechanical, chew and swallow. AND he tells me. His father just died, headed out on Emergency Leave. Wounded in Korea, in a wheelchair, drank way too much--he and Ski didn't get along. But, partner-one Dad is all you're issued. Said he'd try'n stop by before he left. <Ski, Ski, Ski--find a way not to come back. You ain't ard enuff--stickin' with me on religious horseshit kept you out of a body bag.>

Tape goes blank...

<Old ladies and kids under crushing weight of pope's bricks--had sandbag wall 'on me' one time--every breath, you lose a little more.>

"YOU DIDN'T SEE ME."

... woke in my rack around 0730 and then again at 0900. Grabbed canteen and M-16, went to one of the fighting posts. Sat, smoking joints--gazing out over barbed-wire and sand. Lowe dead, Boone dead, LeHoullier dead, Donaldson dead. Burned up that rope, tryin' to make IT go away.

Willie Lee--nicest kid I've ever known--I don't care WHO your father is--no, lightning hadn't got me--they nailed you.

Boone--had had a few brews with Doc, had liked him...

LeHoullier, came up on 'em in the bush-new boots and one good attitude. "Where you from?" "New Hampshire." "No shit, my mother's family has a grocery store in Sommersworth." "My father built it." Store had gone from Mom 'n Pop to Big 'n Modern-his dad was contractor. I'd talked to him that once-liked him, we'll get together later in the rear. Buried across the street from my grandparent's house--so, I'm told-never went over.

274

Donaldson, a Lieutenant, his people felt about him the way we did about Coarser--he took care of/looked out for 'em. Had had a couple 'ol Mils with him. Standing in the dark, he said, "No bars." (no rank) and we just rapped.

Had to have worked on some of 'em and/or loaded their bodies--cannot remember.

A Dude went by, "You didn't see me." Maybe 45 seconds, Bang-Bang and he-came-back up the-trench-movin'-quick. Waited a couple minutes and followed.

In tent, people were cleaning their weapons--passing rods, swabs, WD40 around. Joined 'em.

Call it ten minutes, a couple officers walked in. Dipshit major was shot in the back while walking across the Z. Anyone know anything?

See what, Man? If we'da split up and pinched/in--it'da been us kickin' Charlie's ass. Didn't even kill the prick--just popped a lung. chickenshit bastard never even came across the river. Stayed over there talkin' on his goddamn radio.

Dead dead dead dead-wounded and fuckin'dead-smokin'dope-repops comin', "So long, Willie." dead on that ragged edge. See what, Man?

NOTES:
The following members of India-3-1 were killed in action on April 5, 1969:
Boone, William E. HM3 B325937; 25; Tuskegee, Alabama
Daniel, Stephen A. LCpl. 2427872; 19; Waco, Texas
Donaldson, Steven E. 2dLt. 0106781; 22; Peabody, Massachusetts
Geary, William S. 2dLt. 0107385; 23; Winona, Mississippi
Le Houllier, Paul R. LCpl. 2336494; 21; Somersworth, New Hampshire
Leavell, Richard T. Pfc. 2447052; 21; Jeffersonville, Indiana
Lira, Robert C. 2413969; 19; Indio, California
Lowe, Willie L. Pfc. 2405682; 20; Louisville, Kentucky
Ozger, Islam Pfc. 2467699; 21; New York, New York
Perdue, Richard W. Pfc. 19690405; 19; Rocky Mount, Virginia
Smith, Steven J. HM3 B527575; 19; Viroqua, Wisconsin

The following member's of India-3-1 were wounded that day:
Baylark, Johnnie Jr. LCpl. 2356297; 20; Little Rock, Arkansas
Beckner, Duane L. Pfc. 2481809; 19; South Charleston, West Virginia
Brubach, John P. LCpl. 2407074; 19; Cincinnati, Ohio
Brymer, Bruce F. LCpl. 2444714; 19; Baltimore, Maryland
Cage, Leroy E. Jr.LCpl. 2391746; 20; Nashville, Tennessee
Carter, Robert L. LCpl. 2449547; 20; Dallas, Texas
Carter, Tommie L. Cpl. 1975128; 26; Richmond, Virginia
Chester, Larry LCpl. 2448513; 19; Hornell, New York
Duke, Griffith P. LCpl. 2504988; 19; Kansas City, Missouri
Holmes, Lyle A. LCpl. 2478818; 19; Cleveland, Ohio
Latham, Terry D. Cpl. 2015190; 25; Denver, Colorado
Lewis, James H. Jr. LCpl. 2421009; 20; Detroit, Michigan

Martin, Bruce R. Cpl. 2377979; 20; Boston, Massachusetts
McHenry, John K. LCpl. 2483955; 19; Detroit, Michigan
Meadows, Ronald J. Cpl. 2339670; 20; Raleigh, North Carolina
Moore, Anthony D. Cpl. 2396749; 20; New Haven, Connecticut
Moore, Morris J. LCpl. 2473172; 19; Jackson, Mississippi
Paul, Hugh D. LCpl. 2423960; 19; Buffalo, New York
Peirce, Larry R. LCpl. 2427537; 19; Kansas City, Missouri
Peters, George D. LCpl. 2417341; 20; Baltimore, Maryland
Phillips, Willie R. LCpl. 2441232; 19; South Charleston, West Virginia
Porter, Herman P. LCpl. 2439468; 19; Baltimore, Maryland
Shepard, Bobby R. Pfc. 2455994; 19; Oklahoma City, Oklahoma
Surprise, James M. Pfc. 2466532; 19; Boston, Massachusetts
Taylor, Claude E. LCpl. 2406962; 20; Cincinnati, Ohio
Wright, Curtis D. Cpl. 2226002; 21; Louisville, Kentucky

The Silver Star Medal was earned by Second Lieutenant Eric D. Shafer, First Lieutenant Kamille K. Shibley and Lance Corporal Paul R. Lehoullier. The award to Lehoullier was posthumous.

First Lieutenant Kamille K. Shibley was the commanding officer of India Company on April 5, 1969. When the lead unit was halted by intense enemy fire he moved the company to a vantage point. Shibley assessed the situation and commenced a fierce assault against the North Vietnamese emplacements. During the initial moments of the attack the company sustained many casualties including the deaths of two platoon commanders, Second Lieutenant Steven E. Donaldson and Second Lieutenant William S. Geary, who were shot down by the murderous fire. Shibley paused to reorganize his unit, evacuate the wounded, and direct supporting arms fire.

During the evacuation of the wounded Lance Corporal Paul Le Houllier, fully exposed to the enemy fire moved to their aid. He moved across the battlefield from one wounded Marine to another and rendered first aid and words of encouragement. Le Houllier carried one of the casualties to a covered position and returned for another. On this trip he was shot down and killed by an enemy sniper.

Shibley then launched a savage attack against the flank of the North Vietnamese emplacements, which allowed the trapped Marines to move to better positions. He and his Marines inflicted heavy casualties upon the enemy and were instrumental in his unit's killing of seventy two North Vietnamese soldiers.

Corporal Tommie L. Carter had been a Marine since 1961 and was on his second tour in Vietnam. He participated in Operation Starlite in August of 1965 as a member of Hotel Company, Second Battalion, Fourth Marines. Carter was wounded on August 18 and received his first purple heart. He was to go on to serve twenty years in the Marine Corps and retired as a

Master Sergeant in 1981.

Dennis Moffett paid this compliment to Steven Donaldson. "Donaldson stood tall and proud as a Marine Lieutenant we hardly got to know before he fell leading a charge into enemy territory on April fifth. He fell with many of our heroes that dreadful day. We never had a chance against so many yet he was an exemplary product of the good that makes Marines."

Dick Bienvenu remembers Daniel. "Tex was the toughest Marine that I knew in Vietnam. He was good at what he did and truly loved the Corps. Tex was a good friend to all and taught us much about what it meant to be friends and brothers. There were five of us that were close, Tex, A.T., Marshall, Tennessee and me. We all made it home but Tex. He died doing one of the things he did best firin' a LAAW at the NVA."

Private First Class Duane Beckner who suffered a fragmentation wound to his right arm went on to serve 30 years in the Marine Corps and retired a Sergeant Major in 1998. He later served as a drill instructor at Parris Island and participated in both Operations in Iraq.

Lance Corporal Larry D. Ballance was a competent professional who tirelessly trained his men. He later distinguished himself on May 12, 1969 when Company I came under heavy fire from well entrenched North Vietnamese soldiers. Ballance disregarded his own personal safety and repeatedly exposed himself to hostile fire as he made certain that an isolated platoon from his company was adequately supplied with ammunition.

Other Marines mentioned in the story are as follows:
Ballance, Larry D. LCpl. 2424292; 20; Buffalo, New York
Bienvenu, Richard Cpl. 2401036; 20; Bridgeton, Missouri
Buster, Robert D. LCpl. 2405635; 20; Bowling Green, Kentucky
Courser, Charles B. 1stLt. 0107335; 22; Detroit, Michigan
Cherry, David E. LCpl. 2339257; 20; Woodland, North Carolina
Harwood, Michael H. LCpl. 2396670; 20; New Hampshire
Moffett, Dennis Pfc. 2463576; 19; Philadelphia, Pennsylvania
Opolski, Frederick Pfc. 2451655; 19; Meridian, Idaho
Watson, James L. LCpl. 2376824; I-3-1; 20; Nashville, Tennessee
Cherry was killed in action on May 12, 1969.

* * * *

In the late fifties and into the Vietnam War years, my hometown, Opelousas, Louisiana, was a hotbed for Marine Corps recruiters. In January of 1960 five members of the 1958 Opelousas High School football team decided to join the Marine Corps. Back then the old downtown section was thriving and I was walking through it on my way home from school when I encountered the five new recruits. They were excited about the new adventure that they were about to undertake and in their zeal they tried to get me to join with them. I really wanted to but I was only sixteen at the time and had yet to graduate. These guys were popular, big time, and I was honored that they would ask me to go with them. James Doucet, Paul David, Clarence "Boo Boo" Brinkman and Larry Lalonde were really good football players and Lalonde had won a state boxing championship. Whitney Cropper, the fifth recruit, was an affable, roly-poly, good guy. The seed was planted. I would join a few months later as would John Adolph Haas, one of the most unforgettable characters to ever walk the streets of my hometown.

"In Vietnam being strange is not looked upon as any sort of aberration; it is just seen as part of one's personality---like a sense of humor; Being strange is not only acceptable---it is viewed with great relish. Being strange fills a need. Strange is a simple humor. Macho humor"

First Lieutenant Ernest Spencer
D-1-26
From *Welcome to Vietnam, Macho Man*

15

"Mo" Haas

I had known Johnny Haas since we began school at The Academy of the Immaculate Conception in 1948. AIC, or "The Convent" as it was sometimes called, was the Catholic school in Opelousas.

It was located within the city block that is now Roy Motors. The campus was a collection of aged, wood framed, clapboard buildings that were painted a drab gray color.

Johnny and Bill Edwards were the two biggest kids in the first grade. Even then he was unique in being extremely intelligent and a little strange. Well, maybe real strange. The bubble on his level was more than a tad off center. He was a dynamo on the playground and was always the dirtiest kid there. By dirty, I mean that Johnny worked at recess. It was his passion and he was difficult to restrain. Now, he could hurt you and he often did, with relish. He was in constant motion, jumping and rolling on the grassless playground. When recess ended Johnny was sweaty and filthy and often sent home to clean up. It happened to me once, to Johnny many times.

As I recall, Johnny was an adopted child. He had the good fortune of being adopted by Johnnie Haas and into one of the wealthiest families in the Opelousas area. In this area the name Haas is pronounced "Hays." Mr. Haas owned and managed a hardware store that was housed in an old wooden building that was located where the Checker's restaurant now sits. The place was run down and was gone long before we finished high school. The family wealth was in property owned. They lived in a house on the corner of Cherry and Union Streets. It is still there and now painted a grey color.

When I reached the fifth grade, AIC moved to its present location on Prudhomme Lane. That year we switched to the public school and Johnny and I went to school together every year thereafter. We were not always in the same class but always in the same grade. I graduated in 1960 and he did not. Johnny's behavior and general lack of effort did not allow him to do so. I joined the Marine Corps on a Monday, graduated that Thursday night and left for boot camp in San Diego the following Monday. The recruiter was Staff Sergeant Robert Alley, who had the good fortune to be assigned to the Lafayette area. The name, Acadiana, had not yet been given to this region of the state.

Alley really scored with the Opelousas boys. In addition to the five previously mentioned, and Haas and I, he also enlisted Rod Bernard, Kermit Folks, Billy Lawless, Bucky McLeod and John Tujague. From AIC he got Phillip LeBlanc, Teddy Hebert and Zack Mills. I am sure there were others; I just cannot recall the names.

Needless to say, Marine Corps boot camp was difficult and by the end of the first week I suffered a severe stress fracture to one of my legs. It was placed in a cast and I was sent to a casual company which was anything but what the name suggested. We had drill instructors and were given no slack. On August 1, 1960 I was assigned to a new platoon (363) and started on the first day of the 12 week training cycle.

Every Sunday the drill instructors made us attend the church service of our choice. Attendance was mandatory. I was raised a Catholic so I went to church with the "Mackerel Snappers". It was the name given to all Catholic's because at that time we didn't eat meat on Friday. It was alright to eat fish. On one such morning during mid-August I was sitting in church when I felt a tap on my back. I turned around and there, surprisingly, sat Johnny Haas. I had no idea that he had joined the Corps. His head had recently been shorn of all hair, and he had that shocked look, on his face, that all recruits have during the first few days of recruit training. Most of the photographs in this book are boot camp "mug shots" that were taken shortly after the recruit's arrival. All of the Marines whose photos are posted there have that, "What have I gotten myself into?" expression. Johnny and I only exchanged a few words and I tried to tell him that the harsh treatment would continue but he'd get used to it.

It was the only time that I saw him during the four years that we served. During the spring of 1962 the Marine Corps conducted maneuvers in the Far East. There, we practiced the assault and occupation of one of the Philippine Islands. We were at sea for a period of seventeen days as we sailed in a World War II type convoy. Aircraft carriers and troop ships were located in the middle with Destroyers patrolling the flanks as protection. Johnny was aboard the USS Hancock, a World War II era carrier. I was aboard LST 1167, the Westchester County, which carried troops and supplies. I saw the Hancock everyday but, of course, I never saw Haas.

I next saw Haas at the Southern Club a few days after I was discharged from the Corps. He had asked for and received an early release for the purpose of attending college. In fact, Johnny had gotten out a couple of days before I did. Well, we raised hell for several months and both ended up at Northwestern State College in Natchitoches. While there we were both excellent students.

As I mentioned before, Johnny Haas was a strange individual. Unique. One of a kind. After the Lord created Johnny he threw away the pattern. Had to. He lived until the ripe old age of 52 and died of stomach cancer. The incubus of this cancer was believed to be exposure to Agent Orange. As far as I know, Johnny never had another job other than the approximate ten years that he served in the Marine Corps. Not before and certainly not after.

He was a charismatic leader and could get the younger students to do anything he desired. They were at Johnny's beck and call. He was a smoker but I don't think he ever bought a pack of cigarettes. Haas was a beer drinker that I never saw buy a beer. He called a beer an "Oil" and his favorite expression was "Mo needs an oil". To him food was "Grease".

Johnny never had a driver's license and never drove a car. He never had a girlfriend. I don't think he ever went out on a date with one.

I don't know where he picked up the sobriquet "Mo". When he returned to Opelousas after his first hitch in the Corps he demanded to be called that. Maybe someone in the Marine Corps gave it to him but I suspect that he probably just decided to hang that moniker on himself. I never did call him "Mo". He was always Johnny to me.

The following is a letter written by Johnny Haas to Charlie Kessinger sometime during 1969.

PFC J.A. Haas 1928368
HQ Btry 11th Marines
1st Mar. Div. FMF
FPO San Francisco, Calif. 96602

Hi Glenn,
   I tried to find you the weekend I came home, but "Happy" said you were gone. "Frog" is really a Non-Hacker.

   I skated tonight. I am sitting by myself in a miserable little Hooch about 50 yards from the perimeter. This place is so damn screwed. The bastards have been probing our wire the last couple of nights and there have been rockets in the area. Comes a Ground Attack, Goodbye 11th Marines and Goodbye "Mo".

   Robert Sylvester is with 1st BN-5th Marines, about 25 miles from here and they have been catching some Shit. I will go down to see him if I can find a convoy heavily armed enough. This place is about as safe as the old Acadian Inn on Friday night. If only some asshole from Dasahn doesn't decide to pot you while you are using the head. Dasahn is a Ville about 400 yards from this perimeter. There are some good looking wenches there, but much VD. OH! OH! Flares on our side of the hill. Let me know what is happening at home.

   So long! I will now lock and load!

   "Mo"

   Don't join this "Green Bitch" (USMC)
   'Mo' Haas

   Send your care packages to the troops in the field. Fuck It!

   In this letter, Haas appears to be serving with Headquarters Battery of the 11th Marines but I believe, either at that time or some point during his tour, he was with a CAP unit which was concerned with the pacification of the populace around An Hoa.
   Glenn, in the letter, refers to Charles Kessinger a friend of ours who was then playing football at Louisiana College. "Happy" and "Frog" is the same person. "Happy Frog" being the nickname that Johnny bestowed on the late Benny Childs. "Frog," the non-hacker, was soon drafted into the Army.

281

Robert Sylvester was another Opelousas lad doing a hitch in the Marine Corps. He was the son of, then, Clerk of Court Howard Sylvester. Robert was a squad leader in a heavy weapons platoon serving with Charlie Company of the 1st Battalion, 5th Marines near the combat base at An Hoa. Sylvester was wounded by shrapnel in the summer of 1969. He was in Vietnam from February of 1969 to February of 1970.

The Acadian Inn was the local hangout. It was a beer and hamburger joint that everybody from about age fifteen through college age gathered every Friday night. Every pretty girl in town would be there. And so would be all the local dudes, posturing. You had better not be a visitor from another town. If you were your life expectancy could be greatly diminished to about that of a machine gunner in a firefight. When beer and male testosterone is combined in large amounts fighting usually occurs. The fights were clean, no guns or knives, no kicking, just fists, but many times extremely violent. Many an eye was blacked, many a nose busted and a lot of blood was shed.

Haas was a big guy and, if he chose too, could be very capable in a fight. I saw him tangle with Charlie Kessinger in the parking lot behind the Acadian Inn on a week day afternoon. I was sitting in a car on the corner at the old Burger Chef restaurant which is where Oupac now sits. I had a clear view of the action and saw them walk outside and square off. Johnny attacked, swinging wildly and Kessinger hit him square on the jaw with a straight right hand punch. A one punch knockout! Johnny was unconscious for quiet sometime.

On another occasion, I saw Haas knock Benny Blundell senseless with a vicious sucker punch. Blundell was walking toward the entrance of the Acadian and never saw the punch coming. Benny was a muscular, very athletic guy who had tried out with the New Orleans Saints during their first year of existence in 1967. He was the only one to survive the tryout camp and for a while was some what of a celebrity. Blundell went to training camp with the team but was cut soon after. I can't remember the two having a disagreement; Johnny just didn't like him and didn't appreciate his notoriety. When you were around "Johnny Boy," as his mother called him, you had to be careful.

After my hitch in the Marine Corps I was doing very well in college. I attribute this fact to the things that I learned in the Corps, mainly self-discipline and the knowledge that if I persevered that I could do anything. At the beginning of the fall semester of 1967 Johnny rode with me to Natchitoches in my old, lime green, 1957 Plymouth with the big tail fins. It was the exact same car that was in the Stephen King movie Christine, with the exception being the color. Christine was red. It was almost a three hour drive on the old two lane highways that existed back then. Now, Johnny was doing well in school also but it was bothering him that I was receiving a little attention for my accomplishments. Probably the same thing that led to him "bopping" Blundell. As we drove he proposed a wager. The bet was that he would have a better grade point average than I did at the end of the semester. We agreed to wager a case of beer on the outcome. The only rule, as I understood, was that we had to take at least 16 hours of classes.

At the end of the semester I had made straight A's. It was the first time in my life that I had ever done that on any level of school. Haas had given the letter grade A the name "Bullet" because it's shape resembled that of a rifle round. When he asked me what I had made I responded with a one word answer, "Bullets". This didn't sit too well with Johnny who had made a couple of B's. Obliviously I had won the bet, or so I thought, but he refused to pay claiming that I had cheated. I had studied and worked my tail off to make those grades and he knew it. When I asked, "How did I cheat?" He answered, "You studied." I didn't realize that one of the rules was that no studying could be involved. Johnny, who never cracked a book, just assumed that I knew this. He was brilliant and very, very strange.

There was another event that greatly influenced Johnny. In the summer of 1966 a former Marine, Charles Whitman a student at the University of Texas at Austin, went on a killing spree that resulted in the deaths of 16 people and injury to 32 others. He first killed his mother and wife as they slept by stabbing them in the chest. Whitman then went to the campus tower where he killed a receptionist with a butt stroke from one of several guns that he carried. He barricaded the stairs leading to the observation deck and when a group of people encountered it Whitman opened fire on them with a sawed-off shotgun killing two and badly wounding two others. Once he reached the observation deck he randomly began to pick off people, with a high powered rifle, as they moved around the campus. Whitman was killed later that day by Austin police. During a later autopsy, suggested by Whitman in his suicide note, it was discovered that he had a brain tumor that probably contributed to his violent attack.

Johnny was greatly impressed by Whitman and the fact that he was a former Marine. He came up with a plan that had Haas, himself, climb the Opelousas city water tower with a rifle. It was the one located downtown just behind the building owned by the Opelousas Women's Club which was then used as the city library. Once there he would begin to pick off people from a hit list that he had written down. Someone who saw the list told me that my name topped it. I was number one. Of course, he never carried out this plan but just the idea that he might was downright scary.

Shortly thereafter, Haas rejoined the Marine Corps. He wanted to go to Vietnam and asked me to go with him. I had served my time and though tempted to rejoin I decided not to. I was doing very well in school and had a beautiful girlfriend. Going back into the Corps would have been a difficult thing to do.

Before he left, Johnny gave me that strange look that he had and called me a nonhacker.

NOTES:

Robert Pitre, also from Opelousas, was a member of Marine Air Support Squadron-3 (MASS-3) located on Hill 327 near DaNang. At that time he was a First Lieutenant and the senior air director in charge of controlling all-weather bombing missions in that area.

I remember Robert and his older brother, Henry, who was an excellent boxer for Opelousas High School and a maestro on the dance floor. In our part of Louisiana the name Pitre is pronounced "Pete."

Pitre recalls, "I received a letter from home telling me that Robert Sylvester was somewhere in the bush near where I was and that he was not doing well physically. According to the letter he was losing weight and was suffering from dysentery. I took a couple of days off and flew to his location on a helicopter and brought him a bunch of stuff that he could not get in the field. He was very thin and did appear to be losing weight. In fact, soon after my visit he was brought to the rear for a while so that he could regain his health and recoup the weight loss."

Sometime after that Sylvester and Johnny Haas visited me on Hill 327. The three of us, along with a Navy corpsman named Mike O'Brien, had a photo taken. O'Brien was a squadron corpsman from nearby Eunice, Louisiana. I sent a copy of the photo to my mother and our neighbor Althea Galyean had it printed in the Daily World.

Pitre writes about a harrowing experience that occurred in Vietnam shortly after his arrival in country in a story entitled *A Night to Remember*.

"I had only been in Nam since January 29, 1969 when on a mid-February night a combined force of VC and NVA attacked Hill 327 where I was based with a Marine Air Support Squadron (MASS). Hill 327 overlooked the DaNang airbase and was the location of the 1st Marine Division HQ at the base of Hill 327. Hill 327 was also known as Freedom Hill.

Due to my recent arrival in Nam, I was being indoctrinated that night as a senior air director for the MASS-3 unit controlling all-weather bombing missions against insurgent VC and NVA forces. Our position on Hill 327 was on the reverse slope on the DaNang side with our perimeter on the forward slope being guarded by MASS marines in fighting holes dug above and along the barbed-wire perimeter. We had claymore mines and warning devices along the perimeter wires, but the enemy did their recon well and penetrated our perimeter that night by coming up a ravine being guarded by a fighting hole at the top of the ravine. The ravine depth gave concealment to the enemy force on both sides of the ravine allowing them to move up unseen on the fighting hole to take it out with a satchel charge. The story, as I was later told, was that one of the men in the two-man fighting hole had left to relieve himself away from the fighting hole leaving the other Marine to watch the ravine. I will tell you that urinating consistently near a fighting hole is not cool as it will leave a very unpleasant smell for anyone manning the hole. Evidently, the remaining Marine either fell asleep or did not observe the enemy approach his fighting hole and the position was subsequently overrun.

After slipping through the gap in our perimeter, the enemy force proceeded to the top of Hill 327 where they destroyed our unit's equipment used to control fixed wing Marine, Navy and Air Force attack aircraft in all weather bombing missions. As noted previously, our MASS unit conducted all-weather bombing strikes against insurgent enemy forces in support of Marine grunts in the field. This was accomplished with the use of radar that would track fixed wing aircraft to bomb enemy targets supplied by Marine Intelligence (G2). The tracking radar was powered by a generator that also supplied power to a control module that housed various computers and the MASS air controller. By knowing the coordinate position of an enemy target, our own coordinate position and the position of an attack aircraft via the tracking radar, the MASS air controller through computer triangulation computations could direct the aircraft pilot via course heading, altitude and air speed corrections to the designated target to drop the aircraft's ordinance. What made this system effective was that the MASS air controller could direct a bombing aircraft flying at 25,000 feet to a target in fair or foul weather without the pilot or the MASS air controller ever seeing the target;, or the enemy ever knowing they were about to get hit. The maximum range of the system was approximately 50 miles, but the entire system was mobile via helicopter to any secure or semi-secure position making our MASS unit even more effective in support of Marine grunts in the field.

Because our MASS unit used a radar system to track and control bombing aircraft, the radar had to always be located at the top of hills to extend its 50-mile radar wave range. Radar waves don't penetrate solid objects, but bounce off solid objects which is the principle of the Doppler Effect. This Doppler effect of bouncing radar waves off attack aircraft provided continual data to the system computers allowing the air controller to direct and control aircraft to designated targets. Because we were so visible at the peaks of hills, we were a target to our enemy; especially after they became aware of our capability to hit them in any weather when they least expect being hit. On that night in February 1969, we were a target.

After the second explosion taking out our power generator and radar, our control module equipment shut down. About that time, the technician who was manning the tracking radar and power generator at the top of Hill 327 came running into our control module with his right arm nearly severed by an RPG used to take out the radar and generator. I quickly put a sling around his neck to support his right arm and sent him down the hill to our main base and bivouac area. Subsequently, the Marine officer who was indoctrinating me that night also received a shrapnel wound to his back and he too vacated the control module area leaving me and a sergeant in the control module on the reverse slope approximately 100 meters from the main base/bivouac area. For some reason, we did not vacate the area soon enough and got trapped at the control module area. Stupid us!

So there we were my sergeant and I facing a force of unknown size with me having only my M1911 .45cal. pistol with two clips of seven rounds each and my sergeant with an M16 and one clip of 18 rounds. All he was wearing was his camo pants, Ho Chi Minh sandles and his M16; not exactly combat gear to get into a firefight. The sergeant was leaving the squadron to return to the U.S. within the week. At least I was wearing my camo pants, shirt and boots. With the limited firepower we had, there was no way we were going to take on a force of unknown size. So, we did the most logical thing. We ran like scared rabbits out of the control module area on the reverse slope and hid in some waist high monkey grass approximately 75 meters from the control module area. Meanwhile, the enemy force evidently had seen us vacate the control module area and go into the grass as they began to spray the grass with automatic weapons fire. Fortunately, we were low enough, and I mean low into mother earth, that we were not hit. There in the monkey grass we stayed for approximately 20 minutes until we saw a chance to escape down Hill 327 when a Marine rifle platoon from the First Marine Division came up Hill 327 to counter attack the enemy position at the top of the hill. When the rifle platoon laid down some counter fire onto the enemy positions and the enemy was distracted, my sergeant and I took the opportunity to crawl and eventually run through the monkey grass that concealed our movement down to the safety of the base camp area. During that time in the monkey grass, I must have promised God everything I could think of; from going to Church every holy day to not ever taking a drink again. For a Marine to not ever take a drink, that certainly was a desperate measure. A promise that later in the day was simply impossible to keep. The only injury I had to show for my escape from the razor-sharp monkey grass was some irritating cuts on arms, face and neck. My sergeant didn't fare so well without a shirt and socks. For those in Nam who were exposed to monkey grass, they know what I mean.

So I found myself safe at base camp with my MASS unit fellow Marines, but still under the threat of the enemy force of unknown size holding the high ground at the top of Hill 327. Their position was a threat to us and had to be removed. When I got to the squadron base camp, I was surprised to find the rifle platoon that came up to help eliminate the enemy force was commanded by a First Lieutenant that I went through TBS (The Basic School) at Quantico, Virginia. TBS is the Marine Corps' advance officer training program after completion of OCS. After some exchange of pleasantries, he went about his work with his platoon and with their help and the firepower of a Marine Huey Cobra gunship, we were able to eliminate the enemy position and chase the remaining enemy force from the hill. We did capture their CO who stated under interrogation that his force numbered 35 personnel with only a few escaping. The Cobra pretty much wiped out his unit and severely wounded him which is why he did not escape. It was commendable that he stayed behind to cover the escape of the surviving personnel knowing that his end would probably not be pleasant. He was pretty banged up, but proud of his leadership. I doubt he survived long after his capture as he was taken away by Marine and ARVN intelligence personnel who especially were rough on prisoners. As for the Lieutenant who saved my butt that night, I told him that if we ever met again, I'd buy him his favorite fifth of liquor. As it turned out, we met again in 2008 at our 45th TBS reunion in D.C. and I did just that. I bought him a fifth of Glenlivet 60 year-old single malt scotch and again thanked him profusely.

Four Bronze Stars were handed out by our squadron CO that week to squadron marines as a result of that night, including the sergeant I was with. I turned down the medal as I felt I did nothing to deserve it. I was only concerned that night with survival and not in any mind set to attack a superior force with no effective means to fight. Handing out medals in Nam seemed to be a means to enhance moral in a crappy conflict where you never knew where Charlie would come from. In hindsight, a Bronze Star would have looked good on my resume after leaving the Corps. Couldn't do it!

I vividly recall that night and the shock of the initial attack and not knowing what to expect; the fear of being pinned down in the monkey grass without sufficient cover and not having any effective means to defend myself and my sergeant; the adrenalin rush and relief of dashing to the safety of our base camp; and the exhilaration of knowing that we along with the Marine rifle platoon and Cobra gunship had smashed the enemy force thereby eliminating any threat to our MASS unit. While other Marine field units encountered such attacks in the I-Corp area on a consistent basis, we were a Marine air support squadron that was supposed to be isolated and protected from such bullshit. We had heard the rumors that our area was on high alert due to the annual Tet holiday, especially after the 1968 Tet enemy offensive, but no one suspected an attack of such veracity on our supposedly "rear" non-combat unit. Maybe some probing and harassing fire, but no full-out attack was expected. I learned quickly that night in February 1969 on Hill 327 that there were no secure areas in Nam and that we and I needed to be more vigilant about our defenses. Believe me, after that night, we were."

✶ ✶ ✶ ✶

On January 30, 1970 the Marines of Alpha Company, Third Platoon, First Battalion, First Marines were involved in a sweep and clear operation near the Sung Vu Gia River in central Quang Nam Province. The unit was in light contact with enemy forces when one squad crossed into an old overgrown minefield setting off a mine and injuring several Marines.

> "God damn it! Stop shooting at those gooks.
> They are trying to surrender!"
>
> Warren G. Cretney
> Commanding Officer VMO-2

> "Such is the life of a Marine. Covered with glory one minute,
> Covered with shit the next."
>
> An old Marine Corps axiom

In compiling this story, interviews with the participants, the awards citations of many of the participants, the rosters and diaries of the participating units, the command chronology of the First Battalion, Third Marines for April 1966, and information obtained from the National Personnel Records Center were used.

16

Blood, Sweat and Tears

On January 31, 1970 Lieutenant Colonel Walter R. Ledbetter, Jr. was a Marine Helicopter Pilot in Vietnam and the commanding officer of Marine Medium Helicopter Squadron 263. They were flying the CH-46 assault helicopter out of Marble Mountain.

On that day HMM-263 carried all 178 Officers and men of Alpha Company, First Battalion, First Marines into one hell of a fight. Shortly after inserting all three platoons of Alpha Company, into three different positions, the squadron had to start Medical evacuation missions.

Private First Class Eddie West wrote: Marines were dying. They had hit an explosive device, and the Marines of Alpha Company, 3rd Platoon, 1st Battalion, 1st Marine Division closed ranks. They had been sweeping an area on the left flank of their insertion point. They had already engaged the enemy twice, had tagged a booby trap just moments earlier, and some of them were in a running firefight as they pursued enemy forces. The sound of that explosion told the Marines someone had just found another booby trap, the hard way.

The Marines closest to the blast site moved to help the wounded. Some of them had to go through concertina wire to get to them, and others simply ran into the large clearing where the wounded lay. The remaining Marines down the line circled in to set up a defensive perimeter. There, they held positions to provide cover for the Med-Evac they knew would probably already be en route. They didn't know they had just taken up positions in a minefield and were now standing at Death's door. In a matter of minutes this reality would come crashing down on them. They would soon be giving their lives trying to save each other, and one Marine; Private First Class Mike Clausen would be trying to save them all. As Lieutenant Colonel Ledbetter barreled his CH-46 aircraft into the mine field, the three Marines at his one O'clock disappeared in an eruption of earth filled with searing metal and with the blood sweat & tears of young Marines. The crew of the Colonel's aircraft was going to "Gopher Broke", collect that precious Blood Sweat & Tears, save the rest of those Marines, and get the hell out of there! But the dying wasn't over. The stage was set. Lieutenant Colonel Ledbetter's aircraft, Blood Sweat and Tears, had just landed at Death's door, joining the Marines of Alpha Company, third platoon. What happened next may be viewed by some as a waste of humanity. It would be recognized by others as representative of a Marine's commitment to his Fellow Marines and would answer two questions: How many Marines does it take to save a Marine? As many as it takes. How many times do they try? As many as it takes.

It was the morning of January 31, 1970. I was a radio operator (known then as Delivery Boy 1-4 / Lance Corporal West) serving on the Tactical Air Control Party (TACP), Head Quarters Company, First Marine Division on Hill 55 just south of Da Nang. Another of the "Kingfisher" Patrols was already underway as Lieutenant Cruikshank and I left the command bunker and ran down the hill to the lower landing zone. 1st Lieutenant Cruikshank, an A-4 Skyhawk Pilot, was serving as an Air Liaison officer. Major James W. Rider, was the Regimental Air Liaison Officer in charge of the TACP and a Gun Ship Commander with HML-367. Two Ch-46 Sea Knight Helicopters sat with rotors turning at a high rpm ready to lift off, and the Marines of 3rd Platoon, Alpha Company were already in line at the ramp of the Sea Knight we reached first. As we joined the end of their ranks, we all advanced up the ramp and into the orifice of the CH-46. I was amazed at how quickly we had placed ourselves elbow to elbow in the long canvas seats given such a confined space. The ramp closed. The whine of the engines increased and there was an audible change in the pitch of the rotors and the large Sea Knight lifted off, leaving Hill 55 with the Marines of 3rd Platoon, Alpha Company, First Battalion, First Marine Division. Most of them would never see Hill 55 again.

Alpha Company's second Platoon had been inserted earlier and encountered a sizeable enemy force; third Platoon was now being inserted as a blocking force to cut off any possibility of retreat by the enemy. The order to, "Lock and Load", was given, and I could feel the aircraft flare as the ramp came down. Lieutenant Cruikshank and I exited the aircraft and turned to our right taking up a position along a small rice paddy dike followed by the Marines of third platoon. I could not believe it. Right there in front of us was Charlie, wearing black pj's, running diagonally across the rice paddy in a southwesterly direction. I raised my weapon to my shoulder and squeezed off about four rounds when I felt Lieutenant Cruikshank's hand on my shoulder. He gave me a stern, "That's not what we're here for," and then was back on the radio. I couldn't believe my ears. Hell, I was thinking, "How did I miss that guy?" as I realized he was still getting it across that paddy. There must have been 14 Marines to my right along that dike, all firing at him. Still he just kept trucking, heading for the far right corner of the paddy. The Cobra Gun Ship glided in at treetop level right on the heels of Charlie. That's why we were there, TACP. The Scarface pilot of the Cobra positioned the nose of his aircraft directly above Charlie and fired his automatic grenade launcher. Charlie disappeared in a circle of explosions only to emerge looking as if he was running a 50-meter dash. He had picked up the tempo and now disappeared into the jungle at the far right corner of the paddy. He was one dedicated individual, who was probably dead by now in the jungle. The Marines to my right formed a line and moved down the paddy dike, we followed suit. As we reached the corner at the end of the paddy, a Marine warned, "Booby Trap." That's why Charlie was so set on his direction across the paddy, not zigging or zagging to escape our fire. He was leading us to a place of death even as we were killing him. We didn't know it yet, but his buddies were doing the same thing. "Booby Trap." My senses tightened as the words echoed in my head. I strained to see where the Marine was talking about, but we were already on the move again. Turning east we moved across the end of the paddy and into a clearing. In front of us was a drainage ditch with a narrow tree line parallel to it. The lead elements were already through the ditch and trees, and there was sporadic firing in the distance beyond.

As I waded into the chest deep murky water of the drainage ditch, I couldn't stop from thinking, "Booby trap; please don't let me get blown up in this crap and drown." Up out of the ditch and through the tree line into a brushy area. Where the hell did everyone go? It seemed to be just the four of us there at the end of the column. There was high scattered brush everywhere and a tree line in the distance to the south. The Marine with the M-79 must have seen something move in that tree line and squeezed one off. There was the distinctive thoop followed by an explosion in the distance. I looked at him, and as he turned and looked at me, we heard that same distinctive thoop, and then the bush between us exploded. We both cringed and dropped to the ground. I'm sure the expression on my face said, "Don't do that again." Then we heard that first deafening explosion. It came from our 8:00 O'clock, behind us in a northwesterly direction approximately 50 yards away behind a group of small trees and brush. The words "Booby Trap" still echoed in my head. There was a rifleman, Lieutenant Cruikshank, myself and the Marine with the M-79 grenade launcher taking up the rear. We turned and ran to an opening in the brush just to the left of the small clump of trees. It was there we hit the barbed wire and, as fast as you can go through wire, we went right through it. The small group of trees was now on our right; in front of us was a large open area with a tree line in the distance. On our left was more high brush jutting out into the clearing. We skirted the small tree line to our right, moving to the group of Marines now in front of us. There were a lot of wounded Marines lying there. Others were on top of them as if in some type of football huddle. We stopped just short of them. Lieutenant Cruikshank was now between me and the wounded Marines. The two other Marines with us had stopped at about seven yard intervals and were covering the rear. I leaned past Lieutenant Cruikshank and looked into the group of Marines. This face glanced up at that same time, and I had eye contact with a young Corpsman. He was all I could see in that maze of green. The look on his face told me I was lucky. Lucky I hadn't seen what he had. What he now turned his back to. There are no words to describe the sound of that second explosion as it killed one Marine and wounded two others. The Marines in front of us cringed as dirt and shrapnel from that second explosion brought home the realization we were in a mine field. It had come from behind the first group of wounded Marines, further down the tree line and a bit further into the clearing. Lieutenant Cruikshank and I had cringed also, and I was now smacking at my right knee trying to stop the burning sensation from the tiny piece of metal that had made it past the wall of Marines in front of us and was now embedded in my right knee. It was nothing, but it had gotten my attention as if someone had stuck a lit cigarette to my flesh. Lieutenant Cruikshank was busy on my radio as some of the Marines in front of us shifted position to help the second group of Marines just hit.

 I turned to my left and looked into the large clearing. To the left of us, about 60 yards away were three Marines. They seemed to be too far from us to be part of the third platoon. They may have been, but I suspect they were elements of the second platoon which had been moving to link up with us for an assault on the distant tree line. Whoever they were, they were in a very bad place, and they knew it. The tree line in the distance was full of the Vietcong and North Vietnamese Army elements that the second platoon and the lead elements of 3rd Platoon had chased there just moments earlier. They continued to lay down sporadic fire at us and at the second platoon. The Cobra gunships overhead were making them think twice before giving up their positions. These three Marines were in a position to have witnessed that second explosion and must have known what we knew--to move was to die. To remain where they were was to ask for a sniper to end their dilemma. The center Marine appeared to kneel and reached down as if to probe ground. Then they were gone. They had disappeared in an eruption of earth filled with searing metal and with the blood sweat & tears of young Marines. Still I thought I'd be OK. Lieutenant Cruikshank was on the radio and looking around as if giving a situational report. The two Marines behind us remained in position, and the look on their faces told me they too understood our situation. Our world filled with the sound of the large CH-46 as it came in from behind us, seemingly skimming the tree tops. They must have seen those three Marines disappear. It had just happened. Lance Corporal Bish had seen it. He was a radio operator assigned to Alpha Company's Commanding Officer, Lieutenant Purdy who circled above in the Command Huey. Bish had been moving with lead elements of the second platoon to link with the third platoon for an assault on the distant tree line. He had seen that first devastating explosion as it cut down five Marines. Bish had called for a "Corpsman Up" and had been moving toward our position. He now stopped and was on the radio with Lieutenant Colonel Ledbetter, Command Officer, and the pilot of the aircraft, HMM-263, "Blood Sweat & Tears." We had become his third Med-Evac mission since "Kingfisher" began earlier that morning. The Colonel barreled his aircraft down into the mine field as Bish advised, "Popping Smoke." The Corpsman had just run past Bish heading for the first group of wounded Marines as Bish threw his smoke grenade. They both hit mines. I never saw Bish or the explosion the smoke grenade set off. I never saw the explosion that had just killed the Corpsman as he ran to help wounded Marines. They were just out of line of sight behind the tall brush to our left. I never heard them either. These two mines detonated almost simultaneously with the one that was about to send me airborne. Maybe it had already happened and the sound had been drowned out by the rotors and engines of "Blood Sweat & Tears." They would have been the fourth and fifth explosions since the sound of that first one that had summoned us to the mine field moments earlier.

  The sound of that aircraft dropping down in front of us brought my attention back to Lieutenant Cruikshank and the remaining two Marines with us. Lieutenant Cruikshank told the these two Marines to go to the wire as the large Sea Knight, now at eye level, moved directly across from us to the site where the three Marines had disappeared. They made it to the wire and Lieutenant Cruikshank motioned for us to join them. As I turned to go, I noticed a crew member of "Blood Sweat & Tears" run down the ramp and into the minefield. The aircraft had moved across the mine field and now appeared to be straining to maintain a hover in a nose high attitude. In fact, it wasn't hovering at all. PFC Mike Clausen had directed Colonel Ledbetter to a precautious landing in the minefield. To reduce the chance of hitting mines, they had worked together to place the main wheel mounts on craters left by detonated mines. They would repeat this process three times. Bish saw him too. It was Clausen, disobeying the direct orders of Colonel Ledbetter not to leave the aircraft on his first of what would total six trips into the mine field. Bish had yelled, "You dumb SOB, you're in a mine field." Of course Clausen knew that, but he was busy trying to retrieve the body of one of those Marines that he had seen disappear in the dirt and smoke from the third explosion. We didn't get very far. We took one step, and Lieutenant Cruikshank hit the mine that had been right there between us that whole time. Not much room considering I had the radio and Lieutenant Cruikshank had the handset. Defying the laws of gravity, I was flying in slow motion through the air. It seemed an eternity before I felt the earth bring my flight to an end. Stunned and confused I realized my rifle was gone and noticed the blood at my right elbow. I felt as if I had been hit by a Mack truck, and there was a strange numbness accompanied by an intense burning sensation. I pulled my helmet off and tried to puke in it. I couldn't and realized I wasn't trying to puke after all, I was trying to breathe. I was in fact breathing, but that first breath after the mine was forced and hard, not the usual unconscious effort associated with breathing. I threw my helmet down, a really stupid thing to do, considering we were in a mine field. Fortunately, I got away with it, unlike Bish's smoke grenade.

I looked behind me for Lieutenant Cruikshank and saw him lying on his back with a large smoking hole between us. Through the ringing in my ears, I heard this loud hissing noise. The large column of yellow smoke coming from the radio strapped to my back told me it was from the smoke grenades I had attached there earlier. I had just popped another smoke, the hard way. I put my hands on the ground and attempted to get up to go to Lieutenant Cruikshank's aide when a new reality confronted me. My right pants leg dangled in the air, and my right leg was gone. I looked at my left leg to find it severely mangled with my left foot lying at some unnatural angle barely attached. Suddenly, one of the two Marines that had made it to the wire was in front of me knocking me back to the ground. He began to apply tourniquets to my legs. The other Marine was now on top of me also. He was busy slipping the radio off my back and then knelt there with his hands on my shoulders, as if to hold me down. I looked down at the Marine working on my legs and asked, "How are my legs?" to which he replied "don't worry about them…they're gone." He was definitely one "Born Again Hard" Marine. I looked back up at the Marine on top of me and said, "Check the Lieutenant." He seemed to release me and was gone before I finished getting the words out of my mouth.

Bish hadn't seen or heard the explosion that had just claimed both of my legs and one of Lieutenant Cruikshank's. His attention had been torn between his radio, the rescue efforts in front of him at the site of that third detonation, and the lifeless body of the Corpsman. The Corpsman was obviously gone. He had covered a lot of minefield before fate caught up with him and his lifeless body now rested close to the site of that first mine. Lance Corporal Bish and the Marine with him remained where they were as Colonel Ledbetter air taxied the large aircraft backwards across the mine field to the site of the first, second and sixth detonations. It was here that the majority of the wounded Marines lay. Here Private Clausen would again direct Lt. Colonel Ledbetter to landings at detonation sites and would exit the aircraft another five times, helping the wounded and retrieving the dead. I don't know how many Marines were left at this position that had not been injured. There were the two who had been with Lieutenant Cruikshank and me when we entered the mine field. There must have been several at the site of the first explosion. Lieutenant Cruikshank and I watched as they formed that huddle over wounded. I never saw the ones that hit that second mine beyond the first group. I was flat on my back, facing the opening in the brush that we had come through. A Marine placed a stretcher on the ground next to me. It may have been Clausen; I just don't know. This Marine and I believe the one that had tied off my legs, moved me to the stretcher. I could feel the torn muscle and broken bones for the first time. There is no gentle way to quickly move a mangled limb. The pain had become excruciating as the Marines carrying my stretcher ran to the ramp of "Blood Sweat and Tears" and deposited me on the deck.

Lance Corporal Bish watched in disbelief as the large aircraft wavered, first pitching and then yawing, just inches from disaster. He almost took off running despite the mines. It looked for a moment like it was going to roll over in the minefield and come after him. The rotor stabilized and Lieutenant Colonel Ledbetter regained control of his now damaged aircraft chock full of dead and dying Marines. Clausen, Colonel Ledbetter's crew chief, was still outside the aircraft. Lieutenant Cruikshank and the Marines carrying him had been hit by shrapnel. A Corpsman with them had been killed. Clausen had been knocked down by that blast. He got up and continued helping the wounded Marine with him to the aircraft. He then returned to the mine field for a fifth and then sixth and last time to help Lieutenant Cruikshank, the wounded Marines with him and to retrieve the body of the Corpsman. The helicopter, "Blood Sweat and Tears", was finally able to leave the mine field. Bish watched as the aircraft lifted off and turned away from him heading north east toward Da Nang. He would be going south west back to the second platoon and the pursuit of the Vietcong and North Vietnamese

Army elements still in the distant tree line, still lying down sporadic fire. All he had to do was turn in a direction as instructed by Lieutenant Purdy and walk out of the mine field. I am glad he hadn't seen our attempt at that. He had already seen enough to tell him what this meant. Bish and the Marine with him turned and left the mine field. He cringed with every step. It was agonizing. There was no wire in his end of the mine field--No discernable barrier or border to offer sanctuary, to offer acknowledgment of their successful passage through the killing field. His thoughts, when he had them, kept returning to a snapshot of the dead Corpsman. In his own words, "It had turned into, just another bad day in Viet Nam." Bish, Lieutenant Purdy's radio operator, he had always been in the thick of things, and this was his ninth or tenth "Kingfisher Patrol." The second platoon had been "kicking ass" all morning. They now waited for Bish to join them. Bish continued his walk out of the mine field and the Marine with him, possibly a Sergeant, a lifer from the looks of him, took point. They made it clear of the mine field and continued with the mission. Guess they were some more of those "Born again Hard," Marines.

I was so cold I was starting to shiver. There was a crewman moving back and forth passed me as I lay there on the deck in my stretcher. As he passed by I looked up and said, "I'm cold." He stopped and turned behind him to the flak jackets on the deck at the feet of the door gunners. Retrieving these, he turned back and covered me with them, pushing them in close to me as if trying to tuck me in for the night. Then he was gone. I think it was Clausen. I looked up and my eyes were locked with those of the starboard door Gunner. Colonel Ledbetter would later smile when I mentioned this and say, "That was Sergeant Major Martin S. Landy." His eyes left my gaze and would scan the carnage that filled the aircraft. Then he would turn and stare down the barrel of his weapon. His expression said, "Die you SOBs, die." He did this the remainder of the flight to the hospital. As bad as things had been, nothing could compare with what happened next. When they dropped into the minefield, Lieutenant Colonel Ledbetter had directed his flight crew to take off their flak jackets and place them on the deck. It was a wise decision. Again, there are no words to describe that seventh and last detonation. Clausen was returning from the left front of the aircraft with a wounded man. Lieutenant Cruikshank, now on a stretcher carried by two Marines was just steps from the ramp of the aircraft. I had cringed during that second detonation, and experienced a personal encounter with the sixth. It still wasn't over. There was a horrific noise, a combination of high explosive, rotor wash and the whine of the engines. Metal and debris smacked into the rear rotor and left rear of the aircraft. I rolled to my right side towards the center aisle and moved my left arm and hand to cover my head. I remember thinking, "God, please don't blow me up again." There was the sound of metal hitting metal, and I held tightly to the stretcher.

We were down, and people filled the cramped interior of the aircraft. Two of them grabbed my stretcher, and I was now in the bright sunlight of the landing zone. They were running and each step they took jarred my broken bones. The pain was immediate and unbearable. I tried to rise to my elbows and was yelling at them to stop. All I could do was hold on. We were now in a room, a room filled with saw horses. They placed my stretcher across two of those saw horses, and about five people pounced on me. Some were cutting my clothes off; others were working on my legs and taking my pulse. They all seemed to be in a hurry. One real irritating Marine knelt at my head with a clipboard and kept asking me for my name, rank, and serial number; how many times did I need to tell him? Whatever they were doing to my legs hurt. I moaned in agony. It felt as if they had just closed bear traps on them. A voice came from behind the Marine with the clipboard. It said "Hang in there West." The Marine beside me moved, and I could see Lieutenant Cruikshank. He was on the saw horses on my left and like me, surrounded by people working on him. We made eye contact and he closed his eyes, passing out. Why the hell was I still awake? All I wanted was to go to sleep and escape the unrelenting pain. I didn't care or think about waking up. I just wanted it to be over. It occurred to me that if I was still this aware of things, I should just get up and leave this place. I went to rise up on my elbows, made it about half way and fell back to the stretcher, drained from the effort.

The people over me shifted position and he pushed his way between them. He rubbed something on my chest and pushed the long needle between my ribs and into my heart. It may as well have been his fist. The sensation was that of a crushing pressure as if he was trying to push my heart out of my back. He never spoke a word, and then he was gone. It seemed like a lull in a storm. There were fewer people over me, and a back door adjacent to me opened up. Someone stuck his head in the room and said the "O. R. is ready." Hell, everyone knows they put you to sleep in the operating room. Again I tried to lift my head, turned to the person in the door and shouted, "West is ready!" The people at my side grabbed my stretcher and took me through that door to a room for an X-ray. As soon as the button had been pushed, they moved me down the hall into the O. R. I looked up through the blinding bright light and made eye contact for the last time that day. It was the face of a kindly, older man. His short, graying hair bordered his head cap like a halo. He reached down placing his right hand on my forehead and said, "It's OK son, you're going to sleep now."

"It's OK Son." I had heard these same words so many times as I sat in the safety of the command bunker handing off and monitoring the traffic between ground forces and air crews. "It's OK Son,' calm down and listen to me, I can't see your heat tab. You're going to have to do something else, or I can't find you."

"It's OK Son.' It will be all right. I know you have many wounded. I don't know where your two o'clock is. Where are you in relation to me?"

Even Bish lost it once as he was targeted by automatic weapons fire from a well entrenched position in a tree line. Evidently the degree of your speech impediment is directly related to the proximity of the fire you are receiving. The Scarface pilot came back with a "Calm down. It's OK Son," in response to the unintelligible babble that came across his radio. "Where you from, Son?" the pilot asked. "Oh yeah, I know where that is. OK Son" what are we doing here?" Bish directed the pilot to the position in the tree line where Charlie was deep under a large fallen tree. Several bursts of the mini gun and a few rockets later, Scarface had put an end to that crap.

First Lieutenant Bruce Cruikshank writes: I have no memory of getting on or off the helicopter. The scene starts with Eddy West and I moving at a good clip with several grunts to our right, We were heading east from our landing, across a rice paddy, sort of line abreast, when suddenly some Viet Cong are flushed out of some high grass some distance ahead of us. The men to our right, and maybe our left, start firing at the retreating enemy. We all give chase and I pull out my .45 side arm and fire a few rounds. I'm amazed that none of the Viet Cong seem to be hit and they disappear into some distant vegetation. The pursuit and firing stops and the group turns left and moves slowly in a northerly direction.

The radio traffic between someone on the ground and the lead pilot in the circling helicopters picks up. There's a group of Marines to the west in trouble and we start looping around further in that direction. I don't think this looping maneuver was more than 100 yards across. About this time I decide that I should say something on the radio and I am told that things are under control, shut up. I am following Eddy as we turn or I may have been ahead of him because when we reversed course I was then following him. We turn in a south westerly direction and approach the troubled squad. We get to within about 25 feet before we realize what I think was the platoon commander waving us back. We stop. The commander is standing in a spread out group of six to nine men. As we watch, on the far side maybe 60 to 90 feet away there's an explosion and something that looked like a sack or large rag but was probably one of the grunts is blown six to eight feet into the air.

Eddy and I turn and start back the way we came heading north east. I am following three feet behind him. I recall making no effort to follow in Eddy's foot steps. On about the third step I see a white flash where my left foot was. There's a deafening bang and I'm blown up in the air a bit. The first thought that crosses my mind is, "I'm going home." When I come to rest I am lying on my back with my feet to the southwest. Eddy is above my head on the ground to the northeast. The corpsman with our group is removing my belt to use as a tourniquet. About this time the 46 lands right in the area of this mine field to the west of me where all the other Marines were. The corpsman is working at my feet and facing me. One or two of the grunts were crouched down on my right side doing something like putting me on a poncho, they are facing the corpsman with their butts toward my head. There is another bang. When I come to, I'm in the same position and the helicopter is in the same position. It has mud splattered on its side with a big gob stuck to the side behind the cockpit window below the rotor mast. I see nothing of the rescue of Eddy. I think I'm the last one on the ground. I raise up on my right elbow and look over at the helicopter. I see the crew chief, Clausen standing on the rear ramp waving at me with both hand to come on board. I raise my hand in a helpless gesture and he trots out some 30 to 50 feet to me. Clausen hooks on to my arm pits and drags me back into the helicopter. As I am dragged up the ramp I remember seeing my feet dangling at very odd angles. Clausen props my head up but I tell him that I want it down and he lets it hang. I figure what blood I have left should be in my head.

When we reach Danang Eddy and I are placed together and we gave words of encouragement to each other. The corpsman kept asking for my service number and I got very annoyed having to repeat it so often. It was probably done just to keep us awake."

NOTES:
The following members of Alpha 1/1 were killed in action on January 31, 1970:
Doronzo, Paul F. HM3 B619922; 23; Denver, Colorado
Harrell, Raymond D. Cpl. 2477515; 19; Fries, Virginia
Rozell, Edward A. LCpl. 2538134; 18; West Seneca, New York

For heroism that day the Medal of Honor was earned by Private First Class Raymond M. Clausen, Jr.
The Navy Cross was awarded to Lieutenant Colonel Walter R. Ledbetter, Jr.
The Silver Star was earned by First Lieutenant Paul D. Parker, II and the Bronze Star by Major Alan B. Kehn, First Lieutenant Bruce W. Cruikshank, Lance Corporal Walter A. Gillin and Private First Class E.A. West.

Captain James W. Rider, a pilot flying with HML-367 wrote: I served two tours in I Corps, flying 0-1's and UH-1E's (slicks and guns) in 1965-66, and OV-10's and AH-1G Cobras in 1969-1970. An area I got very familiar with was just South of DaNang, bordered on the North by the Song Tuy Loan/Dong Cau Do, East by the South China Sea, West by Charlie Ridge and the South by the Song Tra Bon/Song Ky Lam. The area was spotted by many small villages surrounded by rice paddies, lightly vegetated except for rice paddies, and generally flat. Our squadron, VMO-2 flew all over I

Corps, but that unnamed area was as familiar as my home town by 1970. I had received two Purple Hearts for actions in that all too familiar area.

In 1970 I was assigned against my wishes as Air Liaison Officer with the First Marine Regiment, the "First of the First." I had served as a BAR man and Squad Leader (MOS 0311) with E-2-1 in Korea.

The Headquarters for the First Marines was on Hill 55, a place referred to as "Double Nickel", and one that I was well acquainted with. The First Marines was commanded by Colonel Lloyd Wilkerson (Major General USMC (Retired). Wilkerson, one of the most respected infantry officers I have ever known, was a Sergeant at Guadalcanal and a true leader, in every sense of the word.

My new sleeping quarters were in a South East Asia "Hooch" with 3 other Majors. The Regimental S-3 was Major Dave "Skag" Ramsey, a heavy .30 caliber water cooled machine gunner in Korea and a Bronze Star winner. Dave was serving with a battalion from the Third Marines in 1965, when they made a helicopter assault on Hill 55. The Marines encountered very light contact after landing. Ramsey was a Captain at that time, in H & S Company. The first morning after the landing the battalion command post was blown away by a suspected bobby trapped 155 round. The battalion commander was a KIA, along with several other Marines. I flew as the gunship escort for the VMO-2 Huey Med Evac flight that brought out the bodies. The med Evac bird was flown by our squadron commander, who was a classmate and close friend of the dead battalion commander.

Dave Ramsey later wrote a Reader's Digest "My most unforgettable character" article on the battalion CO. Dave showed me a small monument fastened to a rock on Hill 55, shortly after I checked in with the First Marine Regiment.

In 1970 in our hooch on Hill 55, over a couple of beers, Dave Ramsey outlined a derivative of the Sparrow Hawk mission. Dave emphasized that this was not that of a Sparrow Hawk reaction force, but of a helicopter patrol that went looking for the enemy and initiating action. We talked long into the night. Our ideas went up the chain of command, and what was christened "Kingfisher" was born.

Kingfisher was made up of:
1. A reinforced rifle platoon, which included a forward air control (FAC) team, and one Kit Carson Scout (former VC).
2. Four CH-46 helicopters.
3. Two AH-1G Cobra gunships.
4. A UH-1E Command and Control ship with the ground company commander aboard, piloted by a Tactical Air Control Pilot.
5. An OV-10 Bronco spotter, air control and light armed (M-60's and 2.75 rockets) reconnaissance aircraft.
6. An Army light fixed wing airplane with a rebroadcast air to ground

capability for radio transmissions which was useful for propaganda and recommendations for surrender by enemy soldiers (but never by U.S. Marines).

7. At least two Marine attack aircraft (usually A-4's) armed only with napalm and 20 mm. machine guns. The inclusion of the attack aircraft was for mandated by III MAF. Ramsey and I felt this was more to include fixed wing attack aircraft in the game for inter service political reasons, than for tactical necessities. Dave and insisted on using Napalm instead of high explosive bombs, because of concern for the civilian population. We had both seen the mediocre effects of Napalm in the damp, lush vegetation in our areas, and felt there was limited risk for anyone on the ground that was not physically struck by the canister, during a "nape scrape."

Sparrow Hawk was highly effective. On one notable occasion the helicopters landed to check the ID cards of two "farmers" strolling down a road. Without a shot fired, an NVA (not a VC) Lieutenant Colonel and an NVA private were apprehended and given a free ride to DaNang, in time for lunch. Our intelligence officer (S-2) reported in the next morning's brief, that when interrogated separately, both men said they were looking for a chance to surrender, but were afraid to tell the other one.

Other Kingfisher Missions were less amusing and more violent! After a few months of Kingfisher operations we received a message from III Marine Amphibious Force, congratulating us on over 100 kills and captures with only one Marine WIA during Kingfisher operations. All that changed on January 31, 1970.

My Tactical Air Control Party consisted of an acting TACP Chief Lance Corporal Springer.

Bruce Cruikshank and Eddie West volunteered to serve as the Forward Air Control team in support of the Kingfisher platoon which had drawn the mission for that day.

The Kingfisher patrolled around our area. There was a little radio chatter, but not much going on. An Army Light Observation Helicopter (LOH) reported taking fire a couple of clicks Northwest of Hill 55 on the east side of the Song Tra River. The fire was coming from a village that we knew was strongly infiltrated by the Viet Cong. I shifted the Army helicopter to another radio and told the pilot to check in with the Kingfisher command and control bird. The pilot told me that he wanted to engage the Viet Cong immediately. I told him to switch to the Kingfisher frequency, get clear of the target or leave our area.

When the Army observation helicopter came up on the Kingfisher frequency I asked him to fire a smoke rocket on the area from which they took fire, and then get out of the way. The Fire Support Coordinator gave the Command and Control Huey clearance to fire, in less than a minute the Cobras had prepped the zone with rockets, Machine guns and automatic grenade fire. The Kingfisher CH-46's landed right behind the gun run, the troops were off and chasing the fleeing enemy and the fight was on!

I called the command and control ship and asked them to get me some prisoners to Hill 55, as soon as possible. Our Kit Carson Scout was advising the enemy in Vietnamese to surrender, and they would not be harmed. Then I recognized the distinct voice of the Cobra Squadron Commander, "G__ damn it! Stop shooting them, they are trying to surrender!"

Within minutes an H-46 landed at our pos, and began taking prisoners off, most of which had been wounded. Our medics began working on the wounded immediately. Then a Marine shoved a young, healthy looking North Vietnamese soldier towards me. The enemy soldier held a freshly opened Coke in his hand. The Coke was so cold that water vapor was condensing on the sides of the can. Up to that point we had thought we were engaging the local the Viet Cong forces. I told the Kit Carson Scout to have the prisoner broadcast to the Army's rebroadcast frequency asking the enemy to surrender or die.

The Kit Carson Scout pointed his pistol over the head of the North Vietnamese soldier, smiled and spoke to the prisoner in Vietnamese. The enemy soldier was surprised to hear his voice transmitted over head.

I looked down and saw a hard looking, deeply tanned, weathered, tired looking enemy soldier on a stretcher near me. He was obviously in pain, covered with blood and lying in the hot sun. I called the Navy Regimental Surgeon over to the stretcher. The doctor took one look at the North Vietnamese and said, "He's dying", and turned to leave. I grabbed his arm and said, "Couldn't you do something to make him comfortable. The Surgeon looked me in the eye and said, "I don't have time for gooks."

I had adopted a three year old boy in Viet Nam, and taken him home, during my first tour (1965-66) in Nam. The word "gook" infuriated me, and was forbidden by Lieutenant General Lewis Walt, commander of the III Marine Amphibious force. I snarled "So much for the Hippocratic oath!" and the surgeon walked away.

To be cautious I gave my pistol to Springer, bent down and offered the soldier my canteen. He stared intently at me and took some of the water. We moved him into some shade near the command bunker. I asked the Kit Carson Scout what the prisoner's name was and he told me, "Lieutenant Colonel." I looked the wounded man in the eyes, then placed my hand on his shoulder and said "I'm sorry." We stared at each other for a couple of seconds. I said a silent prayer for him. He struggled and also made the sign of the cross. Then I saw what looked like a Roman Catholic "Miraculous Medal" around his neck. I turned away and went back to my radio. I looked for this man later, but never found him, but assumed that he died on Hill 55.

Springer told me that a helicopter was landing to pick up Marine casualties. As the helicopter made his approach I recognized the voice of Lieutenant Colonel Walter Ledbetter, the pilot. I knew Walt through two Viet Nam tours and loved the guy. He was the epitome of a combat helicopter commander.

I had not heard Eddie West or Bruce Cruickshank on the radio for some time. It did not concern me. Sometimes people have to switch frequencies, or are temporarily off the net.

Springer told me that he thought Cruikshank had called for a medical evacuation. As the CH-46 hovered there were several explosions in the zone. There were a lot of explosions as gunships fired rockets. Marines threw grenades and a mortar team had been flown into the fray.

It occurred to me that the NVA had fled towards the DaNang Barrier, a clearly marked mine field, that circled around the South side of the city. We called it the McNamara line.

Other helicopters landed. The reports of the number of wounded climbed. There were explosions all around Walt's 46, and for some time it seemed like it was taking forever to get the casualties on board.

Reports came that the NVA had fled through the DaNang barrier, that the warning signs had been removed, and that large numbers of Marines had been wounded.

Eventually all firing had ended. Prisoners and casualties were evacuated. The helicopters left for fuel and quickly returned. Finally the ground force was flown back to the base camp of Second Battalion, First Marines.

I still couldn't get land line phone from 2/1. I was told that both Cruikshank and West had been wounded and evacuated, and that there were a lot of seriously wounded Marines. The officer who called added that there were several amputees.

I got on the radio and land lines and learned that most of the casualties had been taken to the Naval Support Activity Hospital, across the road from Marble Mountain Air Facility.

I found Colonel Wilkerson and told him that my guys were hurt and I wanted to go to DaNang. The colonel told me that a number of Marines had been hurt in the DaNang Barrier and that I could ride with him, in his jeep, to the hospital. I grabbed the AK-47 I had picked up on patrol. Wilkerson rode in the passenger seat, with a 12 gauge pump shotgun across his lap. The Colonel's radio operator sat next to me. We took off for DaNang.

When we rolled into the hospital parking area I was right at home. On my first Vietnam tour I took many seriously injured casualties to NSA. I got caught by a mortar attack when eating dinner at the hospital one night and took shelter in a sand bagged sentry post.

I knew where Triage was located and I stepped up to a small wooded shelter outside of triage.

A Navy Chief set in the small shack that looked like a hot dog stand. I told the Chief that I was looking for Lieutenant Cruikshank and Lance Corporal West, who had been evacuated that day. The sailor grunted, "Cruikshank is a double amputee." Then in a more gentle voice, "West is also a double amputee. Both of them lost both legs. Cruikshank's are both above the knees. West's are below.

I felt like I had been hit by a sledge hammer. I followed the Chief's directions to Bruce's ward, and found him awake. He recognized me immediately and appeared alert. The conversation went something like this.

"Hello, Bruce. I am awfully sorry about what happened."

Bruce looked melancholy. He forced a smile and said, "Thank you for coming in." We attempted small talk about when he would be sent out of country. Bruce talked about treatment, where he would go, etc.

Before I left I told Bruce that I was sure that he would be able to handle the difficult times ahead, and then I mentioned another pilot and that I did not think that man would have been able to overcome the hardship if it was he that was wounded. It seemed like a real dumb thing to say. The Bruce brightened up and said, "Well, …at least they didn't get my balls!"

I stumbled down the hospital passageway, tears in my eyes, to find Eddy West's bed. I prayed silently for those two. Young horribly wounded Marines. When I walked up to West's bed he was awake and alert. He saw me. I couldn't speak. Eddy said "Gee, Major thanks for coming to see me!" I told him how sorry I was.

Eddie told me enthusiastically that he was being awarded a Bronze Star and a Purple Heart. He was upbeat and smiling. I was in tears. He told m that some of the other amputees talked about killing themselves, but that he knew he would be alright. He rattled on, and I grew more morose, forcing myself to appear upbeat and encouraging.

When I got back to 55 Springer and Jake were in the command bunker. I told them that I saw both the Lieutenant and Eddie West, and that they were both double amputees. I reported that Eddie was doing pretty well, under the circumstances, but that the Lieutenant had even more serious wounds, was weaker, and more dispirited. I told them to pray for both of these guys.

These two Marines and some other younger troops talked quietly about the whole affair. When I told them that both our guys would be awarded the Bronze Star and Purple Heart, One of the radio operators said to the other, "Would you give up your legs to get a Bronze Star and a Heart, the answer was "Probably." I blew my stack. I told them what a stupid thing to say. I knew I was reacting to all the horrible things that had happened so far that day. I later apologized.

I made notes on everything that had happened, and wrote a recommendation that both Cruikshank and West be awarded the Silver Star. Then a morbid and morose fog settled over me. My heart ached for our two casualties. I thought about how the whole rest of their lives would be changed. When I stumbled back to the hootch I sat staring out past the barb wire at the rice paddies and trees that surrounded us, and finally crashed onto my cot, praying for West and Cruikshank.

The next morning I attended the Commanding Officer's daily brief, and gave a short account of what I knew about the previous day. Our regimental briefings were usually short, upbeat and enthusiastic. That morning the place was like a funeral parlor. We took every casualty seriously, but the difference this morning was that the staff all knew and respected the two young amputees.

I checked in at the TAC Center, read the log, set my chin and slapped Springer on the back. I said, "Well Corporal Springer, you've got your work cut out for you. Tell the Com Chief that we want a radio operator, not a wireman, and I want him to be junior to you. It's up to you rebuild the spirit of the TAC party. Tell me what I need to do to make it easier for you."

Then I went to Bruce's hootch that he shared with other company grade officers. I inventoried and recorded the contents of his footlocker and other personal belongings. I had performed this grim task several times in Korea and my two tours in Viet Nam. We always removed the things that might be embarrassing to the casualty's family. Bruce's foot locker contained a neat pile of civilian and military aviation periodicals. In other inventories I had usually found stacks of Play Boy and other "girlie" publications. From looking at Bruce's stuff I could see that the biggest passion in his life was aviation.

I told Springer to shuffle the watch schedule and for he and Jake to try to get to the NSA hospital, at separate times.

The regiment's awards board met a couple of days later. I was a member. We were involved in a soul searching debate about an infantry platoon sergeant from the third battalion, this staff NCO had literally thrown himself on a Chi Com grenade during a brief fire fight near Hill 37. When the grenade did not go of he had pushed himself up and, with a sheepish expression, threw the frag explosive as far as he could. His men reported what he had done and we were debating about a recommendation for the Medal of Honor. We decided to reconvene the Board later, and personally interview the witnesses.

I raised my recommendation that both Cruikshank and West receive the Silver Star. The regimental executive officer, who headed up the awards board, dismissed this, because we really had no eye witnesses available. They had either died or been evacuated out of country. I told the Board that after I had left the regiment, I would reopen the case with Lieutenant Cruikshank's squadron in DaNang.

After he awards board broke up the XO told me to come to his little shelter, left over from the French who had occupied Hill 55 when fighting the Viet Minh. The Lieutenant Colonel sharply criticized me for sending Cruikshank and West on the mission. I explained to him that we had included a forward Air Controller, and TAC party radio operator on all the Kingfisher missions, because of the involvement of many aircraft, and the fact that the ground component could find itself upon landing in close contact, not always sure of their exact location, and without other supporting arms. I told the XO that we had carefully spelled out the whole operation in selling it to the Division and Wing, and that the tactical air control element was critical.

The XO listened carefully and said he understood the necessity of the air-ground communication. His attitude seemed to soften and he sympathized with me over the loss of our two men, and all of the battalion's casualties.

Within a week we received a letter from Bruce Cruikshank's mother. Bruce had been evacuated to Guam and the government was flying her to visit her son. This was grim news. The fact, that Bruce was in Guam meant that he was not strong enough to be flown to the continental United States. Also, we knew that dependents were not flown at government expense to visit casualties, unless the doctors felt that here was a medical necessity in which the presence of a loved one might help the patient to recover. I later learned from other letters from Bruce's mom, and years later from Bruce himself, that his mother's presence did make the difference. In Bruce's words, "it gave him the will to live."

Kingfisher continued to fly. My orders had specified six months of temporary duty with the First Marine Regiment. I returned to the Wing, where I chose to fly with a Cobra Squadron, HML-367. I was with the squadron for seven days, when I was shot down over the "Tennis Courts" in the mountains West of An Hoa. My co-pilot Dennis Grace was killed either by the impact of the crash, or was shot by the North Vietnamese who were all over our aircraft. I was rescued by a CH-46, evacuated to NSA Hospital, hen to USNH, Yokosuka and eventually Bethesda NNMC.

My recovery was slow. As soon as I could drive I went to Philadelphia Naval Hospital, where most Marine amputees from the East Coast, were sent. A major from HML-367, his roommate, a lieutenant from my home town, the major's roommate, and Eddie West were patients there. I had called and left a message for Eddie, that I was coming. I found him in a wheel chair along with a whole bunch of young, enlisted Marine amputees. They were bunched up at the top of a long, concrete ramp that ran from one story to the next. These Marines and their wounded Navy Corpsmen buddies were having races down the ramp, and then pushing themselves back up to the starting area. I once again brushed a few tears from my eyes when I saw the undefeatable spirit of these young heroes.

A couple of years later I was back on duty, and stationed at Headquarters Marine Corps (HQMC). Eddie was going to college in nearby Maryland. On night he and his girl friend were visiting me and my family. Eddie told me he had never received his Bronze Star. I raised the issue at HQMC, and Eddie was scheduled to receive the award from Brigadier General "Smoke" Spanjer. General Spanjer had pinned my third Purple Heart on my bed clothes when I was in the NSA Hospital at DaNang. When I talked to the general at length about Eddie, he said that he was sure that he had pinned the medal on West when Eddie was in the NSA Hospital, but that it wouldn't hurt to do it again.

On the day of the ceremony I met Eddie in the parking lot in front of HQMC, where I had arranged for handicapped parking. He struggled out of his car and onto his feet. He was in his forest green service uniform, probably the first time since he had left for Viet Nam. He did not have his cover. I helped him switch his Marine Corps collar emblems to the correct sides on the greens. His sideburns were way to long among the high and tight cuts of the Marines around him. As Eddie struggled along the sidewalk and ramped climb to the Navy Annex, a Colonel started to pass us. The Colonel took a look at Eddie and said, "Hold on there, Marine." I quickly stepped inside of Eddie and said in a low voice, "Colonel. This man is a double amputee He has been medically discharged and is here to receive a combat award." He colonel's face softened. He shook Eddie's hand, and said, "Son. It is an honor to meet you. Thank you for what you did for our country in Viet Nam.

Eddie's parents and younger brother arrived at General Spanjer's office. While pleasantries were being exchanged (Smoke could charm the stripes off a tiger), Mr. West, Edie's father somewhat belligerently complained about the delay in Eddie's award. General Spanjer promised he would take immediate action on the matter. The Bronze Star and Purple Heart were formally awarded in a dignified and somewhat sober manner. The General went to each person and shook their hand, and expressed a sincere thank you for Eddie's paying the heavy price in defense of our country.

After the ceremony my wife Sue and I took Eddie and his girlfriend to a restaurant with a spectacular view of the Potomac River and Washington DC. It was a warm and satisfying evening. Eddie and his girl friend visited us several times before we were transferred to Marine Corps Air Station, New River, in North Carolina. Eddie would occasionally call us and ride down on his motor cycle. He fascinated my boys when he told them about all the times when one or the other of his artificial limbs would fall victim to a motor cycle mishap. Then we did not see Eddie again, until the ceremony commemorating Blood, Sweat and Tears was conducted in Charlotte.

Bruce Cruikshank and I wrote each other, emailed, and talked on the phone several times. He was in a battle to get his private pilot's license back. After a long hard fight he got his flight status reinstated. I heard how he had built another plane, and one year he sent me a Calendar with thrilling pictures of him in flight.

Finally I received word about a Squadron Reunion for Walt Ledbetter's Viet Nam squadron at Charlotte North Carolina's airport. Included in the reunion was a presentation of the actual aircraft that had rescued the Marines from the Viet Nam minefield, dedicated to an aviation museum at the airport. The aircraft "Blood, Sweat and Tears" had been damaged by the

U.S. Army in Iraq, and then repaired and restored by Marine Volunteers at MCAS, Cherry Point.

Eddie rode his motor cycle to the event. Bruce and his charming wife flew in from California. After the weekend Eddie and the Cruikshanks visited us at Topsail Island.

I feel deeply privileged to have served with these heroes, and to have been a witness to what Walt Ledbetter and Mike Clausen did that day near Hill 55.

God bless the Marines!

Major Rider, was flying as a gunship commander with HML-367 on March 11, 1970 when his aircraft was shot down. He remembers the incident and writes: "I launched on an extract of a combined force of recon Marines and ARVN soldiers. It was the final episode of an extract that went bad on 2 March. Denny Grace was my co-pilot-gunner. I had just joined the squadron after six months as the First Marines ALO, and I hardly knew Denny. We had met at a discussion group being run by the Catholic chaplain at MMAF. We talked a little while we flew, rearmed and refueled. It turned out that we had both gone to small rival Catholic colleges in Western New York. Denny was from Niagara Falls and I was from the suburbs of Buffalo. He was on his second tour in Vietnam.

Ted Solliday had experienced a cobra tail rotor loss a few days before, and Denny and I briefed on the emergency procedure. The white board in the squadron ready room warned of air bursts in the Arizona area. We were operating near the "Tennis Courts" across the River, West of An Hoa.

On the second sortie after rearming at An Hoa we were hit. While climbing out from a gun run at about 600 feet AGL I felt a lurch. In my peripheral vision I saw a red or pink cloud. It looked like an airburst, but could have been hydraulic fluid. The aircraft immediately began to spin wildly. I rolled the power off and we were in an almost completely inverted, nose down attitude. Once the power was off the aircraft responded nicely to control input, but we could only fly in a rapidly descending large, right hand circle. I got out a May Day. As the ground came rushing up I could see we were going to hit on a rocky, steep slope of a mountain. I tried to squeeze on a little power but the aircraft started to spin again. I wasn't absolutely sure up to that point that we had lost the tail rotor. It could have been loss of tail rotor control, or a stabilizer computer malfunction. I flared and pulled collective as we went in, facing up the mountain.

With the beginning of the spin when I started to add power we had gotten over some trees. I had briefed Denny to challenge me on "Fuel" if we were going to impact during the flight. He did and I told him "Secure Fuel." The engine was winding down. We were in an extremely nose high attitude and I could not see what was in front of us. I gave Denny the controls, with a "You've got it." He took the controls with a kind of incredulous "I've got it!" I saw we were settling into the tops of trees and then remembered nothing until I woke up on the ground. I think the cyclic hit my bullet bouncer, which in turn smacked me in the jaw and knocked me unconscious.

I came to on the ground, sitting in the cockpit. I had all the wind knocked out of me and could hardly talk. There was AK-47 fire very near to us. I tried to call Denny but could barely speak. I passed in and out of consciousness, several times. My helmet was over my face with a big vertical crack down the middle of it. Through that crack I saw two people, carrying AK-47's and speaking Vietnamese. My right arm had a compound fracture and the only way I could move it was to grasp my right wrist with my left hand and move my arm around. I could not get to my survival radio, which was under my bullet bouncer. I got my .38 out. It was loaded with some green star cluster rounds that Major Dan Shar had given me. I fired the pistol through the canopy a couple of times and Shar saw the rounds and knew someone was alive.

I came to at one point and there was a crewmember from a 46 next to the cockpit. He dragged me out of the cockpit. I asked him about Denny and he said he looked pretty bad. I ran on my broken leg to the hoist and while ascending to the CH-46, a round hit the hoist seat, broke up and fractured my lower right arm, putting a deep laceration there. This wound gave me my third purple heart, which meant a free trip home. In the CH-46 I talked to a crew chief or gunner and asked them to get the Catholic Chaplain from MAG-16 over to the NSA hospital to administer last rites to Denny.

My health record was pretty screwed up. I had a broken jaw and lost some teeth. All of the ribs on my left side were broken as were the tibia and fibula of my right leg. I also had a large hole in my right forearm and my spleen and gall bladder were removed. I spent a long time in hospitals at DaNang, Yokuska, Yokota and finally Bethesda. Denny's brother, Ed came to see me in the hospital, and his parents eventually visited my family and me in Virginia.

I felt that his mother never really recovered from Denny's death. I remember Denny and his family every week in my prayers. I still have dreams about various parts of this incident. It was a shame that a young officer like Denny had to die. I sometimes feel a little guilty that he died so young, and I lived on. The whole incident still stands out vividly in my memory, but there are a lot of unknowns that I remain curious about."

Walter Ledbetter wrote: "In 1969 and 1970 I was a Marine helicopter pilot in Vietnam. I was a Lieutenant Colonel, and the commanding officer of HMM-263 at Marble Mountain, Vietnam, we were flying the CH-46 assault helicopter. January 31, 1970 on a mission code named Kingfisher, my crew chief that day was Private First Class Mike Clausen. On that mission he earned the Medal of Honor, the only aviation enlisted crewman to be awarded that medal, and I received the Navy Cross.

Thirty five years later that same helicopter crashed in Iraq. The Marine Corps declared it an historic aircraft and returned it to the States to be rebuilt as it was that day in Vietnam. The historic branch of Headquarters Marine Corps called and told me about the crash and what was going to be done to the aircraft. Truthfully, I didn't think much about it but about the same time my crew chief passed away. I gave the eulogy at his funeral and it was written up in a few newspapers.

A short time after the funeral I got a call from a man, named Ed West. He told me that he had read about Mike Clausen's funeral and that he had been looking for him for years. He went on to say that on that day he was a nineteen year old Lance Corporal laying in that mine field with both of his legs blown off; and if we had not landed when we did and picked him up he would be dead. He said he was calling just to say thank you. We talked for a long time and I told him about the aircraft.

We arranged to meet each other at the helicopter; it was rebuilt by the Carolina Aviation Museum in Charlotte, North Carolina. We met about a month later. It was a real emotional experience for both of us. He showed me where he was lying on the aircraft and told me what he could remember of that day. He asked me a lot of questions about that day and other kingfisher missions that I had flown. His questions made me think about things that I had forced out of my mind a long time ago. We determined that we would find all of the other people involved in that mission.

On that day we carried all 178 Officers and men of Alpha Company First Battalion, First Marine Regiment, First Marine Division into one hell of a fight. Shortly after inserting all three platoons of Alpha Company, into three different positions, we had to start medevac. I made two emergency medevacs for the first two platoons we had landed. I picked up six wounded in those first two medevacs, one of the wounded, Walter A. Gillin, had lost his foot. Every landing that day was into a hot zone. We were all being shot at.

Then we were called and told about the nineteen marines in a mine field. They had inadvertently walked into the mine field pursuing the NVA. The mine field was a U.S. minefield (the Danang barrier) that was supposed to be tended by the ARVN forces. They didn't do their job. The NVA took one side of the fence down and had a path across the field. The NVA crossed the field on their path with the marines in pursuit. Once the marines entered the minefield they were taken under fire and if they moved they stepped on mines. I landed three times and got them out. Mike Clausen, against my orders, got out of the aircraft six times to carry in wounded and dead marines. If he had not done what he did we probably would not have been successful. I would have had to land three or four more times and the odds were such that I would probably have landed on a mine and crashed. That would have been a real problem for everyone. There were four killed, eleven wounded, and four that made it out unhurt. Three of the wounded were double amputees. All together that day my crew and I picked up seventeen wounded and four dead. We took moderate damage to the aircraft when a mine went off and sent shrapnel up through the rotor system and through the aft pylon but luckily the aircraft was still flyable.

The platoon commander in the mine field, a young First Lieutenant, Joseph A. Silvoso, had a pretty severe head wound, but when I got them to the medevac hospital he refused treatment and took the four that were not wounded and returned to the fight. The last man that we have found was a young Private First Class, Authur Trujillo, at that time, he stepped on a mine and was wounded thirty three times in his legs, arms, and back, plus he took another three shrapnel wounds to his head.

When I talked to him he said he had been trying to find the others involved for many years. All he could remember from the incident was coming to after the explosion and seeing two marines coming toward him to help. They stepped on a mine and just disappeared. He had always felt responsible for their deaths. I had them call him. They were two of the amputees.

The aircraft was finished being rebuilt and refurbished in September 2007 and was dedicated on October 20, 2007. We tried to have everybody that was involved in Kingfisher Missions present. We got a good number to attend. Clausen had named his aircraft and painted the name on the armor plate next to the door. He named the aircraft Blood, Sweat, and Tears, after his favorite rock group.

We are still trying to finding the others. All I have is a company roster with their names and old Marine Corps serial numbers, we did not use social security numbers back then, and where they joined the Marine Corps. I have found my aircrew, the copilot, First Lieutenant Paul Parker, and the two gunners, Sergeant Major Martin S. Landy and Corporal Steve M. Marinkovic, and seven of the wounded and about 10 others. Last summer I found one more of the wounded men from the mine field.

That day I witnessed unbridled courage. A young corpsman blown up running toward the wounded in what he knew was a minefield. I saw young marines moving across that field to help their comrades and going up in smoke. My crew chief walked out of that helicopter six times into the killing field, to help others. He knew, full well, that each step could be his last. He served three years in Vietnam and flew over one thousand nine hundred and ninety combat missions. It may not be a record but it is a damn good average.

If you have read this far, there is a lot more to this story. Kingfisher mission were made up of three Ch-46's transport aircraft carrying a reinforced infantry platoon, from Alpha Company, First Battalion, First Marine Regiment. There was a lead Huey that would scout a designated area looking for things (people) that should not be there. We followed at low altitude accompanied by four Cobra gunships. There was also an OV-10 observation aircraft to control any fixed wing support that was needed in the fight, the OV-10 was armed with rockets and machine guns.

When the Huey spotted something out of place, and he always did, he called me and I would speed up going toward him, I would stay low to preclude the enemy from observing my approach. The Huey pilot would then mark the spot with a smoke grenade or rocket. My job was to land right on them, not fifty yards away but right in the middle of the enemy force; and the fight was on. When the fight started we would then pick up the other two platoons and put them into the battle as blocking forces. Some days we brought in a lot more troops than that. It all depended on the size of the enemy force. The Kingfisher missions were given credit for disrupting the enemies plan for a 1970 Tet offensive.

I led the first thirty two of these missions. To illustrate the ferocity of these fights, on an earlier mission in January, I took 57 rounds in the cockpit and numerous hits all over the aircraft. That type of intense fire was routine. During that time the various aircraft that I flew on Kingfisher missions took more hits than all of the other aircraft in the first marine air wing combined; and the young men of Alpha Company, in a forty to forty five day period were in constant fights for thirty two of those days.

In 1970 the Marine Corps was reducing its force in Vietnam. From our section around Danang the Twelfh Marine Regiment was withdrawn and sent home. They reduced the forces in Northern I Corps. They also sent a lot of the long timer's home and mix mastered the force. The Marines that were involved in Kingfisher were mostly new guys in a mix mastered unit. Very few of them knew each other, and they were short on staff NCO's. The company commanders were First Lieutenants. It was a brutal assignment for young inexperienced Marines and they performed like Marines of the old corps.

First Lieutenant Paul D. Parker was Ledbetter's co-pilot that day and he writes, "My memory about that day is somewhat jumbled after these many years. We made several insertions this day and on one of these we saw a VC running across the field. The Marines were shooting at him but he kept running. One of the Cobras rolled in on him with machine gun fire and kicked up a lot of dust. The guy went down and we figured he must have been hit. As the cobra rolled off the guy got up and started running. Then the second cobra made a run on him with the same results. When he got up this time he made it to the tree line so it must have been his lucky day.

As far as specific events that day I don't have many memories. It was a very fast paced 6 hours. As copilot I was busy assisting the pilot and working the radios, working the ramp, watching the engine gages, and clearing the aircraft as we air-taxied between spots."

Clausen was far from being a model Marine and had several brushes with the UCMJ system. He was thought of by many as being a somewhat disagreeable person. He was on a second tour in Vietnam having been with HMM-364 from December of 1967 to January of 1969. He returned to Vietnam in November of that year and was at the end of his second tour when he earned the Medal. Walt Ledbetter was always a kind of concerned, hands on, kind of leader and it is believed that Clausen was his crew chief so that he, Ledbetter, could keep an eye on him. Clausen did disobey Walt Ledbetter's orders, and walked down that ramp into a minefield, which he knew was there. Probably most of those Marines who he rescued would have soon bled out where they lay. Clausen deserves the recognition of us all, and our corps.

Lance Corporal Walter Scott Brank, III, a helicopter crew chief who earned the Silver Star and two Bronze Stars in Vietnam knew Clausen well. He attended helicopter mechanic school with him and made this observation, "Clausen was a wild one who liked to drink and raise hell. He used to tell us that he was a Catholic and that he could do whatever he wanted, then go to confession and all would be forgiven. I heard that he was about to be processed for an undesirable discharge when the mine field incident took place. Scuttlebutt is that he was separated, called back to active duty for a few days, awarded the Medal and separated again, immediately. I heard that he asked to return to active duty in the Marine Corps but his request was refused. He was a hard guy to like."

After action reports indicate that two other helicopters from HMM-363 participated that day. One was piloted by First Lieutenant Carroll L. Wright with First Lieutenant Peter A. Mazurak as copilot. The crew chief was Lance Corporal Arthur E. Peckitt, Jr. and the gunners were A. Leary and A.T. Lomers. The pilot of the other aircraft was First Lieutenant Robert W. Watkins. His copilot was First Lieutenant Thomas E. Heifner and Sergeant Donald E. Stueve was the crew chief. Flying as gunner was Corporal John H. Mora

During the course of the action Lance Corporal Walter A. Gillin had his left foot blown off and suffered intense pain. Although weak from loss of blood and unable to continue his combat efforts, his calm presence of mind and unwavering courage was a constant inspiration to the Marines near him.

Also among the wounded was Lance Corporal James K. Singleton with a fragmentation wound in the neck and Lance Corporals Chris Nick, Edward W. Sanderson and Larry Wynn, and Privates First Class Arthur L. Forbes, Larry E. Lasater and Frank G. McKeever with multiple shrapnel wounds.

Nick had carried the first wounded aboard the aircraft out of the minefield. He then went back out and was badly wounded.

Others mentioned in the story are as follows:
Andrus, Kermit W. LtCol. 058689; 41; Jamestown, New York
Chill, John E. Pfc. 2565205; A-1-1; 19; St. Louis, Missouri
Clausen, Raymond M. Pfc. 2258929; HMM-263; 22; Hammond, Louisiana
Cretney, Warren G. LtCol. 062107; HML-367; 47; Kansas City, Missouri
Cruikshank, Bruce W. lstLt. 0102799; FAC-1-1; 24; San Luis Obispo, California
Forbes, Arthur L. Pfc. 2594700; A-1-1; 19; Minneapolis, Minnesota
Gillin, Walter A. LCpl. 2536861; A-1-1; 19; Baltimore, Maryland
Grace, Dennis F. 1stLt. 0104205; HML-367; 26; Niagra Falls, New York
Heifner, Thomas E. 1stLt. 0104484; HMM-263; 22; Ruston, Louisiana
Kehn, Alan B. Maj. 080877; H&S-1-1; 28; Troy, New York
Landy, Morton S. SgtMaj. 621217; HMM-263; 40; Curwinsville, Pennsylvania
Lasater, Larry E. Pfc. 2569768; A-1-1; 18; Evansville, Indiana
Ledbetter, Walter R. Jr. LtCol. 063973; HMM-263; 40; Shreveport, Louisiana
Marinkovic, Steve M. Cpl. 2445390; HMM-263; 20; Pittsburg, Pennsylvania
Mazurak, Peter A. 1stLt. 0105185; HMM-263; 25; Milwaukee, Wisconsin
McKeever, Frank G. Pfc. 2588815; A-1-1; 18; Oakland, California
Mora, John H. Cpl. 2404421; HMM-263; 21; New Mexico
Nick, Chris LCpl. 2491792; A-1-1; 19; Jacksonville, Florida
Parker, Paul D. II 1stLt. 0102654; HMM-263; 24; Austin, Texas
Peckitt, Arthur E. Jr. LCpl 2439983; HMM-262; 20; Chicago, Illinois
Purdy, William R. 1stLt. 0107823; A-1-1; 21; Raleigh, North Carolina
Ramsey, David A. Maj. 070095; H&S-1-1; 37; San Francisco, California
Rider, James W. Maj. 077451; HML-367; 33; West Seneca, New York
Sanderson, Edward W. LCpl. 2409456; A-1-1; 20; Jackson, Mississippi
Shar, Danny A. Maj. 072730; HML-367; 35; Unknown
Silvoso, Joseph A. II 1stLt. 0108780; A-1-1; 23; Columbia, Missouri
Singleton, James K. LCpl. 2510279; A-1-1; 19; Little Rock, Arkansas
Soliday, Theodore D. 1stLt. 0104790; HML-367; 23; Wheeling, Ohio
Spanjer, Ralph H. BGen. 011919; AWC-1-MAW; 47; East Orange, New Jersey
Stueve, Donald E. Sgt. 1306860; HMM-263; 36; St. Louis, Missouri
Trujillo, Arthur Pfc. 2504775; A-1-1; 18; San Antonio, Texas
Waltrip, Steve W. Capt. 089632; HML-367; 27; Herculaneum, Missouri
Wilkerson, Herbert L. Col. 045163; CO-1MarDiv; 50; Charleston, South Carolina
Wright, Carroll L. 1sLt. 0104821; HMM-263; 28; Pinellas, Florida
Wynn, Larry LCpl. 2545635; A-1-1; 19; St. Louis, Missouri

★ ★ ★ ★

I have been working on a list of combat medals earned by Marines in Vietnam since 1998, when I discovered that no such list existed. I was fortunate to have acquired a copy of Jane Blakeney's rare book, *Heroes: U.S.Marine Corps 1861-1955*, early in my post Marine Corps life. Jane Blakeney had worked at the Marine Corps Award Branch for many years and had meticulously compiled and listed every combat award earned by Marines since 1861. When she retired, around the time her book was published, no one chose to continue this valuable compilation.

When I made my first visit to the Awards Branch it was located in Clarendon, Virginia. I mistakenly thought that a list of Marine combat award winners was on file and all I had to do was go there and obtain a copy. Much too my chagrin I discovered that no such list existed. Undaunted I was determined to compile such a list and I have doggedly attempted to do so since then.

The list that appears in this book is complete for the Medal of Honor and the Navy Cross. The Silver Star list is incomplete but close to being so. If I erred it was on the side of omission. I much prefer the name of a deserving person be omitted than an undeserving one included. That said, this is the best that I have been able to do to this point. The Silver Star list remains a work in progress. Thanks to my friend C. Douglas Sterner, I recently added the names of three Navy corpsmen just days before going to press. I'll keep working.

> "Medals are very difficult to deal with as any honest person knows that a whole lot of people have done more and gotten less and some people have done less and gotten more."
>
> First Lieutenant Karl A. Marlantes
> Charlie Company 1/4
> Navy Cross

In compiling this story, interviews with the participants, the awards citations of many of the participants, the rosters and information obtained from the National Personnel Records Center were used.

17

Medals

James Rider wrote: "I was the secretary of the awards board in VMO-2 both times I was in Vietnam, and on the awards board for the First Marine regiment. While doing this I became aware that there was never anything fair about awards, and no matter what you had done to be recommended for an award, your best friends would say, "Well he really didn't deserve it." People would lie, cheat and steal to get awards.

While with VMO-2 in 1965 and 1966, the commanding officer, Rigor Bauman had a policy that no one would be recommended for awards from within the squadron. That made my life easy as secretary of the awards board. Dobie Gillis, the executive officer of the squadron landed in the middle of some Viet Cong to pick up American wounded on Harvest Moon. General Walt called him over to his headquarters and General Victor Krulak pinned a Silver Star on him. I had to write it up. It was just like doing an investigation. You got statements from the witnesses, took facts out of the statements, and put together a recommendation based on the facts.

A few weeks later, Gillis got a call to go to General Walt's office. As he left, Major Gillis said, "I wonder if I am going to get another Silver Star." When he returned he looked defeated. That day, he had refused to control attack aircraft on a village from which heavy caliber fire was blasting the helicopters in an assault landing zone. Major Gillis could see women and children in the village. He had words with General Walt over this incident, and as usual, the General prevailed. Gillis sat down on the bench in front of the ready room with me, and said something like this, "Johnny, what happens when you've been doing something for seventeen years and all of a sudden find out that you hate it?" I think he retired at his first opportunity and went to New Zealand to live.

When Abshire, Doc Mayton, Dick Drury and I flew that god awful medevac, (*Honor the Warrior* page 18) I was aware that a lot of the enlisted men in the squadron were disgruntled over the fact that the officers got all the awards, and all they ever got were Purple Hearts an single mission air medals. That was part of my motivation for recommending them for the Navy Cross. I told the awards board this. They actually asked me if I should get the Medal of Honor. I told them no, and that Jack Enockson, my gunship wing man, deserved a higher award than I did, because he went in and snatched the radio operator from under the noses of the North Vietnamese. I couldn't get in to that spot because every time I tried I took hits in the air. Jack asked to try and I radioed, yes! He made a gun run to a quick stop, a hard landing and pick up. He got the DFC.

I was recommended for the Silver Star but I told them that I would rather have the DFC. Doc Mayton was the only one who initially received the award he was recommended for. I chalked that up to the fact that he was in the Navy. I wrote Headquarters Marine Corps about Bob Abshire's award and they later upgraded his Silver Star to the Navy Cross. Much later, my award was upgraded, I think at the instigation of Bob Abshire after he left the Marine Corps.

I came to the conclusion that most awards were forthcoming after fiascoes. When everyone did his job correctly there was no need for heroics. When people were in a long sustained fight, there was no thought about decorations. The worse the battle, the less likely that awards were forthcoming.

Many people never receive awards they deserve. A very few may receive awards they don't deserve. Usually, successful combat operations are a result of a team effort. One or two people get recognized for what the whole team did, usually those who were the leaders, took the most risk, inflicted the most enemy casualties, or were hurt the worst. Some time ago I read an article in the *Marine Corps Gazette*, recommending a series of small unit awards, for squads, fire teams, air crews, and other small units, for heroic action. That makes a lot of sense to me. I wish everyone on my aircraft in the Ashau valley had received a small unit Silver Star Medal, instead of just me.

Combat decorations are an enigma. Not very long after the action, all but a few have remembered what you did to get the award; they only remember that you have a certain award."

Richard O. Culver wrote, "I have seen truly deserving lads, mostly enlisted folks, actually turn down even the suggestion that he or they might have done something worthy of an award, especially when so many acts of bravery and self sacrifice went unnoticed.

 Now the kind of Marines I am/was used to dealing with would rather take a fanny kicking than ask anybody to write them up for any sort of medal – it comes with being a professional who simply thinks he is or was simply doing his job."

On the list of medal winners that follow in some cases the Marines age at the time the medal was earned is an approximation. Where hometown is concerned in many cases the place of entry into the Marine Corps is used as I was unable to determine a hometown. I did the best I could with the information that I was able to obtain.

MEDAL OF HONOR

Name	Service #	Unit	Date	Age	Hometown
Anderson, James Jr Pfc	2241921	F-2-3	2/28/67*	20	Compton, California
Anderson, Richard A LCpl	2420891	E-3-Recon	8/24/69	21	Houston, Texas
Austin, Oscar P Pfc	2472757	E-2-7	2/23/69*	21	Phoenix, Arizona
Barker, Jedh C Cpl	2207369	F-2-4	9/21/67*	22	Franklin, New Hampshire
Barnum, Harvey C Jr 1stLt	084262	H-2-9	12/18/65	25	Waterbury, Connecticut
Bobo, John P 2dLt	092986	I-3-9	3/30/67*	24	Niagra Falls, New York
Bruce, Daniel D Pfc	2485891	H&S-3-5	3/1/69*	18	Beverly Shores, Indiana
Burke, Robert C Pfc	2359360	I-3-27	5/17/68*	18	Monticello, Illinois
Carter, Bruce W Pfc	2511589	H-2-3	8/7/69*	19	Miami Springs, Florida
Clausen, Raymond M Jr Pfc	2258929	HMM-263	1/31/70	22	Hammond, Louisiana
Coker, Ronald L Pfc	2452732	M-3-3	3/24/69*	21	Alliance, Colorado
Conner, Peter S SSgt	1280832	F-2-3	3/8/66*	33	East Orange, New Jersey
Cook, Donald G Col	072794	POW	12/64-12/67	45	New York, New York
Creek, Thomas E LCpl	2403648	I-3-9	2/13/69*	18	Amarillo, Texas
Davis, Rodney M Sgt	1978754	B-1-5	9/6/67*	25	Macon, Georgia
De La Garza, Emilio A LCpl	2532368	E-2-1	4/11/70*	20	East Chicago, Illinois
Dias, Ralph E Pfc	2421279	D-1-7	11/12/69*	19	Shelocta, Pennsylvania
Dickey, Douglas E Pfc	2199321	C-1-4	3/26/67*	20	Rossburg, Ohio
Foster, Paul H Sgt	1903536	H-3-12	10/14/67*	28	San Francisco, Cal
Fox, Wesley L 1stLt	096702	A-1-9	2/22/69	37	Round Hill, Virginia
Gonzalez, Alfredo Sgt	2142473	A-1-1	1/31/68*	21	Edinburg, Texas
Graham, James A Capt	088847	F-2-5	6/3/67*	26	Frostburg, Maryland
Graves, Terrence C 2dLt	0101090	3-F-Recon	2/17/68*	22	Groton, New York
Howard, Jimmie E GySgt	1130610	C-1-Recon	6/16/66	37	Burlington, Iowa
Howe, James D LCpl	2462231	I-3-7	5/6/70*	21	Liberty, South Carolina
Jenkins, Robert H Jr Pfc	2428700	3-Recon	3/5/69*	20	Interlachen, Florida
Jimenez, Jose F LCpl	2472949	K-3-7	8/28/69*	23	Red Rock, Texas
Johnson, Ralph H Pfc	2356797	A-1-Recon	3/5/68*	19	Charleston, S C
Keith, Miguel LCpl	2517987	CAP-1-3-2	5/8/70*	18	Omaha, Nebraska
Kellog, Allen J Jr GySgt	1927666	G-2-5	3/11/70	26	Bridgeport, Connecticut
Lee, Howard V Capt	069961	E-2-4	8/8-9/66	33	Dumfries, Virginia
Livingston, James E Capt	084449	E-2-4	4/30/68	28	McRae, Georgia
Martini, Gary W Pfc	2217825	F-2-1	4/21/67*	18	Portland, Oregon
Maxam, Larry L Cpl	2141892	D-1-4	2/2/68*	20	Glendale, California
McGinty, John J III	1602718	K-3-4	7/18/66	26	Boston, Massachusetts
Modrzejewski, R J Capt	073356	K-3-4	7/15-18/66	32	Milwaukee, Wisconsin
Morgan, William D Cpl	2337025	H-2-9	2/25/69*	21	Mount Lebanon, Pennsylvania
Newlin, Melvin E Pfc	2229466	F-2-5	7/4/67*	18	Wellsville, Ohio
Noonan, Thomas P Jr LCpl	2292900	G-2-9	2/5/69*	25	Maspeth, New York
O'Malley, Robert E Cpl	1972161	I-3-3	8/18/65	22	Brooklyn, New York
Paul, Joe C LCpl	2033358	H-2-4	8/19/65*	19	Dayton, Ohio
Perkins, William T Jr Cpl	2296240	C-1-1	10/12/67*	20	Sepulveda, California
Peters, Lawrence D Sgt	2004158	M-3-5	9/4/67*	20	Binghamton, New York
Phipps, Jimmy W Pfc	2412145	B-1-Eng	5/27/69*	18	Culver City, California
Pittman, Richard A Sgt	2119979	I-3-5	7/24/66	21	San Joaquin, California
Pless, Stephen W Capt	079156	VMO-6	8/19/67	27	Decatur, Georgia
Prom, William R LCpl	2421504	I-3-3	2/9/69*	20	Pittsburgh, Pennsylvania
Reasoner, Frank S 1stLt	085378	A-3-Recon	7/12/65*	27	Kellogg, Idaho
Singleton, Walter K Sgt	2056158	A-1-9	3/24/67*	22	Memphis, Tennessee
Smedley, Larry E Cpl	2274116	D-1-7	12/21/67*	18	Orlando, Florida
Taylor, Karl G SSgt	1862790	I-3-26	12/8/68*	29	Avella, Pennsylvania
Vargas, Jay R Capt	083768	G-2-4	5/2/68	29	Winslow, Arizona
Weber, Lester W LCpl	2323793	M-3-7	2/23/69	20	Hinsdale, Illinois
Wheat, Roy M LCpl	2242728	K-3-7	8/11/67*	20	Moselle, Mississippi
Williams, Dewayne T Pfc	2420506	M-3-9	9/18/68*	19	St. Clair, Michigan
Wilson, Alfred M Pfc	2421744	M-3-9	3/3/69*	21	Odessa, Texas
Worley, Kenneth L LCpl	2230824	L-3-7	8/12/68*	20	Modesto, California

NAVY CORPSMEN

Ballard, Donald E HM2	B604255	M-3-4	5/16/68	22	Kansas City, Missouri
Caron, Wayne M HM3	B116083	H&S-3-7	7/28/68*	21	Middleboro, Mass
Ingram, Robert R HM2	7719275	C-1-7	3/28/66	21	Clearwater, Florida
Ray, David R HM2	B308634	H&S-2-11	3/19/69	24	McMinnville, Tennessee

CHAPLAIN

Capodanno, Vincent R Lt	656197	H&S-3-5	9/4/67*	38	Honolulu, Hawaii

NAVY CROSS

Abrams, Lewis H Col	053788	VMA-242	10/25/67*	48	Montclair, New Jersey
Abshire, Bobby W Cpl	1979928	VMO-2	5/21/66	22	Fort Worth, Texas
Abshire, Richard F Sgt	2125809	G-2-4	5/2/68*	23	Abbeville, Louisiana
Adams, John T LCpl	2033889	C-1-Recon	6/16/66*	22	Covington, Oklahoma
Adams, Laurence R III Capt	092937	HMM-165	1/12/69	25	Seattle, Washington
Alfonso, Vincent Pfc	1984618	A-1-3	7/20/66	24	Tohatchi, New Mexico
Allen, Yale G Cpl	2406509	C-1-4	3/5/69	20	Cincinnati, Ohio
Almeida, Russell V LCpl	2083359	C-3-Eng	12/20/65*	19	South Dartmouth, Massachusetts
Ambrose, Gerald D LCpl	2458230	M-3-1	1/8/70	20	Iowa City, Iowa
Amendola, Willet R Cpl	2303053	M-3-7	11/2/67*	19	Deposit, New York
Anderson, John J Sgt	1646153	I-3-9	10/27/65	26	Broad Channel, N Y
Armstrong, Russell P SSgt	1970190	I-3-26	9/7-8/67	24	Ft. Calhoun, Nebraska
Arquero, Elpidio A SSgt	1694536	B-1-3	5/10/67*	27	Honolulu, Hawaii
Aston, James M Pfc	2450092	H-2-26	3/19/69*	18	Wichita Falls, Texas
Ayers, Darrell Sgt	2341301	1-Recon	3/19/70*	32	Alderwood Manor, WA
Badnek, Samuel J Pvt	2012032	H-2-4	8/18/65	20	Youngstown, Ohio
Baggett, Curtis F SSgt	1384825	K-3-5	2/6/68*	31	Raleigh, North Carolina
Bailey, Walter F Sgt	2135341	E-2-5	3/21/70	24	Unknown
Barnes, Robert C LCpl	2083790	B-3-Recon	12/17/66	21	Nashville, Tennessee
Barnett, Robert L Jr Cpl	2259704	A-1-5	5/9/69	19	Minneapolis, Minnesota
Barrett, James J Cpl	2228382	I-3-26	9/19/67	19	Jacksonville, Florida
Barrett, John J Capt	085356	HMM-263	2/26/69	30	Minneapolis, Minnesota
Baskin, Richard W Sgt	2051026	H&S-1-26	6/6/67	24	Philadelphia, Penn
Batcheller, Gordon D Capt	080672	A-1-1	1/31/68	28	Hingham, Mass
Bateman, Kent C Maj	073614	VMA-533	10/25/67	34	Salt Lake City, Utah
Beaulieu, Leo V Pfc	2162804	E-2-5	5/16/66*	21	Lengby, Minnesota
Bell, Van D Jr LtCol	044563	CO-1-1	6/6/66	48	Portsmouth, Rhode Island
Bendorf, David G LCpl	2247275	L-3-9	5/20/67*	20	Livingston, Wisconsin
Benoit, Ronald R 2dLt	096153	D-1-Recon	2/25/67	36	Brunswick, Maine
Berger, Donald J 2dLt	064100	HMM-163	3/12/66	34	Williamsville, New York
Binns, Ricardo C LCpl	2031505	C-1-Recon	6/16/66	20	Bronx, New York
Bird, William C Pfc	2449630	E-2-5	5/15/69	19	Mansfield, Texas
Blann, Stephen L Cpl	2391078	E-2-9	2/16/69*	23	Pompano Beach, Florida
Blevins, Thomas L Jr Cpl	2255036	C-1-26	5/31/69*	21	Middletown, New Jersey
Bogan, Richard E LCpl	2357980	C-1-5	4/12/68	19	Lebanon, Indiana
Brady, Eugene R LtCol	051664	HMM-364	5/15/69	41	York, Pennsylvania
Brandtner, Martin L Capt	080625	D-1-5	9/3/68	31	Minneapolis, Minnesota
Brandtner, Martin L Capt	080625	D-1-5	9/11/68	31	Minneapolis, Minnesota
Brantley, Leroy Cpl	2392175	G-2-5	3/28/69*	20	Charleston, South Carolina
Brindley, Thomas D 2dLt	0101761	I-3-26	1/20/68*	24	St. Paul, Minnesota
Brown, Charles E Cpl	2288651	CAP-H-8	1/31/68	20	Atlanta, Georgia
Brown, David H Sgt	2056298	L-3-26	9/10/67*	21	Saltillo, Tennessee
Browning, Randall A Cpl	2151453	A-3-ATanks	9/10/67	22	Cincinnati, Ohio
Bryan, Charles W Cpl	2307311	B-3-Recon	1/20/68*	20	McKinney, Texas

Name	Service #	Unit	Date	Age	Hometown
Bryant, Jarold O LCpl	2061015	K-3-1	3/21/66	22	Columbus, Ohio
Buchanan, Richard W LCpl	2113260	M-3-27	5/24/68	23	Moraga, California
Burke, John R Cpl	2200152	H&S-1-26	6/6/67*	23	Clearwater, Florida
Burnham, Thomas R Cpl	1426163	F-2-5	10/1/67	31	Pennsylvania
Burns, Leon R SSgt	1487956	B-1-9	7/2/67	30	Portland, Maine
Caine, Lawrence B III Cpl	2131757	I-3-5	5/13/67	20	Salt Lake City, Utah
Calhoun, John C LCpl	2328321	CAP-H-6	1/7/68*	20	South Boston, Massachusetts
Campbell, Joseph T 1stLt	0101784	I-3-5	6/15/68*	23	Stoneham, Massachusetts
Canley, J L GySgt	1455946	A-1-1	2/6/68	31	Little Rock, Arkansas
Carroll, James J Capt	079583	K-3-4	10/5/66*	29	Miami Beach, Florida
Carter, Marshall N Capt	085375	C-1-1	1/14/67	27	ColoradoSprings, Colorado
Casebolt, Henry C Cpl	1933907	F-2-1	2/28/66*	24	St. Joseph, Missouri
Casey, Michael J 2dLt	095659	C-1-1	3/18/68*	23	Dalton, Massachusetts
Casey, Thomas J Jr LCpl	2422191	D-1-7	2/16/69*	22	Milton, Massachusetts
Castillo, William Pfc	2404648	E-2-4	2/25/69	19	Albuquerque, New Mexico
Cavanaugh, Thomas J Cpl	2136906	F-2-7	9/19/68*	22	Waterbury, Connecticutt
Cheatham, Ernest C Jr LtCol	058120	CO-2-5	2/3-3/3/68	38	Long Beach, California
Cheatwood, Paul R LCpl	2345578	B-1-5	2/16/68	19	Tallapoosa, Georgia
Christensen, Paul K Cpl	2105464	3-CAG	11/19/67	21	New York, New York
Christman, William J III 2dLt	0106531	A-1-9	2/22/69*	23	Gaithersburg, Maryland
Christmas, George R Capt	085447	H-2-5	2/5/68	28	Yeadon, Pennsylvania
Christy, Kenneth L Jr 2dLt	099991	L-3-4	1/18/68	23	Lutz, Florida
Cisneros, Roy Cpl	2341534	B-1-3	9/11/68*	19	San Antonio, Texas
Claybin, Edward A Pfc	2127649	D-2-11	6/20/66	19	Los Angeles, California
Cobb, Paul F 2dLt	0103410	A-1-7	5/16/68*	24	Roanoke, Virginia
Cochran, Robert F Jr 2dLt	089648	A-1-Amph	8/18/65*	23	Poplarville. Mississippi
Coffman, Clovis C Jr GySgt	1106891	C-1-Recon	10/10/66	35	Summit, Mississippi
Collins, Bryant C Cpl	2009643	A-3-Recon	7/12/65	21	Gloversville, New York
Cone, Fred J Maj	071439	VMA-242	10/24/67	34	Prescott, Arizona
Conklin, Richard F Cpl	2251197	D-2-13	5/10/68	23	Seattle, Washington
Coolican, James J Capt	079762	HMAC	1/31/68	28	Carbondale, Pennsylvania
Corsetti, Harry J Cpl	2350487	3-Recon	8/15/68	20	Baltimore, Maryland
Cousins, Merritt T LCpl	2295242B	1-12	7/8/67*	19	Clinton, Iowa
Covella, Joseph F GySgt	1001220	Advisor	1/3/66*	37	New York, New York
Cover, Robert L MSgt	1101469	VMO-2	3/17/67	36	Pickerington, Ohio
Crockett, Joseph R Jr Sgt	2257776	1-FRecon	4/23/69	21	Richmond, Virginia
Cummings, Roger W Pfc	2470680	K-3-7	4/20/69*	21	Bloomington, Indiana
Curley, Ronald T Sgt	2168667	F-2-26	5/16/67	20	Las Vegas, Nevada
Curtis, Russell W GySgt	1428625	FFSR	8/21/67	28	Zanesville, Ohio
Dabney, William H Capt	080399	I-3-26	4/14/68	33	Gloucester,Virginia
Dalton, Robert G Cpl	2298765	K-3-3	5/25/69	21	New York
Danner, David J Sgt	2113629	A-3-Tanks	5/8/67	23	Sioux City, Iowa
Darnell, Dana C LCpl	2208738	B-1-9	4/24/67*	19	Greenville, South Carolina
Davis, Dennis D Pfc	2567690	K-3-7	8/28/69*	20	Capitol Heights, Maryland
Dawson, John R 2dLt	0100987	G-3-12	10/27/67*	23	Adrian, Michigan
Day, Edward L Cpl	2283719	L-3-3	8/26/68*	19	Philadelphia, Pennsylvania
De Bona, Andrew D Capt	083217	M-3-26	9/10/67	30	Ebensburg, Pennsylvania
De Planche, Mark B Cpl	2042758	G-2-9	1/14/66	20	Flint, Michigan
Devries, Marvin H 1stLt	0107350	E-2-3	8/10/69	26	Detroit, Michigan
Dickson, Grover L Cpl	1508294	K-3-3	11/11/66*	28	New Orleans, Louisiana
Dillard, Henry C Cpl	2076789	M-3-4	5/29/67	23	Cadiz, Kentucky
Dittman, Carl R Cpl	2344874	IS-3	11/14/68	22	Houston, Texas
Donaldson, Billy M SSgt	1461338	1-3-Recon	8/8/66	30	Valliant, Oklahoma
Donovan, Joseph P 1stLt	0101005	HMM-364	2/22/69	27	Waxahachie, Texas
Donovan, Joseph P 1stLt	0101005	HMM-364	4/21/69	27	Waxahachie, Texas
Dorris, Claude H SSgt	1547450	CAP-H-6	1/7/68*	29	Louisville, Kentucky
Dowd, John A LtCol	059807	CO-1-7	8/13/69*	37	Elizabeth, New Jersey
Downing, Talmadge R GySgt	1176720	M-3-1	3/5/69	33	Middletown, Ohio
Driscoll, Thomas B Cpl	2135382	D-1-5	9/4, 6/67	22	Richmond, Virginia
Duff, Barry W Cpl	2098848	C-1-9	5/21/66*	21	Baltimore, Maryland

Name	ID	Unit	Date	Age	Hometown
Duncan, Richard W Cpl	2327745	M-3-5	11/8/67*	19	Livermore, California
Eades, Lawrence M LCpl	2231362	CAC-P-3	2/2/68	21	San Diego, California
Ebbert, Terry J 1stLt	091101	E-2-5	12/24/66	24	Belvidere, Illinois
Edwards, Craig A 1stLt	0110222	H-2-7	8/20/70	21	Albuquerque, New Mexico
Estrada, Manuel A LCpl	2361066	A-3-Recon	8/16/68	20	Phoenix, Arizona
Evans, Richard A Jr Pfc	2427562	D-1-5	8/29/68*	18	Independence, Missouri
Fairfield, Rupert E Jr Capt	085242	VMO-6	8/19/67	27	New Orleans, Louisiana
Fante, Robert G Cpl	2175781	F-2-5	8/6/68*	20	Roseville, Michigan
Federowski, Robert A Cpl	2204671	D-1-5	5/24/68*	19	Lansing, Illinois
Feerrar, Donald L LCpl	2252390	G-2-1	3/20/67	20	Baltimore, Maryland
Felton, Samuel L Jr Pfc	2479014	C-1-5	6/11/69	19	Cleveland, Ohio
Finley, Michael P LCpl	2204950	A-1-4	5/8/67*	20	Big Rock, Illinois
Fisher, Thomas W LCpl	2244800	M-3-5	9/4/67*	20	Allentown, Pennsylvania
Floren, Jimmy E Cpl	2325565	H-2-5	11/6/67*	21	Klamath Falls, Oregon
Fowler, Earl W Cpl	1984379	C-1-7	3/28/66	22	Albuquerque, New Mexico
Frederick, John W Jr CWO	082847	POW	6/67-8/68*	48	Manito, Illinois
Fryman, Roy A SSgt	1356690	1-Recon	4/29/68	34	Paris, Kentucky
Fuller, John L Jr 2dLt	094257	E-2-5	3/23/67*	22	Atlanta, Georgia
Galbreath, Bobby F Capt	070303	VMO-6	2/16/68*	37	Amarillo, Texas
Gale, Alvin R Pfc	2393859	G-2-26	1/28/69*	21	Hyde Park, Massachusetts
Gallagher, Patrick LCpl	2170557	H-2-4	7/18/66	19	Lynnbrook, New York
Gauthier, Brian J Cpl	2007664	A-1-3	7/11/65*	20	Mansura, Louisiana
Getlin, Michael P Capt	086661	I-3-9	3/30/67*	27	Lagrange, Illinois
Gibson, George R Cpl	2010487	E-2-4	8/8/66	20	Pasadena, Texas
Gilleland, Richard M Sgt	2260688	M-3-7	2/23/69	22	Nashville, Tennessee
Gillingham Richard K Cpl	2118093	H-2-9	5/19/67*	20	Detroit, Michigan
Gomez, Ernesto LCpl	2151708	HMM-262	1/25/68	20	Brownsville, Texas
Gonzales, Daniel G Cpl	2126237	B-1-7	6/7/69	22	Corpus Christi, Texas
Goodsell, William J Maj	055282	VMO-6	6/16/66*	37	Seattle, Washington
Gray, George E LCpl	2259206	E-2-9	2/2/68	19	Shreveport, Louisiana
Green, John S GySgt	1275527	F-2-5	6/2/67	31	Waco, Texas
Green, Maurice O V 1stLt	0103516	IO-1-5	10/27/68	24	Birmingham, Alabama
Gregory, Robert B Jr LCpl	2411976	CAP-2-1-2	2/25/69	20	Los Angeles, California
Gresham, Michael E Sgt	2378513	D-1-5	9/11/68	21	Macon, Georgia
Gresko, Richard W LCpl	2516983	CUPP-3-5	3/11/70	21	Philadelphia, Pennsylvania
Grimes, Paul E Jr LCpl	2264945	F-2-7	3/4/67	19	Corapolis, Pennsylvania
Grosz, Nicholas H Jr 1stLt	086541	H&S-2-7	12/18/65	25	River Edge, New Jersey
Guarino, Timothy S Pfc	2484801	G-2-9	6/1/69	19	Coldwater, Michigan
Guay, Robert P Maj	060940	HMM-261	3/19/66	25	Guilford, Connecticut
Guerra, Victor J SSgt	2111973	L-3-1	10/27/69	25	Buffalo, New York
Halstead, Lee M 1stLt	094751	HML-167	8/29/68*	23	Lansing, Michigan
Harrington, Myron C Capt	081869	D-1-5	2/23/68	27	Augusta, Georgia
Hartsoe, David E Pfc	2279897	L-3-9	5/20/67*	20	Coatesville, Pennsylvania
Hayes, Daniel J LCpl	2422326	L-3-5	11/28/66	20	Cambridge, Massachusetts
Hazelbaker, Vincil W Maj	063157	VMO-2	8/8/66	39	Grangeville, Idaho
Helle, Ronald B Sgt	2146818	G-2-5	1/28/71	23	Toledo, Ohio
Henderson, Billy K LCpl	2470976	H-2-1	7/17/69	20	Philadelphia, Tennessee
Hendricks, Robert L Cpl	2235429	H-2-7	2/19/68	20	Los Angeles, California
Herrea, Felipe LCpl	2341619	A-1-Recon	9/20/68*	20	San Antonio, Texas
Herring, Alfred J Jr LCpl	2478681	H-2-1	9/20/70	19	Mullins, South Carolina
Herron, Lee R 1stLt	0102874	A-1-9	2/22/69*	23	Lubbock, Texas
Hilgers, John J W Capt	074432	OO-2-4	8/24/66	31	Virginia Beach, Virginia
Hill, Lamont D Pfc	2229088	I-3-4	3/6/67*	20	Massillon, Ohio
Hoapili, John GySgt	1764533	K-3-26	5/9/68	29	Honolulu, Hawaii
Hodgkins, Guy M SSgt	1490687	I-3-9	9/3/66*	30	Los Alamos, New Mexico
Hoff, John R Jr 2dLt	0111172	E-1-Recon	4/7/70	24	Boston, Massachusetts
Holmes, Walter C Sgt	2050005	B-1-9	11/27/65	22	St. Louis, Missouri
Honeycutt, James E Pfc	2356099	3-Recon	2/16/68*	20	Haynes, Arkansas
Hopkins, Michael E Pfc	2037164	K-3-9	7/4/66*	21	Norfolk, Virginia
Houghton, Kenneth J Col	013965	CO-5	6/5/67	46	La Jolla, California

Name	Service #	Unit	Date	Age	Hometown
House, Charles A LtCol	026690	HMM-163	3/10/66	44	West Plains, Missouri
Howard, Billy GySgt	554586	E-2-4	3/21/66*	39	Hialeah, Florida
Howell, Gatlin J 1stLt	093190	IO-1-9	7/2-7/67*	31	Colma, California
Hubbard, Robert W Capt	088752	Advisor	2/4/68*	27	Auburn, Alabama
Huffcut, William H II Maj	076752	VMO-6	9/28/69	33	Talahassee, Florida
Huggins, Michael A Pfc	2470841	G-2-4	6/27/69	21	Indianapolis, Indiana.
Hughes, Stanley S Col	012654	CO-1	2/3/68	49	Rome, Pennsylvania
Hunnicutt, Hubert H III Cpl	2277992	C-1-9	4/18/68	20	Suwanee, Georgia
Jaehne, Richard L 2dLt	0106958	K-3-7	8/28/69	22	Salt Lake City, Utah
Jmaeff, George V Cpl	2436055	C-1-4	3/1/69*	23	Oliver, BC, Canada
Johnson, James L Jr Cpl	2288355	E-2-9	2/17/69	20	Plymouth, Michigan
Johnston, Clement B Jr LCpl	2116397	D-1-Recon	4/28/66*	18	Pittsburgh, Pennsylvania
Jones, Phillip B 2dLt	0106589	G-2-26	1/28/69*	21	Slater, Iowa
Joys, John W SSgt	1624442	A-1-4	8/26/66*	27	San Francisco, California
Judge, Mark W Pfc	2235869	E-2-4	9/21/67*	20	Torrance, California
Kaler, Richard D Cpl	2031739	H-2-4	7/21/66*	22	Patchogue, New York
Kaufman, David M 1stSgt	1016850	E-2-3	6/15/69*	42	Plaquemine, Louisiana
Keck, Russell F Cpl	2201914	A-1-3	5/18/67*	20	Okmulgee, Oklahoma
Kelley, Edwin C Jr 1stLt	0104244	M-3-4	3/13/69	25	Shillington, Pennsylvania
Kelly, James R III Cpl	2040353	I-3-3	3/24/67*	22	Sacramento, California
Kelly, Robert A 1stLt	088103	I-3-9	9/3,5/66	25	Woonsocket, Rhode Island
Kemp, Marwick LCpl	2017255	E-2-4	3/21/66	23	Pensacola, Florida
Kenison, Benjamin A LCpl	2098437	D-1-4	9/16/66*	20	Jefferson, New Hampshire
Kennedy, Johnnie M GySgt	598516	Advisor	4/19/65	38	Miami, Florida
Keys, William M Capt	079544	D-1-9	3/2/67	29	Fredericktown, Pennsylvania
Koelper, Donald E Maj	060953	Advisor	2/18/64*	32	Northbrook, Illinois
Koontz, Leonard Cpl	2395551	M-3-4	5/27/68	19	Johnstown, Pennsylvania
Korkow, Kenneth A Cpl	2258125	B-1-26	3/30/68	21	Blunt, South Dakota
Kowalyk, William LCpl	2122016	G-3-12	8/26/66	19	Detroit, Michigan
Kuzma, Marc J Pfc	2391847	A-1-4	4/26/68*	19	South Hadley Falls, Massachusetts
La Pointe, Alvin S Pfc	2083443	C-1-7	3/28/66	19	Sudbury, Massachusetts
La Porte, Alfred P Jr Sgt	2076173	H-2-4	2/25/69	25	New York, New York
Lain, Bobby D Capt	077695	B-1-1	2/19/67	29	Round Top, Texas
Lankford, Albert J III 1stLt	0106303	G-2-9	4/21/69	24	Montgomery, Alabama
Laraway, William D LCpl	2210088	CAC-H-4	4/11/67*	23	Seymour, Indiana
Lauer, Charles R Cpl	2214200	A-3-Amph	6/18/67*	20	Downey, California
Lazaro, Lawrence J Cpl	1950010	E-2-7	9/19/66	21	Boston, Massachusetts
Lebas, Claude G LCpl	2141933	B-1-3	4/2/66	22	Long Beach, California
Ledbetter, Walter R Jr LtCol	063973	HMM-263	1/31/70	40	Shreveport, Louisiana
Lefler, Alan C LCpl	2264470	VMO-2	3/17/69	23	Davenport, Iowa
Leftwich, William G Jr Maj	061154	Advisor	3/9/65	39	Germantown, Tennessee
Lineberry, Jerry E SSgt	1993557	B-1-7	2/12/70*	28	Wadesboro, North Carolina
Livingston, Lawrence H Capt	0107447	Advisor	7/11/72	31	Bonsall, California
Long, Melvin M Sgt	2226408	F-2-5	6/2/67	22	Toledo, Ohio
Lopez, Jose G Sgt	2261270	B-3-Recon	9/2/67	20	Fort Douglas, Utah
Lopez, Steven D Pfc	2312521	A-3-Recon	5/9-10/67	18	Silver Spring, Maryland
Loweranitis, John L Cpl	2052170	I-3-9	3/30/67*	22	Dubois, Pennsylvania
Lowery, Steven M Cpl	2198568	C-3-Recon	3/5/69	23	Las Vegas, Nevada
Lownds, David E Col	015530	CO-KSCB	11/67-3/68	47	Westerly, Rhode Island
Luca, Joseph Sgt	2239172	F-2-7	3/5/67	20	Meriden, Connecticut
Lumbard, Donald W LCpl	2117497	B-1-Amph	7/4/66	20	Thomaston, Connecticut
Lunsford, Glen T Sgt	2101197	D-1-7	2/3/68*	20	Danville, Virginia
Mac Vane, Matthew C Cpl	2239397	CAP-T-P-3	8/14/67	22	Portland, Maine
Malone, George M Jr 2dLt	0105740	A-1-9	2/22/69	22	Portland, Oregon
Mann, Bennie H Jr Maj	060180	HMM-163	3/31/65	37	Yuma, Arizona
Marlantes, Karl A 1stLt	0103269	C-1-4	3/1-6/69	24	Seaside, Oregon
Martin, Raymond C Sgt	1511600	F-2-9	3/18/66*	28	Pomona, California
Mc Afee, Carlos K Capt	067830	Advisor	6/12/66	32	Oklahoma City, Oklahoma
Mc Cauley, Bertram W Maj	058997	HMM-263	9/5/66	37	New Albany, Indiana
Mc Cormick, Michael P 2dLt	0107462	D-1-4	3/20/69*	24	Wellston, Ohio

Name	Service #	Unit	Date	Age	Hometown
McDaniel, James V 2dLt	SS#	G-2-9	5/15/75	24	Unknown
Mc Donald, Thomas C Capt	092570	HML-167	3/28/71	29	Cincinnati, Ohio
Mc Henry, William D Cpl	2253419	H&S-1-5	2/1/68	20	Cincinnati, Ohio
McRae, Arthur G Cpl	2254285	F-2-7	3/16/67	21	Newark, New Jersey
Mc Whorter, James E LCpl	2492266	L-3-3	8/22/69*	20	Beaverton, Oregon
Meier, Terrance L SSgt	1999759	M-3-3	7/21/67*	22	Portland, Oregon
Mendez, Angel Sgt	2030367	F-2-7	3/16/67*	20	New York, New York
Meuse, John R Pfc	2249495	E-2-3	5/3/67*	18	Malden, Massachusetts
Miller, Cleatus A Jr Pfc	2530192	K-3-7	10/25/69	20	Raleigh, North Carolina
Mitchell, Robert G Maj	070988	HMM-165	11/29/68	34	Alameda, California
Moe, Robert F SSgt	1096741	L-3-3	12/9/65	34	Mobile, Alabama
Moffit, Richard E Sgt	2127413	G-2-26	5/17/67	23	Omaha, Nebraska
Monahan, Frederick G LCpl	2195404	E-2-3	5/3/67	20	Holland, Pennsylvania
Monahon, Robert E LCpl	2121721	D-1-9	5/28/67*	20	Swedesboro, New Jersey
Montgomery, Robin L 2dLt	0107468	MP-1-5	6/8/69	23	Brookfield, Connecticut
Moore, Freddie L LCpl	2494568	B-1-7	8/12/69	20	Chicago, Illinois
Moore, Ronald A LCpl	2241747	I-3-7	7/19/67*	20	Manhattan Beach, Cal
Mosher, Christopher K LCpl	2130200	FAC-K-3-5	5/13/67	20	Dowigac, Michigan
Muir, Joseph E LtCol	049816	CO-3-3	8/24/65*	37	Meadow Bridge, W.V.
Mulloy, James E Jr Sgt	1288230	H&S-3-3	8/18/65	30	Jeffersontown, Kentucky
Murphy, James E Capt	091159	AO-2-4	10/26/67	24	Midland, Michigan
Murray, John D Capt	091848	M-3-5	9/4/67	28	Pittsburgh, Pennsylvania
Myers, William H Pfc	2315999	D-1-5	5/12/67*	20	Salem, Indiana
Neil, Michael I 2dLt	0101333	D-1-7	12/20/67	27	San Diego, California
Noel, Thomas E 2dLt	0106151	C-1-4	3/1/69	22	Tulsa, Oklahoma
Noon, Patrick J Jr Sgt	2016255	F-2-4	8/24/66	22	Cheverly, Maryland
Norris, James A LCpl	2458214	I-3-5	8/13/69*	19	Des Moines, Iowa
Norton, John J Capt	077989	E-2-5	4/21/68	31	Foster City, California
Norwood, George O LCpl	2060343	G-2-7	3/4/66	23	Holy Grove, Arkansas
O'Brien, Joseph J Capt	076414	OO-2-9	9/23/67	29	Bronx, New York
O'Conner, Martin E Maj	072430	Advisor	11/5/69	36	Rahway, New Jersey
Oakley, Johnny L LCpl	2197945	G-2-3	8/22/66	20	Ridgecrest, California
Panian, Thomas C Sgt	2098929	I-3-5	9/6/67	21	Essex, Maryland
Parrott, Lee R GySgt	1533280	MP-2-3	8/10/69	31	Noblesville, Indiana
Paskevich, Anthony Jr 1stLt	094196	VMO-2	3/17/69	25	Chagrin Falls, Ohio
Peczeli, Joseph S SSgt	1427514	1-AAmph	3/24/67	30	Utica, New York
Peters, William L Jr 1stLt	0102208	HMM-165	4/12/69*	26	Fort Dodge, Iowa
Peterson, Dennie D 2dLt	0100590	FO-2-11	9/6/67*	24	Los Angeles, California
Phelps, John J LCpl	2079635	VMO-6	8/19/67	23	Louisville, Kentucky
Piatt, Louis R 2dLt	0106014	M-3-7	2/23/69	23	New Orleans, Louisiana
Pichon, Louis A Jr GySgt	605087	I-3-3	3/24/67*	39	Slidell, Louisiana
Pierpan, Herbert E Maj	082433	OO-1-4	3/20-22/69	30	North Adams, Mass
Pitts, Roy E Pfc	2425604	G-2-9	2/17/69	18	North Highlands, California
Popp, James A Pfc	2246112	F-2-4	4/8/67*	21	Hemlock, Michigan
Poulson, Leroy N GySgt	1209285	VMO-6	8/19/67	35	Newell, Iowa
Powell, Charles T Sgt	2085259	I-3-9	5/31/68*	23	Columbus, Georgia
Quick, Robert L Pfc	2395468	K-3-3	2/7/68*	18	Wallaceton, Pennsylvania
Ralya, Warren H Jr Cpl	2287779	A-1-Amph	1/20/68*	21	Shalimar, Florida
Rash, Donald R Pfc	2230103	B-1-26	3/30/68*	19	Pocahontas, Virginia
Ray, Darrell T Pfc	2108299	C-2-1	2/28/66*	18	Olympia, Washington
Reid, John M Cpl	2145615	C-1-3	5/10/67*	20	Magnolia, New Jersey
Reilly, Donald J Maj	061230	VMO-2	12/9/65*	36	Ladue, Missouri
Reilly, James R Pfc	2042806	G-2-3	3/17/66*	20	Esopus, New York
Reis, Tiago Cpl	2209245	F-2-4	9/21/67*	20	New Bedford, Mass
Richards, Thomas A Cpl	2417681	H-2-9	6/5-6/69	22	Madison, Wisconsin
Rienschke, Harold A SSgt	1880764	B-3-Tanks	3/24/69	27	Beatrice, Nebraska
Ripley, John W Capt	081239	Advisor	4/2/72	32	Radford, Virginia
Rivera Sotomayor, Miguel A Cpl	2174871	F-2-9	7/29/67	20	Philadelphia, Pennsylvania
Rivera, Jose L LCpl	2447515	L-3-5	3/26/69	20	Waukegan, Illinois
Rivers, Jettie Jr 1stSgt	1300239	D-1-9	5/15/67*	34	Nashville, Tennessee

Name	Service #	Unit	Date	Age	Hometown
Roberson, James J 2dLt	097738	1-AAmph	3/24/67	35	Boise, Idaho
Rodrigues, Joe G Jr Sgt	2058221	L-3-4	3/3/69*	22	Dallas, Texas
Rogers, Raymond G Jr 1stSgt	583291	I-3-9	3/30/67	39	Cincinnati, Ohio
Roller, Robert T Sgt	1998691	F-2-5	10/13/66	23	South River, New Jersey
Rollings, Wayne E 1stLt	017099	1-Recon	9/18/69	28	Elloree, South Carolina
Romine, Richard E LtCol	058482	HMM-165	6/3-4/67	36	Palm Harbor, Florida
Rosenberger, Roger D Pfc	2484828	M-3-3	6/17/69*	18	Swartz Creek, Michigan
Ross, David L Maj	074837	VMO-2	9/4/67	35	Cabool, Missouri
Rusher, Robert C Cpl	2278308	CAP-H-6	1/7/68*	25	Tracy, California
Russell, Timothy W Cpl	2229575	D-1-4	2/2/68	21	Cleveland, Ohio
Russell, William E Capt	081670	E-2-3	5/28/68	31	Omaha, Nebraska
Rusth, John E Cpl	2274554	C-1-5	5/10/67	20	Klamath Falls, Oregon
Sadler, Charles D Cpl	2130908	A-1-9	5/21/66	22	St. Louis, Missouri
Sampson, Gerald H Capt	089284	B-1-3	8/28/69*	32	Williamsport, Pennsylvania
Sanders, Thomas Cpl	2169627	C-1-3	5/10/67*	22	East Elmhurst, New York
Sargent, George T Jr LtCol	051686	CO-1-4	3/20/69*	39	Auburn, Alabama
Schley, Robert J Cpl	2162020	M-3-3	4/30/67*	23	Fitchburg, Wisconsin
Schreiber, Klaus D 1stLt	0100653	C-1-Recon	10/14/67	25	Dortmund, Germany
Schunck, Henry M Cpl	2248186	D-2-13	5/10/68	20	San Francisco, California
Scott, Donald W Sgt	1539245	D-1-26	9/18/66	28	Birmingham, Alabama
Seath, Ned E LCpl	2026608	K-3-4	7/16/66	25	Reed City, Michigan
See, Roger D Cpl	2382260	A-3-Recon	6/8/69	20	Indianapolis, Indiana
Sexton, Charles T Cpl	2461865	3-Recon	2/5/70	20	Columbia, South Carolina
Sexton, Harry E LtCol	064225	HML-367	9/11/70	38	Anaheim, California
Sexton, Merlyn A Capt	091890	I-3-4	7/8/68	28	Chamberlain, South Dakota
Sherman, Andrew M 2dLt	097870	E-2-4	8/8/66*	32	Doylestown, Ohio
Sipple, Conrad A Cpl	2043729	C-2-4	3/5/66*	22	Salem, Indiana
Sirousa, Michael A Pfc	2569447	C-1-7	2/12/70*	25	Chicago, Illinois
Skibbe, David W 2dLt	0108861	C-1-Recon	3/2/70*	23	Des Plaines, Illinois
Skweres, Jeff C Cpl	2428288	HMM-364	6/1/70	20	Madera, California
Slater, Albert C Jr Capt	084435	A-1-9	7/6-7/67	26	Venice, California
Slater, Robert M S 1stLt	099864	Advisor	1/5-11/68	32	Buffalo, New York
Sleigh, Duncan B 2dLt	0105657	M-3-7	11/6/68*	23	Marblehead, Massachusetts
Sliby, Dennis M LCpl	2209054	A-1-5	3/30/68	21	Boston, Massachusetts
Smith, Ray L Capt	0102290	Advisor	4/1/72	26	Shidler, Oklahoma
Snyder, Stephen F 2dLt	093446	F-2-4	8/24/66*	23	Sunbury, Pennsylvania
Soliz, Thomas Cpl	2180067	A-1-Amph	9/6/67*	19	Bakersfield, California
Spark, Michael M Col	049041	CO-3-3	1/15/69*	41	New York, New York
Spicer, Jonathan N Pfc	2390916	C-3-MB	3/8/68*	19	Miami, Florida
Srsen, Steve A Pfc	2196925	A-1-3	1/27/67*	20	San Lorenzo, California
Stahl, Mykle E Sgt	2055308	K-3-26	1/21/68	24	Abilene, Texas
St. Clair, Clarence H Jr Cpl	2460676	K-3-7	8/28/69	20	Pensacola, Florida
Starrett, Edward F LCpl	2573512	G-2-5	12/9/70	19	Baltimore, Maryland
Stewart, Michael E LCpl	2135524	A-1-9	5/13/67*	18	Culpepper, Virginia
Stockman, Robert D Sgt	2203740	3-MPB	1/14/70	22	Kansas City, Missouri
Stuckey, James L LCpl	2114059	C-1-9	7/6/67	21	Seminole, Florida
Sullivan, Daniel F Jr Cpl	2067034	L-3-4	4/11/66	21	Wakefield, Massachusetts
Sullivan, George R 1stLt	094356	L-3-4	3/17/67	23	Lynnbrook, New York
Thatcher, Charles D LCpl	2178769	A-3-Tanks	5/8/67	20	Chicago, Illinois
Thomas, Michael H 2dLt	0102330	I-3-26	1/20/68*	25	Pawnee, Oklahoma
Thompson, Brock I Cpl	2304778	E-2-7	10/19/67	20	Denver, Colorado
Thompson, Clinton W Cpl	2434938	M-3-4	3/14/69	20	St. Louis, Missouri
Thompson, Jerrald R Cpl	1892012	C-1-Recon	6/16/66*	24	Pataskala, Ohio
Thompson, John C Sgt	1425383	HMM-364	4/30/64	27	Corinth, Texas
Thompson, Robert H LtCol	062346	CO-1-5	3/3/68	37	Corinth, Mississippi
Thoryk, Barry L Cpl	2316814	A-1-9	4/4/68	21	Mantua, Ohio
Thouvenel, Armand R Pfc	2272239	M-3-4	5/29/67*	24	Wheatridge, Colorado
Timmons, James M Pfc	2389485	M-3-7	11/6/68*	20	Groveport, Ohio
Tonkyn, Michael S LCpl	2255465	C-1-5	6/11/69	19	Mendham, New Jersey
Trent, William D LCpl	2381048	M-3-5	5/9/68*	19	East Peoria, Illinois

Name	Service #	Unit	Date	Age	Hometown
Tycz, James N Sgt	2082767	A-3-Recon	5/9/67*	22	Milwaukee, Wisconsin
Tyrone, Willie D SSgt	1221421	Advisor	5/31/65*	32	Carbon, Texas
Underwood, David F Capt	088999	HMM-163	2/16/68	27	Waynesville, North Carolina
VanCor, Norman W LCpl	2418325	C-3-Recon	5/7/69	22	Ashfield, Massachusetts
Vasquez, Jesus R Sgt	2126694	FFSR	1/30/68*	20	El Paso, Texas
Verheyn, David A LCpl	2117334	A-1-Recon	2/3/67	21	Lenoxdale, Massachusetts
Wallace, Ernie W Cpl	2034491	H-2-4	8/18/65	22	Wayne, West Virginia
Ward, James C Cpl	2439597	D-1-5	5/9/69*	21	Alexandria, Virginia
Warren, Roger O LCpl	2274526	F-2-5	2/3/68	22	Klamath Falls, Oregon
Webb, Bruce D Capt	068857	I-3-3	8/18/65*	31	Wheaton, Illinois
Webb, James H Jr 1stLt	0106180	D-1-5	7/10/69	23	Saint Joseph, Missouri
Weise, William LtCol	057704	CO-2-4	5/2/68	39	Philadelphia, Pennsylvania
Widger, Robert I Cpl	2380489	K-3-1	6/7/69	20	Syracuse, New York
Williams, Robert S 1stLt	092019	I-3-5	7/25/66	26	St. Genevieve, Maryland
Wilson, Willis C 1stLt	087454	B-1-3	4/2/66	24	Sharpsville, Pennsylvania
Wirick, William C Cpl	2386310	I-3-26	12/8/68*	20	Toledo, Ohio
Woods, Lloyd Cpl	2245531	F-2-5	6/2/67	21	Grand Rapids, Michigan
Wynn, Edward H Pfc	2424476	E-2-4	5/25/68	20	Napa, California
Yarber, Vernon L LCpl	2366678	L-3-3	8/26/68*	18	Jacksonville, Florida
Yates, John C 2dLt	0108133	B-1-Amph	10/17/68*	26	Fergus Falls, Minnesota
Yordy, Charles R Pfc	2420003	K-3-27	5/24/68	20	Fruitport, Michigan
Young, William H Cpl	0000000	A-1-3	3/7/68	00	Unknown

NAVY CORPSMAN

Name	Service #	Unit	Date	Age	Hometown
Ashby, James W HM3	9140268	L-3-9	6/1/67*	23	Park Rapids, Minnesota
Barber, William B HM3	B793357	I-3-4	11/25/68	21	Austin, Texas
Benoit, Francis A. HN	9140346	E-2-9	3/16/67*	22	Red Lake Falls, Minnesota
Burns, Dewey R Jr HM3	B716750	CAP-1-3-9	9/13/69*	20	Sulphur Springs, Texas
Braun, Kenneth R HM3	9188839	I-3-9	3/30/67	18	Hopkins, Minnesota
Casey, Robert M HM3	B111377	G-2-7	5/16/68*	19	Guttenberg, New Jersey
Clay, Raymond D HN	2318817	G-2-7	9/24/66	20	Atlanta, Georgia
Crawford, Charles H HM3	6874759	M-3-4	5/29/67*	26	Batavia, Ohio
Cruse, James D HN	1392506	M-3-4	6/15/68*	22	Paducah, Kentucky
Gerrish, Alan R HN	B116456	3-MPB	9/7/68*	19	Woburn, Massachusetts
Gillespie, Martin L Jr HM2	9027756	D-1-4	3/21/66*	26	East Boston, Mass
Grant, Gollie L HN	6833808	B-1-26	9/19/66*	22	Old Fort, North Carolina
Hancock, Eugene S HM2	B317923	I-3-7	2/24/69*	22	Gainesville, Florida
Henry, Daniel B HM1	2365535	1MB	8/27/67*	32	Olean, New York
Hickey, William L HM2	5848365	K-3-9	7/4/66	22	Dallas, Texas
Holmes, Billie D. HM3	6848245	C-1-Recon	6/16/66	23	Madison, Tennessee
James, Alan C HM2	7894913	B-1-3	9/9/68	22	Lisle, Illinois
Kierznowski, Terrence E HM2	B509521	K-3-3	9/12/69	21	Crete, Illinois
Leal, Armando G Jr HN	B704027	M-3-5	9/4/67*	20	San Antonio, Texas
Lewis, David H LtCdr	633562	1MD	9/24/69	32	Martinsville, Virginia
Mack, Francis W HM3	5950010	F-2-4	8/17/66*	24	Jersey City, New Jersey
Mayton, James A. HM1	2973461	VMO-2	5/21/66	33	Manchester, Tennessee
McKeen, Gerald C HN	9190058	G-2-7	9/24/66*	22	Sac City, Iowa
Mercer, William I. HM2	B980863	M-3-4	6/15/68*	21	Los Angeles, California
Orlando, Samuel G HN	5910304	H-2-7	3/4/66*	21	Birmingham, Michigan
Phillips, John C HN	B673934	C-1-7	12/19/68	20	Des Moines, Iowa
Powell, Richard L HN	B418314	L-3-7	8/29/68*	20	Youngstown, Ohio
Rudd, Donald L HM2	B502831	L-3-4	3/3/69*	23	Tecumseh, Michigan
Strode, Gerald M HM3	B809190	B-1-5	9/4/67	20	Moses Lake, Washington
Taft, David A LtCdr	711699	1-MB	8/27/67	33	Denver, Colorado
Valdez, Phil I HN	9997731	B-1-1	1/29/67*	20	Dixon, New Mexico
Wilhelm, Mack H BM3	B713921	D-1-9	2/19/69*	23	Rockport, Texas
Willeford, Franklin P HN	3537852	F-2-4	12/14/68*	25	Lawton, Oklahoma
Work, Warren A Jr HM3	9158408	F-2-4	4/8/67	20	Tulsa, Oklahoma

DISTINGUISHED SERVICE CROSS

Barrow, Robert H Col	023471	CO-1	3/18/69	47	St. Francisville, Louisiana
Crowley, Fred R 1stLt	094635	Advisor	9/18/68	24	Philadelphia, Pennsylvania

SILVER STAR

Aaron, Robert J Jr1stLt	093840	A-1-9	3/24/67	24	Philadelphia, Pennsylvania
Abely, John LCpl	2249793	CAP-2-1	4/19/67	19	Chelsea, Massachusetts
Achterhoff, James P LCpl	2385864	E-2-4	3/18/68*	18	Muskegon, Michigan
Adams, George W III Cpl	2337404	C-1-26	6/7/69	20	Altoona, Pennsylvania
Adams, Laurence R III Capt	092937	HMM-165	12/15/68	24	Seattle, Washington
Adams, William C Cpl	2082965	C-1-5	6/2/67	22	Concordia, Kansas
Addesso, Peter F Cpl	2281600	H-2-5	2/3/68	19	Lynnbrook, New York
Aguilar, Vicente Jr Pfc	2334659	G-2-4	5/1/68	20	Tulia, Texas
Akers, Jerry R 2dLt	094503	L-3-4	2/28/67	24	Sheffield, Alabama
Albers, Vincent A Jr LtCol	057287	CO-2-7	9/11/70	40	Houston, Texas
Albrecht, John A 1stLt	086107	D-1-4	3/24/66	24	Columbia Falls, Montana
Albright, Edward J Sgt	2334347	A-3-Tanks	4/29/69	21	Jacksonville, Florida
Alderette, Arnold L Cpl	2233722	A-1-9	2/8/68	22	Valinda, California
Alexander, Robert B Capt	075013	B-1-4	3/26/66	30	McKeesport, Pennsylvania
Alexander, Robert B Maj	075013	OO-1-7	8/12/69	33	McKeesport, Pennsylvania
Allen, Carl Jr Capt	096016	L-3-26	2/28/69	36	Camp Lejeune, N.C.
Allen, David C Pfc	000000	Unk	2/28/66	00	Unknown
Allen, Don W Cpl	2083773	D-1-3	7/15/66	20	Arab, Alabama
Allen, Donald W LCpl	2337744	K-3-7	4/10/68	20	Pittsburgh, Pennsylvania
Allen, Fred A LCpl	2520196	I-3-1	1/31/70	20	Atlanta, Georgia
Allen, Joe E 2dLt	0103089	D-1-5	5/24/68*	22	Bay St. Louis, Mississippi
Allen, Melvin L LCpl	2232029	D-1-3	5/4/67*	21	Beulah, Mississippi
Allen, Paul D C Pfc	2090875	F-2-I	2/28/66	23	Indianola, Iowa
Allen, Ronald L Capt	080654	F-2-4	6/27/66	24	Sabula, Iowa
Allen, Terence M LtCol	057053	PRU	8/7/69	28	St. Louis, Missouri
Allison, Buryl R Sgt	2128865	I-3-27	5/28/68	21	Cincinnatti, Ohio
Allord, Gary C 1stLt	0107842	E-1-Recon	3/28/70	21	Madison, Wisconsin
Almanza, Ricky J Cpl	2379991	M-3-5	9/3/68*	20	Moline, Illinois
Alo, Ta'Aloga S SSgt	1300719	C-3-ATanks	8/28/67	32	Long Beach, California
Altazan, Kenneth A Sgt	2303933	HMM-364	5/9/69	23	Baton Rouge, Louisiana
Althoff, David L Maj	064955	HMM-262	2/2/68	35	Chandler, Arizona
Althoff, David L Maj	064955	HMM-262	5/13/68	35	Chandler, Arizona
Althoff, David L Maj	064955	HMM-262	5/18/68	35	Chandler, Arizona
Alvarado, Jose J SSgt	1259372	B-3-Tanks	8/20/66	33	San Juan, Puerto Rico
Alvarez, Enrique Cpl	2192883	A-1-4	5/7/67	20	El Paso, Texas
Alvarez, Robert Pfc	2193505	C-1-4	3/24/67*	20	Clint, Texas
Ambrose, Frank LPfc	2366901	M-3-5	2/7/68	18	Orlando, Florida
Anasiewicz, Richard J Pfc	2169441	M-3-5	7/18/66*	20	New Brunswick, New Jersey
Anderson, Andrew E Jr Maj	061855	Advisor	8/6-7/66	35	Unknown
Anderson, Clinton H Jr 2dLt	0100013	K-3-9	5/20/67	22	El Segundo, California
Anderson, Donald F Pfc	2183557	B-3-Tanks	9/19/66	20	New Waverly, Massachusetts
Anderson, John J Sgt	1646153	I-3-9	10/27/65	22	New York, New York
Anderson, Joseph S Pfc	2348018	E-2-4	3/13/68	20	Los Angeles, Caifornia
Anderson, Lee H 2dLt	091901	L-3-5	7/24/66	22	Concord, New Hampshire
Anderson, Ralph C 1stLt	0100817	HMM-262	7/31/69	26	Portland, Oregon
Anderson, Ralph T LCpl	2366871	H-2-4	5/2/68*	18	St. Petersburg, Florida
Anderson, Ronald D Pfc	2320067	H&S-2-26	4/6/67*	18	New York, New York
Anderson, Terry C Pfc	2421839	A-3-Recon	2/17/69	19	Dallas, Texas
Anderst, James L Capt	082185	VMA-242	10/25/67	29	Plankton, South Dakota
Andrews, James F III 1stLt	0902684	HMM-265	3/13/69	24	Westchester, New York
Angle, Peter F Capt	081994	HMM-161	7/8/68	28	Sewickley, Pennsylvania

Name	Service #	Unit	Date	Age	Hometown
Angle, Peter F Capt	081994	HMM-161	8/18/68	28	Sewickley, Pennsylvania
Anzaldua, Alberto T Sgt	2019091	A-1-3	8/10/69*	23	Santa Rosa, Texas
Apodaca, Ramon LCpl	2044969	G-2-1	3/4/68	23	Salt Lake City, Utah
Atkins, William LCpl	2309299	D-1-26	6/9/68	23	Philadelphia, Penn
Araujo, Espiridion Jr LCpl	2423491	A-1-4	11/21/68	22	Harlingen, Texas
Arbogast, Allen G Cpl	2290277	A-1-27	4/13/68	20	Middlesex, Massachusetts
Arenas, Lorenzo T Cpl	2054727	B-1-4	1/27/66	21	El Paso, Texas
Ariss, David W Capt	082292	HMM-265	2/8/68	28	Pomona, California
Arizmendez, Daniel M Cpl	2202057	B-1-26	5/29/68*	22	Holland, Michigan
Armer, Billy R GySgt	1196747	G-2-4	4/27/68	35	St. Louis, Missouri
Armes, Willard P 2dLt	0106503	M-3-9	2/10/69	22	Hague, New York
Armstrong, Elton Pfc	2424349	H-2-5	5/11/69	20	Rochester, New York
Armstrong, John C Cpl	2377114	G-2-9	4/21/69	22	Seattle, Washington
Arroyo, Joseph U Capt	090455	I-3-9	1/28/69	26	San Francisco, California
Arthur, Lawrence K Pfc	2380637	F-2-3	5/28/68*	18	Lowville, New York
Ashby, James C Pvt	2124853	M-3-1	4/14/67	20	Dallas, Texas
Ashley, Maurice C Jr LtCol	045707	CO-1-Tanks	11/17/68	43	Poughkeepsie, New York
Atkins, William LCpl	2309299	D-1-26	6/9/68	23	Philadelphia, Pennsylvania
Austin, Randall W LtCol	075337	CO-2-9	5/15/75	38	Glenside, Pennsylvania
Avalos, Manuel Jr Cpl	2080450	B-1-3	4/2/66	20	Tracy, California
Avery, John M GySgt	1518220	B-3-1	10/25/68*	31	Cottondale, Alabama
Axley, Lawrence A LCpl	2513991	L-3-3	8/22/69	19	Phoenix, Arizona
Ayers, Gerald J 2dLt	0107747	D-1-5	6/9/69	21	St. Louis, Missouri
Babb, Wayne A Capt	085006	D-2-11	3/19/69	29	Brevard, North Carolina
Babitz, Donald M Maj	073760	HMM-165	11/14/66	32	Cicero, Illinois
Bach, Albert W Sgt	1994442	B-3-Recon	1/18/67	22	Hialeah, Florida
Bachta, Thomas E Cpl	2051749	1-F-Recon	8/8/66	21	Chicago, Illinois
Backeberg, Bruce B Pfc	2100837	D-1-26	9/19/66*	19	Helena, Montana
Baez, Jose M LCpl	2243475	D-1-4	10/26/67	20	Adjuntas, Puerto Rico
Bailey, Gene E Maj	076205	HML-367	3/28/71	34	Santee, South Carolina
Bailey, Thomas B Cpl	2273846	D-3-Recon	9/2/68	22	Houston, Texas
Baird, John R Jr LCpl	2423144	A-1-9	2/22/69*	19	Oak Lawn, Illinois
Baker, Billy R LCpl	2542949	H-2-7	4/24/70*	19	Leesville, Texas
Baker, Clyde L 2dlt	096075	D-1-4	2/25/67	31	Seagoville, Texas
Baker, David R Cpl	2032366	L-3-7	3/22/66	22	Macon, Georgia
Baker, Fred J LCpl	2147111	A-3-Recon	4/25/67	21	Piedmont, Alabama
Baker, Harvey L Pfc	2288653	L-3-4	2/26/69	19	Detroit, Michigan
Baker, Herbert G SSgt	1557100	D-1-1	7/2/68	31	Gary, Indiana
Baker, Kenneth A SSgt	1471002	D-1-1	11/22/68*	32	New Bedford, Mass
Baker, Sam R II Sgt	1973607	C-1-5	8/10/66	23	Bronx, New York
Baker, Steven D Cpl	2151521	F-2-3	5/28/68	21	Rigby, Idaho
Baker, Sydney A 1stLt	0105860	HML-367	9/11/70	24	San Antonio, Texas
Baker, William H Sgt	2036045	G-2-7	3/4/66	20	Louisville, Kentucky
Balanco, John J Sgt	2113184	CAP-0-1	1/21/68	24	Florence, Oregon
Balfanz, Duane A Capt	096079	Advisor	11/6/69	36	Minneapolis, Minnesota
Balignasay, Pedro L GySgt	1348870	E-2-4	2/25/69	41	Philippine, XP
Ballew, Donald L GySgt	649328	Advisor	2/7/66	42	Staunton, Virginia
Ballew, Henry Jr LCpl	2584132	A-1-3	7/27/69*	18	Atlanta, Georgia
Ballin, Joe M Jr LCpl	2126236	D-1-4	9/16/66*	19	Fresno, California
Banks, Adam J Jr SSgt	1579708	M-3-5	11/8/67	29	New Orleans, Louisiana
Banks, Andrew B Jr Maj	081351	HML-167	11/29/68	31	Los Angeles, California
Banks, Edward J Capt	076822	H-2-1	1/31/67	30	South Bend, Indiana
Banks, Johnny L Pfc	2491653	D-1-7	2/14/69	18	Thomasville, Georgia
Banning, John J 2dLt	0102452	A-1-Tanks	5/25/68	25	Joplin, Missouri
Barber, Russell M Sgt	1866636	K-3-4	7/18/68	25	Boston, Massachusetts
Barclay, Boyd L Capt	087876	VMO-3	6/8/67	27	Oklahoma City, Oklahoma
Barden, Roosevelt Jr LCpl	2490809	K-3-7	10/25/69	19	Aliquippa, Pennsylvania
Barents, Brent J Capt	089369	HMM-263	6/21/69	28	Long Beach, California
Barham, Robert L II Sgt	1960007	D-3-Recon	1/4/67	23	Santa Monica, California
Barnard, Roger H LtCol	052696	CO-3-7	5/15/68	38	Lincoln, Nebraska

Name	Service #	Unit	Date	Age	Hometown
Barnes, Alan R LCpl	2496022	B-1-7	8/12/69	19	Houston, Texas
Barnes, Eric M 1stLt	089806	C-1-26	9/24/66	24	Windsor, Connecticut
Barnes, Louis Cpl	2150373	F-2-5	1/2/67	22	Chicago, Illinois
Barnes, Robert C LCpl	2083790	B-3-Recon	10/3/66	21	Nashville, Tennessee
Barnes, Robert L 1stLt	0105456	HML-367	12/8/69	23	Wichita, Kansas
Barnett, Meredith L Sgt	2584200	I-3-1	11/6/70*	20	Belpre, Ohio
Barr, James B Capt	090775	HML-167	6/10/71	29	Pittsburgh, Pennsylvania
Barrett, Clarence A Pfc	2387775	L-3-5	3/1/68	19	Redlake, Minnesota
Barrett, James A LCpl	2137396	C-1-Recon	10/10/66*	21	Pittsburgh, Pennsylvania
Barrett, John J Capt	085356	HMM-263	2/10/69	30	Minneapolis, Minnesota
Barrett, John J Capt	085356	HMM-263	4/3/69	30	Minneapolis, Minnesota
Barrett, John J Capt	085356	HMM-263	5/26/69	30	Minneapolis, Minnesota
Barron, Jeffrey M Cpl	2296415	C-3-12	2/25/69*	20	La Puente, California
Barrow, Thomas M Jr Pfc	2308150	K-3-3	4/26/67*	18	Atlanta, Georgia
Bartlett, Dan M LCpl	2115490	E-2-5	5/16/66	21	Artesia, New Mexico
Bartlett, Gary R Sgt	1940379	F-2-3	7/20/66	24	Irwin, Pennsylvania
Bartolotti, Richard J LCpl	2365941	L-3-7	2/7/68	20	St. Louis, Missouri
Bartusevics, John Sgt	1897519	C-1-Tanks	1/15/67	26	Philadelphia, Pennsylvania
Basel, John M 1stLt	0102818	H&S-3-12	4/13/69	22	McLean, Virginia
Bateman, James A Sgt	2102290	CAP-2-8-1	5/15/69*	23	Mundelein, Illinois
Bates, Robert A 1stLt	091299	G-2-3	10/22/66*	24	Lake Forest, Illinois
Bathurst, Sheldon J Sgt	1820763	E-2-1	7/18/66	25	Baltimore, Maryland
Battista, Anthony J 2dLt	091530	H-2-9	4/16/66*	23	Owego, New York
Bauer, William A LCpl	1884761	K-3-5	7/24/66	24	New York, New York
Beans, James D Maj	072931	Advisor	5/2/71	37	Annapolis, Maryland
Beard, Lary G LCpl	2343333	E-2-1	5/17/68	21	Memphis, Tennessee
Beaver, Donald E GySgt	1201018	HMM-164	9/2/66	32	Selinsgrove, Pennsylvania
Becham, Gary V LCpl	2378778	G-2-9	2/17/69	20	Macon, Georgia
Beck, Hugo T 1stLt	2241449	HMM-262	3/6/70	25	Albany, New York
Beckman, Bruce E 1stLt	095492	VMA-242	2/23/68	23	Minneapolis, Minnesota
Beebe, William A II 1stLt	0100848	HMM-364	4/8/69	25	Asheville, Alabama
Beeler, Robert A Capt	089734	I-3-5	8/13/69	27	Louisville, Kentucky
Belko, Laurence E Sgt	1842683	I-3-3	5/26/68	28	West Haven, Connecticut
Bell, Earl W SSgt	1670116	H&S-3-Rec	3/24/67	29	Claymont, Delaware
Bell, James E Sgt	1083156	G-2-4	3/22/66	34	Washington, Pennsylvania
Bell, Marcus R Cpl	2197635	M-3-5	5/13/67	20	Seattle, Washington
Bell, Thomas J Sgt	1973855	F-2-4	6/25/66	22	Montgomery, Alabama
Bell, Van D Jr LtCol	044563	CO-1-1	2/22/67	47	Portsmouth, Rhode Island
Below, Jack W SSgt	1889549	A-1-Rec	3/20/67	24	Wyatt, Missouri
Belser, Joseph H Jr 1stLt	0101738	K-3-27	5/24/68	22	St. Louis, Missouri
Bem, Walter P Pfc	2490068	B-1-7	4/30/69*	18	Indiana, Pennsylvania
Bench, Arnold E LtCol	051325	CO-2-4	7/24/66	41	Corvallis, Oregon
Bench, Edmund Jr LCpl	2478200	G-2-9	6/1/69	19	Salt Lake City, Utah
Bendell, Lee R LtCol	050744	CO-3-4	1/24/68	40	Melrose Park, Illinois
Bender, Lawrence J II Capt	095847	Advisor	5/15/70	33	Cleveland, Ohio
Benet, Peter E Maj	066168	HMM-263	6/14/70	36	Austin, Texas
Benfatti, Raymond C 1stLt	096148	L-3-9	2/17/69	39	Paterson, New Jersey
Benigo, Ronald 1stLt	089391	C-1-5	8/10/66	24	Detroit, Michigan
Benjamin, John Cpl	2095699	H-2-7	3/4/66	20	San Francisco, California
Bennett, Billy J Pfc	2327147	F-2-9	7/29/67*	19	Chattanooga, Tennessee
Bennett, Jesse D Jr Maj	081265	I-3-26	6/21/68	34	Little Rock, Arkansas
Bennett, Kenneth D Cpl	2370807	F-2-5	1/25/69	20	Nashville, Tennessee
Bergeron, Robert H SSgt	1375888	K-3-4	9/27/66	30	Uxbridge, Massachusetts
Bergerson, John F 2dLt	092624	B-1-1	1/29/67*	23	Mercer Island, Wash.
Bergman, Carl E Capt	092979	HMM-163	2/16/68	24	Kenmare, North Dakota
Berman, Stuart C 2dLt	094544	C-1-ATanks	3/21/67	20	New York, New York
Bernard, Henry W Jr LCpl	2322441	H&S-2-9	7/29/67*	19	Willimantic, Connecticutt
Berry, Johnny K GySgt	2006360	I-3-1	12/22/67	23	Cincinnati, Ohio
Berry, Larry J Cpl	2091073	A-1-3	7/6/67	23	Mobile, Alabama
Betts, Albert L SSgt	1982176	D-1-7	2/6/68	24	West Palm Beach, Florida

Name	Service #	Unit	Date	Age	Hometown
Beyerlein, David Sgt	2217783	A-1-9	2/22/69	21	Portland, Oregon
Bianchino, Richard L Capt	088193	HMM-364	4/14/69	28	Albany, New York
Bibeau, Arthur K Cpl	2378231	C-11-Eng	3/24/69	22	Boston, Massachusetts
Biber, Joseph F Sgt	2198273	E-3-Recon	9/19/68*	22	Lompoc, California
Bickert, Edward T Jr LCpl	2279268	2-CAG	2/7/68	20	Philadelphia, Pennsylvania
Bickford, Thomas E 2dLt	096171	FO-3-11	9/25/66	36	Flandreau, South Dakota
Bickley, Leroy A Maj	077461	OO-1-5	2/26/70	33	Sandusky, Ohio
Biddulph, Stephen G 1stLt	0115801	1-Anglico	7/13/72	26	Rexburg, Idaho
Biehl, Michael C Cpl	2353385	C-1-4	3/5/69	22	Newark, New Jersey
Biggers, Archie J 1stLt	0107246	C-1-9	2/20/69	25	San Diego, California
Biggs, Jimmy D Pfc	2451993	A-1-7	12/7/68*	19	Kansas City, Missouri
Billings, Roger L LCpl	2361338	D-1-5	2/22/68	20	Kansas City, Missouri
Billups, Josh LCpl	2471975	K-3-9	2/12/69	19	Chicago, Illinois
Bingenheimer, James Sgt	2145060	A-1-Recon	3/15/71*	23	Atlantic City, New Jers
Bird, Loren W SSgt	1517379	H-3-7	1/11/68	31	Kansas City, Missouri
Biskey, Robert A Pfc	1871095	I-3-1	3/5/66	24	Hill City, Minnesota
Black, Charles H Maj	062737	OO-1-5	9/6/67	40	Dumfries, Virginia
Black, Robert A Capt	087392	B-1-1	7/6/68	26	Honolulu, Hawaii
Blackburn, Glenn J Pvt	2160441	D-1-3	5/4/67	19	Dover, Massachusetts
Blackman, Thomas J Pfc	2398632	D-2-13	5/10/68*	19	Racine, Wisconsin
Blackwell, Kenneth G LCpl	2400695	I-3-26	2/23/69*	20	Tucson, Arizona
Blades, Arthur C 1stLt	090430	C-1-5	8/10/66	24	West Hampton Beach, N.Y.
Blair, Frank S III 1stLt	092681	F-2-4	9/21/67	24	Albuquerque, New Mexico
Blair, Frank S III 1stLt	092681	F-2-4	10/14/67	24	Albuquerque, New Mexico
Blair, John D Cpl	2208761	I-3-4	1/27/68	21	Columbus, Georgia
Blair, Lawrence K III Sgt	1643102	B-1-26	9/16/66	27	Worcester, Massachusetts
Bland, Richard P L Capt	080411	HMM-364	6/4/68	30	Kansas City, Missouri
Blankenship, Dennis R SSgt	1690600	I-F-Recon	5/20/66	27	Coalwood, West Virginia
Blankenship, Sidney H SSgt	1547310	Advisor	4-1/68	29	Huntington, West Virginia
Blanton, Charles G Sgt	1938124	IF-3-Recon	12/16/65	22	Little Rock, Arkansas
Bleacher, Ronald T Cpl	2283664	H-2-5	10/29/68*	19	Marshallton, Delaware
Blizzard, David W Capt	0100870	Advisor	11/9/71	30	Ocean City, New Jersey
Blocker, Eugene SSgt	1904253	D-1-9	2/21/67	25	Adah, Pennsylvania
Bloomberg, Richard N Capt	092595	VMA-121	5/10/68	28	Los Angeles, California
Blough, David A Pfc	2378332	E-2-3	12/28/67*	19	Spencer, Massachusetts
Blunk, Harold R Pfc	2358969	F-2-3	5/28/68	21	Chicago, Illinois
Boatman, Michael L Pfc	2373653	G-2-7	5/7/68	19	Lakewood, Colorado
Bobak, Raymond W Pfc	2479097	C-1-7	8/12/69	19	Cleveland, Ohio
Bobian, Ralph D LCpl	2304829	D-1-7	12/20/67*	18	Denver, Colorado
Bodenweiser, Alec J 2dLt	0103347	E-2-26	2/5/68	22	Salem, Oregon
Boeck, Gary R LCpl	2616935	L-3-5	1/6/71*	20	Braham, Minnesota
Bogart, John G 2dLt	0106510	L-3-3	6/16/69	24	Chappaqua, New York
Bohn, Robert D Col	037498	CO-5	12/27/67	46	Alexandria, Virginia
Boillot, David A 2dLt	0102669	A-1-27	6/19/68	22	Rye, New York
Boldes, James M LCpl	0000000	0-0-0	10/3-5/66	00	Unknown
Bolding, Benjamin F Cpl	2257470	M-3-26	2/15/69*	20	Moore, Oklahoma
Bolduc, George J 2dLt	0103348	L-3-7	2/7/68	23	New York, New York
Bolduc, George J 2dLt	0103348	L-3-7	5/18/68	23	New York, New York
Boles, James M LCpl	2128736	I-3-9	4/9/66	19	Irontown, Ohio
Bollman, Henry C III Capt	085842	HMM-165	7/11/72	31	Dallas, Texas
Bolton, Gilbert H SSgt	1892020	M-3-7	11/2/67	26	Portsmouth, Ohio
Bolton, Michael L Cpl	2235185	D-1-3	9/18/68	20	Los Angeles, California
Bonnelycke, Clyde L Sgt	1806021	C-1-3	1/19/68	28	Honolulu, Hawaii
Bonsper, Donald E Capt	091908	Advisor	1/9/68	23	Portville, New York
Book, Floyd G Jr LCpl	2130395	CAP-D-2	6/5/67	19	Hammond, Indiana
Boomer, Walter E Capt	079957	H-2-4	2/5/67	29	Rich Square, North Carolina
Boomer, Walter E Maj	079957	Advisor	4/3/72	34	Rich Square, North Carolina
Boone, Samuel Jr Cpl	2294933	B-1-26	3/30/68	20	Baltimore, Maryland
Booty, Larry O J Cpl	2062303	F-2-9	1/8/66*	21	Greenwell Springs, Lousiana
Bosley, Charles W Pfc	2122523	C-1-Recon	6/16/66	19	Greencastle, Indiana

Name	Service #	Unit	Date	Age	Hometown
Boss, Otis E Jr LCpl	2363521	F-2-4	5/2/68	20	Richmond, Virginia
Bosser, Johnny S LCpl	2452803	K-3-7	9/28/69*	19	Fort Lupton, Colorado
Bost, Barry N Cpl	2315872	A-1-7	5/17/68	20	Kannapolis, North Carolina
Botello, Alfred L Sgt	1961587	E-2-5	12/19/66	23	Austin, Texas
Bott, Daniel J SSgt	1694088	D-1-1	9/25/68	29	New York, New York
Bourne, Frank L Jr LtCol	052717	CO-3-4	6/18/68	39	Savannah, Georgia
Bower, Burton K Pfc	2238843	L-3-27	5/18/68	20	Springfield, Massachusetts
Boyer, Robert L Cpl	2234967	E-2-3	12/28/67*	21	Long Beach, California
Braddock, Harold T Jr Cpl	2083160	E-2-5	5/16/66	21	Holden, Massachusetts
Braddon, John R Maj	057518	HMM-364	4/27/64	34	Wellsville, New York
Bradley, Gerald G Cpl	2394208	E-2-1	1/15/69*	21	Braintree, Massachusetts
Bradley, John M LCpl	2306247	M-3-1	12/27/68	20	Buffalo, New York
Brady, Eugene R LtCol	051664	HMM-364	2/26/69	40	York, Pennsylvania
Brady, Phillip O 1stLt	082115	Advisor	12/31/64	24	Quantico, Virginia
Brandon, Jack A Maj	066095	VMO-6	7/31/69	39	El Paso, Texas
Brandon, Wayne H 2dLt	0100069	K-3-5	9/6/67	23	Martin, Tennessee
Brank, Walter S III LCpl	2212179	HMM-165	5/5/68	21	Jonesborough, Tennessee
Branscombe, Robert A Sgt	2157195	CAP-B-1	2/6/68	20	Glendale, California
Breeding, Earle G Capt	080698	E-2-26	2/5/68	35	Roswell, New Mexico
Brennan, John L 1stLt	0100072	HMM-364	2/12/69	24	Woodbridge, Virginia
Brenno, Wesley C Pfc	2310818	C-1-1	3/20/67	22	Minot, North Dakota
Brent, Lawrence J Jr Cpl	2079230	A-1-3	7/20/66	21	Chicago, Illinois
Brewer, Herman R Jr Sgt	2055993	A-1-Amp	3/6/67	22	Miami, Florida
Brick, George E Cpl	2217775	M-3-5	8/17/68	20	Prineville, Oregon
Brickey, Billy J SSgt	1657083	I-3-5	7/22/66	27	Hickory Ridge, Arkansas
Bright, Robert III Cpl	1692324	3-FRecon	10/21/66	27	Saddle River, New Jersey
Brisco, Joseph O Cpl	2450884	H-2-4	8/20/69	20	New Orleans, Louisiana
Britt, Ted D Pfc	2378809	B-1-26	3/30/68*	19	Decatur, Georgia
Britton, Albert K SSgt	1820832	G-2-4	3/30/68	27	Baltimore, Maryland
Broadtman, Henry R Jr Pfc	2384975	L-3-3	1/26/69*	18	Waggaman, Louisiana
Brodrick, Steven P 1stLt	0103916	F-2-4	12/11/68*	25	Selma, California
Bronars, Edward J Maj	050746	Advisor	7/22/65	37	Chicago, Illinois
Brooks, Robert P Capt	073934	K-3-9	7/4/66	37	San Diego, California
Brophy, Daniel R 1stLt	082441	AO/11	2/23/69	29	Los Angeles, California
Broquist, Steven A 2dLt	094500	D-1-9	5/14/67*	22	Champaign, Illinois
Brown, Calvin F SSgt	1446627	C-1-1	1/14/67	32	Chicago, Illinois
Brown, Charles L Sgt	1998034	L-3-9	2/24/69	24	Nashville, Tennessee
Brown, David B Capt	082465	F-2-5	9/30/68	29	Shillington, Pennsylvania
Brown, Gary E 1stLt	088179	E-2-4	2/9/66	24	Clinton, South Carolina
Brown, Gary E 1stLt	088179	E-2-4	3/21/66	24	Clinton, South Carolina
Brown, James G LCpl	2080108	H-2-5	4/21/66*	21	Gatesville, Texas
Brown, James H 1stLt	0103368	HML-367	12/2/69	22	Baltimore, Maryland
Brown, James S Jr 1stLt	094575	C-1-12	6/11/68	25	Jackson, Mississippi
Brown, Jerome W Maj	064917	CO-7-MTB	2/13/69	35	Cincinnati, Ohio
Brown, Joseph C Cpl	2047608	B-3-Tanks	7/30/66*	20	Pasadena, Maryland
Brown, Lawrence R LCpl	2089250	I-3-9	9/5/66	19	Cincinnati, Ohio
Brown, Leslie E Col	019930	MAG-12	10/28/65	45	Compton, California
Brown, Marc A Pfc	2297413	M-3-1	4/21/67*	19	Long Beach, California
Brown, Michael A LCpl	2453984	G-2-4	3/11/69	19	Savannah, Georgia
Brown, Michael R Sgt	2233563	H&S-3-5	8/9/68	22	Ashland, Kentucky
Brown, Richard J Sgt	2031415	K-3-5	9/6/67*	23	Pine Beach, New Jersey
Brown, Terry R Pfc	2307478	A-1-Recon	10/28/68	20	Dallas, Texas.
Browning, Michael L LCpl	2251278	B-1-7	10/24/67*	19	Fullerton, California
Broyer, Clifton L Cpl	2249149	C-1-4	2/28/69*	23	Cataumet, Massachusetts
Bruggeman, David C 1stLt	0113783	1-Anglico	4/1/72*	25	Pittsburgh, Pennsylvania
Bruinekool, Dewayne G Cpl	1975618	M-3-3	7/21/67	23	Kansas City, Missouri
Bruno, Edward LCpl	2243160	F-2-3	8/6/69*	21	Woodridge, New York
Bryan, Byron E Jr 1stLt	0100083	K-3-9	4/30/68	23	Wayne County, NC
Bryant, David P LCpl	2206916	FO-3-9	10/1/67	20	Freeport, New York
Bryant, Donald L Sgt	2009710	G-2-26	10/4/68	24	Pittsburgh, Pennsylvania

Name	Service #	Unit	Date	Age	Hometown
Buchholz, Edward A LCpl	2088095	H-2-4	7/23/66	23	Borger, Texas
Buchs, Christopher J LCpl	2107157	I-3-3	8/18/65	20	Fraser, Michigan
Buckles, William T LCpl	2333245	M-3-5	2/6/68*	19	Hollywood, Florida
Buckley, Alton LCpl	2260763	E-2-9	2/2/68	20	Nashville, Tennessee
Buckner, Gordon H II Maj	065539	HMM-362	11/18/65	33	St. Clair, Minnesota
Budd, Talman C III Maj	057146	Advisor	7/31/67	37	Rochester, New York
Bulger, Thomas E LtCol	050747	CO-3-1	4/4/69	42	Staten Island, New York
Bull, Edward D Cpl	2306688	C-1-4	3/1/69	21	Memphis, Tennessee
Bunda, George J 1stLt	093008	HMM-265	5/10/68	26	Chicago. Illinois
Burch, James M 1stLt	090802	H-2-3	8/20-26/66	24	Meridian, Mississippi
Burchett, Donald D Cpl	2139595	G-2-4	7/23/67	20	South Charleston, WV
Burchett, Johnny R Pfc	2226011	H-2-1	1/29/67	20	Louisville, Kentucky
Burgess, Glenn F 1stLt	0100907	HMM-364	8/26/69	25	Waukegan, Illinois
Burgett, George S Capt	088652	F-2-5	10/13/66	25	Chandler, Arizona
Burghardt, James E LCpl	2215780	I-3-9	3/30/67	20	Belmont, California
Burke, Francis M Maj	080135	I-3-5	9/6/67	30	New York, New York
Burke, Nicholas E Capt	074930	VMO-6	5/10/67	32	Niagra Falls, New York
Burkhardt, Douglas K Pfc	2365259	A-1-Recon	3/5/68	19	Detroit, Michigan
Burleson, Eugene B Jr Capt	076404	C-1-3	10/11/67	35	Citra, Florida
Burns, John R Jr 2dLt	0101776	M-3-4	1/18/68*	24	St. Louis, Missouri
Burns, Raymond M Capt	091609	VMA-242	10/27/67	24	Glynco, Georgia
Burns, Roy T Pvt	2264776	B-3-Recon	9/2/67	20	Garyville, Florida
Burns, Terry P Capt	086169	M-3-5	3/3/69	28	Little Rock, Arkansas
Burr, Stewart S Pfc	2255287	E-2-9	4/23/69*	20	Passaic, New Jersey
Burton, Ronald E LCpl	2367230	E-2-7	5/12/68	19	Mansfield, Ohio
Burtsell, Ronald L 1stSgt	1075210	A-1-27	8/23/68	35	Santa Monica, California
Butler, Christopher P SSgt	1403463	H&S-2-12	10/20/66	30	Galveston, Texas
Butler, James T Capt	090634	HMM-164	2/12/68	25	Tampa, Florida
Butler, John C SSgt	0000000	CAC-1-1	5/9/68	00	Unknown
Butler, Robert H Jr Sgt	1968622	M-3-5	7/18/66*	22	Sheridan, Indiana
Butt, Thomas E LCpl	2172737	I-3-9	3/30/67	20	Rockville, Maryland
Buttry, Robert L Jr LCpl	2307924	D-1-1	8/1/67	20	Fort Jackson, SC
Butts, John D Sgt	2038912	CAC-1	1/10/68	29	Newport News, Virginia
Byers, James N IV 2dLt	0104543	D-1-5	9/11/68	22	Providence, Rhode Island
Byler, Earl D 2dLt	0106933	G-2-4	1/30/69	22	Denton, Texas
Byrne, Gerald J Jr Pfc	2208155	H-2-5	9/10/67	22	Catskill, New York
Byrne, Joseph L Jr Pfc	2116537	I-3-4	7/20/66*	19	Roscoe, Pennsylvania
Byron, Michael J 1stLt	088284	A-1-3	7/1/65	23	Santurce, Puerto Rico
Cabrera, Alfredo Sgt	2105940	A-1-7	7/6/70	25	New York, New York
Caceres, Edgardo LCpl	2108262	B-1-9	5/12/66*	21	Tacoma, Washington
Cahoon, Curtis R LCpl	2455194	I-3-7	6/19/69	19	Columbia, North Carolina
Cain, Jerome F Pfc	2361727	B-5-Tanks	6/15/68	20	Buffalo, New York
Calabria, Joseph R Sgt	1830887	11-Eng	5/8/67	27	Youngstown, Ohio
Calderone, Thomas Cpl	2207193	B-1-1	5/31/68	20	New York, New York
Caldon, David L Capt	089878	TAC-2-26	4/6/68	25	Cheshire, Connecticutt
Caldwell, Robert P Pfc	2236993	C-1-1	3/18/68	21	Macon, Georgia
Calhoun, Lester R Jr LCpl	2197670	M-3-3	4/30/67	19	Los Angeles, California
Call, John G LCpl	2308315	B-1-5	9/4/67*	19	Columbus, Georgia
Callaham, Robert LCpl	2101141	E-2-3	12/28/67	20	Richmond, Virginia
Callahan, Melvin R LCpl	2101566	A-1-1	10/30/65	20	Tuscaloosa, Alabama
Callaway, Johnny P Pfc	2474595	B-1-Recon	11/29/68	19	Baltimore, Maryland
Camp, Aubie Pfc	2083796	A-1-1	10/30/65	19	Apison, Tennessee
Campbell, Donald A Cpl	2128722	B-1-3	9/9/68*	21	Glouster, Ohio
Campbell, Harold J Jr Capt	085928	HMM-265	1/27/67	28	Somerville, Mass
Campbell, Kenneth C Sgt	1919370	1-Anglico	2/1/67	25	Seattle, Washington
Campbell, Thomas E Capt	083152	Advisor	6/29/66	28	Dallas, Texas
Campbell, Wallace L 1stLt	092078	VMCJ-1	5/1-6/1/67	25	Pearl Harbor, Hawaii
Candelario, Rafael A Sgt	2030045	A-1-5	5/9/69	23	San Juan, Puerto Rico
Canney, George T LCpl	2266543	B-1-3	9/17/67	19	Boston, Massachusetts
Cannon, James R 2dLt	093569	E-2-3	5/3/67	31	Fredericksburg, Virginia

Name	Service #	Unit	Date	Age	Hometown
Cannon, James T Jr SSgt	1662890	F-2-3	8/7/69	29	Columbus, Georgia
Cantieny, John B LtCol	060027	CO-1-13	2/23/69	37	Minneapolis, Minnesota
Cantu, Andres Jr Cpl	2276912	I-3-3	9/11/68	21	San Antonio, Texas
Capers, James Jr 2dLt	095892	3-F-Recon	4/3/67	29	Baltimore, Maryland
Caple, Edward F Pfc	2255277	K-3-3	9/15/69	19	Newark, New Jersey
Capozzoli, Orlando S Cpl	2022090	M-3-7	4/21/66	22	Chicago, Illinois
Capraro, Claud W Cpl	2373669	K-3-7	12/3/68*	19	Penrose, Colorado
Carey, Michael D 1stLt	091723	I-3-5	7/22/66	25	Torrance, California
Carey, Michael D 1stLt	091723	I-3-5	7/24/66	25	Torrance, California
Carey, Thomas W LCpl	2252396	F-2-9	5/20/67	19	Baltimore, Maryland
Carl, Daniel R Sgt	2479982	H-2-7	7/13/70	21	Cleveland, Ohio
Carlin, John P Capt	092415	E-2-3	6/3/68	24	Wooster, Ohio
Carlisi, Ignatius Pfc	2170881	C-1-Recon	6/16/66*	20	New York, New York
Carlisle, Jimmy D Sgt	2164281	I-3-5	8/29/70	25	Zillah, Washington
Carlisle, Richard P Maj	080714	Advisor	3/25/71	33	Bethlehem, Pennsylvania
Carlson, Clarence R LCpl	2192592	A-3-Recon	5/9/67	21	Milltown, Wisconsin
Carlson, Gary E Capt	087394	Advisor	6/29/66	25	Annapolis, Maryland
Carlton, David J Cpl	2259576	C-1-1	8/1/67*	20	St. Paul, Minnesota
Carlton, John D Maj	071434	MCRS-1	4/16/72	38	Austin, Texas
Carmean, Paul R Jr Cpl	2136365	MABS-36	3/17/68	21	Raleigh, North Carolina
Carmody, John J 2dLt	0101793	D-1-4	12/6/67	23	Albany, New York
Caro, Alexander J LCpl	2242963	M-3-4	1/27/68	20	Brooklyn, New York
Carolan, Fredrick A Capt	072088	VMCJ-1	5/6/67	34	Minneapolis, Minnesota
Carpenter, Peter B 2dLt	0106231	C-1-Recon	1/16/69	23	Springfield, Missouri
Carr, John D Capt	083158	L-3-4	1/18/68	34	Belfast, Maine
Carr, Richard W Maj	068300	HMM-161	4/18/69	36	Anderson, South Carolina
Carr, William D 1stLt	083966	VMA-513	2/21/68	35	New York, New York
Carrel, Jeffery L LCpl	2311797	I-3-5	5/9/67	19	Kansas City, Missouri
Carroll, Peter R LCpl	2375967	H-2-1	1/15/69*	20	Winters, California
Carter, Coy C Sgt	1385867	FSPB	2/16/67	31	Fort Wayne, Indiana
Carter, Johnnie III Pfc	2488459	1-FRecon	6/18/69	20	New Orleans, Louisiana
Carter, Thomas C 1stLt	088603	B-1-7	9/10/65	24	Winslow, Arizona
Carter, Timothy G LCpl	2482858	F-2-1	7/1/69*	20	Ely, Nevada
Carter, Wallace M SSgt	1526745	G-2-5	12/19/66	29	Centerville, Iowa
Caruolo, Richard A LCpl	2011788	K-3-7	3/22/66*	21	Providence, Rhode Island
Cashman, Cornelius J LCpl	2479699	D-1-7	8/12/69*	19	North Royalton, Ohio
Cassidy, John J 1stLt	090418	B-3-Recon	3/15/66	23	Central Islip, New York
Castagnetti, Gene E Capt	083162	B-1-5	6/9/69	30	Needham, Massachusetts
Castaneda, Robert L Sgt	2435039	B-1-Recon	7/23/70	22	San Antonio, Texas
Castania, Donald W LCpl	2428331	C-3-Recon	2/17/69	22	Fresno, California
Castillo, Alfredo R LCpl	2198359	A-1-1	4/9/67	20	Detroit, Michigan
Castillo, Charles R LCpl	2311620	H&S-1-26	6/5-6/67	19	Detroit, Michigan
Castor, James W Cpl	2452355	I-3-5	8/13/69*	20	Natoma, Kansas
Caswell, Russell J Capt	075626	C-1-5	5/10/67	31	North Hollywood, Cal
Catalogne, Paul R Capt	088796	C-3-Eng	7/30/67	26	St. Louis, Missouri
Cauble, John D Jr SSgt	1511999	HMM-262	1/25/68	31	Brenham, Texas
Caudle, Jerry W 1stLt	089540	HMM-361	10/30/66	24	Winston-Salem, North Carolina
Cecil, Richard D Sgt	1127963	C-3-Tanks	5/20/66	32	San Diego, California
Centers, Jack Cpl	2139630	E-2-1	5/14/67	20	South Charleston, WV
Centers, Norman B 1stLt	091328	A-1-26	6/7/67	30	Olmstead Falls, Ohio
Centers, Ronald L Pfc	2128537	H-2-4	8/18/65.	19	Cincinnati, Ohio
Cerda, Rene Cpl	2318335	B-3-Tanks	5/19/68	22	Fresno, California
Cerna, Narciso R Jr LCpl	2110587	K-3-4	7/15/66*	18	San Antonio, Texas
Chacon, David A LCpl	2409589	A-1-9	2/22/69*	20	Gilcrest, Colorado
Chacon, Richard S Jr Sgt	1371893	K-3-3	7/7/68	34	Houston, Texas
Chadwick, Leon G III Capt	087306	PMO-6	10/31/66*	26	Raleigh, North Carolina
Chaisson, John R Col	07244	G-3-3	4/9/66	50	Swampscott, Massachusetts
Chamberlain, Gene Pfc	2217725	I-3-9	3/30/67	19	Milwaukie, Oregon
Chambers, John C Cpl	2178941	A-3-Tanks	5/19/68	20	Evanston, Illinois
Chambers, Lester E Pfc	2336047	A-1-ATanks	2/24/68*	21	Dallas, Texas

Name	Service #	Unit	Date	Age	Hometown
Chambliss, Carl S LCpl	2532187	A-1-Recon	4/23/70	19	Chicago, Illinois
Champe, Charles R 1stLt	0102666	1-FRecon	8/18/69	24	Albuquerque, New Mexico
Chapa, Richard E 1stLt	093882	C-1-3	5/10/67	24	San Antonio, Texas
Chapman, Darrell H LCpl	2459791	A-1-9	2/22/69	20	Claremont, New Hampshire
Chapman, Harlan P LtCol	071437	POW	7/66-10/67	32	Elyria, Ohio
Chapman, William W SSgt	1920454	B-1-9	8/22/68	26	San Francisco, California
Charlie, Peter LCpl	2591299	B-1-5	8/8/70*	21	Farmington, New Mexico
Charlton, Albert K LtCol	066768	VMO-6	9/21/69	37	Spokane, Washington
Chase, Jerry D 2dLt	095902	I-3-3	3/24/67	31	Vestal, New York
Chavarria, Manuel T SSgt	1638209	B-1-4	3/22/67	29	San Antonio, Texas
Chaves, Allen F Pfc	2269791	D-3-Recon	5/4/67*	19	Winslow, Arizona
Cheatham, Charles W 1stLt	089036	I-3-12	7/29/66	24	Burlington, NC
Cheff, Stanley W 1stLt	088234	C-1-Recon	10/16/65	24	Grand Rapids, Michigan
Chesnut, Robert E LCpl	2238708	G-2-26	4/7/68	21	Springfield, Mass
Chidgey, Donald J LCpl	215 1762	C-1-1Rec	10/10/66	21	Bakersfield, California
Chiofolo, Vincent N Pfc	2417922	K-3-27	5/17/68	19	Jacksonville, Florida
Chism, Samuel M Sgt	2200443	B-1-13	6/17/68	22	Charleston, Illinois
Christensen, Harry C Cpl	2328989	B-3-Tanks	1/24/68	20	Marblehead, Massachusetts
Christy, Howard A Capt	069814	A-1-9	5/21/66	29	Provo, Utah
Cincotta, Thomas A Pfc	2483094	L-3-9	6/28/69*	18	San Rafael, California
Cisneros, Philip C Capt	081499	I-3-9	6/27/69	37	Los Angeles, California
Cisneros, Tony A LCpl	2327825	K-3-3	2/7/68	19	Oakland, California
Clancy, John J III 1stLt	085172	A-1-4	3/20/66	25	Norcross, Georgia
Clark, Delbert E Cpl	2036461	C-1-3	6/5/65	21	Jackson, Missouri
Clark, Frank A LtCol	058931	CO-1-7	11/12/69	38	Portsmouth, Rhode Island
Clark, John L Jr Maj	079737	VMO-2	5/5/68	31	Omaha, Nebraska
Clark, Johnnie M Pfc	2397358	A-1-7	8/3/68	18	St. Petersburg, Florida
Clark, Read M Capt	078778	I-3-7	1/30/67	31	Winston Salem, North Carolina
Clark, Robert N Jr 2dLt	093883	D-1-9	3/4/67*	24	Indianapolis, Indiana
Clark, Stephen W Capt	091131	VMA-235	5/3/68*	26	Plymouth, California
Clary, William B Capt	088003	Advisor	6/3/70	26	Atlanta, Georgia
Clay, Henry H Pfc	2403575	D-1-1	11/22/68	19	Albuquerque, New Mexico
Cleary, Robert E GySgt	1223540	G-2-1	6/25/66	35	Tewksbury, Massachusetts
Clements, Charles G Sgt	1938124	1F-3-Rec	12/16/65	22	Little Rock, Arkansas
Clements, Marvin W Cpl	2357427	L-3-7	8/19/68	20	Boston, Massachusetts
Cleven, Wayne R 1stLt	0100125	H&S-12	5/5/68	22	Hewlett, New York
Clute, Michael A Pfc	2454839	G-2-9	2/17/69*	19	Hinsdale, New York
Coachman, Albert H Cpl	2350228	M-3-1	2/2/68	21	Washington, D.C.
Coats, Charles T LCpl	2352780	G-2-9	4/19/68*	18	Klamath Falls, Oregon
Cobb, Daniel J III Capt	085599	VMO-6	12/22/67	26	Santa Barbara, California
Cody, Richard L Capt	085109	A-1-5	10/7/68	30	Richmond, California
Coe, Michael G 2dLt	0101819	A-1-7	2/6/68	23	Del Rio, Texas
Coffman, Clovis C Jr 1stLt	1106891	3-FRecon	8/31/69	38	Summit, Mississippi
Colasanti, Robert J Sgt	2027058	G-2-4	5/2/68	24	Grosse Point Woods, Michigan
Colby, Brian L LCpl	2261912	F-2-7	2/28/67*	19	Lansing, Michigan
Coleman, David M Cpl	2200633	G-2-3	4/30/67	21	Dunbar, West Virginia
Coleman, George W 2dLt	0103411	D-1-7	3/17/68*	25	Wildwood Crest, New Jersey
Coleman, Matthew Pfc	2258782	L-3-1	1/1/67	20	Baton Rouge, Louisiana
Collins, David L Cpl	2127450	F-2-5	2/1/68*	20	Carson City, California
Collins, Edward Jr Pfc	2404021	B-1-26	3/30/68	20	Drew, Mississippi
Collins, James A Pfc	2086051	I-3-26	1/20/68*	22	Broadwell, Illinois
Collins, Patrick G Capt	076230	D-3-Recon	5/14/65	22	Grosse Isle, Michigan
Compton, James L Maj	071749	B-3-Recon	3/15/66	30	Kansas City, Missouri
Conatser, Bernis B Jr Capt	087397	Advisor	1/22/70	28	Annapolis, Maryland
Conger, William A Capt	082209	K-3-9	2/15/68	29	Cincinnati, Ohio
Connell, George M 1stLt	089342	C-1-9	5/21/67	23	Cheverly, Maryland
Connelly, Edward W Jr Capt	093033	HMM-165	5/5/68*	24	Agawam, Massachusetts
Connelly, William 1stLt	0101824	C-1-9	4/16/68	25	Cleveland, Ohio
Conner, Samuel D LCpl	2122469	G-2-9	1/13/66	24	Indianapolis, Indiana
Conti, Robert F 1stLt	0105592	E-2-5	11/24/69*	23	Arlington Heights, Illinois

Name	Service #	Unit	Date	Age	Hometown
Contreras, Anselmo Cpl	2110338	K-3-3	3/24/67	22	San Antonio, Texas
Convery, Joseph F Jr Pfc	2347401	F-2-3	2/6/68*	19	Chester, Pennsylvania
Cooke, Harold T Pfc	1686983	F-2-9	1/30/66*	27	San Diego, California
Coomer, Richard R Cpl	2397845	A-1-1	10/27/68*	21	Placentia, California
Cooper, David H 2dLt	095915	H&S-3-3	3/25/67	25	Abington, Pennsylvania
Cooper, James L Maj	067148	E-2-5	6/21/66	37	Penhook, Virginia
Cooper, Thomas M Capt	085558	D-1-7	2/14-16/69	30	Elizabethtown, Kentucky
Copeland, Randolph G Capt	080139	G-2-1	3/4/68	28	Jacksonville, Florida
Copeland, Shalor II LCpl	2307499	B-1-3	8/17/67	20	Dallas, Texas
Corbett, Wayne Jr Pfc	2134448	B-3-Recon	12/17/66	20	Jacksonville, Florida
Corliss, Gregory A Maj	058138	HMM-262	2/1/67	26	Arlington, Virginia
Cornett, Charles B Cpl	2307003	VMO-2	1/6/69	21	Athens, Tennessee
Corsino, Eddie N LCpl	2478857	I-3-4	9/17/69*	20	Lorain, Ohio
Costello, William F LCpl	2097787	F-2-7	3/4/66	21	San Diego, California
Cothran, Terry E Cpl	2010356	D-3-Recon	5/14/65	21	Houston, Texas
Cotter, Richard L 2dLt	0105059	-1-Amph	2/10/69*	23	Peabody, Massachusetts
Courtney, Paul H Capt	088591	HMM-363	4/13/67	25	Leavenworth, Kansas
Cowart, David L LCpl	2281133	A-1-1	9/25/67*	20	Aliquippa, Pennsylvania
Cox, Eugene P Capt	018940	TF-77	4/16/72	26	San Francisco, California
Cox, George W Jr Maj	066082	HMM-364	5/7/69	38	Overland Park, Kansas
Cox, Lester E Jr Cpl	2637650	HMM-165	7/11/72	20	Jacksonville, Florida
Cox, Robert K Pfc	2477312	E-2-4	3/4/69	20	Shelton, Connecticut
Craft, David J Sgt	2360609	F-2-7	7/22/70	21	Albany, New York
Cramer, Robert M Maj	076778	HMM-362	1/8/68*	32	Stoutland, Missouri
Cravens, Michael J Pfc	2235732	L-3-3	9/7/67	19	Kansas City, Missouri
Crawford, Curtis E Cpl	2127011	G-2-3	2/28/67*	19	Dunkirk, New York
Crawl, Daniel Jr LCpl	2178449	FLSGA	9/3/67	22	Cleveland, Ohio
Creech, Jimmie A Maj	077879	VMO-2	7/26/69	33	Jenkins, Kentucky
Creed, Jerry L Capt	0110455	Recon	11/5/70	22	El Dorado, Kansas
Creel, John B Jr 2dLt	089621	A-1-9	9/9/65	23	Thomaston, Connecticut
Creel, John B Jr Capt	089621	H&S-1-13	2/23/69	26	Thomaston, Connecticut
Creelman, Malcolm W Jr Pfc	2393161	H-2-1	9/16/68	20	Providence, Rhode Island
Crews, Curtis T Capt	090827	HML-367	2/23/71	29	Burke, Virginia
Croft, Harold A Pfc	2249792	K-3-3	3/25/67	23	Malden, Massachusetts
Crooks, Ronald L Cpl	2214965	B-1-26	6/7/67*	21	Hickory, North Carolina
Cross, Robert J Jr 1stLt	094633	I-3-4	8/16/68	26	Los Angeles, California
Crouch, George M SSgt	1959468	A-1-Recon	4/15/69	27	New York, New York
Crouch, Jack E Jr LCpl	2115687	C-1-26	9/22/66*	18	Mount Vernon, Illinois
Crowder, David L Cpl	2400990	M-3-4	5/27/68	19	Atlanta, Georgia
Cruz, Luis A LCpl	2541704	B-1-5	11/4/70*	20	New York, New York
Cuddy, Francis J Jr Capt	094240	HMM-265	6/8/69	29	Samford, Connecticut
Cudnik, Edmund V Pfc	2421050	M-3-7	12/26/68*	19	Detroit, Michigan
Cullen, Terence M 2dLt	0100156	L-3-9	6/1/67	25	Ithaca, New York
Culver, Richard O Jr Capt	075696	H-2-3	7/21/67	31	Hopewell, Virginia
Cumbie, William T Cpl	2308561	I-3-3	2/9/69*	19	Jacksonville, Florida
Cummings, Charles H Pfc	2359128	C-3-Recon	9/14/68	18	Washington, Illinois
Cummings, John D Capt	083974	VMFA-122	8/30/68	32	Olathe, Kansas
Cunningham, Frederick R 1stLt	0100158	M-3-1	2/2/68	25	Glen Ellyn, Illinois
Curd, James H R Capt	070637	C-1-9	3/4/67	30	Golden, Colorado
Curd, James H R Capt	070637	C-1-9	3/24/67	30	Golden, Colorado
Curry, Jimmy D Cpl	2196774	F-2-4	9/21/67*	21	San Jose, California
Curry, Richard T Pfc	2131932	A-1-1	8/12/67	19	Waltham, Massachusetts
Cushman, Earl D Jr Cpl	1949791	C-3-Eng	4/10/68	26	Portland, Maine
Cutbirth, Richard E LCpl	2377740	F-2-3	5/28/68*	19	Marionville, Missouri
Cutshall, David W LCpl	2217976	M-3-3	3/6/68*	21	Rapid City, South Dakota
Dabney, William H Capt	080399	I-3-26	1/20/68	33	Gloucester, Virginia
Daerr, Richard L Jr 1stLt	095297	L-3-9	5/20-21/67	22	College Station, Texas
Dagger, Carl R Cpl	2304358	I-3-4	5/17/68*	20	Urbana, Ohio
Daigle, Isreal J Sgt	1637938	G-2-1	12/16/65	25	Vinton, Louisiana
Dalbey, Rolland M 1stLt	057291	HMM-165	5/26/67	38	Omaha, Nebraska

Name	Service #	Unit	Date	Age	Hometown
Daley, Edmund W Pfc	2148768	H&S-3-3	3/24/67	20	Hull, Massachusetts
Daley, Richard F LtCol	056483	CO-1-5	3/19/69	41	Madison, Wisconsin
Dalhouse, John D 1stLt	089075	D-1-3	2/14/66	26	Montgomery, Alabama
Dally, David J Sgt	2195759	L-3-4	6/15/68	21	Blakeslee, Pennsylvania
Danielson, Darrell C Maj	056511	XO-1-9	7/2/67	38	Burlington, Iowa
Danis, John A Sgt	1888311	G-2-9	6/14/66	24	Medford, Massachusetts
Darden, Ronald C Cpl	2300409	L-3-3	9/7/67	20	Bradenton, Florida
Darlington, Earl G 2dLt	096498	D-1-Recon	11/22/66	36	New Braunfels, Texas
Darrh, Floyd J GySgt	1297336	F-2-3	3/14/68	34	Constantia, New York
Dartt, Robert J Cpl	2049979	C-1-9	12/25/65	20	Maunie, Illinois
David, George J Cpl	2186692	D-1-Recon	9/13/67	19	Seekonk, Massachusetts
Davis, Alan F Capt	092159	HMM-262	9/20/68	25	Albany, New York
Davis, Allan D Maj	065655	HMM-165	12/31/68	37	Gulf Breeze, Florida
Davis, Bruce Cpl	2035575	F-2-7	3/4/66	20	Gilbert, Louisiana
Davis, Bruce W Cpl	2384439	A-1-Tanks	2/23/69	20	Houston, Texas
Davis, Earl R SSgt	1375531	B-1-9	5/12/66	31	Baltimore, Maryland
Davis, Edward D Pfc	2307471	C-1-9	7/6/67	20	Dallas, Texas
Davis, Garry D Cpl	2203969	A-1-9	5/21/66	19	Springfield, Oregon
Davis, Gene F SSgt	1460528	C-1-1	7/24/66	30	Indianapolis, Indiana
Davis, George B 1stLt	0107803	D-1-7	11/12/69	22	Boston, Massachusetts
Davis, James R Capt	090147	Advisor	5/20/67	27	Jacksonville, Florida
Davis, Jerry M LCpl	2486554	A-1-7	2/24/69	19	Walla Walla, Washington
Davis, Michael H Pfc	2236733	E-2-5	2/7/68	21	Greer, South Carolina
Davis, Stanley Col	024203	CO-5	9/4-15/67	45	Niagara Falls, New York
Davis, Thomas R Cpl	2416988	C-1-3	1/31/69*	20	Frostburg, Maryland
Davis, William J LtCol	049480	CO-1-7	2/14/68	45	Lynn, Massachusetts
Day, James L Maj	056003	CO-1-28	3/2-4/67	41	East St. Louis, Illinois
Day, Wesley D Pfc	2365134	L-3-7	11/26/67*	19	Grand Rapids, Michigan
Deal, Earl M LCpl	2351262	H&S-1-5	1/18/68	20	Farmville, North Carolina
DeAtley, Hillmer F LtCol	053925	CO-3-1	2/12/67	42	Wood River, Illinois
DeAtley, Hillmer F LtCol	053925	CO-3-1	4/21/67	42	Wood River, Illinois
DeBlanc, Daniel J 1stLt	0103432	VMO-2	3/17/69	24	Slidell, Louisiana
Deegan, Gene A Capt	075623	F-2-1	4/21/67	30	St. James, Minnesota
DeForest, Roy E Maj	082183	OO-3-3	6/17/69	30	San Antonio, Texas
Deichman, Jack E Capt	0101857	G-2-4	4/30/68	21	Freemansburg, Pennsylvania
Delong, Earl R LtCol	048884	CO-2-3	5/9/67	47	Jackson, Michigan
Demartino, Pasquale W Capt	062915	CO-3-3	12/9/65	34	Cumberland, Maryland
DeMille. Roy W Pfc	2269614	H-2-3	5/3/67	19	Phoenix, Arizona
Dennis, Dan M Cpl	2053890	M-3-5	5/13/67*	21	Houston, Texas
Dennis, Phillip V Pfc	2433936	D-3-Recon	12/17/68	22	Toms River, New Jersey
Denton, Charles L Cpl	2073195	C-3-Tanks	8/18/65	20	Detroit, Michigan
Dermody, Denis J Cpl	2242960	VMGR-152	3/17/68	22	Brooklyn, New York
Derryberry, Abraham R III LCpl	2259201	A-1-3	3/7/68*	18	Shreveport, Louisiana
Deschuytner, Victor R LtCol	057310	AOO-9	11/17/65	37	Chelsea, Massachusetts
Desmond, Lawrence E Cpl	2240370	C-1-5	2/13/68	21	Kansas City, Missouri
Desselle, Thomas W Pfc	2434898	G-2-7	11/20/68*	20	San Antonio, Texas
Detora, Ernest F Jr Cpl	2222067	L-3-3	8/21/67	21	Glendora, New Jersey
DeTrempe, Barry V Cpl	2231831	-1-Amph	2/10/69*	19	Peoria, Illinois
Dewey, Henry C Maj	063819	VMA-242	10/27/67	34	Peckville, Pennsylvania
DeWilde, Peter F Pfc	2484172	C-3-Recon	3/5/69*	18	Lansing, Michigan
Dewlen, Michael L 2dLt	0102756	C-1-12	6/11/68*	24	Amarillo, Texas
Dextraze, Richard P LCpl	2416530	E-2-9	4/23/69*	21	Buckingham, Quebec
Dias, Raymond R III Pvt	2190384	A-1-9	4/16/68	20	Honolulu, Hawaii
Diaz, Frank E Sgt	2110609	E-1-Recon	6/14/70	26	Hephzibah, Georgia
Dickey, Dwight R Maj	066224	O-1-Tank	2/6/68	35	Eagle Point, Oregon
Dickey, Thomas R Pfc	2465956	L-3-9	2/10/69*	19	Concord, Massachusetts
Dillberg, David W Cpl	2296859	G-2-1	1/22/68	19	Los Angeles, California
Dinota, Dennis T SSgt	1858560	M-3-5	2/1/67	25	Clearwater, Florida
Dirr, Michael A LCpl	2086576	D-3-Recon	9/3/69	20	Independence, Kentucky
Dito, Raymond E 2dLt	0103445	F-2-3	5/28/68	24	San Francisco, California

Name	Service #	Unit	Date	Age	Hometown
Doan, Thomas L 2dLt	0105493	L-3-4	12/8/68	28	Danville, Illinois
Dobbins, Kent E 1stLt	0102429	C-1-Tanks	2/23/69	24	Lawrence, Kansas
Dobbs, John W Jr LCpl	1910349	D-E-Recon	5/14/65	22	Tuscumbia, Alabama
Dockendorff, Gary D 2dLt	093066	M-3-9	8/20/66	23	Danville, Iowa
Dodge, Harvey J Pfc	2129286	G-2-9	1/13/66	20	Fruitland, New Mexico
Dodson, James A Cpl	2235324	CAP-2-3-2	3/17/69	22	Los Angeles, California
Dodson, Robert G 1stLt	0101865	C-1-1	7/7/68*	25	Bloomfield, New Jersey
Doherty, Jerome J Jr Capt	088119	H-2-5	1/26/67	25	Denver, Colorado
Doherty, John B 2dLt	091916	L-3-4	7/18/66	22	Norfolk, Virginia
Dolan, David L Pfc	2192786	G-2-26	11/17/66	20	Brooklyn, Minnesota
Dombrova, Louis A 1stLt	0101867	E-2-4	8/2/68	22	Jacksonville, Florida
Doneghy, James M SSgt	1858526	C-1-9	6/18/69	28	Norfolk, Virginia
Donnelly, Alan C LCpl	2391481	B-1-Eng	11/12/68*	20	Gloversville, New York
Donnelly, William R Jr 2dLt	0101619	A-1-1	2/4/68	24	Minnville, Oregon
Donofrio, Ernest Jr LCpl	2237518	HMM-165	3/3/67	20	Enola, Pennsylvania
Donovan, Joseph P 1stLt	0101005	HMM-364	5/9/69	27	Waxahachie, Texas
Dooley, Dennis D 2dLt	096559	C-1-4	3/26/67	34	Mankato, Minnesota
Dopko, Theodore G Capt	079129	FLSG	1/30/66	29	Taylor, Pennsylvania
Doss, James G Jr Maj	051888	-3-Tanks	8/17/66	40	Phoenix, Arizona
Doss, Larry D LCpl	2370576	G-2-7	9/20/68*	19	Nashville, Tennessee
Doublet, Alvin J Capt	060073	H&S-2-7	12/18/65	37	Pittsburgh, Pennsylvania
Dougherty, James E Sgt	2096769	B-1-5	2/7/68	22	Conshohocken, Pennsylvania
Dougherty, Steven J Cpl	2356913	C-1-1	7/7/68	21	Philadelphia, Pennsylvania
Douglas, William L Jr Cpl	2103244	K-3-4	7/15/66*	19	Canal Fulton, Ohio
Douthit, Roger W SSgt	2131930	G-2-7	9/8/70	25	Boise, Idaho
Dowd, John A LtCol	059807	CO-1-7	4/21/69	37	Elizabeth, New Jersey
Dowd, Lawrence K LCpl	2466300	E-2-3	8/10/69*	19	Bridgewater, Massachusetts
Downing, William K Cpl	2201839	E-2-3	5/3/67*	20	Fort Gibson, Oklahoma
Downs, Edward J Cpl	2350684	K-3-7	7/28/68*	19	Washington, DC
Downs, Michael P Capt	081727	F-2-5	2/3-7/68	27	Oak Bluffs, Massachusetts
Downs, Vernon L Jr Cpl	2306783	H&S-3-9	2/14/68*	19	Huntsville, Alabama
Drake, Benjamin A Cpl	2385832	H-2-1	3/3-4/69	18	Grand Rapids, Michigan
Drake, Warren G Sgt	1519600	HMM-165	1/28/67	30	Cambridge, Ohio
Draude, Thomas V Capt	089211	M-3-7	4/21/66	25	Kankakee, Illinois
Draude, Thomas V Capt	089211	M-3-7	7/15/66	26	Kankakee, Illinois
Drollinger, Harry B 2dLt	0101014	B-1-5	2/1/68	24	Dallas, Texas
Drone, Frank L Cpl	2305373	M-3-3	3/24/69	20	Evansville, Indiana
Drury, Paul S LCpl	2376703	D-1-5	8/4/68	20	Knoxville, Tennessee
Dube, Marcel J. Capt.	076408	Advisor	3/20/67	34	Twentynine Palms, California
Dulude, Daniel D LCpl	2186420	HMM-265	4/24/67	20	Cumberland, Rhode Island
Dulude, Daniel D LCpl	2186420	HMM-265	4/25/67	20	Cumberland, Rhode Island
Dunbar, Charles L SSgt	1565358	H-2-5	1/13/69	30	Cleveland, Ohio
Duncan, Billy R LtCol	060076	CO-2-1	5/19/68	37	Amarillo, Texas
Dunham, Richard F Cpl	2077038	K-3-4	3/23/67	22	Rochester, New York
Dunn, John F 1stLt	096578	Advisor	8/19/68	35	Providence, Rhode Island
Dunn, John H LtCol	059941	POW	65-68	28	Glendive, Montana
Dunn, Vincent Cpl	1924381	L-3-4	7/15/66	24	Elizabeth, New Jersey
Dunning, Clifford R Capt	081620	Advisor	11/18/67	29	Berkeley, California
Duphiney, Randall W Maj	061345	HMM-361	10/28/65	36	Carlsbad, California
Durand, Hershal D Cpl	2259228	H-2-7	8/22/66	20	Shreveport, Louisiana
Durham, Larry W Cpl	2113914	H-2-4	3/31/68	21	Houston, Texas
Durham, Thomas W Pfc	2486190	E-2-5	3/27/69*	19	Richmond, Virginia
Dustin, Charles R Sgt	2157384	B-1-5	4/20/68	21	Las Vegas, Nevada
Duvall. Doyle W SSgt	1442752	F-2-7	12/18/65	31	Clifty, Kentucky
Dye, Andrew T SSgt	1864613	USNMP	8/27/67	27	Los Angeles, California
Early, Richard J 2dLt	0105077	E-2-5	10/12/68	24	Bronx, New York
Easter, Boyce E Pfc	2354407	H-2-4	3/18/68	20	Charlotte, North Carolina
Eaton, Robert F LCpl	2333634	C-3-Recon	6/23/68	20	Bedford, Indiana
Eberhardt, Edward V Sgt	2163646	G-2-26	5/16/67	22	Hartford, Connecticut
Ector, Jerry Pfc	2406562	C-1-3	7/5/68*	20	Cincinnati, Ohio
Edgar, Gene E Pfc	2337757	B-1-26	3/30/68	19	Pittsburgh, Pennsylvania

Name	Service #	Unit	Date	Age	Hometown
Edwards, Daniel L Pfc	2291073	B-1-7	8/10/68*	18	Ceredo, West Virginia
Edwards, Michael J Pfc	2337706	E-2-9	2/2/68	20	Pittsburgh, Pennsylvania
Edwards, Ted W 2dLt	0100215	K-3-1	2/2/68*	23	Charlotte, North Carolina
Edwards, Thomas B III 1stLt	0103468	K-3-7	8/28/69	25	Louisville, Kentucky
Egger, Charles H F Maj	063656	HMM-361	9/10/67	37	West Palm Beach, Florida
Eggers, Michael A Cpl	2313037	I-3-27	5/17/68	21	Houston, Texas
Eichler, Thomas J Cpl	1838959	E-2-26	2/5/68	27	Chicago, Illinois
Eisenbach, Charles R II 2dLt	0101622	D-1-Recon	7/4/68	22	Annapolis, Maryland
Eisenson, Henry L Maj	078516	HMM-163	10/12/67	29	Pensacola, Florida
Elder, William F Cpl	2014043	A-1-9	9/9/65	20	Fort Wayne, Indiana
Elkins, Roger L Pfc	2387144	G-2-9	2/17/69*	23	Fort Gibson, Oklahoma
Ellefson, David J LCpl	2492657	M-3-7	4/22/70*	19	North Bend, Oregon
Eller, Franklin P Jr Capt	067808	Advisor	12/31/64	31	Middle River, Maryland
Eller, John A LCpl	2351913	D-1-MPB	8/23/68*	24	Norfolk, Virginia
Ellis, Frank B Maj	054468	HMM-265	1/14/67	38	Austin, Texas
Ellis, Gerald L Maj	070766	MABS-12	9/27/67	34	Cherry Point, North Carolina
Emberger, George J SSgt	1336158	L-3-1	4/5/69	35	Philadelphia, Pennsylvania
Emerick, George A Sgt	1863759	I-3-3	8/18/65	23	Chesterfield, Michigan
Emerson, William Capt	093908	HMM-265	2/3/68*	26	Concord, Massachusetts
Emery, Lawrence W Cpl	2012328	D-1-3	6/29/65	21`	Carrollton, Ohio
Emmons, Michael A Pfc	2385269	B-1-Eng	12/18/68	20	La Crosse, Wisconsin
Emrick, Ervin J SSgt	1889464	K-3-5	5/11/69*	29	St. Louis, Missouri
Enedy, Robert J Sgt	2151824	H-2-4	4/30/68*	22	San Diego, California
Engel, Donald F Capt	092342	HMS-16	11/26/67	21	Springfield, Ohio
Englade, Bertly V Jr Cpl	2384912	E-2-3	4/4/69	21	Belle Chasse, Louisiana
Epps, Robbie L LCpl	2334217	A-1-7	2/6/68	21	Gainesville, Florida
Erickson, Eric C Sgt	2386023	E-2-9	9/9/68	22	Orion, Ohio
Erickson, Roger A 1stLt	092924	MAG-11	9/17/68	24	Minneapolis, Minnesota
Eriksen, Darrell G Sgt	2348367	C-3-Recon	3/5/69	25	Los Angeles, California
Eron, John R LCpl	2277659	M-3-3	3/16/68	21	Fort Jackson, South Carolina
Erpelding, Wayne C LCpl	2143564	B-1-1	2/9/68	21	Waterloo, Iowa
Eshelman, William P Maj	077473	Advisor	2/2/68	30	Silver Spring, Maryland
Esmond, Donald V 1stLt	095503	HMM-364	11/17/69	25	Akron, Ohio
Espenes, Olav J 2dLt	0105504	G-2-5	8/18/68	22	Minneapolis, Minnesota
Espinola, Robert J LCpl	2289764	I-3-4	1/27/68	24	Boston, Massachusetts
Esslinger, Dean E LtCol	045012	CO-3-5	4/21/67	41	Sheffield, Iowa
Estes, Edward S LCpl	2329731	C-1-5	2/13/68*	23	Dallas, Texas
Eubank, Gareth D Pfc	2420441	L-3-3	8/26/68	19	Detroit, Michigan
Eustis, Michael S 1stLt	0116851	AO-2-9	5/15/75	24	Harvey, Illinois
Evans, Eddie L Cpl	2094918	L-3-9	6/1/67	22	Dallas, Texas
Evans, Gary G Pfc	2269577	L-3-5	5/26/67	20	Phoenix, Arizona
Everett, Lucious L Sgt	2106822	M-3-3	12/9/66*	20	Detroit, Michigan
Everett, Michael T LCpl	2372400	HMM-165	4/15/69	19	Meridian, Mississippi
Ewers, Norman G LtCol	028152	HMM-163	3/31/65	41	Tujungai, California
Fales, William E SgtMaj	1039761	H&S-1-5	3/19/69	45	Norfolk, Virginia
Falk, Frederick J Jr LCpl	2322062	I-3-1	1/26/68*	19	Torrington, Connecticut
Fanter, Stephen A Cpl	2250623	C-1-3	10/11/67	20	Chicago, Illinois
Farley, James C Jr Cpl	1937825	HMM-163	3/31/65	21	Tucson, Arizona
Farlow, Gary A LCpl	2263163	D-1-5	5/12/67*	19	North Olmstead, Ohio
Fassett, Randolph D Sgt	1963302	I-3-4	7/20/66	24	Sacramento, California
Federowicz, Michael J LCpl	2347563	C-1-9	4/16/68	20	Philadelphia, Pennsylvania
Feille, William B 1stLt	088825	F-2-7	12/18/65	25	Fort Worth, Texas
Fellers, Robert M 2dLt	090862	F-2-1	5/26/66	24	Tulsa, Oklahoma
Fenton, Larry W Capt	0102803	HMM-262	6/7-11/69	23	Aurora, Missouri
Ferracane, Louis J Jr Capt	090153	VMA-224	5/29/72	30	Braintree, Massachusetts
Festa, Donald Capt	076922	A-1-9	3/3/67	34	Stafford, Virginia
Fiala, Joseph B Cpl	2231752	1-AAmph	3/1/68	23	Great Falls, Montana
Fields, Joseph P Cpl	2080918	E-2-5	12/19/66	22	Lakeland, Florida
Figueroa, Dennis C LCpl	2197951	B-1-4	3/22/67	21	Concord, California
Fillon, William LCpl	2132370	F-2-4	2/2/67	21	Weymouth, Massachusetts

Name	Service #	Unit	Date	Age	Hometown
Finkel, Charles LCpl	2282715	I-3-4	9/4/67*	19	New York, New York
Fisch, David A LCpl	2197161	F-2-26	5/16/67*	20	Remsen, Iowa
Fisher, Jeffrey M 1stLt	093917	G-2-4	9/21/67	25	Williamsport, Pennsylvania
Fisher, Joseph R LtCol	045857	CO-2-4	8/18-24/65	43	Westwood, Massachusetts
Fisher, Larry P SSgt	1957045	L-3-3	11/7/68	28	Kansas City, Missouri
Fisher, Thomas F LCpl	2145555	F-2-9	7/5/67	21	Philadelphia, Pennsylvania
Fitch, William H LtCol	055351	VMA-533	2/21/68	38	Fort Meade, Florida
Fite, William C III 1stLt	094694	Advisor	2/24/68	25	Clearwater, Florida
Fitzgerald, Arthur L GySgt	663891	E-2-5	6/3/67	37	Portland, Maine
Flanagan, Lawrence J Maj	057065	HMM-362	5/10/67	36	Sayville, New York
Fleischer, Charles W Jr 1stLt	0104098	HMM-262	3/6/70	25	Alameda, California
Fleming, William B LtCol	050585	MAG-11	12/2/66	38	Putnam, Alabama
Flood, Denis W Pfc	2282676	M-3-5	5/13-14/67	19	Brooklyn, New York
Flores, Librado L SSgt	1166569	3-FRecon	8/0/66	33	San Antonio, Texas
Floyd, David A LCpl	2259076	I-3-5	3/1/69*	20	Minden, Louisiana
Floyd, Donald Cpl	2269120	A-1-1	2/9/68	20	Hamilton, Ohio
Foley, Gifford T Cpl	2447541	M-3-3	3/24/69	22	Winnetka, Illinois
Fontaine, Edward L SSgt	1810940	D-1-1	7/2/68	28	Carthage, Missouri
Ford, Vernon J LCpl	2055451	M-3-5	7/22/66	20	Monroe, Louisiana
Forgea, Bernard LCpl	2289312	B-1-Amph	8/30/67	20	Hartford, Connecticut
Forness, Richard W LCpl	2360345	7-Comm	6/16/68	19	Coral Gables, Florida
Forte, Ollie SSgt	1803556	L-3-5	6/11/68*	30	Raleigh, North Carolina
Foster, Douglas G Cpl	2221643	CAP-1-31	9/14/68*	21	Beaver, Washington
Foster, Gregory P Pfc	2467323	G-2-4	3/23/69	19	Bronx, New York
Foster, Jerome C LCpl	2145219	B-1-26	3/30/68	21	Philadelphia, Pennsylvania
Foster, Larry E Pfc	2376944	M-3-7	12/26/68*	18	Lenoir City, Tennessee
Fox, George C LtCol	047459	CO-2-9	3/18/69	45	Chapel Hill, North Carolina
Foy, Thomas L Pfc	2277787	B-1-3	5/10/67*	20	Coolidge, Georgia
France, Phillip S LCpl	2294742	G-2-4	10/14/67*	20	Baltimore, Maryland
Francis, Glen A Pfc	2148555	D-1-7	9/7/67	20	Boston, Massachusetts
Francis, Larry J LCpl	0000000	CAP-4-2-3	7/1/70	00	Unknown
Frank, Harvey E Sgt	2220889	E-2-4	8/2/68	22	Hartford, Connecticut
Frash, Wilford E Cpl	1957702	I-3-7	8/25/69	25	Pennsville, Ohio
Frasier, Ronald R Pfc	2385302	F-2-5	2/3/68	20	Hennepin, Minnesota
Frazier, Jerry R Pfc	2457855	L-3-4	2/26/69*	18	Fairfield, Iowa
Frazier, Melvin D Pfc	2366536	B-1-26	3/30/68	19	Jacksonville, Florida
Frederick, John W Jr CWO	082847	POW	5/66-4/67	48	Manito, Illinois
Frenier, David E LCpl	2517744	CAP-1-3-2	5/8/70	20	Langhorne, Pennsylvania
Fretwell, John B 1stLt	092067	M-3-5	5/26/67	24	Coral Gables, Florida
Frey, Kevin J LCpl	2422000	D-1-9	2/27/69	20	Boston, Massachusetts
Fricker, Jerrell T Capt	079130	E-2-7	9/19/68	33	San Diego, California
Friel, Joseph A Pfc	2209039	C-1-26	9/21/66*	21	West Roxbury, Massachusetts
Friese, Laurence V Capt	091579	POW	2/11/68	25	Huron, South Dakota
Frisina, JohnT 1stLt	094708	Advisor	6/1/68	25	Cleveland, Ohio
Frost, Carlton A Pfc	2261252	D-1-4	5/19/68*	20	Winslow, Maine
Fryman, Roy A GySgt	1356690	Advisor	8/24/69	36	Paris, Kentucky
Fulford, Carlton W Jr 2dLt	092863	D-1-5	9/4/67	23	Newnan, Georgia
Fullerton, Donald G Sgt	1532460	H-2-4	12/26/65	28	Chicago, Illinois
Fults, Lawrence A Jr Pfc	2198691	E-2-27	3/13/68*	20	Tucson, Arizona
Furman, James T Sgt	1999793	Advisor	5/10/69	25	Powell, Wyoming
Gaboury, Laurence R Capt	068458	Advisor	8/14/66	33	Taunton, Massachusetts
Gaffney, Kenneth M Cpl	2098483	H-2-1	1/29/67	22	Rochdale, Massachusetts
Gagnet, Ronald W Pfc	2425054	B-1-4	5/25/69	20	San Francisco, California
Gaines, Byron A Jr Pfc	2308724	E-2-3	12/28/67*	20	Jacksonville, Florida
Galiana, Rudolph S Pfc	2297083	I-3-1	4/21/67*	20	Los Angeles, California
Galindo, Benjamin L Cpl	2040332	A-1-1	10/30/65	21	Acampo, California
Galindo, Herman Cpl	2222759	G-2-9	4/19/68	20	El Paso, Texas
Gallegos, Carlos E Cpl	2058382	K-3-9	3/7-20/66	22	Leadville Lake, Colorado
Galyardt, Robert G GySgt	1409591	H&S-1-1	1/31/68	32	Russell, Kansas
Gammack, Gregg L Capt	089587	HMM-361	5/10/67	26	Seattle, Washington

Name	Service #	Unit	Date	Age	Hometown
Gandy, William G SSgt	2083735	CAG-1-1	5/24/68	24	Nashville, Tennessee
Garber, Charles P 1stLt	096730	H-2-4	2/25/69	33	Sarasota, Florida
Garcia, Richard Pfc	2251755	I-3-7	1/30/67*	19	Galveston, Texas
Gardner, Donald R Capt	079807	C-3-Recon	8/18/66	28	Memphis, Tennessee
Garner, Cecil A Sgt	1306811	HMM-163	3/31/65	29	Tacoma, Washington
Garner, Ricky D Pfc	2359605	L-3-1	3/5/66*	18	Dallas, Texas
Garrett, Charles SSgt	1489226	A-1-7	12/8/66	29	Washington, D.C.
Garringer, David F 2dLt	0104040	H-2-7	8/22/68*	22	San Francisco, California
Gartman, Jerald B Capt	091496	HMM-165	3/5/68	26	Lamar, Missouri
Garvey, Vaughn D Pfc	2135957	F-2-9	1/30/66	18	Redondo Beach, California
Gates, Phillip R Cpl	2326768	HMM-361	8/10/69	20	Georgetown, Ohio
Gatlin, Thomas Pfc	1993968	A-3-Recon	7/12/65	22	Atlanta, Georgia
Gaugash, Jeffrey A 2dLt	091920	K-3-9	9/5/66	23	San Francisco, California
Gawlinski, Stanley A Jr Pfc	2323533	L-3-26	8/24/68	20	Calumet City, Illinois
Gayer, Kenneth E Sgt	2248157	A-3-Recon	1/25/69*	22	Fresno, California
Gelinas, Timothy K. Cpl	2163569	B-1-26	11/10/68	19	Woonsocket, Rhode Island
Gentile, Wayne M Cpl	2322369	G-2-1	3/4/68	22	New Haven, Connecticut
Gentry, George H Jr Maj	053634	XO-2-7	12/18/65	36	Baytown, Texas
Geatherall, Joseph T Jr Sgt	2092339	M-3-26	12/22/66	20	Boston, Masschusetts
Gibbons, Donald R GySgt	1377206	B-1-Tanks	12/9/67	32	Detroit, Michigan
Gibbs, Joseph W III Capt	080147	L-3-1	8/17/67	29	Montrose, Colorado
Gibson, Peter Cpl	2366795	HMM-262	5/10/69*	19	Orlando, Florida
Gibson, Richard LCpl	2370606	D-1-7	3/16/69	19	Nashville, Tennessee
Giddens, LaVon Cpl	2288707	E-2-27	3/3/68*	20	Detroit, Michigan
Giles, Jerrald E Capt	077857	K-3-9	7/4/67	31	Moscow, Idaho
Gilley, William B Sgt	2090023	F-2-9	5/20/67	23	Baltimore, Maryland
Gillis, James E Maj	058243	VMO-2	12/8/65	35	Long Beach, California
Giordano, Andrew M LCpl	2206251	M-3-5	9/4/67*	19	Smithtown, New York
Giretti, Anthony A 2dLt	0106806	G-2-9	3/19/69*	22	Baldwin, New York
Glaenzer, George W Sgt	2065302	C-3-Recon	1/12-14/67	23	Baltimore City, Maryland
Glass, Billy W LCpl	2509049	C-1-1	4/19/70*	21	Covington, Tennessee
Glawe, Thomas D Pfc	2130228	C-1-Recon	6/16/66*	18	Rockford, Illinois
Gleason, Robert J Sgt	1525006	M-3-5	3/21/66	28	Santa Rosa, California
Glenn, Clifford D LCpl	1960069	D-2-11	6/21/69	22	Bakersfield, California
Glover, Bobby Sgt	1938144	G-2-1	2/28/66	24	Little Rock, Arkansas
Glover, Leo P 1stLt	088701	K-3-4	7/18/66	24	Slippery Rock, Pennsylvania
Goble, Woodruff C 1stLt	095262	HMM-262	8/24/68	25	Livingston, New Jersey
Godwin, Solomon H CWO2	043252	1-C-T	1/31/68*	33	Hot Springs, Arkansas
Golub, Richard M Sgt	2099586	1-Amp-11	12/6/68	22	San Antonio, Texas
Gomez, Harold Cpl	2178660	A-1-9	2/21/67*	20	East Chicago, Illinois
Gonzales, David Cpl	2348336	VMO-2	12/18/68	18	Ventura, California
Gonzales, Edwardo J LCpl	2227946	H-2-3	5/3/67*	20	Sinton, Texas
Goodman, Edward M Sgt	2221603	HMM-165	1/12/69	23	Seattle, Washington
Goodrum, Lance A Capt	093140	HMM-164	11/5/68	24	Paris, Illinois
Goodwin, Forrest 1stLt	091778	L-3-3	3/2/67*	25	Tylertown, Mississippi
Goodwin, Paul B Capt	086261	K-3-3	6/11/69	31	Portsmouth, Virginia
Goodwyn, Ben R 1stLt	085210	C-1-7	3/28/66	25	Ennis, Texas
Gordon, James T Maj	069896	HMM-362	2/19/67	34	Odell, Illinois
Gorney, Jerry E 1stLt	093926	A-1-9	3/3/67*	24	Hudson, Ohio
Gorton, Ralph S III 2dLt	0103511	D-1-3	5/27/68*	25	Boise, Idaho
Gosch, Thomas C 1stLt	096807	M-3-1	2/21/67*	34	Oceanside, California
Gose, Elvin W Pfc	2357960	E-2-4	3/18/68*	19	Cumberland, Indiana
Goss, Jeffery A LCpl	2261627	M-3-5	2/7/68*	19	Orem, Utah
Grammar, William M 1stLt	091923	Advisor	5/20/67*	25	Oklahoma City, Oklahoma
Grantham, James T Sgt	1951293	L-3-4	2/23/67	25	New Roads, Louisiana
Graves, Floyd A Cpl	2115496	A-1-4	8/26/66	20	Shreveport, Louisiana
Graves, Joel E Cpl	2451592	B-1-5	5/3/69	20	Detroit, Michigan
Graves, Paul L LCpl	2448323	I-3-26	12/8/68	19	Coral Gables, Florida
Gray, Alfred M Maj	056067	COCAB	5/14/67	36	Point Pleasant Beach, NJ
Gray, Andrew D Cpl	2352191	G-2-5	5/7/68	21	Waterford, Michigan

Gray, Danny R Pfc	2382785	M-3-9	2/10/69	20	Pinto, Texas
Gray, Gary G Pfc	2490289	E-2-7	8/26/69*	22	Rimersburg, Pennsylvania
Gray, Ruzell Pfc	2384529	E-2-3	5/31/68*	20	Crocket, Texas
Green, Maurice O V 1stLt	0103516	D-1-5	2/3/68	23	Birmingham, Alabama
Green, Terry R LCpl	2152200	E-2-5	12/20/66	20	Indianapolis, Indiana
Green, William F 2dLt	0102882	G-2-4	6/29/68	24	Boston, Massachusetts
Greenberg, Gerald Cpl	2105360	I-3-5	9/6/67	20	Bradley Beach, New Jersey
Greenberg, Leroy F Jr Pfc	2121024	H-2-9	9/6/67	20	Scranton, Pennsylvania
Greene, Bruce A Maj	080216	OO-1-26	3/23/68	34	Birmingham, Alabama
Greer, Jesse R Maj	065804	HMM-361	4/3/67	36	Pearl, Mississippi
Greer, Matthew E Cpl	2142211	D-1-27	5/5/68*	21	East Palatka, Florida
Gregory, Simon H 1stLt	087543	L-3-1	3/5/66	24	Utica, New York
Gress, Howard K Jr Maj	063053	HML-367	6/8/69	37	Berlin, Pennsylvania
Griego, Cresenciano Jr Pfc	2184799	F-2-3	5/28/68	21	Albuquerque, New Mexico
Griffin, Albert Sgt	1441273	A-1-1	10/30/65	32	Pittsburgh, Pennsylvania
Griffin, Walter J L Pfc	2450222	C-1-9	6/18/69*	20	Hawkins, Texas
Grignon, Andrew L Cpl	2253368	C-1-26	2/23/68	21	Pittsfield, Maine
Grim, Thomas M SSgt	1938316	E-2-5	12/0/66	24	Norfolk, Virginia
Grimes, Michael B LCpl	2242452	B-1-4	3/22/67*	20	North Hollywood, California
Grinalds, John S Capt	077793	Advisor	9/2-4/66	28	Macon, Georgia
Grist, William A LCpl	2337813	F-2-3	5/28/68*	19	Bessemer, Alabama
Griswold, Gary C Pfc	2289655	G-2-4	10/14/67*	20	Bethel, Connecticut
Gross, Edmund K Jr 1stLt	0105913	K-3-Recon	6/7/69	24	Jacksonville, Florida
Gross, Wayne W 2dLt	0105109	M-3-9	8/19/68*	27	Carroll, Iowa
Gross, William H E Jr GySgt	602829	K-3-9	9/5/66	37	Bakersfield, California
Grove, Robert W GySgt	306455	Advisor	2/16/65*	47	Casper, Wyoming
Groves, Donald D LCpl	2409669	C-1-4	1/14/69	21	Denver Colorado
Grover, Lewis Pfc	2500960	H-2-5	12/9/69	18	Los Angeles, California
Gruerman, Gary R Pfc	2268725	B-3-Recon	12/17/66	21	Kansas City, Missouri
Grunewald, Bruce W Cpl	2286472	3-Recon	3/8/68*	19	Springfield, Illinois
Gudjonsson, Gunnar Capt	093148	Advisor	6/27/71	29	Perrysburg, Ohio
Guinee, Vincent J Jr Maj	064735	HMM-361	7/10/66	34	New York, New York
Guinee, Vincent J Jr Maj	064735	HMM-361	8/10/66	35	New York, New York
Gulash, David J Sgt	1898797	L-3-5	9/12/67*	25	Flint, Michigan
Gull, Rawlin C Sgt	1803308	HMM-363	1/1/66	25	Salt Lake City, Utah
Gum, William E SSgt	1812620	Advisor	4/18/69	28	Albuquerque, NM
Gump, Jackie A Pfc	2090264	G-2-9	1/13/66	20	Coburn, West Virginia
Gundlach, Richard M LCpl	2479035	L-3-4	10/9/69	19	Huron, Ohio
Gunning, Roger O 1stLt	0101115	E-2-3	12/13/67	23	West Lafayette, Indiana
Gurrola, Michael A Capt	086269	I-3-5	2/19/69	29	San Diego, California
Gustafson, Gordon A Pfc	2259807	F-2-26	5/16/67	20	St. Paul, Minnesota
Gutzwiler, Norman P Cpl	2251344	HML-367	9/24/68	22	Seattle, Washington
Guy, John W Maj	081309	Advisor	2/25/69	35	Cheyenne, Wyoming
Haaland, John D 1stLt	096849	AO-3-26	9/10/67	31	Daly City, California
Hackett, David S 1stLt	092088	H-2-3	4/30/67*	23	Ligonier, Pennsylvania
Hackett, William R LCpl	2250676	B-1-4	10/27/67*	21	Chicago, Illinois
Hadnot, Otha N SSgt	1868918	K-3-9	9/5/66	30	Jasper, Texas
Hadsock, William A Pfc	2334155	E-2-7	8/18/68*	19	Tarpon Springs, Florida
Haga, Gene H 2dLt	0108674	A-1-7	5/29/68	28	Jacksonville Beach, Florida.
Hagan, John R 2dLt	0102650	G-2-9	4/19/68	22	Savannah, Georgia
Hagemann, Richard L Cpl	2353171	A-3-Recon	2/17/69	20	Omaha, Nebraska
Hagen, Jerome T LtCol	062126	VMA-311	2/27/71	38	North Dakota
Hagerty, Roy H 1stLt	094171	HMM-263	3/19/69	24	Greenville, North Carolina
Haggerty, Ancer L 1stLt	0100493	B-1-3	9/11/68	24	Portland, Oregon
Hailston, Earl B 1stLt	0108635	E-1-Recon	6/16/70	23	Albany, New York
Hainsworth, John J Capt	078253	Advisor	2/1-2/68	33	Philadelphia, Pennsylvania
Hair, John C III Sgt	2145572	B-1-3	9/10/68	21	Philadelphia, Pennsylvania
Haley, Harrison L Capt	080836	D-1-1	5/26/68*	31	Martinez, California
Haley, William J Jr Cpl	2563754	CAP-1-3-9	7/9/70	19	Seattle, Washington
Hall, Edward M Pfc	2403963	A-1-26	8/18/68	19	Detroit, Michigan

Name	Service #	Unit	Date	Age	Hometown
Hall, Rickey W LCpl	2471142	E-2-3	8/10/69*	20	Indianapolis, Indiana
Hamblett, Robert B LCpl	2486513	D-1-7	11/12/69*	19	Roanoke, Virginia
Hamilton, David A Pfc	2326751	G-2-4	10/14/67*	19	Springfield, Ohio
Hamilton, Don E 1stLt	082187	HMM-163	3/31/65	25	Eastchester, New York
Hamilton, Robert E Capt	078253	Advisor	7/30-31/65	30	Oklahoma City, Oklahoma
Hamm, John R Pfc	2384231	B-1-1	5/31/68	21	Houston, Texas
Hammel, Alfred A Jr SSgt	1691303	A-1-Eng	6/26/66	26	Hartford, Connecticut
Hammonds, Roy L LCpl	2450003	L-3-7	1/4/70*	21	Waxahachie, Texas
Hammons, Herbert D LCpl	2332783	D-1-5	2/15/68*	20	Pocasset, Oklahoma
Hancock, David Capt	083348	A-1-3	3/7/68	29	Los Angeles, California
Hanifin, Robert T Jr LtCol	048995	CO-2-1	2/27/66	40	Boston, Massachusetts
Hanley, Patrick J LCpl	2443880	E-3-Recon	6/7/69	19	Portland, Oregon
Hanna, Rocky W Pfc	2486550	H&S-3-M	1/11/69*	18	Addy, Washington
Hannah, Samuel J 2dLt	0101388	D-1-26	6/7/68*	25	Lincoln, Arkansas
Harbin, David D LCpl	2238359	H-2-5	2/4/68	20	Albuquerque, New Mexico
Hardin, Richard P 1stLt	094172	HMM-364	2/7/69	25	Richmond, Virginia
Hardy, Warren Jr Cpl	2171504	E-2-7	6/15/67*	23	Montgomery, Alabama
Harrell, Roger P Capt	085965	HMM-361	10/31/66*	28	Jacksonville Beach, Florida
Harrington, Myron C Jr Capt	081869	D-1-5	2/15/68	27	Augusta, Georgia
Harris, Charles E LCpl	2312798	1-FRecon	1/14/68*	20	Norfolk, Virginia
Harris, John A SSgt	1518077	D-1-1	7/23/66	30	Birmingham, Alabama
Harrison, Bruce W Cpl	2226354	L-3-9	2/24/69	21	Louisville, Kentucky
Harrison, Clyde L LCpl	2366560	C-1-4	8/25/68	20	Jacksonville, California
Harrison, James R Cpl	2349898	G-2-4	3/9/69	18	Reseda, California
Harshman, Richard L 1stLt	091599	G-2-4	9/21/67	25	Denver, Colorado
Hart, James W Jr Cpl	2239759	G-2-26	5/16/67	21	Topeka, Kansas
Hart, Michael K Pfc	2512333	D-3-Recon	5/16/69	19	Louisville, Kentucky
Hartley, William L Capt	087550	A-1-4	8/23/68*	27	Memphis, Tennessee
Hartney, Alan H Capt	083357	L-3-4	3/2/67	28	Ashland, Ohio
Harvey, Michael A Pvt	2366216	H&S-2-4	2/25/69*	18	Milwaukee, Wisconsin
Harville, Lawrence GySgt	1640020	M-3-5	5/9/68*	30	Princewick, West Virginia
Haskell, William C 2dLt	0106577	K-3-3	3/9/69	23	Quantico, Virginia
Hatfield, John C GySgt	665452	K-3-9	7/4/67	36	Wilkinson, West Virginia
Hathcock, Carlos II Sgt	1873109	Sniper	9/16/69	27	Little Rock, Arkansas
Hatzfeld, William G LCpl	2242155	M-3-5	5/14/68	21	Los Angeles, California
Hausrath, Donald A Jr 2dLt	0103550	F-2-5	2/3-4/68*	22	Villa Park, California
Hawley, Jack E Capt	091575	VMA-242	2/18/68	32	Moffett Field, California
Hayden, Thomas R LCpl	2317751	HMM-262	5/18/68	20	Buffalo, New York
Hayes, Fred J SSgt	1822754	D-1-27	4/13/68*	25	Walnut Creek, California
Hayes, Ivey J Cpl	2236650	CAP-4-1-10	4/4/69*	19	Jesup, Georgia
Hayes, James M 2dLt	094769	B-1-11	4/21/67	25	Wauwatosa, Wisconsin
Hayes, Leonard C 1stLt	084167	D-1-3	7/17/65	25	Whittier, California
Hayes, Ray A Sgt	1942851	A-1-5	9/17/68*	24	Knoxville, Tennessee
Hayes, William A Pfc	2359755	E-2-26	6/6/68*	19	Chicago, Illinois
Haynes, John L 2dLt	079127	A-FLC	9/2/66	36	Sarasota, Florida
Hazelwood, John E Cpl	2268715	I-3-27	5/28/68*	24	Gypsum, Kansas
Hazen, Donald L LCpl	2353111	C-1-3	1/19/68	20	Omaha, Nebraska
Head, Glen R Pfc	2052341	A-3-Eng	7/25/66	22	Lexington, Kentucky
Head, William H Jr SSgt	1973115	B-1-3	8/17/67	25	San Marcos, California
Heald, Ronald E Maj	073991	VMO-6	2/8/67	31	Adams, New York
Heeter, James R Pfc	2374329	D-1-3	8/18/68*	19	Spencer, West Virginia
Helton, John K Sgt	2046079	E-2-4	7/6/67*	24	Griffin, Georgia
Helton, Michael LCpl	2218163	C-1-4	3/26/67	19	Cincinnati, Ohio
Helton, Michael P Cpl	2323276	E-2-4	2/4/68	21	Cincinnati, Ohio
Hembrough, John A Pfc	2465384	H&S-2-4	2/25/69	20	Jacksonville, Illinois
Hemingway, Joseph E Sgt	2114023	H-2-9	11/16/66	22	Richmond, Virginia
Hemp, Stuart F Pfc	2511850	L-3-3	8/22/69*	19	Richmond, Virginia
Henderson, Earl W 1stLt	093935	E-2-9	4/6/68	25	Los Angeles, California
Henderson, Jim A Jr Cpl	1984516	D-1-5	6/2/67	24	Albuquerque, New Mexico
Henderson, William B SSgt	1588608	F-2-5	3/24/67	27	Memphis, Tennessee

Name	ID	Unit	Date	Age	Hometown
Henderson, William T 2dLt	089766	A-3-Recon	7/12/65	23	Princeton, New Jersey
Hendricks, Robert L LCpl	2235429	H-2-7	2/24/68	20	Los Angeles, California
Henebury, Joseph D LCpl	2114415	L-3-3	12/9/65	21	Belmont, Massachusetts
Hengevelt, William G Cpl	2142331	I-3-11	11/4/68	21	Jacksonville, Florida
Henning, Thomas L Sgt	2264035	C-3-Recon	6/1/69	20	Canton, Ohio
Henricks, Charles D 1stLt	0102443	HMM-161	3/23/69*	26	Solana Beach, California
Henry, Clark G Maj	060399	XO-3-4	7/18/66	39	Oceanside, California
Henry, Howard B Maj	085097	VMO-2	10/22/69*	36	Baltimore, Maryland
Henry, Richard H 1stLt	0100319	E-2-5	5/15/69	27	Chapel Hill, North Carolina
Henry, Walter M Pfc	2388813	D-1-MPB	8/23/68*	19	Seattle, Washington
Henson, Clark L Jr Sgt	2063752	D-1-26	6/7/68*	21	Joplin, Missouri
Henson, Fred G Jr Cpl	2175080	I-3-1	4/21/67	21	Philadelphia, Penn
Herlihy, Patrick E Pfc	2209977	D-1-5	8/29/68	20	Wilmington, Delaware
Herman, Charles W Cpl	2329901	E-3-Recon	5/17/69	22	Montgomery, Alabama
Herman, Steven G Cpl	2149874	S1-1-4	5/8/67	21	Munster Lake, Indiana
Hernandez, Daniel L Pfc.	2135637	M-3-1	3/5/66	20	Boyle Heights, California
Hernandez, Jose F Cpl	0000000	1-Anglico	7/16/72	20	El Paso, Texas
Hernandez, Leonardo Cpl	2238532	M-3-1	4/21/67	21	Midland, Texas
Herrera, Manuel Pfc	2106574	F-2-4	3/21/66*	22	Pueblo, Colorado
Herrera, Phillip A LCpl	2140517	G-2-4	9/8/66*	21	Selma, California
Herring, Charles L Cpl	1993881	F-2-9	1/30/66	21	Austell, Georgia
Herringer, Charles L SSgt	1969142	L-3-7	1/17/66	24	Springfield, Oregon
Herrington, Richard L 1stLt	0103562	HML-167	12/31/69	23	San Leandro, California
Herrod, Randell D Pfc	2503459	K-3-3	7/28/69	20	Calvin, Oklahoma
Herron, Mark P LCpl	2273873	H&S-2-4	2/23/68	19	St. Louis, Missouri
Hertberg, Edward C LtCol	064604	HMM-164	5/11/72	42	New York, New York
Hescock, Donald W Pfc	2422793	D-1-9	2/26/69	20	Walpole, Massachusetts
Hesser, Peter M 1stLt	092873	G-2-3	2/6/68	24	Stillwater, Oklahoma
Hester, James R Cpl	2132683	F-2-5	6/2/67	22	Louisville, Kentucky
Hester, Michael P LCpl	2434020	A-1-9	2/22/69	20	Dallas, Georgia
Hester, William W Pfc	2399347	I-3-4	7/1/68*	19	Philadelphia, Pennsylvania
Hettinger, John R Pfc	2323783	A-1-1	9/25/67	19	Chicago, Illinois
Hewitt, John R 2dLt	0101979	K-3-5	2/6/68	23	Cleveland, Ohio
Heyward, Robert LCpl	2236996	H-2-9	9/1/68	19	Macon, Georgia
Hickey, Lloyd R Cpl	2344181	FPD-3-12	2/8/69	20	Oklahoma City, Oklahoma
Hicks, Larry D Pfc	2184537	D-1-4	2/22/67	18	Farmington, New Mexico
Higgins, Jerome Pfc	2457197	D-1-7	11/12/69*	19	Springfield, Ohio
Higgins, John F LCpl	2420162	E-2-4	12/8/68*	19	Lincoln Park, New York
Higgins, Martin C Capt	089589	E-2-7	12/30/66	24	Fall River, Massachusetts
Hignight, Daniel J Cpl	2380117	H&S-3-5	8/9/68	26	East Moline, Illinois
Hildreth, Raymond S LCpl	2147409	C-1-Recon	6/16/66	19	Tulsa, Oklahoma
Hilgartner, Peter L LtCol	051942	CO-1-5	9/4-6/67	39	Newport, Rhode Island
Hill, Byron E 1stLt	094785	F-2-11	5/13-14/67	24	Sparta, Illinois
Hill, John M Sgt	2106248	B-1-Tanks	8/22/69*	23	Middletown, Kentucky
Hill, Thomas R Sgt	2145860	C-1-1	3/18/68	21	Buffalo, New York
Hill, Twyman R LtCol	045538	1-MPB	1/30/68	38	Denton, Texas
Hilliard, George S Jr LCpl	2356108	H-2-27	7/18/68	19	Pine Bluff, Arkansas
Hiltbrunner, Donnal E Capt	080851	VMA-533	10/29/67	29	Bushton, Kansas
Hilton, Judson D Jr 1stLt	080851	G-2-4	5/1/68	23	St. Paul, Minnesota
Hinkle, Dennis A 2dLt	0100327	G-2-4	6/11/67	22	Warren, Pennsylvania
Hinkle, Thomas F Capt	086682	M-3-9	2/9/69	28	Wilmington, Delaware
Hinojosa, Juan N Jr LCpl	2110369	B-3-Tanks	9/26/65	20	San Antonio, Texas
Hinson, Amos B 1stLt	086290	K-3-3	8/18/65	26	Montgomery, Alabama
Hinson, Don Pfc	2466761	K-3-3	12/22/68*	19	New York, New York
Hinson, Donald SSgt	1474530	G-2-4	6/11/69	32	Miami, Florida
Hitzelberger, Daniel A Capt	087413	G-2-9	2/5/69	29	Glassboro, New Jersey
Hoare, Thomas J Jr 2dLt	095789	L-3-4	1/7/68*	23	Bellerose, New York
Hoban, John J Cpl	2029200	K-3-9	7/4/66	25	Hartwell, Nebraska
Hobbs, Thomas A 1stLt	0102952	K-3-1	4/17/69	25	Roseburg, Oregon
Hoch, Larry D LCpl	2438147	G-2-1	5/3/69*	21	Topton, Pennsylvania

Name	Service #	Unit	Date	Age	Hometown
Hodges, Homer L Jr LCpl	2094758	K-3-9	9/9/68*	22	Mineral Wells, Texas
Hodges, Richard P LCpl	2172138	A-1-9	2/22/69*	20	Fulton, Georgia
Hoeck, Richard J LCpl	1960766	H-2-4	8/18/65	22	Wyoming, Minnesota
Hoff, John R Jr 2dLt	0111172	E-1-Recon	3/26/70	24	Boston, Massachusetts
Hoffert, Duwayne W Maj	062272	HMM-363	2/6/68	37	St. Peter, Minnesota
Hoffman, Richard C 1stLt	0105132	M-3-3	6/17/69	25	Seattle, Washington
Holden, Thomas J 1stLt	0849419	G-2-3	8/23/66	25	Hasbrouck Heights, NJ
Holden, Thomas J 1stLt	0849419	G-2-3	10/22/66*	25	Hasbrouck Heights, NJ
Holler, Donald D SSgt	1845128	E-2-9	2/2/68	28	Indianapolis, Indiana
Holloway, James O Jr LCpl	2355107	A-1-9	4/4/68*	25	Newark, New Jersey
Hollowell, Leon R Pvt	2073469	M-3-4	6/4/65	22	Detroit, Michigan
Holmgreen, John C Jr 1stLt	094422	K-3-4	3/17/67	23	San Antonio, Texas
Holt, Robert M GySgt	1059760	L-3-3	8/21/67	34	Pearl Harbor, Hawaii
Holtzclaw, Gary E 1stLt	0101637	F-2-4	4/10/67	24	Corbin, Kentucky
Holycross, Richard L CWO	085755	H&S-3-26	9/7/67	36	Columbus, Ohio
Holzmann, James C Cpl	24791919	1-FRecon	8/4/70	22	North Olmstead, Ohio
Honeycutt, James Q LCpl	2037848	G-2-3	10/22/66	22	Fayetteville, North Carolina
Honsinger, Arthur W Pvt	2300745	G-2-5	2/7/68	20	Troy, New York
Hooten, Richard J Jr Capt	088161	HMM-363	2/22/69	27	Pensacola, Florida
Hoover, Ronald E 1stLt	096988	I-3-26	11/20/68	33	Harrisburg, Pennsylvania
Hopkins, John I Capt	070652	Advisor	2/22/66	31	Brooklyn, New York
Hopkins, Julius B SSgt	1940257	E-2-7	4/12/67	24	Bishopville, South Carolina
Horak, Frank J Jr Capt	079105	MAG-16	3/3-6/67	37	Pensacola, Florida
Horcajo, Robert A Pfc	2425989	K-3-7	2/23/69*	19	Milpitas, California
Hord, Raymond A 1stLt	0103281	C-1-7	8/12/69	24	Dale City, Virginia
Horn, James M 1stLt	0104465	G-2-9	4/21/69	24	Stillwater, Oklahoma
Horner, Richard L 2dLt	0101999	F-2-5	2/1/68	26	Los Angeles, California
Horton, Charles R GySgt	662527	D-11-Eng	5/8/67	38	Clute, Texas
Hoskins, Sheldon D Cpl	2261445	G-2-5	10/7/68*	20	Blackfoot, Idaho
Hossack, Donald A Pfc	2275733	M-3-3	4/30/67	19	Kalispell, Montana
Houghton, John P LCpl	2121819	1-Anglico	1/10/67	22	Philadelphia, Pennsylvania
Houghton, Kenneth J Col	013965	CO-5Mar	4/25/67	46	San Francisco, California
Houze, Benjamin C Jr LCpl	2304536	E-2-11	8/19/69	21	Cincinnati, Ohio
Hovanesian, Daniel G 1stLt	0106803	AO-11	12/31/70	24	Watertown, Massachusetts
Howard, Ralph Pfc	2359925	M-3-5	2/6/68	18	Chicago, Illinois
Howe, Leroy C Cpl	2153778	F-2-9	5/13/67*	19	Holland Patent, New York
Hoyez, James K Pfc	2252555	M-3-5	9/11/68*	19	Albany, Oregon
Huckins, Raymond L LCpl	2241312	K-3-3	4/25/67	20	Lubec, Maine
Huckstep, Gordon W LCpl	2458910	1-FRecon	9/18/69	19	Cape Girardeau, Missouri
Hudson, Thomas J Jr1stLt	0103726	VMO-2	10/20/70	23	Denver, Colorado
Huebner, Anthony C Capt	074791	HMM-363	10/10/66	30	Carlsbad, California
Huff, Edwin L Capt	076402	VMA-155	9/12/66	30	Spokane, Washington
Huffman, Donald R Maj	075231	MAG-18	2/23/69	33	Bentonville, Arkansas
Huffman, James W Jr Capt	097004	B-1-7	4/5/69	38	Dallas, Texas
Huffman, James W Jr Capt	097004	B-1-7	4/21/69	38	Dallas, Texas
Hughes, David M LCpl	2451969	CAP-1-1-6	12/7/68	19	Kansas City, Missouri
Hughes, Robert C 1stLt	091695	HMM-163	3/5/67	27	Camp Hill, Pennsylvania
Humphrey, Gilbert W Jr 1stLt	0101169	E-2-7	9/19/68	23	Cleveland, Ohio
Humphries, Claude Jr Cpl	2085643	G-2-7	9/24/66	21	Columbia, South Carolina
Hundley, David D 1stLt	0101583	HMM-165	12/8/68	22	St. Petersburg, Florida
Hunt, Homer S SSgt	2083285	I-3-4	9/17/69	22	Burnham, Maine
Hunt, William H 2dLt	0106804	E-2-4	2/25/69*	22	Merritt Island, Florida
Hurtado, Albert S LCpl	2072324	E-2-7	9/19/66*	20	National City, California
Huskey, James H Sgt	2363681	C-1-9	4/4-6/68	22	Albany, New York
Hutchins, Norman I LCpl	2267716	HMM-164	8/21/68	20	Portland, Maine
Hutchinson, Edward L Jr Capt	090290	C-1-9	7/2/67	26	Merritt Island, Florida
Hutchinson, Franklin G Jr Maj	067488	Advisor	3/16/69	34	Lawrence, Kansas
Hutchinson, Kenneth P Jr LCpl	2290977	C-1-27	6/5/68*	18	Huntington, West Virginia
Hutchison, Joseph F LCpl	2475781	M-3-9	6/2/69	22	Freedom, Maine
Hutton, William R Cpl	2141953	K-3-4	9/28/66	19	Canoga Park, California

Name	Service #	Unit	Date	Age	Hometown
Hynes, Michael M 1stLt	092589	HMM-363	6/7/67	23	Albany, New York
Inks, Earl L LCpl	1986563	L-3-7	3/21/66	22	Salem, Oregon
Inman, Linwood J LCpl	2147102	A-1-4	8/26/66	19	Corunna, Michigan
Inscore, Roger V Cpl	2234781	M-3-5	5/26/67*	21	Santa Ana, California
Irvin, Stephen L Cpl	2305659	H-2-5	9/10/67*	20	Columbia, Missouri
Irwin, David L Cpl	2383039	HMM-161	6/11/70	22	Dallas, Texas
Jackson, Curtis R LCpl	2010458	G-2-3	6/4/65	22	Pollock, Texas
Jackson, Kenneth E Pfc	2441265	G-2-9	2/5/69*	19	Beckley, West Virginia
Jackson, Marlin W Cpl	2153526	M-3-27	5/26/68	19	Beaverton, Oregon
Jackson, Richard D Capt	080223	M-3-4	2/22/67	30	Huntington, West Virginia
Jackson, Terry K LCpl	2400904	K-3-9	2/13/69*	21	Hawkinsville, Georgia
Jacobs, Quiles R Cpl	2235586	B-1-26	6/5/68	22	Compton, California
Jacques, Raymond J Jr Sgt	1913045	B-1-9	5/12/66	24	Waterville, Maine
Jadlow, Robert L 1stLt	087562	B-1-9	5/12/66	26	Topeka, Kansas
Jakovac, Donald L Pfc	2337678	D-1-4	2/2/68	19	Pittsburgh, Pennsylvania
Jakubowski, Walter P Sgt	640095	C-3-Eng	5/29/66	37	New York, New York
James, Buddy L Sgt	1579538	I-3-4	7/20/66	27	South Charleston, WV
James, Danny D SSgt	1178515	E-2-3	8/22/65	33	Chester, Illinois
James, Lee R GySgt	669040	Advisor	9/5/65	35	Louisville, Kentucky
James, Richard D Cpl	2382319	L-3-26	12/8/68*	20	Shelbyville, Virginia
Jandik, Frank Jr Sgt	2132892	D-1-5	2/15/68	21	Miami, Florida
Jarboe, Bernard K Pfc	1952880	L-3-7	3/21/66	23	Columbus, Mississippi
Jarrell, Eugene E Pfc	2457391	B-1-Recon	11/29/68	19	Chicago, Illinois
Jasinski, Daniel Sgt	1914115	CAD-A	2/2/68	26	Toledo, Ohio
Jenkins, Carl E LCpl	2091760	C-1-7	9/8/65	19	Jacksonville, Florida
Jenkins, Charles E Sgt	2116407	D-1-5	9/4/67	21	Pittsburgh, Pennsylvania
Jenkins, Homer K 1stLt	085550	H-2-4	8/18/65	26	Hughes Springs, Texas
Jenkins, James A Cpl	2339189	C-1-3	5/20/69*	18	Marion, North Carolina
Jenkins, Robert M SSgt	1498750	F-2-5	1/26/67	29	Ft. Myers, Florida
Jensen, Duane S Maj	076524	HMM-364	11/8/70	36	San Diego, California
Jensen, Paul A 1stLt	095672	VMO-6	2/16/68*	23	Asheville. North Carolina
Jensen, Robert R SSgt	1641814	A-1-9	2/22/69	29	Janesville, Wisconsin
Jesperson, Marshall W Sgt	2168889	L-3-26	9/10/67	22	Atascedero, California
Jessup, Daniel G Sgt	2134362	I-3-26	1/20/68	21	Scotia, California
Jester, Herbert A Pfc	2369954	K-3-27	5/24/68	19	Cincinnati, Ohio
Jette, Peter L Cpl	2163516	D-1-4	5/8/67	19	Lisbon, Connecticut
Jinkerson, Victor Cpl	2286460	K-3-3	2/7/68	19	Farmington, Missouri
Joeckel, Charles E Cpl	2294975	A-1-7	1/23/68	21	Annapolis, Maryland
Johann, John C Sgt	1132242	E-2-7	5/12/66	32	Greenwich, Connecticut
John, Richard H Pfc	2142092	A-3-MTB	3/21/66	21	Keams Canyon, Arizona
Johnson, Charles H Sgt	2062309	D-3-Recon	6/13/69*	25	Baldwin Park, California
Johnson, Dennis G LCpl	2406779	F-2-9	3/28/69*	20	Cincinnati, Ohio
Johnson, James R SSgt	1859910	D-3-Recon	9/2/68	27	Highland Falls, New York
Johnson, Joe V Cpl	2350594	HMM-364	3/8/70	23	Torrence, California
Johnson, Julius C Cpl	2075579	C-1-26	9/24/66	20	Keyport, New Jersey
Johnson, Ken H Capt	084925	VMO-2	1/18/69	29	Blacksburg, Virginia
Johnson, Larry A Sgt	2232731	A-1-Recon	9/20/68*	21	Varna, Illinois
Johnson, Michael J Cpl	2291712	G-2-4	3/19/69	20	Brooklyn, New York
Johnson, Robert F II Sgt	2068483	K-3-5	9/6/67	00	Unknown
Johnson, Terry W Sgt	0000000	Advisor	2/23/69	00	Unknown
Johnson, William D Jr Cpl	2357895	B-1-Recon	7/3/68*	20	Kokomo, Indiana
Johnson, William S Cpl	2208849	E-2-26	12/5/68	22	Brunswick, Georgia
Jones, Andrew L Pfc	0000000	1-Recon	4/23/69	00	Unknown
Jones, Barry T 1stLt	0103599	F-2-27	8/23/68	22	Mentor, Ohio
Jones, Charles D 1stLt	088170	C-1-7	12/5/65	23	Huntsville, Texas
Jones, Clarence W SSgt	1460601	3-Tanks	8/26/66	29	Richmond, Virginia
Jones, David R 1stLt	0101182	E-2-4	5/1/68	24	Silver Spring, Maryland
Jones, Douglas A 1stLt	085267	Anglico	9/23/65	25	Seattle, Washington
Jones, Edward R Cpl	2253584	G-2-5	7/30/68	20	Fairlawn, New Jersey
Jones, James H LCpl	2366047	E-1-Recon	8/1/68*	21	Beltsville, Maryland

Name	Service #	Unit	Date	Age	Hometown
Jones, James L Jr 2dLt	0102030	G-2-3	5/27/68	24	Washington, DC
Jones, Joe N Sgt	1994040	H-2-4	5/2/68	24	Unknown
Jones, John H Sgt	2152507	C-1-3	8/16/67*	19	Nashville, Tennessee
Jones, John J LCpl	2347659	HMM-164	11/7/68	20	Wilmington, Delaware
Jones, Joseph J Sgt	2147920	B-1-Recon	7/3/68*	21	Scotland Neck, North Carolina
Jones, Larry G Sgt	1940762	I-3-9	7/6/68	25	Pittsburgh, Pennsylvania
Jones, Otis R Cpl	2512511	B-1-7	6/26/70*	20	Donerail, Kentucky
Jones, Wayne E GySgt	1442915	M-3-26	3/4/69	25	Pittsburgh, Pennsylvania
Jordan, Charles F Cpl	2445603	F-2-7	4/12/69	20	Jacksonville, Florida
Jordan, Kenneth D Capt	0108043	3-FRecon	1/16/67	29	Houston, Texas
Jordet, Ronald G LCpl	2156757	K-3-4	9/26/66*	21	Reedpoint, Montana
Joyce, Michael J Jr Cpl	2446486	L-3-9	2/17/69	20	Pittsburgh, Pennsylvania
Judkins, Paul G Maj	076747	VMGR-152	6/13/66	28	Buffalo, New York
Judson, Donald H SSgt	1927194	E-24	5/25/68	26	Waterbury, Massachusetts
Julian, Paul M Pfc	2225380	B-1-3	8/17/67	20	Coral Gables, Florida
Justis, Donald E SSgt	1458636	L-3-5	6/11/68	32	Cincinnati, Ohio
Juul, John F Capt	090402	D-1-4	5/8/67	25	New Orleans, Louisiana
Kabeller, George R Cpl	2043283	H-2-7	3/4/66	22	Parkersburg, West Virginia
Kahabka, James G LCpl	2408062	G-2-7	2/25/69	20	Minneapolis, Minnesota
Kalm, Raymond W Jr Capt	081597	M-3-4	1/24/68	35	Los Angeles, California
Kappmeyer, Paul J Sgt	2013858	M-3-9	8/20/66*	22	Indianapolis, Indiana
Karkos, Norman F Sgt	1866644	1-F-Recon	1/23/69	28	Lisbon Falls, Maine
Kasson, Leonard C Cpl	2118060	K-3-4	9/28/66	20	Cincinnatus, New York
Kaus, Harry L Jr Pfc	2077042	H-2-4	8/18/65*	18	Dunkirk, New York
Kaylor, James N LCpl	2349911	E-2-26	6/7/68	19	Costa Mesa, California
Kean, Billie O SSgt	1851628	C-1-4	3/16/68*	25	Alexandria, Ohio
Keaveney, John P Cpl	2207230	A-1-Amp	1/20/68	23	Brooklyn, New York
Keaveney, Paul S LCpl	2476234	3-FRecon	2/7/70	27	New Smyrna Beach, Florida
Keckler, Richard C Capt	093224	HMM-364	1/11/69	26	Irvington, New York
Keddy, Michael L Sgt	2027455	D-1-Recon	5/6/68	23	Boston, Massachusetts
Keefe, Floyd M GySgt	1332594	H&S-1-5	3/19/69*	35	Montgomery, Alabama
Kegel, John C 1stLt	0104243	C-1-4	3/5/69	26	Pittsburgh, Pennsylvania
Kegley, Walter M Sgt	2221435	CAC-2-3	8/6/68	22	Jacksonville, Florida
Keif, David J Pfc	2378228	F-2-5	2/4/68	19	Milton, Massachusetts
Keilty, Kevin P LCpl	2422647	M-3-9	2/19/69	20	Peabody, Massachusetts
Keister, Roger P Cpl	2180551	E-1-Recon	8/1/68	21	Phoenix, Arizona
Keith, Archibald K Pfc	2241803	K-3-4	9/28/66*	23	Bainbridge, New York
Keleher, Michael K Sgt	2002102	D-3-Recon	5/4/67	24	San Bernardino, California
Kellams, Glennis R Sgt	1858569	L-3-3	4/13/68*	25	New Albany, Indiana
Kelleher, Richard A SSgt	1556718	9-MAB	4/30/68	29	Aurora, Illinois
Keller, Albert W Maj	059625	Advisor	9/15/66	38	Seminole, Oklahoma
Kelley, Paul X LtCol	050603	CO-2-4	3/21/66	37	Boston, Massachusetts
Kelly, John A 1stLLt	088887	I-3-3	8/18/65	25	New York, New York
Kelly, John A Capt	088887	C-1-9	1/20/69	28	New York, New York
Kelly, Ronald J Cpl	2182506	C-7-Eng	9/4/67	20	Jacksonville, Florida
Kelly, Vincent J Jr SSgt	1517827	F-2-7	2/25/66	28	Boston, Massachusetts
Kelly, Vincent J Jr GySgt	1517827	L-3-26	4/10/69	31	Boston, Massachusetts
Kendall, George P Jr GySgt	1103292	H&S-3	2/4/68*	37	Missoula, Montana
Kendley, Patrick D Capt	095388	Advisor	8/13/66	32	Boothbay, Maine
Kennedy, Bruce T Pfc	2430740	L-3-3	8/26/68*	19	Espanola, Ontario
Kennedy, Robert R Jr LCpl	2271588	CAP-E-4	2/8/68	20	Albany, New York
Kennedy, Thomas J Jr Capt	069955	Advisor	6/12/66*	36	Erie, Pennsylvania
Keppen, Thomas R 1stLt	0103608	B-1-3	7/6/68*	22	Evansville, Indiana
Kerrigan, Thomas G 2dLt	0106816	L-3-1	5/5/69	23	Newark, New Jersey
Keshner, Keo J Pfc	2481163	A-1-5	6/8/69*	21	New Florence, Missouri
Kettering, Alvah J Capt	067506	HMM-263	12/9/65	32	Riverside, California
Keys, William M Capt	079544	D-1-9	3/5/67	30	Fredricktown, Penn
Kiely, Denis J Maj	077488	VMF-235	3/16/68	33	Boston, Massachusetts
Killian, Mark A Sgt	2096371	D-3-Recon	6/7/69	21	Sacramento, California
Kimener, Robert P 2dLt	0110642	B-1-7	6/26/70	23	Cincinnati, Ohio

Kines, Charles G 2dLt	0109043	C-1-7	8/12/69	25	Detroit, Michigan
King, Charles W 2dLt	0102049	L-3-26	4/14/68	22	Ben Wheeler, Texas
King, Cleveland Jr Cpl	2431840	C-1-4	3/1/69	20	Youngstown, Ohio
King, George R Jr Cpl	2384884	B-3-Recon	2/2/69	20	Fairfax, Virginia
King, Samuel L LCpl	2184066	B-1-1	2/19/67	21	Waco, Texas
Kinnaman, George M Sgt	1986364	G-2-3	10/22/66	22	Springfield, Oregon
Kirk, Alexander Capt	089986	HMM-164	8/3/67	24	Keansburg, New Jersey
Kisala, Walter Pfc	2079832	B-3-Recon	2/13/66*	18	Chicago, Illinois
Klages, Robert J 1stLt	094840	B-3-Amph	8/1/67*	23	St. Louis, Missouri
Kline, Gary D Sgt	2351248	A-1-Recon	5/26/69	20	Manitowoc, Wisconsin
Kline, Robert J LCpl	2288879	L-3-7	4/10/68*	20	Bay City, Michigan
Knapp, William F Jr LCpl	2427918	B-7-MTB	3/26/69	19	Houston, Texas
Knight, Jerome G Pfc	2181371	G-3-12	2/25/69	21	Lakeport, California
Knight, John E Jr Capt	088479	H-2-4	2/25/69	27	Washington, DC
Knight, Terry W Cpl	2353535	E-2-4	4/11/69	20	Birmingham, Alabama
Knight, William A Capt	093963	HMM-362	9/26/68	26	Ellsworth, Maine
Koehler, David J Pfc	2537986	I-3-3	6/4/69*	20	Clarence Center, New York
Kolakowski, Henry Jr Capt	081631	I-3-5	1/30/68	29	Farmington, Michigan
Kolter, Bruce 2dLt	0107436	M-3-3	6/17/69*	23	Wapakoneta, Ohio
Konn, Raymond LCpl	2309615	CAP-C-1	11/8/67	20	Wilkes-Barre, Penn
Konrady, Lester W Cpl	2124523	L-3-5	7/18/66	20	Lakeworth, Florida
Kopka, John R 1stLt	090184	G-2-3	10/22/66	26	Auburn, Massachusetts
Kosoglow, Joseph J Pfc	2074108	C-1-Recon	6/16/66	20	Irwin, Pennsylvania
Krages, Bert W. Maj	076354	HML-367	3/28/71	36	Valley Junction, Kentucky
Krec, Frank LCpl	2309152	M-3-1	3/2/68*	24	Williamson, New York
Kreh, Gary H Cpl	2175644	A-1-4	5/8/67*	19	Flint, Michigan
Kretzschmar, William Capt	082939	VMA-533	10/29/67	29	Brooklyn, New York
Krogh, Richard O Cpl	2266195	L-3-3	8/26/68*	21	Mercer Island, Washington
Kruger, Earl A Capt	0101214	Advisor	4/30/72	29	Bloomington, Indiana
Krulak, Charles C 1stLt	089423	L-3-3	6/3/69	27	Exeter, New Hampshire
Kubik, Kenneth A 2dLt	0109525	C-1-Recon	10/22/69*	25	Hollywood, Florida
Kuci, Richard A Maj	058735	HMM-361	8/18/66	35	Coraopolis, Pennsylvania
Kuci, Richard A Maj	058735	HMM-361	10/12/66	35	Coraopolis, Pennsylvania
Kuehlmann, Werner G LCpl	2438305	K-3-5	11/24/68	19	Philadelphia, Pennsylvania
Kufeldt, Edward Capt	089361	VMO-6	2/7/68	27	Homestead, Florida
Kupcho, Thomas G Pfc	2408039	B-1-26	5/29/68	19	Minneapolis, Minnesota
Kupec, Keith L Pfc	2113873	E-2-1	4/16/66	18	Houston, Texas
Kurtz, Norman F Cpl	2389280	D-1-7	12/20/67	19	Montgomery, Alabama
Kuske, Gregory W Sgt	2087698	HMM-165	3/6/68	22	Limestone, Tennessee
Kustaborder, Thomas W Cpl	2421708	K-3-9	2/14/69*	22	Juniata, Pennsylvania
Lacey, Fred E Jr Maj	064297	HMM-364	10/17/68	38	Monroe, Louisiana
Lackey, Ira E Sgt	1933997	E-2-4	7/8/67	24	Dallas, Texas
Lagrone, Leonard Jr LCpl	2297090	G-2-9	7/29/67	21	Marshall, Texas
Laine, Elliot R LtCol	050691	CO-3-9	3/15/69	41	Arlington, Virginia
Lake, Bruce R 1stLt	0101584	HMM-265	12/6/68	21	Harrisville, New Hampshire
Lala, Nolan J Pfc	2135240	C-1-MTB	1/31/68	19	Denver, Colorado
Lamb, Allan W Capt	065948	RLT-7	8/24/65	37	Ellensburg, Washington
Lambrecht, Donald M Sgt	1931632	L-3-4	9/25/66	24	Rockford, Illinois
Lamontagne, Edward J LtCol	054688	EO-3-MD	1/31/68	39	Manchester, New Hampshire
Lampo, Stephen F 1stLt	0101221	Advisor	1/31/68	26	Neosho, Missouri
Landes, Burrell H Jr Capt	079715	B-1-3	7/6/67	28	Topeka, Kansas
Landreth, Alfred F Jr LCpl	2529967	E-2-5	9/9/70	19	Greensboro, North Carolina
Landry, Daniel J Cpl	2098154	C-1-Recon	2/25/67	19	Newburg, Massachusetts
Landry, Eddie L LCpl	2035493	H-2-4	8/18/65*	20	Gonzales, Louisiana
Langdon, Howard W Jr 1stLt	0102068	C-1-Recon	4/30/68	23	Raleigh, North Carolina
Langevin, George Pfc	2290354	C-1-Recon	10/14/67	19	Framingham, Massachusetts
Langley, David F Sgt	2130678	D-1-1	7/5/68*	23	Jacksonville, North Carolina
Langley, William D Cpl	2447716	D-1-3	5/1/69	19	Chicago, Illinois
Lanham, Daniel L LCpl	2379976	D-1-5	8/29/68	21	Des Moines, Iowa
Lantry, Merrill L Cpl	1993291	B-1-7	9/2/65	22	Council Bluffs, Iowa

Name	Service #	Unit	Date	Age	Hometown
Laramy, Robert E Capt	092373	TAC	2/22/68	26	Stores, Connecticut
Larson, Edward D Pfc	2138412	D-1-26	9/19/66	21	St. Louis Park, Minnesota
Larson, Stephen P 1stLt	0101222	C-3-Tanks	2/3/68	23	Buffalo, New York
Lasseter, Lee T Maj	074063	VMFA-333	9/11/72	37	Lake Wales, Florida
Latting, Charles W Capt	077176	M-3-1	3/5/66	31	Cordona, Tennessee
Lau, James 1stLt	085372	H-2-7	3/4/66	27	West Point, New York
Lawrence, Gearie E Pfc	2460838	D-1-7	4/16/69	19	Falkville, Alabama
Lawrence, John F LCpl	2455299	B-1-7	4/30/69*	21	Norfolk, Virginia
Lawrence, Larry E LCpl	2307915	RLT-26	6/6/67*	19	Brunswick, Georgia
Lawrence, Michael J Pfc	2255903	H-2-7	8/23/69*	19	Maplewood, New Jersey
Laws, Billy W Pfc	2156134	K-3-4	9/26/66*	23	Grandview, Missouri
Lawson, Archie H LCpl	2241083	E-2-4	3/18/68	20	Little Rock, Arkansas
Lawson, Charles A Sgt	1912373	B-1-7	6/6/69	27	South Charleston, WV
Lawson, Curtis G Maj	065073	VMA-533	7/25/68	37	Elberta, Alabama
Leaf, Stephen D Cpl	2389051	A-1-1	3/7/69	21	Hobart, Washington
LeBaugh, Christopher M LCpl	2247724	C-1-7	6/26/67	21	Daly City, California
LeBlanc, Ross A LCpl	2309963	I-3-3	9/7/67	19	Schenectady, New York
LeCornu, John 1stLt	081634	Advisor	7/13/66	26	Nashville, Tennessee
Ledin, Jerry W Capt	071293	E-2-4	12/5/65	32	Los Gatos, California
Lee, Alex Capt	070686	H&S-2-7	3/4/66	32	Hemet, California
Lee, Cleo E 1stSgt	1023518	F-2-5	6/2/67	34	Dayton, Ohio
Lee, Douglas W Cpl	2112997	C-1-4	3/26/67*	20	Winston-Salem, N C
Lee, Fielding J Cpl	2384119	D-1-1	2/23/69	22	Indianapolis, Indiana
Lee, George N Jr Cpl	2270446	H-2-1	5/17/68	20	Los Angeles, California
Lee, Gregory W Capt	082477	HMM-362	8/10/66	28	King of Prussia, Pennsylvania
Lee, Johnnie P Sgt	2135311	C-3-Recon	9/14/68	23	Richmond, Virginia
Lee, Vincent B 2dLt	0109436	I-3-5	11/23/69*	24	Lawrence, Massachusetts
Lefefe, Anthony T Sgt	1976884	C-1-9	7/6/67	30	San Francisco, California
Leftwich, William G LtCol	061154	CO-2-1	8/4/70	39	Germantown, Tennessee
Legas, David S 1stLt	0104263	HMM-364	6/1/70	27	Ogden, Utah
Legaux, Merlin P Cpl	2144012	K-3-4	7/18/66	19	New Orleans, Louisiana
Lehoullier, Paul R LCpl	2336494	I-3-1	4/5/69*	21	Somersworth, New Hampshire
Lenna, Frank Sgt	2076091	E-2-5	8/17/67	22	New York, New York
LeNoue, Bruce V Cpl	2162858	A-1-1	3/26/67*	20	Anoka, Minnesota
Leon, Mario R Pfc	2398690	F-2-9	3/29/69*	18	Milwaukee, Wisconsin
Leonard, William Cpl	2241751	F-2-7	1/14/70*	21	Marlborough, Massachusetts
Leshow, William F Cpl	2073488	C-1-5	5/10/67	22	Detroit, Michigan
Lewin, Lanny K LCpl	2337311	H-2-1	10/22/68	19	Pittsburgh, Pennsylvania
Lewis, Frederick E Maj	081635	VMFA-542	12/27/69	28	Annapolis, Maryland
Lewis, Willie R Pfc	2390887	C-1-4	6/6/68	19	Coral Gables, Florida
Libutti, Frank 2dLt	0100427	C-1-9	7/2/67	23	Huntington, New York
Lifred, Hubert M LCpl	2355625	L-3-9	9/15/68	20	Newark, New Jersey
Limones, Jesus M Pvt	2256502	D-1-4	5/8/67*	19	Del Rio, Texas
Linde, Richard V Pfc	2218294	C-1-1	12/21/66*	26	Cleves, Ohio
Lindsay, Paul T 2dLt	0105251	F-2-7	11/12/68	23	Chicago, Illinois
Lindsey, Henry J LCpl	2344236	D-1-1	6/28/68	21	Oklahoma City, Oklahoma
Linkous, Noah R Pfc	2238101	C-1-7	11/2/69	21	Richmond, Virginia
Linn, Orie O LCpl	2156017	K-3-9	3/30/67	21	Greenbelt, Maryland
Linn, Robert A II 2dLt	0106604	C-1-4	3/23/69	23	Bloomsburg, Pennsylvania
Lipscomb, Roger D Sgt	2034550	K-3-1	1/15/67	22	South Charleston, WV
Little, James J SSgt	1410063	E-2-1	12/16/65	28	Willow Springs, Missouri
Lively, Stephen G LCpl	2647966	9-Amphib	7/11/72	20	Harrison, Arkansas
Livingston, Alastair J Cpl	2133814	C-1-Recon	10/10/66	20	Willodale, Ontario
Livingston, Bruce B LCpl	2263489	I-3-4	9/2/67*	18	Lorain, Ohio
Livingston, James E Capt	084449	E-2-4	3/18/68	28	Telfair, Georgia
Livingston, Lawrence H Capt	0107447	Advisor	4/12/72	31	Bonsall, California
Livingston, Thomas E Sgt	2094261	C-1-1	2/7/68	23	New Orleans, Louisiana
Lloyd, Edward J 1stLt	088909	G-2-7	2/22/66	28	Dallas, Texas
Lloyd, Lowell R LCpl	2265956	L-3-5	5/26/67*	20	Woodlawn, Illinois
Lochridge, Willard F 2dLt	092140	B-3-Tanks	9/5/66	23	Scarsdale, New York

Lockhart, James S Pfc	2361268	9-Amph	3/2/68	19	Topeka, Kansas
Lockwood, Bryce F SSgt	1859856	Liberty	6/8/67	27	Springfield, Missouri
Lockwood, Robert H Maj	069751	H&MS-36	1/28/67	33	Greenwich, Connecticut
Lofink, Walter F GySgt	563664	F-2-7	3/16/67	41	Staten Island, New York
Logan, Thomas E Pfc	2212878	F-2-26	5/16/67	20	Kansas City, Missouri
Logan, Westley R 1stLt	0100436	VMO-2	12/18/68	25	Fulton, Missouri
Lomen, William W Cpl	2315005	C-1-1	7/7/68	21	Seattle, Washington
Long, Donald E Pfc	2222600	G-2-5	2/5/68	20	Salt Lake City, Utah
Looney, Kenneth L LCpl	2213785	B-1-1	12/15/67	20	Montgomery, Alabama
Looney, Paul T Capt	089775	HMM-164	5/10/67 *	25	Shelburne Falls, Massachusetts
Loop, James S Maj	071973	HMM-161	2/23/69	29	San Marino, California
Lopez, Adrian S Pfc	2375936	3-FRecon	2/16/68 *	19	San Martin, California
Lopez, David LCpl	2341710	L-3-7	9/14/68	19	Austin, Texas
Lopez, Felix R Sgt	1332115	B-1-4	3/20/66	32	Denver, Colorado
Lopez, Joseph Pfc	2344087	C-1-5	2/13/68	20	New York, New York
Lopez, Joseph P Cpl	2135326	E-2-7	9/19/68 *	23	Denver, Colorado
Lopiano, Richard C LCpl	2346189	E-2-4	2/9/68	19	Samford, Connecticut
Lopinto, Frank T LCpl	2253402	B-1-9	4/27/67	19	Lyndhurst, New Jersey
Loucks, Burdette W Jr GySgt	1347464	L-3-5	8/17/68	31	Palm Coast, Florida
Lough, Robert M Jr SSgt	1881709	H&S-11	2/23/69 *	27	Moundsville, WV
Loughrey, Wayne F 1stLt	089856	E-2-9	3/25/66	26	Woodstock, New York
Lounsbury, Theodore W Jr LCpl	2151661	HMM-361	10/30/66	20	Los Angeles, California
Love, Clarence L LCpl	2383315	L-3-5	12/1/68*	19	Tyler, Texas
Love, Donald E Capt	089672	H&MS-16	12/25/67	26	Lafayette, Indiana
Lowder, Charles L 1stLt	0107449	1- Recon	8/10/69	24	Sullivan, Illinois
Loweranitis, John L Cpl	2052170	I-3-9	9/3/66*	22	Du Bois, Pennsylvania
Lowery, William T Sgt	2135395	C-1-7	8/12/69	24	Norfolk, Virginia
Lozano, Samuel H Pvt	2080080	B-1-27	6/17/68	19	Des Moines, Iowa
Lucas, Robert D Pfc	2329712	A-1-Recon	3/18/67	20	Dallas, Texas
Lucus, Gary A Sgt	1859141	HMM-265	7/15/66	25	Cody, Wyoming
Lund, Craig L Pfc	2361128	M-3-9	5/1/68	20	Flagstaff, Arizona
Lund, Dean T SSgt	1572519	A-1-Tanks	2/6/68	30	Syracuse, New York
Lunsford, Robert L SSgt	1608023	A-1-1	2/13/66	28	San Francisco, California
Lyle, Charles R LCpl	2326521	CAP 1-3-5	6/26/68	20	Owensboro, Kentucky
Lynch, Eugene A Capt	079753	I-3-1	4/21/67	29	Geneva, New York
Lyon, Alfred E Capt	086886	E-2-3	5/3/67	27	Cincinnati, Ohio
Lyon, John W 1stLt	087597	H&S- 2-1	12/10/65	25	Glen Ridge, New Jersey
Lyon, Thomas H SSgt	1372440	H-2-7	9/21/68	34	Vancouver, Washington
Lyons, Joseph W Cpl	2315470	1-FRecon	6/5/68*	22	Phoenix, Arizona
Lyons, Paul C Cpl	2122037	G-2-26	5/17/67	22	Detroit, Michigan
Lyons, Roger G Pfc	2369478	E-2-4	2/9/68*	19	Amelia, Ohio
Lyons, Walter J 2dLt	0100447	E-2-7	6/15/67*	21	Jacksonville, Arkansas
Maas, Bertram A LtCol	061155	VMO-6	8/19/68	37	St. Paul, Minnesota
Macalik, Daniel E. SSgt	1856841	B-11-Eng	6/14/67	28	Ennis, Texas
MacCormack, Dana F 1stLt	097272	C-1-7	2/15/68	37	Redwood City, California
Machulda, Thomas E Sgt	1114301	1-Recon	2/7/68	33	Los Angeles, California
MacKenna, James J GySgt	1356083	E-2-1	5/29/66*	37	Denver, Colorado
Madden, Ernest J Cpl	2137377	D-1-5	5/12/67*	20	Wellsville, Ohio
Maddox, Richard G Capt	095664	D-1-7	9/2/68	32	Cupertino, California
Maggi, David R Cpl	2387352	I-3-3	9/10/68	19	San Diego, California
Mahoney, John M Capt	076396	K-3-4	3/17/67	34	Brooklyn, New York
Main, Stanley W 1stLt	076635	H&MS-17	3/23/67	39	Modesto, California
Maki, Roger L LCpl	2454323	L-3-3	1/12/69	20	Columbus, Montana
Maldonado, Robert P Pfc	2270235	I-3-7	1/30/67	20	Los Angeles, California
Mallette, Robert D LCpl	2141497	I-3-3	11/28/67	22	Beltsville, Maryland
Mallobox, Jesse A Pfc	2297064	D-2-12	5/13/69*	21	El Centro, California
Malnar, John M SgtMaj	528234	H&S-2-4	4/30/68*	41	Sawyerville, Illinois
Maloney, William R LtCol	052112	VMO-6	2/1/67	41	Fairfax, Virginia
Manac, Don Pfc	2599169	H-2-7	8/20/70*	20	Fargo, North Dakota
Manfra, Howard T Jr Sgt	2096513	M-3-5	9/4/67	21	Philadelphia, Pennsylvania
Mangrum, Richard G LCpl	2213707	L-3-4	1/7/68*	19	Lynn Haven, Florida

Name	Service #	Unit	Date	Age	Hometown
Mangual, Jose M Pfc	2463614F-2-7		1/14/70*	20	Ponce, Puerto Rico
Mara, Donald E Sgt	1653151	M-3-3	11/11/68	28	Candia, New Hampshire
Marcantel, William E Capt	083491	Advisor	6/22/66	25	Iowa, Louisiana
Marcombe, Stephen G LCpl	2303802	M-3-5	6/2/67*	20	Westwego, Louisiana
Marengo, Anthony H SSgt	1692871	F-2-9	1/26/69	28	Richmond Hills, New York
Maresco, Richard E Capt	082456	K-3-5	1/26/67	28	Suffern, New York
Marino, Jack Jr SSgt	991615	A-1-Amp	8/18/65	41	Mobile, Alabama
Marinos, Nicholas SSgt	1490955	F-2-5	8/6/68	32	Vafe, Greece
Markham, Joel M LCpl	2436595	H-2-4	12/11/68	20	Boteturt, Virginia
Marks, David C Sgt	2044624	B-1-5	6/9/69	28	Venice, California
Marks, David E Capt	083493	F-2-1	2/28/68	30	Moscow, Idaho
Marsden, Robert P Cpl	2404784	K-3-5	1/20/71*	19	Randolph, Massachusetts
Marsh, James W LtCol	050363	CO-3-3	3/6/68	40	New Mexico
Marsh, William C SSgt	1637431	A-1-Tanks	2/24/68*	27	Amarillo, Texas
Marshall, Billy G SSgt	655042	A-1-4	2/26/66	37	Fillmore, California
Martin, Emerson Pfc	2475620	A-1-7	5/29/69*	21	Church Rock, New Mexico
Martin, Ervin P Jr 2dLt	0105180	D-1-5	8/29/68	20	Portland, Oregon
Martin, John A Maj	072618	VMA-242	10/27/67	31	Unknown
Martin, Justin M II 2dLt	0103097	F-2-3	8/15/68	23	Alexandria, Louisiana
Martin, Laurence A SSgt	517200	D-1-3	6/29/65	37	San Rafael, California
Martin, Robert J Maj	073452	Advisor	3/25/71	35	New York, New York
Martin, Rodney J Pfc	2410709	D-1-5	10/26/68	20	Chicago, Illinois
Martin, W L 1stSgt	298134	F-2-7	12/30/66	41	Dallas, Texas
Martinez, Donacano F GySgt	1262320	G-2-26	5/16/67	34	Dallas, Texas
Martinez, Jorge Pfc	2341584	M-3-7	7/28/68*	19	San Antonio, Texas
Martinez, Manuel O Maj	067305	HMM-161	9/18/65	32	Corpus, Christi, Texas
Martinez, Robert LCpl	2083073	C-1-Recon	6/16/66	21	Garden City, Kansas
Marvel, Jerry W LtCol	070296	POW	2/68-8/69	35	Evansville, Indiana
Mascarenas, Alcadio N LCpl	2129336	C-1-Recon	6/16/66*	22	Sapello, New Mexico
Mastrion, Robert J Capt	088497	G-2-4	3/30/68	28	Brooklyn, New York
Mattern, Robert S Sgt	2131996	L-3-9	6/1/67*	25	Worcester, Massachusetts
Matheny, Roy W LCpl	2356155	L-3-4	6/15/68	20	Little Rock, Arkansas
Mathews, Robert A Capt	080639	H&MS-16	3/14/67	29	Lake Villa, Illinois
Mathews, Vincent M LCpl	2207094	B-1-4	10/27/67	19	New York, New York
Mathis, Jack D Sgt	2013119	C-3-Recon	5/23/66	23	Brandon, Florida
Mattern, Larry A Pvt	2103373	D-1-1	7/23/66	19	Saegertown, Pennsylvania
Matthews, Gordon B 1stSgt	097328	B-1-1	2/7/68*	36	Bloomfield, Iowa
Matthews, Harry E Pfc	2321080	I-3-3	2/16/68*	19	New York, New York
Mattie, John G Jr Sgt	1982986	H-2-9	4/16/68	23	Swoyersville, Pennsylvania
Mattingly, Robert E 2dLt	091273	H-3-Tank	8/5/66	24	Preston, Maryland
Matye, Clemence T Cpl	2297648	B-1-Tanks	11/12/68	20	Granada Hills, California
Maxim, Robert J Cpl	2141186	B-1-9	8/24/67	20	Albany, New York
Maxson, Leonard W Pfc	2294681	1-SPB	3/26/68	21	Baltimore, Maryland
May, Dennis P LCpl	2149714	M-3-1	1/10/67	21	Wentzville, Missouri
May, James A Sgt	2090588	B-1-27	4/13/68	23	Mud, West Virginia
Mayer, Brian S Cpl	2282884	H&S-1-5	2/22/68	20	New York, New York
Mayfield, Leonard Pfc	2372525	E-2-5	4/25/68	20	Jackson, Mississippi
McAllister, Robert A Pfc	2359816	Mts-1-13	8/18/68*	20	Tinley Park, Illinois
McCall, Gerald A SSgt	1592696	H-2-7	7/13/68*	28	Atlantic City, New Jersey
McCallum, Daniel P LCpl	2220639	D-1-1	9/24/66	18	Coventry, Rhode Island
McCann, Patrick J Capt	092375	Advisor	6/2/70	26	Philadelphia, Pennsylvania
McCann, Roger D Pfc	2454735	I-3-5	9/11/69	19	Hammondsport, New York
McCardell, Raymond J Cpl	2148831	C-1-Recon	2/29/68	22	Boston, Massachusetts
McCart, Daniel P 2dLt	093280	K-3-4	3/17/67	23	New Castle, Pennsylvania
McCarter, James W Jr Capt	085448	G-2-4	10/14/67*	27	New Orleans, Louisiana
McCarthy, Brian S 1stLt	0103750	SPRC	2/23/69	25	Detroit, Michigan
McCarthy, Peter R Capt	080028	Advisor	7/31/67	28	Providence, Rhode Island
McCarty, James A Capt	069975	Advisor	3/10/65	32	San Diego, California
McCauley, Bertram W Maj	058997	HMM-263	9/17/66	37	New Albany, Indiana
McClain, David L Pfc	2080858	H-2-9	12/18/65	19	Jacksonville, Florida

Name	Service #	Unit	Date	Age	Hometown
McClain, William D Sgt	2024794	K-3-4	5/16/68*	22	Waco, Texas
McClary, Patrick C III 2dLt	0102130	A-1-Recon	3/5/68	26	Pawleys Island, SC
McClintock, Ted E SgtMaj	338430	H&S-3	1/15/69*	47	Seattle, Washington
McCluey, Harry H III Pfc	2497158	B-1-26	9/10/69	20	Seattle, Washington
McConnell, Paul R 1stLt	0101651	E-2-5	4/22/68	23	Portsmouth, New Hampshire
McCord, William C Cpl	2240989	B-1-7	8/10/68	22	Pulaski, Arkansas
McCormick, Charles M Cpl	2290978	H-2-5	9/29/68	20	Ashland, Kentucky
McCormick, James P LCpl	2289865	G-2-26	4/7/68	21	Boston, Massachusetts
McCourt, Edward F 2dLt	097355	I-3-3	11/28/66	30	Long Island, New York
McCoy, James H SSgt	1495782	F-2-5	2/4/68	30	Pittsburgh, Pennsylvania
McCracken, Jack H Capt	092037	HMM-165	6/3/67	25	Austin, Texas
McDonald, Alan V Cpl	2366683	H-2-5	2/5/68	20	Jacksonville, Florida
McDonald, Guy T Sgt	1998534	H-2-5	1/30/68	24	Fitzgerald, Georgia
McDonnell, John M Pfc	2322914	A-1-9	5/13/67	20	Chicago, Illinois
McElroy, James R Jr Capt	080235	M-3-5	5/13/67	29	Birmingham, Alabama
McElroy, James R Jr Maj	080235	Advisor	5/14/71	33	Birmingham, Alabama
McGee, Stephen D LCpl	2231850	M-3-26	5/9/68*	18	Mishawaka, Indiana
McGinley, Gerald G LCpl	2424469	A-3-Recon	12/5/68	20	Concord, California
McGrath, Thomas H LCpl	2381139	G-3-12	2/25/69*	18	Homewood, Illinois
McInturff, David L 1stLt	091629	D-1-5	6/2/67	25	Pettit, Texas
McKee, Thomas E LCpl	2066048	A-1-Tanks	7/6/67*	20	Palm Springs, California
McKeon, Joseph T Jr 2dLt	094899	B-1-3	5/10/67*	23	Chicago, Illinois
McKercher, Paul R LCpl	2358242	M-3-1	3/1/68	19	Johnson City, New York
McKim, Edward A Sgt	2058209	D-3-MPB	1/31/68*	23	Dallas, Texas
McKinny, James O Pfc	2055426	C-1-Recon	6/16/66*	18	Monroe, Louisiana
McLaughlin, Thomas H LCpl	2182042	C-1-26	9/21/66	20	Lowell, Massachusetts
McLean, Ronald W 1stLt	0105587	A-3-Recon	6/8/69*	24	Beverly Hills, California
McMahon, Daniel K Jr 1stLt	087835	B-1-4	3/20/66	24	Long Island, New York
McMahon, Daniel K Jr 1stLt	087835	D-1-4	9/19/66	24	Long Island, New York
McMillin, Donnell D Sgt	1813597	F-2-7	3/4/66 *	25	Mena, Arkansas
McMullin, Robert LCpl	2447184	L-3-5	5/12/69	19	Camp Lejeune, North Carolina
McMurray, Daniel D 2dLt	0102572	B-1-7	8/10/68	22	Toledo, Ohio
McNalley, Ronald Sgt	2149765	C-1-7	1/31/68	22	St. Louis, Missouri
McNally, Paul A LCpl	2121897	D-1-5	5/12/67	20	Philadelphia, Pennsylvania
McQuerry, Thomas O Sgt	2151702	A-1-4	3/6/68	22	Los Angeles, California
McQuown, Max LtCol	053321	CO-1-3	2/2/68	43	St. Helena, South Carolina
McVey, Lavoy D Capt	091424	C-1-Recon	3/2/70 *	32	Lamar, Colorado
McWilliams, James P Jr Maj	073146	H&S-1-9	9/5/69	34	Philadelphia, Pennsylvania
Mead, James M Maj	073315	1-M G3	4/22/68	32	Malden, Massachusetts
Meade, Joseph L Pfc	2450534	M-3-26	1/25/69*	19	Kingsport, Tennessee
Meadows, Charles L Maj	083834	G-2-5	2/6/68	29	Beaverton, Oregon
Mees, William L SSgt	2077837	G-2-1	5/5/69	25	Butte, Montana
Meester, Harold W Pfc	2168010	G-2-9	6/14/66	20	Rock Rapids, Iowa
Meier, Terrance L SSgt	1999759	M-3-3	4/30/67*	22	Portland, Oregon
Meilinger, John J Cpl	2078241	I-3-1	11/12/66	22	Bethlehem, Pennsylvania
Melim, Patrick R F K Sgt	1854914	C-1-Recon	5/28/67	26	Honolulu, Hawaii
Melson, Wilton E Jr Cpl	2378995	D-3-Recon	5/15/69	20	Birmingham, Alabama
Melton, Walter W Jr GySgt	1434949	G-2-3	6/1/69	36	Gulfport, Mississippi
Melton, William R 2dLt	0103896	F-2-5	9/30/68	31	Marysville, California
Melvin, James L Cpl	2182214	A-1-1	11/26/67*	20	East Boxford, Mass
Menagh, Philip S 2dLt	0103686	F-2-7	3/22/68	23	Toledo, Ohio
Mendenhall, William E Sgt	1533344	VMA-214	10/28/65	27	Kaneohe Bay, Hawaii
Menzies, Alexander J N LCpl	2133773	B-1-3	4/2/66*	20	Walworth, New York
Meskan, Donald J Maj	076109	HMM-262	9/15/69	34	Minneapolis, Minnesota
Messer, James M 2dLt	0108392	L-3-3	9/13/69	23	Morgantown, West Virginia
Meydag, Richard H Maj	083832	VMO-6	4/24/69	28	Tulsa, Oklahoma
Meyer, Ronald W 2dLt	091945	C-1-5	6/16/66*	23	Dubuque, Iowa
Michalowski, Raymond J LCpl	2323809	K-3-5	2/2/68*	18	Chicago, Illinois
Michalski, James Pfc	2471622	I-3-3	6/4/69*	21	La Porte, Indiana
Mickelson, Dennis E LCpl	2394453	A-3-Recon	12/5/68*	19	Eugene, Oregon

Name	Service#	Unit	Date	Age	Hometown
Mihalovich, John M Cpl	2108819	G-2-27	3/15/68*	22	Milwaukee, Wisconsin
Mikaele, Puni Sgt	2376013	I-3-9	6/27/69	23	American Samoa
Miller, Allen P Sgt	1985062	D-3-Recon	12/19/66 *	24	Midwest City, Oklahoma
Miller, Dana L Cpl	2263958	D-2-12	5/13/69*	20	Akron, Ohio
Miller, Donald E P Maj	076614	VMO-2	4/23/69	34	Williams, California
Miller, Edwin D Sgt	1805091	Advisor	2/17/71	31	Fenton, Missouri
Miller, Floyd E Cpl	2190910	F-2-1	4/21/67	20	Albany, New York
Miller, Fredrick W Cpl	2012460	M-3-7	4/21/66*	22	Millersburg, Ohio
Miller, James E LCpl	2557592	F-2-5	10/27/70*	18	Lancaster, Pennsylvania
Miller, Richard E 1stLt	0104708	1-FRecon	4/12/69	23	San Francisco, California
Miller, Russell P Pfc	2322957	D-1-3	5/10/67*	19	South Bend, Indiana
Miller, Thomas S 1stLt	0108510	A-1-7	6/11/70	26	New Kinsington, Penn
Miller, William LCpl	2396089	M-3-4	3/14/69	20	Buffalo, New York
Miller, Willie E LCpl	2343048	CAP-1-3-3	8/2/68	19	Cincinnati, Ohio
Mills, Michael T Pfc	2215209	G-2-3	4/30/67	19	St. Louis, Missouri
Millsap, William C Pfc	2135609	F-2-1	2/28/66	20	San Diego, California
Minehart, Russell E Sgt	1522495	B-1-9	7/12/65	29	Callesy, Pennsylvania
Mitchell, Curtis Pfc	2300587	B-1-5	9/4/67*	19	Jacksonville, Florida
Mitchell, Mack E Capt	057303	HMM-262	11/27/68	42	Mishawaka, Indiana
Mitchell, Paul H Jr LCpl	2103805	M-3-4	10/5/66*	20	Mentor, Ohio
Mitchell, Thomas E Jr Sgt	2247666	D-1-5	5/24/68	22	Sacramento, California
Mixon, Michael J Sgt	2062439	B-1-5	9/4/67	23	Plain Dealing, Louisiana
Mixon, William H Cpl	1906658	E-2-3	8/22/65	23	New Orleans, Louisiana
Moguel, Albert Pfc	2384561	B-1-26	3/30/68	19	Houston, Texas
Mollett, Chester A Sgt	2291098	F-2-3	8/7/69*	26	Peytona, West Virginia
Monday, Alvin LCpl	2156748	K-3-9	9/9/68*	23	Eunice, Louisiana
Mondragon, James W Pvt	2344962	A-1-Amp	3/7/68	20	Houston, Texas
Montague, Paul J Capt	081197	POW	3/29/68	33	Anthony, Kansas
Montague, Paul J Maj	081197	POW	6/70-3/73	38	Anthony, Kansas
Montez, Frank J Cpl	2375978	A-1-Recon	9/20/69*	19	Salinas, California
Montjoy, Norris D SSgt	1528788	F-2-1	5/19/68	31	Raleigh, North Carolina
Moody, Jimmy D Pfc	2273558	H&S-3-3	3/24/67*	20	Kennett, Missouri
Mooney, Clifford D 1stSgt	1045500	C-3-Recon	3/16/67	36	Shawnee, Oklahoma
Moore, Brian D Capt	072242	E-2-4	2/9/66	30	Carlisle, Pennsylvania
Moore, Dan E Pfc	2127977	H-3-12	8/26/66	20	San Diego, California
Moore, Kenneth W 2dLt	092890	I-3-5	5/9/67	23	Annapolis, Maryland
Moore, Phillip A Pvt	2269400	A-1-7	4/22/67*	20	Marengo, Ohio
Moore, Willard T LCpl	2016691	A-1-11	4/16/66	22	Tampa, Florida
Moore, William R LCpl	2060008	1-FRecon	12/16/65*	20	Richmond, California
Moores, Kenneth F Pfc	2422372	B-1-4	10/8/68*	19	Waltham, Massachusetts
Morey, Alfred W Sgt	2321746	HMM-463	6/28/70	21	Somers, New York
Morgan, Dennis J LCpl	2292395	M-3-7	9/16/68	20	New York, New York
Morgan, Edward E LCpl	2409040	D-1-9	2/18/69	20	Jackson, Mississippi
Morgan, Henry L Sgt	2034970	E-2-3	12/28/67*	22	Benson, North Carolina
Morgan, Michael R LCpl	2243845	M-3-3	4/30/67*	21	Bethpage, New York
Morgan, Wilkes R 1stLt	0102885	MAG-18	2/23/69	21	St. Helena, California
Moriarty, William S Capt	077897	Advisor	7/30-31/67	30	Winter Park, Florida
Moro, Michael E Pfc	2171692	A-1-4	5/8/67	20	Albany, New York
Morris, Daniel M Cpl	2005641	G-2-5	2/7/68	24	Los Angeles, California
Morris, Phillip W Sgt	2257704	F-2-4	9/21/67	21	Norfolk, Virginia
Morris, Richard A Cpl	2172310	D-1-5	2/22/68	22	Charleston, South Carolina
Morris, Ronald D SSgt	1954054	K-3-7	8/23/68	26	Evansville, North Carolina
Morrisey, John T Jr Cpl	2169854	3-FRecon	8/17/67	25	Eastchester, New York
Morrison, Gary L Pfc	2474926	G-2-4	2/21/69	20	Upper Marlboro, Maryland
Morrison, Lonnie W Pfc	2338568	B-1-26	3/30/68	19	Raleigh, North Carolina
Morrison, Michael K 1stLt	095669	HMM-161	6/9/70	27	Pasadena, Maryland
Morrison, Robert S 1stLt	087345	H-2-4	8/18/65	23	Lake Forest, Illinois
Morrison, Ronald C 1stLt	0101324	A-1-Tank	2/15/68	23	Quantico, Virginia
Morrow, Alfred J Jr LCpl	2171623	HMM-161	3/23/69	22	Albany, New York
Morrow, Harold E GySgt	1136878	M-3-26	7/21/67*	36	Valencia, Pennsylvania

Name	Service #	Unit	Date	Age	Hometown
Moser, Keith M II Pfc	2250629	F-2-5	6/2/67*	20	Lowell, Michigan
Moss, Richard A GySgt	1486097	E-2-3	5/28/68	30	Cincinnati, Ohio
Mower, Kenneth T Jr Sgt	1938315	F-2-7	12/20/66	25	Virginia Beach, Virginia
Moy, William K III Cpl	2219050	M-3-5	9/4/67	21	Philadelphia, Pennsylvania
Moyle, Wesley A Pfc	2336949	B-1-1	12/15/67*	20	Duquesne, Pennsylvania
Much, Gary W LCpl	2082695	L-3-27	5/18/68	20	Marlon, Wisconsin
Mueller, Charles E LtCol	059068	CO-2-7	5/4/68	37	Lyndhurst, New Jersey
Mugler, Charles R LCpl	2204354	C-1-26	9/21/66	20	Richmond, Virginia
Mulhearn, James J Jr 1stLt	0107476	CAC-1-3	9/13/69	24	Los Angeles, California
Mullen, James R 2dLt	0101326	I-3-7	11/19/67	24	Wichita, Kansas
Mullen, Richard W 2dLt	093384	K-3-4	9/28/66	23	Wichita, Kansas
Mullins, James R LCpl	2128488	B-1-26	5/29/68*	20	Franklin Furnace, Ohio
Mulvihill, Daniel K LCpl	2102047	C-1-Recon	6/16/66	19	Chicago, Illinois
Munson, Stephen A 2dLt	0102162	L-3-3	3/3/68	25	New York, New York
Munter, Weldon R Capt	059909	VMO-3	2/23/68	27	Portland, Oregon
Muraca, Patrick J LCpl	2418335	H-2-9	2/17/69*	20	Dalton, Massachusetts
Murphy, Dennis G Pfc	2467070	1-FRecon	6/18/69*	18	Copiague, New York
Murphy, Edmond J Maj	066064	HMM-363	9/4/67	36	Pinehurst, North Carolina
Murphy, Edward T Cpl	2030755	B-1-5	5/21/67*	23	Levittown, New York
Murphy, James K 1stLt	0104702	L-3-9	3/18/69	24	Brooklyn, New York
Murphy, Kenneth R Sgt	0000000	0-0-0	2/3/68	00	Unknown
Murphy, Kevin J 1stLt	0105995	G-2-5	5/15/69	23	Beverly, Massachusetts
Murphy, Michael A Sgt	2058956	CAP-1-3-9	1/3/69	25	Detroit, Michigan
Murphy, Thomas E SSgt	1830087	A-1-1	4/20/68	27	Unknown
Murphy, Thomas H Jr Pfc	2393162	L-3-5	7/31/68	20	Providence, Rhode Island
Murphy, Walter M Maj	075028	H&S-1-1	1/31/68*	31	New York, New York
Murray, David R Cpl	2140769	C-1-9	2/18/69	21	Greenville, South Carolina
Murray, Freddie L LCpl	2065527	A-3-Recon	7/12/65	18	Weirsdale, Florida
Murray, Grover 2dLt	097489	D-1-Recon	1/25/67	35	Boswell, Oklahoma
Mutschler, John L Pfc	2484069	H&S-2-3	8/19/69*	20	Clarksville, Michigan
Myatt, James M 1stLt	087245	A-1-4	3/20/66	25	Monterey, California
Myers, Dale S Jr LCpl	2349067	M-3-3	3/6/68	19	Los Angeles, California
Myers, Donald F Sgt	1277825	C-1-9	2/13/69	34	Indianapolis, Indiana
Myers, Donald F Sgt	1277825	C-1-9	2/22/69	34	Indianapolis, Indiana
Myers, Leo B 1stLt	0102529	H-2-5	2/11/68	24	Alba, Oklahoma
Nachtrieb, Mallery O LCpl	2335562	G-2-7	2/23/69	20	Dallas, Texas
Nappi, Patrick 2dLt	0106382	M-3-5	8/3/69	23	Syracuse, New York
Naugle, Russell W LCpl	2336710	C-1-7	2/15/68*	18	Republic, Pennsylvania
Navadel, George D Capt	075389	M-3-9	6/1/67	30	Buffalo, New York
Neal, Earnest LCpl	2334226	D-1-9	4/29/68	19	Jacksonville, Florida
Neal, Hiahwhanah R Cpl	2168567	M-3-27	5/24/68	20	Talihina, Oklahoma
Neal, Richard I 1stLt	092178	I-3-9	3/31/67	24	Hull, Massachusetts
Neal, Richard I Capt	092178	Advisor	5/22/70	27	Hull, Massachusetts
Neblett, Lynell LCpl	2351619	F-2-7	9/29/67*	20	Blackstone, Virginia
Negron, William P Capt	082954	C-1-4	6/5/68	32	Edison, New Jersey
Nelson, Gregory H 1stLt	0102415	VMO-6	6/17/69	21	Long Island, New York
Nelson, Jacob GySgt	1832510	G-2-4	9/21/67	27	Crowley, Louisiana
Nelson, Jacob GySgt	1832510	G-2-4	10/14/67	27	Crowley, Louisiana
Nesmith, Joseph Q Jr Capt	089927	HMM-165	10/24/67	25	Santa Ana, California
Neu, Melvin P SSgt	1871047	L-3-3	3/2/67	27	Minneapolis, Minnesota
Neuss, William H Capt	092447	C-1-3	10/11/67*	24	Yaphank, New York
Nevarez-Oliveras, Jose A Sgt	2192274	G-2-4	7/2/69	23	Fort Brooke, Puerto Rico
Neville, George G Jr Cpl	2098958	A-3-Recon	9/23/66	20	Rockville, Maryland
Newby, Thomas J Pfc	2293178	CAP-2-1-5	2/23/69	19	Fort Hamilton, New York
Newman, Michael E 2dLt	0107486	G-2-5	3/28/69	21	Chicago, Illinois
Newton, Leonard L Pfc	2387353	I-3-26	1/19/68*	19	Stockton, California
Nichols, Daniel C 1stLt	0104295	D-2-12	2/17/69	25	Westfield, New Jersey
Nichols, Daniel C 1stLt	0104295	D-2-12	5/13/69	26	Westfield, New Jersey
Nickerson, Michael K 1stLt	0100556	HMM-364	4/14/69*	25	Indianapolis, Indiana
Niedopytalski, John J Cpl	2358303	D-2-11	3/19/69	20	Syracuse, New York

Name	Service #	Unit	Date	Age	Hometown
Nielsen, Howard B Capt	0104841	M-3-5	9/11/68	21	San Francisco, California
Niesen, Paul W LtCol	051752	HMM-161	7/17/68	42	Racine, Wisconsin
Niotis, John D Capt	081270	L-3-5	2/22/68	30	Chicago, Illinois
Noakes, David L GySgt	1932904	F-2-5	11/5/70	28	Portland, Oregon
Nolan, James M Capt	073360	F-2-7	12/18/65	30	Swampscott, Massachusetts
Norman, William LCpl	214093	C-1-Recon	6/16/66	19	Sedona, Arizona
Normand, Paul D Pfc	2438799	E-2-12	6/5/69	19	Milwaukee, Wisconsin
Norris, Gleason GySgt	1127120	M-3-26	9/10/67	35	Sells, Arizona
North, Donald R Jr Cpl	2226598	D-1-3	11/11/67	20	Southgate, Michigan
North, Oliver L 2dLt	0106162	K-3-3	5/25/68	26	Philmont, New York
Northington, William C LCpl	2407330	A-1-9	2/22/69	20	Prattville, Alabama
Northrop, Thomas E Cpl	2332091	B-1-1	5/31/68	20	Waverly, New York
Norton, Chandois M LCpl	2335342	HMM-262	5/18/68	20	Dallas, Texas
Norton, John E Cpl	2593738	E-2-1	12/29/70*	19	Fort Oglethorpe, Georgia
Novotny, Stanley J SSgt	1424826	HMM-163	3/31/65	38	Lonsdale, Minnesota
Nowak, Laurance S 1stLt	092686	MAG-16	10/13/67	23	Los Angeles, California
Noyes, David W 2dLt	0102079	M-3-7	11/2/67	23	Birmingham, Michigan
Nulty, Thomas G 1stLt	0104424	K-3-26	11/24/68	26	New York, New York
Nunez, Larry B Cpl	2147189	M-3-5	9/4/67	20	Birmingham, Alabama
Nye, Lindsey H Cpl	2148825	E-2-1	5/14/67	20	Carver, Massachusetts
Nyulassy, Arnold C 2dLt	0109102	K-3-7	8/28/69	23	Saddlebrook, New Jersey
O'Bannon, Robert III Sgt	2002103	E-2-4	3/18/68*	22	San Bernardino, California
O'Brien, Joseph J Maj	076414	G-2-3	7/22/67	34	Minneapolis, Minnesota
O'Connell, George J Jr 1stLt	090982	HMM-163	3/5/67	25	Highland Park, Illinois
O'Connell, John P 2dLt	0106980	E-2-4	8/12/69	22	Redondo Beach, California
O'Connor, Brian R 2dLt	094002	H-2-1	1/29/67*	27	Andover, Massachusetts
O'Donnell, John W Maj	072969	Advisor	8/6-7/66	31	Annapolis, Maryland
Ogden, Howard Jr LCpl	2312138	G-2-7	10/18/67*	19	Omaha, Nebraska
Ohanesian, Victor LtCol	050708	CO-2-3	3/1/67*	40	New York, New York
Okada, Danny G Cpl	2214507	HMM-164	11/5/68	20	Los Angeles, California
O'Leary, Patrick C LCpl	2348223	VMA-211	3/21/69	20	Boise, Idaho
Olenski, Paul F Pfc	2181660	C-1-Recon	7/26/68	21	Chicago, Illinois
Olsen, Spencer F SSgt	1951058	E-2-9	3/16/67	24	Fowler, Colorado
Olson, Bruce G Sgt	1990196	E-2-5	5/15/69	25	Hartford, Connecticut
Orlett, Paul J SSgt	1458705	K-3-5	5/12/67	32	Portsmouth, Ohio
Orsburn, Lyndell M Capt	076293	K-3-3	2/18/66	33	Newport, Arkansas
Ortiz, Jose A LCpl	2466149	I-3-26	8/24/69*	21	Corpus Christi, Texas
Ortiz, Melecio Cpl	2164601	I-3-5	5/13/67*	22	Crystal City, Texas
Osgood, William H Capt	079337	C-1-3	1/19/68	29	Stockton, California
O'Sullivan, Andrew F 1stLt	0105219	FSC-3-12	3/20/69	24	New York, New York
Ott, Theodore A Sgt	2319808	1-FRecon	6/18/69	21	Monroe, Wisconsin
Otto, Charles E 1stSt	650299	E-2-4	3/13/68	40	Union Bridge, Maryland
Owens, Bennett H Jr Sgt	1499308	M-3-9	3/21/66	26	Jacksonville, Florida
Owens, Mackubin T Jr 2dLt	0105300	B-1-4	5/25/69	23	Santa Barbara, California
Pace, David C Pfc	2229358	G-2-26	5/16/67	19	Erie, Pennsylvania
Pace, Simone J 2dLt	091949	A-1-9	5/21/66	23	Teaneck, New Jersey
Pacello, Francis D Capt	082483	M-3-5	8/26/68	28	Wilmington, Delaware
Pacheco, Eugenio Cpl	2256528	F-2-4	9/21/67	20	San Antonio, Texas
Pacholke, Norman L SSgt	1648557	B-1-7	10/10/68	28	Manilla, Iowa
Padilla, Jose M GySgt	1284395	M-3-9	4/30/68	32	Los Angeles, California
Padilla, Rodney J A LCpl	2236044	L-3-7	2/7/68	20	Albuquerque, New Mexico
Page, Douglas B Capt	090578	FAC-1-4	3/20/69	30	Susanville, California
Page, James E Capt	068990	F-2-1	12/10/65	37	Tulsa, Oklahoma
Palacios, Benjamin Jr SSgt	1988589	F-2-5	5/22/68	25	San Francisco, California
Paladino, Robert J LCpl	2284579	L-3-4	2/26/69	20	New York, New York
Palembas, Richard A Sgt	2121528	D-1-9	6/29/67	23	Butler, Pennsylvania
Palmason, Stephen T 1stLt	095621	VMO-6	11/25/68	25	Salem, Oregon
Palmer, David E Cpl	2516133	A-1-Recon	6/11/70	20	Unknown
Palmer, Leonard J Jr LCpl	2447708	K-3-7	3/11/69	19	Berkeley, Illinois
Palumbo, Charles F Cpl	2213161	H&S-3-26	9/10/67	22	Houston, Texas

Name	Service #	Unit	Date	Age	Hometown
Pannell, Horace L Pfc	1998756	C-1-3	9/17/65	20	Warrior, Alabama
Parker, Milton SSgt	1548859	A-1-1	2/5/68	29	Buena Vista, Georgia
Parker, Paul D II 1stLt	0102654	HMM-263	1/31/70	24	Austin, Texas
Parks, Gordon E LCpl	2371272	CAP 2-7-6	9/11/69	19	Nashville, Tennessee
Parlett, Phillip G Cpl	2385188	M-3-4	3/14/69	21	New Orleans, Louisiana
Parmelee, Bruce C Pfc	2249008	I-3-1	4/21/67*	20	Reading, Massachusetts
Parr, William T LCpl	2358150	C-1-9	4/16/68	20	Syracuse, New York
Parrish, Daniel I Cpl	2295719	CAP-L-5	3/16/68	20	San Francisco, California
Parrish, Joseph P Cpl	2397420	1-FRecon	4/25/69	20	Lake City, Florida
Parrish, Robert E 2dLt	0101368	H&S-3-1	6/18/68	23	Quantico, Virginia
Parry, Dale T GySgt	1072827	HMM-165	3/29/68	36	Ogden, Utah
Parsons, James R Maj	067088	HMM-364	2/4/68	36	Santa Ana, California
Parton, Floyd E LCpl	2388524	B-1-4	10/8/68*	19	Asheville, North Carolina
Parton, John E Cpl	2502260	Advisor	6/30/72	24	Douglas, Arizona
Pasieka, Stanley J 2dLt	093348	K-3-1	1/15/67	23	Middletown, Connecticut
Pate, Gerald S Maj	060342	HMM-261	3/15/66	33	Murfreesboro, Tennessee
Patrinos, Charles 1stSgt	410434	E-2-3	5/1/67	40	Salem, Massachusetts
Patten, David D Sgt	1977452	B-1-3	4/2/66	22	Des Moines, Iowa
Patterson, Eldridge Jr Cpl	2159667	L-3-26	4/14/68	21	Dwarf, Kentucky
Patterson, Robert L Pfc	2006885	B-1-12	2/24/66	22	Memphis, Tennessee
Patton, Donald R Cpl	2139361	D-3-Recon	12/19/66	19	Prestonsberg, Kentucky
Patton, Jack L Cpl	2316260	M-3-4	1/25/68	22	Marion, Indiana
Patton, Robert F CWO-3	064572	HMM-364	6/5/64	36	San Antonio, Texas
Paul, James R Sgt	2043910	H-2-4	2/5/67*	20	Huntington, Indiana
Paulin, John T LCpl	2487685	C-1-26	6/7/69*	20	Owensboro, Kentucky
Pavey, Chester R 1stSgt	642952	I-3-3	3/25/67*	36	Anderson, Indiana
Payne, Darnell M Pfc	2474947	E-2-9	4/23/69*	20	Falls Church, Virginia
Payne, Jack S SSgt	1940198	I-3-4	7/1/68	26	Moon, Georgia
Pearce, Donald L SSgt	1385129	K-3-3	6/17/69	34	Jacksonville, Florida
Pearson, Millard L LCpl	1954220	C-3-Eng	4/13/66	23	Indianapolis, Indiana
Peatross, Oscar F Col	07196	CO-7	8/24/65	49	Frogmore, South Carolina
Peavler, Lawrence A Cpl	2225293	B-1-1	5/13/67	21	Coral Gables, Florida
Peck, Garrette W III LCpl	2180320	H-2-9	8/2/66	24	Washington, DC
Peko, Lolesio Pfc	2042244	B-1-5	8/10/66	22	Albany, New York
Pembleton, Elias S SSgt	2036330	E-2-3	6/12/69	26	Buffalo, New York
Pennington, Kenneth E 1stSgt	0102204	VMO-2	4/21/69*	26	Durham, North Carolina
Perez, Ernesto Pfc	2370282	B-1-1	12/15/67*	19	Rake, Iowa
Perez, Jesus R Pfc	2389862	F-2-5	6/23/68*	21	Kingsville, Texas
Perez-Padin, Juan R Cpl	2342741	E-2-7	8/26/69*	21	Quebradillas, Puerto Rico
Perino, Anthony Cpl	2205594	H-2-9	7/29/67	20	New York, New York
Perinotto, Ernest D Pfc	2438154	B-1-7	11/17/68*	20	Allentown, Pennsylvania
Perriquey, Charles D Jr 1stLt	0100588	HMM-367	12/8/69	24	San Gabriel, California
Perry, Henry SSgt	1490605	G-2-7	2/23/69	32	Albuquerque, New Mexico
Perry, Larry E 1stLt	0102206	C-1-1	6/5/68	24	Sherman, Texas
Perry, Richard W Sgt	2035396	E-2-7	9/19/66*	24	Marion, Arkansas
Perryman, James M Capt	067837	VMO-6	6/16/69	33	Washington, DC
Persky, Donald N 1stLt	0106640	HMM-463	9/11/70	22	Delray Beach, Florida
Person, Barnett G GySgt	616367	A-3-Tanks	5/8/67	40	Birmingham, Alabama
Persons, Henry H 1stLt	0101378	Wpns-1-5	2/1/68*	24	Fort Wayne, Indiana
Peters, Anthony J Sgt	1948609	K-3-9	7/4/66	22	Oxford, Pennsylvania
Peters, James E Pfc	2457430	D-1-5	9/11/68	19	Chicago, Illinois
Peters, William J Jr 2dLt	0107002	1-FRecon	4/12/69	24	San Francisco, California
Peters, William L Jr 1stLt	0102208	HMM-165	6/21/69*	26	Fort Dodge, Kansas
Petersen, John W III Cpl	2295074	G-2-1	3/4/68	19	Davenport, Iowa
Peterson, Dennis J Cpl	2311326	HMM-364	5/7/69	21	Detroit, Michigan
Peterson, Michael I 1stLt	0104422	B-1-Amph	2/22/69	24	Nashville, Tennessee
Peterson, Robert V LCpl	2242658	I-3-5	1/30/68*	20	Canton, Mississippi
Petit, Matthew M LCpl	2276696	D-1-7	12/20/67	20	Chicago, Illinois
Petrone, Louis G Jr LCpl	2355206	L-3-4	1/7/68*	22	Morristown, New Jersey
Petrunio, Darryl J Cpl	2337315	I-3-9	5/31/68	20	Pittsburgh, Pennsylvania

Name	ID	Unit	Date	Age	Hometown
Pettengill, Harold D Maj	085115	L-3-5	5/26/67	30	Greely, Colorado
Petteys, David M Capt	086421	HMM-265	6/30/67	25	San Diego, California
Pettigrew, William M III Capt	074829	VMO-6	12/4/65	31	Rome, Georgia
Pfeltz, Albert R III 1stLt	091971	B-3-Recon	8/24/67	22	Pittsburgh, Pennsylvania
Philips, Ralph F Jr Sgt	1423932	B-1-Amph	1/14/66	30	New York, New York
Phillips, Jack W Capt	077504	G-2-4	10/14/67*	32	Mission, Kansas
Phillips, Woodrow N LCpl	2555765	B-1-Recon	7/23/70	19	Pecos, Texas
Phipps, Daniel R 2dLt	0106559	G-2-26	5/17/67	22	San Diego, California
Phipps, Lanny W LCpl	2255099	A-1-7	12/7/68*	23	Rutherford, New Jersey
Piatt, Joseph M Jr Capt	090956	L-3-7	6/18/67	25	Harahan, Louisiana
Picciano, Terrance A LCpl	2366238	E-2-26	6/6/68*	19	Baraga, Michigan
Pierce, Daniel P LCpl	2162927	A-1-1	9/25/66	20	Pine Island, Minnesota
Pilson, Darwin R Sgt	2050793	B-1-5	8/10/66	24	Philadelphia, Pennsylvania
Pina, Ronald C Cpl	2466510	CAP-1-3-2	5/8/70	22	Boston, Massachusetts
Pineiro, Ismael LCpl	235598?	C-1-26	9/24/68	21	Newark, New Jersey
Pinkard, Robert E SSgt	1694158	H-2-5	2/5/68	28	New York, New York
Pinkerton, Robert N Jr Cpl	2142370	C-1-7	3/1/67	21	Brownsville, Texas
Piotrowski, Daniel J LCpl	2122153	B-1-3	4/2/66*	20	Jackson, Michigan
Pipes, Kenneth W Capt	081285	B-1-26	3/30/68	30	Fresno, California
Pitman, Charles H Maj	069426	HMM-265	4/8/67	31	Fort Worth, Texas
Planchon, Randall T II Cpl	2296855	M-3-4	5/16/68*	20	Long Beach, California
Platt, Jonas M BGen	06644	CG-TFD	12/20/65	45	Cranston, Rhode Island
Pless, Stephen W Capt	079156	VMO-6	6/2-4/67	27	Decatur, Georgia
Pohlman, Lyn W Cpl	2033359	A-1-1	10/30/65	20	Cincinnati, Ohio
Poniktera, Stanley F Jr LCpl	2374602	F-2-3	5/28/68*	18	Bethlehem, Pennsylvania
Pontius, John J Cpl	2145210	1-Eng	2/16/67	20	Philadelphia, Pennsylvania
Poole, Henry J Capt	094013	G-2-5	8/18/68	25	Portland, Oregon
Porrello, Richard D 2dLt	0105432	C-1-4	2/28/69	22	East Lyme, Connecticut
Porrello, Richard D 2dLt	0105432	C-1-4	3/1/69	22	East Lyme, Connecticut
Porter, Mervin B Col	023221	MAG-36	3/5/66	45	Bakersfield, California
Porter, Richard L 2dLt	087133	H&S-3-12	8/26/66	32	Seattle, Washington
Porterfield, Charles W Capt	082117	VMA-311	3/24/68*	28	Portland, Oregon
Porteur, Kraig M Cpl	2424993	HMM-236	2/7/70	20	Oakland, California
Post, Gary L Capt	086936	VMA-235	3/16/68	27	Augusta, Missouri
Potocki, Robert A Cpl	2288399	A-1-4	2/2/68	20	Hamtramck, Michigan
Powers, Mark F LCpl	2428638	K-3-9	2/14/69*	20	St. Petersburg, Florida
Powles, Thomas G Pfc	2140524	C-1-Recon	6/16/66	20	Vacaville, California
Prather, Ronald R JR Sgt	2177783	A-1-Tanks	2/7/68*	23	Cave Junction, Oregon
Prescott, Alexander F IV 1stLt	0101397	H-2-4	3/18/68	24	Temple, Texas
Prescott, Alexander F IV 1stLt	0101397	H-2-4	4/30/68	24	Temple, Texas
Prewitt, Robert C Capt	075471	B-1-7	3/4-5/66	30	Altus, Oklahoma
Price, Donald L Maj	082924	Advisor	4/16/72	34	Springerville, Arizona
Price, Thomas H Cpl	2042639	B-1-9	3/2/67	23	Detroit, Michigan
Prichard, John L Capt	081650	I-3-4	1/27/68*	29	Oklahoma City, Oklahoma
Prince, Phillip E Sgt	1531289	CAP-2-7-4	5/11/67	25	Delray Beach, Florida
Proctor, James P LCpl	2124481	H-2-9	7/29/67*	21	Tampa, Florida
Prommersberger, James E Sgt	2012082	C-3-Eng	4/16/66*	21	Youngstown, Ohio
Puliafico, Frederick SSgt	1352120	H-2-5	8/22/69	36	Schenectady, New York
Puller, Lewis B Jr 2dLt	2398118	G-2-1	10/11/68	22	Saluda, Virginia
Pupuhi, Rodney H Sgt	1486978	D-1-Recon	2/19/68	31	Honolulu, Hawaii
Purnell, Richard M 1stLt	086430	I-3-3	8/18/65	26	Berlin, Maryland
Purvis, Bernard G LCpl	2391296	A-1-3	1/29/69*	20	Norfolk, New York
Quinlan, Maurice J 1stLt	0100607	HMM-364	1/2/69	24	Medford, Massachusetts
Quinn, Francis X LtCol	055941	CO-3-7	2/28/69	41	Philadelphia, Pennsylvania
Quinter, Robert H Jr 1stLt	010221	HMM-161	6/11/70	23	Reading, Pennsylvania
Quiocho, Albert Jr. GySgt	1487071	Advisor	6/23/71	35	Honolulu, Hawaii
Rachon, Charles J Pfc	2320778	M-3-1	4/21/68*	22	White Plains, New York
Rackhaus, John P LCpl	2471130	A-1-7	5/29/69*	20	Marshall, Illinois
Radcliffe, Henry J M Capt	081002	A-1-9	7/2/67	29	Charleston, South Carolina
Radcliffe, Henry J M Capt	081002	B-1-9	2/8/68	30	Charleston, South Carolina

Name	Service #	Unit	Date	Age	Hometown
Radics, Emil J Col	021026	CO-1	2/12/67	43	Hightstown, New Jersey
Radish, Danny L Sgt	2454292	HMM-262	2/12/70	20	Eureka, Montana
Ragsdale, Gary W LCpl	2562170	E-1-Recon	6/14/70*	19	Kernan, California
Rainey, Larry L Pfc	2494868	K-3-9	2/27/69*	18	Heyworth, Illinois
Rait, Donald M Jr 1stLt	0102223	M-3-4	5/16/68	23	Fullerton, California
Ralston, Arthur N LCpl	2427660	G-2-1	5/6/69	19	Kansas City, Missouri
Ralston, David J 2dLt	0102823	C-3-Tanks	4/19/68	23	Gary, Indiana
Ramirez, Efrain A LCpl	2075114	I-3-7	3/23/66	21	Bronx, New York
Ramos, Roberto LCpl	2396857	H-2-26	1/15/69*	18	Hartford, Connecticutt
Rangel, Antonio Sgt	1543793	L-3-3	5/14-19/65	31	Gonzales, Texas
Raper, Charles D 2dLt	0103141	L-3-4	5/15/68	23	Long Beach, California
Raper, Robert J Pfc	2401164	M-3-9	6/2/69	19	Atlanta, Georgia
Ratcliffe, Edward K Cpl	2136993	C-1-5	2/19/67	19	Cranston, Rhode Island
Ratliff, Arch Jr Capt	089061	HMM-363	10/22/67	25	Abilene, Texas
Ratliff, Fred A Sgt	2015147	A-1-7	12/7/68*	23	Denver, Colorado
Rau, Arnold B Pfc	2318401	F-2-26	5/16/67	20	Oakland, California
Rawson, William A 2dLt	0100614	I-3-5	6/2/67*	25	Lake Forest, Illinois
Ray, Ronald D Capt	089867	Advisor	6/31/67	24	Louisville, Kentucky
Ray, Ronald D Capt	089867	Advisor	7/12/67	24	Louisville, Kentucky
Ray, Ronald J LCpl	2424199	B-1-7	8/12/69*	20	Greenleaf, Wisconsin
Rayo, Joseph A Sgt	1832266	E-2-1	9/9/66	26	Houston, Texas
Reali, James W Cpl	2423345	K-3-7	4/16/69	21	Glenview, Illinois
Rebelo, Joaquim V LCpl	2254243	B-1-1	1/29/67*	20	Newark, New Jersey
Reck, Ronald L Cpl	2492173	E-2-4	2/25/69	18	Beaver Creek, Oregon
Reed, Louis J SSgt	1383118	G-2-3	10/22/66*	30	White Plains, New York
Reed, Paul M Cpl	2039088	E-2-4	8/23/66	20	Roanoke, Virginia
Reed, Van S Maj	066952	HMM-165	12/8/68	36	Gainesville, Florida
Reese, Clifford E LtCol	063327	HMM-367	3/23/71	44	Greensboro, North Carolina
Reese, Clifford E LtCol	063327	HMM-367	3/28/71	44	Greensboro, North Carolina
Reese, Merle W GySgt	1131770	3-OCIT	3/13/67	31	Johnstown, Pennsylvania
Reeves, Daniel M LCpl	2182308	A-1-Amph	1/20/68	21	Pinellas, Florida
Reeves, Edward R LCpl	2506989	I-3-9	6/27/69	19	New Haven, Connecticut
Reeves, James A Sgt	2080209	C-1-7	3/28/66	22	Kelleysville, West Virginia
Register, Kenneth L Pfc	2408690	A-1-7	5/17/68	19	Montgomery, Alabama
Reid, Sandy R Cpl	2385784	B-3-Recon	12/11/68	21	Hemet, California
Reilly, Donald J Maj	061230	VMO-2	11/18/65	36	St. Louis, Missouri
Reisinger, Richard F Cpl	2127019	M-3-4	9/7/67	22	Buffalo, New York
Renegar, Edwin J 2dLt	097707	FO-2-4	8/23/67	31	Irdell, North Carolina
Renteria, Joe M LCpl	2270433	M-3-9	5/25/67	20	Brawley, California
Resnick, Robert A LCpl	2375393	I-3-4	7/1/68*	19	Upper Darby, Pennsylvania
Restivo, Anthony Jr Pfc	2030200	A-1-9	5/21/66	20	Jersey City, New Jersey
Reynolds, Charles A Maj	074013	VMO-3	5/10/67	31	Nashville, Tennessee
Reynolds, John R Cpl	2308173	E-2-27	8/26/68	20	Atlanta, Georgia
Reynolds, Paul R Jr Sgt	1925704	G-2-3	2/6/68	25	Macon, Georgia
Rhodes, John E 1stLt	0103759	HML-367	9/21/69	25	Glendora, California
Rhodes, John E 1stLt	0103759	HML-367	11/14/69	25	Glendora, California
Ribillia, Mariano Jr Pfc	2388352	K-3-5	1/21/69*	19	Puunene, Hawaii
Rice, Johnny B LCpl	2221113	D-1-3	10/8/66	20	Merritt, Florida
Rice, Ronald R Capt	0100624	Advisor	6/13/71	28	Victor, New York
Rich, Robert C Cpl	2150648	VMO-6	3/24/68	23	De Kalb, Illinois
Richards, Chester C GySgt	1417087	D-1-7	11/12/69	33	Oklahoma City, Ok
Richards, John H Jr LCpl	2354831	A-1-3	1/22/69	20	Wilkes-Barre, Penn
Richardson, Benjamin Cpl	2245536	L-3-5	5/26/67*	23	Detroit, Michigan
Richardson, Jerry W Cpl	2158365	I-3-5	3/24/67	22	Dallas, Texas
Richwine, David A 2dLt	091490	K-3-4	7/16/66	22	Overland Park, Kansas
Rickman, William J Pfc	2237378	L-3-7	2/21/67*	18	Altoona, Pennsylvania
Rider, James W Capt	077451	VMO-2	3/10/66	30	West Seneca, New York
Rider, James W Capt	077451	VMO-2	5/21/66	30	West Seneca, New York
Riely, Daniel L Cpl	2192647	M-3-5	5/13/67	22	Grand Forks, South Dakota
Riggs, William W 1stLt	0103000	K-3-5	11/20/68	22	Shreveport, Louisiana

Name	Service #	Unit	Date	Age	Hometown
Riley, Dennis H LCpl	2553936	CAP-1-4-4	12/31/69*	19	Tell City, Indiana
Riley, Edward F Capt	085111	D-1-9	2/26/69	31	Portland, Maine
Riley, James C Pfc	2278327	B-1-3	5/10/67*	19	Pleasant Hill, California
Riley, James T Capt	095179	HMM-463	2/22/68*	32	Zeigler, Illinois
Rilk, Harlen C Cpl	2254415	B-3-Amph	9/16/67*	21	Dover, New Jersey
Rindfleisch, Jon A Maj	077790	Advisor	12/31/67	31	West Point, New York
Rinehart, Benny D Capt	072921	HMM-361	8/18/65	29	Vandalia, Illinois
Ring, Edwin R GySgt	1201472	C-1-4	3/1/69	35	Hartford, Connecticut
Ringler, Robert L Jr Cpl	2099691	I-3-7	12/17/66*	20	Brackenridge, Pennsylvania
Riordan, James P 2dLt	089549	MB-3-12	9/10/66	24	Seattle, Washington
Riordan, Patrick C Pfc	2359732	F-2-1	5/19/68*	18	Des Plaines, Illinois
Rios, Domingo Jr Cpl	2487108	M-3-7	12/1/69	21	Midland, Texas
Rios, Henry A LCpl	2158833	F-2-7	9/25/66	19	Woodland, California
Ripley, John W Capt	081239	L-3-3	8/21/67	27	Radford, Virginia
Rismiller, Larry J Sgt	2406680	L-3-4	2/17/69	22	Cincinnati, Ohio
Risner, Richard F Maj	067619	MAG-12	4/26/68	36	Temecula, California
Rivera, Antonio G Cpl	2452711	K-3-3	6/17/69	21	Denver, Colorado
Rivera, Virgilio LCpl	2320595	B-1-4	5/22/68	19	New York, New York
Rivers, Frederick M Jr Capt	091634	HMM-163	8/22/68	25	Memphis, Tennessee
Rivers, Robert R 1stLt	094467	VMA-211	3/21/69	24	Rochester, New York
Roark, Bruce J LCpl	2386093	C-1-1	7/7/68	20	Detroit, Michigan
Roath, Louis P III 1stLt	092564	FAC-3-26	9/10/67	23	Hartford, Connecticut
Roberson, James J 2dLt	097738	1-Amp-11	3/24/67	35	Yakima, Washington
Roberts, Derold E SSgt	1659087	B-1-1	12/15/67	28	Sioux City, Iowa
Roberts, Harley R Sgt	2077470	H-2-3	7/7/68*	21	Richville, New York
Robertson, John W P LtCol	060509	CO-3-26	12/9/68	41	Newport, Rhode Island
Robertson, Merle E LCpl	2388054	B-1-3	9/10/68*	19	Campbell, California
Robinette, Randall S 1stLt	0106021	HMM-161	6/9/70	23	Phoenix, Arizona
Robinson, Andrew R Cpl	2415765	E-2-3	8/10/69	22	Chicago, Illinois
Robinson, Jimmie L LCpl	2199792	G-2-7	1/31/67	20	Jacksonville, Florida
Robinson, John C II LCpl	2236557	B-1-7	8/10/68*	21	Savannah, Georgia
Robinson, Robert G SSgt	2006274	A-1-3	3/7/68	24	Cincinnati, Ohio
Robson, Jon R Maj	069499	VMO-2	9/15/66	32	Tucson, Arizona
Rock, George B Sgt	2153923	1-Amp-11	2/21/68	23	Albany, New York
Rockey, William K LtCol	050776	CO-3-5	1/30/68	40	Washington, DC
Rodger, Donald W 1stLt	0102247	I-3-9	3/31/69	23	Williamstown, Mass.
Rodgers, Larry J Pfc	2353911	D-1-7	3/17/68*	20	Ranger, Texas
Rodrigues, David E Cpl	2239582	E-2-1	3/21/68	20	New Haven, Connecticut
Rogers, Harry W 1stSgt	932445	D-3-Recon	5/14/65	40	Portland, Maine
Rogers, Lane Maj	061169	Advisor	4/19/65	37	Annapolis, Maryland
Rohweller, Robert T 1stLt	0100638	K-3-9	2/16/69	25	Jacksonville, Florida
Rollings, Wayne E 1stLt	0107099	1-Recon	4/11/69	28	Elloree, South Carolina
Roman, Joseph G Capt	083648	HMM-265	1/27/67	28	Latrobe, Pennsylvania
Romero, Robert W Capt	093402	HMM-165	4/28/68*	26	San Diego, California
Romig, David J LCpl	2621030	H&S-2-5	3/15/70	19	Witchita, Kansas
Rood, Gary A Pfc	2122346	F-2-7	3/4/66	19	Toledo, Ohio
Rosolie, Walter W Cpl	2282831	A-1-5	2/13/68*	19	Rosedale, New York
Ross, Bruce J 1stLt	091160	VMO-6	11/11/66	24	Erie, Pennsylvania
Ross, Frank M Jr SSgt	1117022	Sniper-1-1	3/18/68*	35	Memphis, Tennessee
Ross, Michael R Capt	088020	D-2-12	5/13/69*	29	Lubbock, Texas
Ross, Reid R Jr Cpl	2395408	I-3-26	12/4/68*	19	Flinton, Pennsylvania
Ross, Ronald J Cpl	2177979	H&S-5	2/7/68	20	Portland, Oregon
Rosser, James E LCpl	2204333	D-1-7	4/6/68	20	Stanley, Virginia
Rosser, Richard C Jr Capt	093408	HMM-164	7/18/68	27	North Hollywood, California
Rosser, Richard C Jr Capt	093408	HMM-164	8/1/68	27	North Hollywood, California
Rostad, Theodore W Cpl	1876996	HMM-163	3/31/65	23	Monterey, California
Roth, Harold B Jr Maj	054026	HMM-364	6/29/68	37	Minneapolis, Minnesota
Roth, Raymond A Jr Sgt	2368049	13-ITT	3/12/71	22	Cleveland, Ohio
Roundtree, Louis GySgt	662089	Advisor	5/30/65	35	Kathwood, South Carolina
Rousseau, Joel Sgt	2104807	I-3-5	9/6/65	22	New York, New York

Name	Service #	Unit	Date	Age	Hometown
Rowden, John W Pfc	2352892	G-2-5	2/9/68*	21	Jacksonville, Oregon
Rowe, Larry E Pfc	2457139	B-1-5	6/9/69	19	Dayton, Ohio
Rowe, Michael M LCpl	2064981	B-1-7	9/8/65	20	Cleveland, Ohio
Rowling, Lamont Sgt	1992019	E-2-4	5/25/68	24	Boston, Massachusetts
Royster, Douglas LCpl	2097047	A-1-Eng	6/26/66*	19	Philadelphia, Pennsylvania
Royston, Joseph E Cpl	2382532	E-2-5	11/1/68	19	Indianapolis, Indiana
Rozanski, Edward C LCpl	2250985	D-1-5	9/4/67*	20	Chicago, Illinois
Rozumniak, David S Cpl	2042331	D-1-3	5/27/65	22	Detroit, Michigan
Rubin, Roy G LCpl	2293045	F-2-9	3/28/69*	19	New York, New York
Ruddick, Morris E Jr 1stLt	093410	F-3-Recon	6/14/67	24	Mahwah, New Jersey
Rudisill, Thomas R Cpl	2400003	K-3-3	3/9/69	21	Philadelphia, Pennsylvania
Ruffer, Jack A 1stLt	091880	C-1-1	10/12/67	26	San Bernardino, California
Ruiz, Jose Cpl	2282540	C-1-9	4/16/68*	24	New York, New York
Runyen, Thomas G 1stLt	095032	I-3-5	9/6/67	22	Riverside, California
Runyon, Marvin T III 2dLt	0107524	E-2-7	2/23/69	23	Detroit, Michigan
Rushing, James M 2dLt	0104763	K-3-7	12/16/68*	23	Pensacola, Florida
Rusnak, Robert J Cpl	2337996	K-3-9	9/9/68*	19	Johnstown, Pennsylvania
Russell, Glenn W Jr Capt	089444	HMM-362	1/22/68	29	Portsmouth, Rhode Island
Russell, Richard L 2dLt	0108776	F-2-7	8/25/69	23	Syracuse, New York
Russell, Verner R Cpl	2225712	CAP-0-1	1/21/68	20	Poplar Bluff, Missouri
Ryan, Richard W LCpl	2281106	A-1-1	8/12/67	19	Hopwood, Pennsylvania
Ryan, William F 1stLt	0104333	C-1-3	5/24/69	24	Buffalo, New York
Sabo, Roy T SSgt	1653906	A-1-Tanks	10/4/66	27	Flint, Michigan
Sachtleben, George W Capt	084855	H-2-9	6/5/69	29	Chicago, Illinois
Sakowski, Eugene J LCpl	2484720	L-3-7	8/13/69	19	Detroit, Michigan
Salles, Daniel R Cpl	2413697	K-3-3	9/15/69	20	Los Angeles, California
Salmon, Christopher B Capt	088969	HMM-361	9/21/67	26	MCAS El Toro, California
Salter, Martin E Jr Maj	062172	HMM-363	12/21/67	40	Greenville, Mississippi
Salvati, Ralph J Maj	071078	XO-2-5	2/4/68	34	Russellton, Pennsylvania
Sampietro, Scott B LCpl	2321303	F-2-5	9/30/68	19	Mahopac, New York
Sampsel, John D Pfc	2473777	H-2-26	1/14/69	18	Lodi, Ohio
Sanchez, Jimmy P Cpl	2072279	L-3-1	3/5/66*	23	Los Angeles, California
Sanchez, Manuel Cpl	2036967	M-3-1	3/5/66	23	San Antonio, Texas
Sanders, James M 1stLt	093764	AOU-1M	2/8/68	29	Carlsbad, New Mexico
Sanders, Samuel C LCpl	2544591	CAP-2-7-1	6/17/70	20	Montgomery, Alabama
Sankey, David H Pfc	2147732	A-1-9	7/6/67	21	Oklahoma City, Oklahoma
Santos, Reuben SSgt	1645553	G-2-3	4/30/67	28	Ponce, Puerto Rico
Sarti, Lawrence E Sgt	2439227	A-3-Recon	6/1/69	21	Baltimore, Maryland
Sasek, Richard J Capt	081030	D-1-9	7/6/67*	30	Topeka, Kansas
Sasser, John R Cpl	2020482	E-2-9	3/25/66	21	San Francisco, California
Sausau, Eli SSgt	1351710	H-2-1	1/29/67	33	Amouli, American Samoa
Sawyer, Kenneth V Cpl	2374130	HMM-364	10/21/69	22	Denver, Colorado
Saxon, Clyde E LCpl	2461616	G-2-7	3/18/69*	19	Waynesboro, Georgia
Scafidi, Vincent A Pfc	2271887	H-2-4	5/2/68	21	New York, New York
Scalici, Dennis F Cpl	2377322	L-3-3	5/8/68	20	St. Paul, Minnesota
Scalici, Dennis F LCpl	2377322	L-3-3	8/26/68	20	St. Paul, Minnesota
Scanlon, Michael J Cpl	2136902	3-FRecon	1/17/67*	21	Norwalk, Connecticutt
Schaefer, Charles H 1stLt	0105643	H&HS-1-1	8/24/69*	24	Streator, Illinois
Schaper, John P Sgt	2348025	HMM-164	8/14/69	23	Los Angeles, California
Schermerhorn, James 1stLt	0105255	B-1-7	8/28/69	27	Niagra Falls, New York
Schley, Harold B Jr SSgt	2010910	A-3-Tanks	7/28/69	25	Montgomery, Alabama
Schmidt, Joseph A Jr Cpl	2049230	H&S-1-5	9/4/67	22	Milwaukee, Wisconsin
Schmidt, Wallace R Pfc	2385350	F-2-4	3/12/68	19	Minneapolis, Minnesota
Schneider, Gerard B Cpl	2365027	D-1-5	9/10/68	20	St. Louis, Missouri
Schneider, Harry W Cpl	2201049	VMO-6	2/16/68*	20	Janesville, Wisconsin
Schrader, Peter A Cpl	2265855	M-3-5	11/7/67*	20	University City, Missouri
Schreiner, Andrew M Jr Sgt	1438592	VMA-214	10/28/65	27	Denver, Colorado
Schultz, Daniel R Cpl	1962874	M-3-9	6/4/65	22	Ypsilanti, Michigan
Schulze, Richard C LtCol	053080	CO-3-3	6/16/69	40	Oakland, California
Schwanda, Rudy T 1stLt	095386	VMA-242	2/23/68	24	Haddon Heights, New Jersey

Name	ID	Unit	Date	Age	Hometown
Scipio, Robert A LCpl	2116526	M-3-7	11/8/66	19	Martins Ferry, Ohio
Scott, Fred S LCpl	2260585	D-1-5	9/5/67	20	Nashville, Tennessee
Scriven, Woodrow Sgt	2300451	G-2-4	5/2/68	22	Jacksonville, Florida
Scully, Patrick R Jr 1stLt	0101464	K-3-9	7/17/68*	23	Chicago, Illinois
Scuras, James B 2dlt	093426	F-2-5	12/20/66	26	Murphysboro, Illinois
Seavy-Cioffi, Lawrence J Pfc	2320621	A-1-9	2/8/68	22	New York, New York
Sekne, Sylvester LCpl	2367097	E-2-7	4/12/69*	19	Cleveland, Ohio
Sellers, Donald T Sgt	2093450	D-1-4	2/2/68	24	Little Rock, Arkansas
Sellers, Wiley J Maj	072880	HMM-262	3/6/70	35	Brooklyn, Mississippi
Semenuk, John M Pfc	2181793	FO-A-1-1	9/25/66	19	Boston, Massachusetts
Seminara, Charles B SSgt	1364958	E-2-3	8/10/69*	33	Syracuse, New York
Serna, Marshall J Jr Pfc	2387619	E-2-4	5/2/68	20	Pittsburg, California
Serrano, Francisco C LCpl	2375819	F-2-1	9/24/68	20	San Jose, California
Serrano, John R Cpl	2320960	K-3-7	7/28/68*	19	Huntington, New York
Setser, Robert E 1stLt	085135	A-1-7	9/8/65	25	Carmi, Illinois
Sexton, Raymond D LCpl	2305797	D-1-5	2/15/68	18	Ullin, Illinois
Seybold, Gerald C 2dLt	0106332	TIO-IS	4/28/71*	36	Concord, New Hampshire
Shafer, Eric D 2dLt	0105649	I-3-1	4/5/69	22	Alliance, Ohio
Shafer, Francis L Jr Capt	083900	D-1-9	3/30/68*	29	Newkirk, Oklahoma
Shafer, Leslie H Pfc	2407202	K-3-3	2/22/69*	18	Dayton, Ohio
Shainline, Thomas E Cpl	2399720	C-3-Recon	6/1/69	21	Collegeville, Pennsylvania
Shainline, Thomas E Pfc	2399720	C-3-Recon	12/12/68	20	Collegeville, Pennsylvania
Shalcosky, Marion S Jr Sgt	2032744	M-3-4	5/27/68	24	Baltimore, Maryland
Shankey, Hugh R GySgt	1863529	HMM-167	3/28/71	31	Bibb, Georgia
Sharp, Oliver R LCpl	2229871	D-3-Recon	1/22/67	21	Marlinton, West Virginia
Sharpe, Thomas E Pfc	2377204	L-3-27	5/18/68*	19	Emmett, Michigan
Shaver, Carl A Maj	082539	L-3-7	5/19/68	30	Plattsburg, Missouri
Shaw, James A SSgt	614499	F-2-9	1/8/66	36	Swartz Creek, Michigan
Shaw, James G LCpl	2356774	D-2-12	8/22/68	20	Oakland, California
Shear, James L Pfc	2127305	3-FRecon	8/4/67	22	San Francisco, California
Sheehan, James P Capt	077100	G-2-3	4/30/67	30	Philadelphia, Pennsylvania
Sheehan, John J Capt	085321	Advisor	9/14-17/68	28	Somerville, Mass
Shehan, Timothy Cpl	2240154	MAG-18	2/23/69	21	Kansas City, Missouri
Shellem, Robert P LCpl	2347256	B-1-7	4/28/68*	19	Atlantic City, New Jersey
Sheridan, Robert F Maj	071807	XO-2-3	3/1/67	36	Isle of Palms, South Carolina
Sherin, Duane V 2dLt	0100672	H-2-5	11/7/68	25	Seattle, Washington
Sherman, Edward J 1stLt	0105653	C-1-26	6/7/69	24	Readville, Massachusetts
Sherrill, Richard W LCpl	2406081	CAP-1-3-9	9/13/69*	18	Hartman, Arkansas
Sherrod, Edward H Pfc	2503409	K-3-7	9/28/69*	20	Oklahoma City, Oklahoma
Shibley, Kamille K 1stLt	0103023	I-3-1	4/4/69	26	Sapulpa, Oklahoma
Shields, John M Capt	077239	VMO-6	6/16/66	30	Paicines, California
Shiffler, George M Jr Maj	064365	HMM-361	10/28/65	25	Pittsburgh, Pennsylvania
Shipley, Jack R Jr Cpl	2272283	A-1-Eng	3/15/68	20	Knoxville, Tennessee
Shivers, Stephen L GySgt	1617603	G-2-5	1/4/71	32	Greenville, South Carolina
Shoemaker, David H Sgt	2040598	M-3-7	11/2/67*	21	San Jose, California
Shore, Samuel E 2dLt	0103795	B-1-27	6/15/68	25	Maryville, Tennessee
Shreve, Ernest L Cpl	2128964	B-5-SPB	6/30/68	22	Seattle, Washington
Shubert, Ronald G LCpl	2005183	M-3-7	10/21/65	21	Hays, Kansas
Sibilly, John R Sgt	2205959	I-3-3	9/7/67*	21	Richmond Hills, New York
Sieloff, Richard J LCpl	2138756	HMM-262	1/25/67	19	Anoka, Minnesota
Siler, Jerry E 2dLt	097889	C-3-Recon	1/1/67	33	Omaha, Nebraska
Silvear, Thomas A Capt	076399	I-3-26	12/23/66	34	Carmel, California
Simmons, Herolin T SSgt	1900256	E-2-1	5/29/66*	24	Ahoskie, North Carolina
Simmons, Jack E Capt	091545	VMA-242	10/31/67	24	Salt Lake City, Utah
Simmons, William N Maj	079498	HMM-161	11/25/68	32	Miami, Florida
Simms, James W 2dLt	0106294	D-1-9	2/11/69	24	Carrollton, Missouri
Simms, James W 2dLt	0106294	D-1-9	2/26/69*	24	Carrollton, Missouri
Simon, Jerry W LCpl	2012697	A-1-9	5/21/66	21	Norton, West Virginia
Simpson, Jerry I Capt	080331	Advisor	7/31/67	29	Charleston, South Carolina
Simpson, Willie L LCpl	2660942	G-2-5	1/26/71	19	San Antonio, Texas

Name	Service #	Unit	Date	Age	Hometown
Sims, John D LCpl	2353778	D-1-5	8/29/68	21	Montgomery, Alabama
Singer, Michael E Pfc	2367488	D-1-1	7/5/68*	19	Canton, Ohio
Sireci, Michael J Cpl	2142269	F-2-7	2/28/67	19	Westerly, Rhode Island
Sisson, Ronald P LCpl	2089489	1-FRecon	12/16/65*	23	Hulberton, New York
Sites, David T Maj	079417	VMA-311	10/9/68	31	Birmingham, Alabama
Siva, Thurlo J SSgt	1139175	B-1-Tanks	5/12/68	34	Warner Springs, California
Sivak, David M LCpl	2107080	L-3-26	6/16/68	22	Saginaw, Michigan
Skalba, John J Cpl	2042737	M-3-3	1/5/66*	20	Detroit, Michigan
Skinner, Robert E Sgt	2316143	B-3-Recon	7/7/68	21	Winchester, Indiana
Skoglund, Dale E Cpl	2310907	D-3-Recon	1/17/69	21	Minneapolis, Minnesota
Slack, Paul D Capt	067841	Advisor	2/7/66	34	Des Moines, Iowa
Sleeper, John A Sgt	2249559	C-1-Recon	5/13/69	22	Boston, Massachusetts
Slocum, Danny M Cpl	2295647	3-FRecon	2/16/68	22	Los Angeles, California
Small, Richard C Sgt	2080326	F-2-7	9/24/66	22	Mission Viejo, California
Smith, Charles L Pfc	2456090	A-1-9	3/4/69*	18	Oklahoma City, Oklahoma
Smith, Clifton B LCpl	2287748	H-2-9	11/30/67*	21	Midland City, Alabama
Smith, Clyde D Maj	092575	VMA-224	4/9/72	34	Randolph AFB, Texas
Smith, Coy E LCpl	2370683	G-2-9	2/5/69	20	Nashville, Tennessee
Smith, Dennis A LCpl	2355739	D-2-12	2/8/68	19	Bayonne, New Jersey
Smith, Frederick W 1stLt	095227	K-3-5	5/27/68	23	Marks, Mississippi
Smith, George W LtCol	050104	CO-1-9	3/18/69	43	Camp Hill, Pennsylvania
Smith, Gilbert E 2dLt	0107541	D-1-9	6/30/69	22	Tampa, Florida
Smith, James L Pfc	2457759	M-3-4	3/13/69*	18	Washington, Illinois
Smith, Joseph N Capt	064240	Advisor	4/25/64	33	Austin, Texas
Smith, Michael S Sgt	1904031	K-4-12	11/10/66	24	Pittsburgh, Pennsylvania
Smith, Paul L LCpl	2450288	I-3-5	9/11/69	19	Fort Worth, Texas
Smith, Ralph E Cpl	2121251	D-1-5	5/12/67*	21	Conyngham, Penn
Smith, Ray L 2dLt	0102290	A-1-1	2/4/68	21	Shidler, Oklahoma
Smith, Ray L 1stLt	0102290	A-1-1	7/7/68	22	Shidler, Oklahoma
Smith, Richard L Capt	091896	ALO-3-26	4/9/68	29	Minneapolis, Minnesota
Smith, Robert W Capt	081949	B-7-MTB	3/24/67	26	DeRidder, Louisiana
Smith, Ronald S LCpl	2438585	G-2-4	3/19/69	19	Omaha, Nebraska
Smith, Stanley D Sgt	1344855	B-1-4	3/22/67	32	Wichita Falls, Texas
Smith, Terry L Cpl	2306818	M-3-26	2/20/68*	20	Nashville, Tennessee
Smith, Timothy J Pfc	2381891	I-3-3	9/9/68*	20	Lake Geneva, Wisconsin
Smith, William N Cpl	2053134	G-2-5	5/3/66	22	Myrtle Beach, South Carolina
Smith, William R Capt	084181	F-2-3	3/14/68	27	New Rochelle, New York
Snyder, Robert A Cpl	2288004	H-2-7	9/27/67	21	Detroit, Michigan
Soard, Charles L Cpl	2128430	C-1-5	1/29/70	22	Portsmouth, Ohio
Soderling, Jerry M SSgt	1835230	A-1-5	5/10/67	26	San Francisco, California
Soldner, Dennis M Cpl	2282431	C-3-Recon	6/1/69	19	Woodside, New York
Sooter, Gary E Pfc	2133934	B-1-7	3/5/66*	19	Independence, Missouri
Southworth, Ronald H Cpl	2102784	H&S-2-4	3/31/67*	21	Massena, New York
Spahn, Jordan A SSgt	1415686	I-3-7	1/30/67	30	Honolulu, Hawaii
Spainhour, Walter J Jr 1stLt	085876	A-5-Recon	9/15/66*	26	Lenoir, North Carolina
Spare, Wayne J Sgt	2312160	M-3-7	2/13/68*	20	Baltimore, Maryland
Sparks, Herbert C LCpl	2077422	F-2-12	12/9/65	20	Butte, Montana
Spawn, Stanley I SSgt	1534052	VMA-223	3/29/68	32	Denver, Colorado
Spencer, James L Jr LCpl	2324335	F-2-5	2/6/68	20	Chicago, Illinois
Spencer, John B Pvt	2076344	L-3-7	10/17/66	22	Brooklyn, New York
Spencer, Thomas E Cpl	2072026	D-1-4	4/26/68	24	Akron, Ohio
Squires, Robert J 1stLt	088986	L-3-4	9/18/65	24	Harrisburg, Pennsylvania
Squires, Robert J Capt	088986	K-3-7	8/23/68	27	Harrisburg, Pennsylvania
Stackpole, Henry C III Capt	076572	I-3-5	5/12/67	26	West Haven, Connecticut
Staggs, Daniel A Pfc	2371706	E-2-26	6/6/68	21	Jackson, Michigan
Stamps, Oliver C SSgt	1561677	G-2-7	1/7/70*	31	Baltimore, Maryland
Stanford, Charles R 1stLt	0110356	D-2-5	12/9/70	23	Conyers, Georgia
Stankiewicz, Kenneth D LCpl	2063870	H-2-4	8/18/65*	19	Buffalo, New York
Stankowski, William N Pfc	2201457	I-3-9	3/30/67	18	Milwaukee, Wisconsin
Stansell, Breck S LCpl	2492219	B-1-5	3/19/69	19	Portland, Oregon

Name	Service #	Unit	Date	Age	Hometown
Staples, Thomas H 2dLt	0103818	H-2-7	1/29/68*	23	Petersburg, Michigan
Starbuck, Robert F Sgt	1939063	A-1-Recon	2/3/67*	25	Montgomery, New York
Starick, Michael G 2dLt	0111710	A-1-7	5/6/70	24	Orchard Lake, Michigan
Steadman, Henry W LtCol	059087	HMM-364	11/18/65	40	Cascilla, Mississippi
Steadman, Henry W Maj	059087	HMM-361	12/19/70	45	Cascilla, Mississippi
Steele, David L Maj	079444	HMM-263	10/12/67	31	Jupiter, Florida
Steinbach, Thomas R LCpl	2384229	B-1-Recon	11/29/68*	21	Texas City, Texas
Steiner, Donald E Cpl	2336548	E-2-7	8/18/68	20	Pittsburgh, Pennsylvania
Stensland, William C Capt	079569	A-1-1	12/15/67	29	San Antonio, Texas
Stern, George E Jr 1stLt	090398	3-F-Recon	11/9/66	26	St. Louis, Missouri
Stevens, Ernest A Pfc	2115400	K-3-4	7/18/66	18	Ruston, Louisiana
Stevens, Michael J 1stLt	095091	B-3-Eng	9/18/67	24	Kansas City, Missouri
Stevens, Walter T Cpl	2244778	A-1-9	9/19/66	20	Scranton, Pennsylvania
Stewart, David T LCpl	2388044	F-2-9	8/31/68	21	Oakland, California
Stewart, Marvin R LCpl	2405463	L-3-3	1/26/69	19	Louisville, Kentucky
Stick, Michael O 2dLt	0102306	D-1-4	2/2/68	25	Harrisburg, Pennsylvania
Stickel, James L Sgt	2013516	D-1-3	4/2/66	22	Springfield, Illinois
Stinson, David T Sgt	1807439	L-3-5	7/22/66	29	Mt. Sterling, Pennsylvania
Stokes, Alvin Pfc	2228293	A-3-Recon	9/23/66	19	Bellville, Georgia
Stokes, Colben B Pfc	2101657	1-MPB	1/30/68*	21	Laurel, Mississippi
Stokes, Thomas M Maj	064904	OO-1-1	2/22/67	35	Charles Town, West Virginia
Stoppa, Michael D Cpl	2122149	M-3-27	5/24/68	24	Detroit, Michigan
Storm, Dennis M 2dLt	0106094	G-2-5	5/15/69	24	Wood Dale, Illinois
Strahm, Robert E Pfc	2263104	E-2-9	11/17/66*	18	Kenton, Ohio
Strange, John B 1stLt	0104793	G-2-4	3/11/69	24	Madison, Alabama
Strange, Richard L Sgt	2039191	G-2-1	6/25/66*	24	Richmond, Virginia
Strassburg, Terry A Sgt	2345236	A-1-1	1/30/68	20	North Tonowanda, New York
Strehle, Ernest W Cpl	2022568	F-2-4	6/25/66*	20	Mundelein, Illinois
Strickland, Charles E SSgt	2036347	I-3-5	6/19/69	24	Towanda, Pennsylvania
Strickland, James M 1stLt	0101519	F-2-9	8/31/68	24	San Antonio, Texas
Stroud, Roger L Cpl	2341655	C-1-1	7/7/68*	18	Corpus Christi, Texas
Suarez, John Pfc	2427816	I-3-4	11/25/68	20	Houston, Texas
Sugg, Robert B LCpl	2228013	D-3-Recon	12/19/66	19	San Antonio, Texas
Sullivan, Brian R 1stLt	0103017	HQ Bty-1	12/23/69	23	New York, New York
Sullivan, David O LCpl	2378304	I-3-9	2/14/69*	20	Quincy, Massachusetts
Sullivan, Hugh J Jr Sgt	1915920	C-1-3	6/5/65*	23	Allentown, Pennsylvania
Sullivan, John P T Capt	084248	B-1-5	8/10/66	29	Chicago, Illinois
Sullivan, Joseph H SSgt	1851528	C-1-7	9/15/66*	24	Cincinnati, Ohio
Sullivan, Thomas J 1stLt	0102448	HMM-364	2/7/69	21	Tappan, New York
Sumner, Norman B Sgt	1805926	HMM-364	12/4/65	24	Seattle, Washington
Sutherland, Reginald J Cpl	2472883	CAP-1-4-6	12/10/69*	21	Hartsdale, New York
Swindle, Orson G III LtCol	078193	POW	8/21-23/67	39	Camilla, Georgia
Swindle, Orson G III LtCol	078193	POW	7/7-30/69	41	Camilla, Georgia
Sykora, Marcus L LCpl	2192997	M-3-5	5/14/67	19	Lubbock, Texas
Szymanski, John S SSgt	1824342	C-1-4	3/23/67*	26	Trenton, New Jersey
Taber, Edward A III 1stLt	091789	A-3-Amph	6/18/67	23	Hanover, New Hampshire
Talone, James R 1stLt	0103834	B-1-9	8/22/68	23	Philadelphia, Penn
Tatum, Harold D GySgt	1102779	B-3-Tanks	9/10/67*	35	Sandy Springs, Georgia
Taufi, Aouliolitau F LCpl	2413656	D-1-1	8/21/69*	23	Los Angeles, California
Taylor, Bayard V 2dLt	0102323	H-2-4	5/2/68	28	Chester County, Pennsylvania
Taylor, Billy J Cpl	2311662	CAP-1-3-3	8/2/68*	20	Wyandotte, Michigan
Taylor, James B Jr LCpl	2333094	A-3-Recon	3/20/69	20	Weirsdale, Florida
Taylor, James C Sgt	2341646	L-3-3	9/17/69	21	Houston, Texas
Taylor, John S Pfc	2369008	C-1-3	5/20/69*	21	Granby, Connecticutt
Taylor, Kenneth T Capt	081806	Advisor	7/31/66	31	Huntsville, Alabama
Taylor, Michael L Cpl	2360274	HMM-364	12/19/70	24	Daytona Beach, Florida
Taylor, Richard B Capt	070569	Advisor	10/31/63	31	Lanesville, Massachusetts
Taylor, Richard H Maj	076381	E-2-4	9/21/67	35	Albany, New York
Taylor, William E Pfc	2372470	F-2-9	11/30/67	18	Florence, Mississippi
Teiken, Dennis M Cpl	2059360	L-3-1	3/5/66	22	Minneapolis, Minnesota

Name	Service #	Unit	Date	Age	Hometown
Telles, Jose A SSgt	1410709	D-1-5	11/20/69	31	Albuquerque, New Mexico
Tennant, Byron L 2dLt	0106684	L-3-4	2/28/69*	24	Farmville, Virginia
Tenney, Joseph R Capt	079574	K-3-5	9/6/67	24	Cape Cod, Massachusetts
Ter Haar, Raymond L Jr Sgt	1803966	F-2-9	11/30/67	26	Rochester, New York
Terhorst, Bernard E Sgt	1896541	F-2-1	4/21/67	25	Paonia, Colorado
Terhorst, Bernard R Maj	068004	HMM-263	2/23/69	37	St. Paul, Minnesota
Terhorst, Bernard R Maj	068004	HMM-263	4/19/69*	37	St. Paul, Minnesota
Terrian, Clyde J Pfc	2146689	B-1-7	3/6/66	18	Flint, Michigan
Terry, Hunter M LCpl	2448732	I-3-26	12/4/68	18	St. Louis, Missouri
Tersteege, Paul F SSgt	1532700	D-1-1	1/24/69*	33	Tuscon, Arizona
Tharp, John J Maj	071855	HMM-165	1/28/68	34	Virginia Beach, Virginia
Theer, Richard E Capt	077292	E-2-7	12/10/65	29	Davenport, Iowa
Theiss, William L Sgt	1369673	A-1-1	1/29/67	29	Newell, West Virginia
Theriault, David G Cpl	2322217	F-2-5	2/2/68	21	Tehachapi, California
Thiewes, Ronald C 2dLt	0106428	L-3-3	1/26/69	22	Winona, Minnesota
Thomas, Daniel G Cpl	2311302	A-3-Amp	1/25/68	21	Detroit, Michigan
Thomas, Velpeau C 2dLt	098055	I-3-7	1/31/67	33	Burton, South Carolina
Thome, Richard J LCpl	2248778	D-1-26	3/18/67	23	Abrams, Wisconsin
Thompson, David B Sgt	0000000	1-Recon	1/11/69	00	Unknown
Thompson, Harvey E Pfc	2077393	D-1-5	1/5/68	23	Albany, New York
Thompson, John R Cpl	2168938	D-1-9	11/7/67	20	Phoenix, Arizona
Thompson, Leslie D Sgt	2366588	M-3-5	3/3/69*	18	Tampa, Florida
Thompson, Robert B Cpl	2275075	B-1-5	4/20/68*	22	Grants Pass, Oregon
Thompson, Stephen M LCpl	2390565	D-2-12	5/13/69*	18	Baltimore, Maryland
Thompson, Wayne W SSgt	1631998	Advisor	2/27/68	30	Hillsborough, Florida
Thoms, Robert L SSgt	1951269	D-1-5	2/15/68	24	Baton Rouge, Louisiana
Thomson, Robert B Cpl	2190293	3-FRecon	2/16/68*	20	Colorado Springs, Colorado
Thuesen, Thomas R LCpl	2375951	M-3-27	5/18/68	19	Seattle, Washington
Tigue, Thomas M 2dLt	0106391	K-3-9	2/15/69	23	Pittston, Pennsylvania
Tilghman, Richard H LCpl	2422175	B-1-3	9/9/68	19	Boston, Massachusetts
Till, Willard H Jr Pfc	2339848	E-2-9	4/23/69*	20	Raleigh, North Carolina
Tillery, Jerry T Pfc	2174609	H-2-5	2/3/68*	22	Philadelphia, Pennsylvania
Tilley, Robert O 1stLt	091071	K-3-5	5/12/67	29	San Diego, California
Timmons, Durward E Jr Cpl	2417274	I-3-5	9/10/68	21	Baltimore, Maryland
Tines, Robert W Pfc	2465645	B-1-7	1/20/69	19	Boston, Massachusetts
Tinker, John G LCpl	2295077	F-2-7	10/28/67*	19	New Boston, Illinois
Tiscia, Joseph R Jr Cpl	2211990	H&S-2-5	2/7/68	24	Memphis, Tennessee
Todd, Gary G 2dLt	0105100	D-1-5	6/24/69	24	Los Angeles, California
Todd, George G Cpl	2306711	G-2-7	5/7/68	20	Parris, Tennessee
Todd, Horace B Cpl	2162410	CAP-1-3-3	8/2/68	21	Wilmington, Delaware
Todd, Larry D Cpl	2215692	S-11-Eng	2/2/68	20	San Francisco, California
Tokarz, Anthony P 1stLt	089467	K-3-7	3/21/66	23	Fairmont, West Virginia
Tolan, Paul D Cpl	2117476	F-2-1	1/5/67	22	Franklin, Massachusetts
Tolbert, Roosevelt Jr Pfc	2386324	C-1-7	12/8/68	20	Detroit, Michigan
Tolentino, Clarence Sgt	1268075	C-3-5	3/5/66	33	San Luis Rey, California
Tolleson, Frederic L Maj	067844	Advisor	3/19-25/71	39	Sisterdale, Texas
Tolliver, Jimmy E SSgt	1813850	VMO-6	2/16/68*	28	Cromona, Kentucky
Tonkin, Terry L 1stLt	0116555	FAC-2-9	5/15/75	24	Grove City, Ohio
Tonucci, Richard L Cpl	2011754	H-2-4	8/18/65	21	Derby, Connecticut
Torres, Felipe Cpl	2320197	I-3-26	12/8/68	18	Bronx, New York
Torrey, Phillip H III Capt	085190	A-1-5	6/9/69	29	Los Angeles, California
Toth, William A Jr Pfc	2457321	H&S-2-4	2/25/69	19	Elgin, Illinois
Townes, Raymond M LCpl	2371093	E-2-26	8/16/69	19	Nashville, Tennessee
Townsend, Gary R LCpl	2424290	F-2-5	9/30/68*	21	Orchard Park, New York
Trautwein, Henry J Jr Capt	083757	C-1-1	7/7/68	32	Austin, Texas
Traylor, Corey S LCpl	2388893	9-Eng	2/26/69	20	Seattle, Washington
Trevino, Elias Sgt	1374493	A-1-4	2/26/66	31	Mercedes, Texas
Trivette, Marion C Jr Pvt	2246889	VMO-6	3/24/68	21	Quinton, Virginia
Trujillo, Gilardo J Sgt	1984214	C-11-Eng	2/2/68	25	Albuquerque, NM
Tubbs, James L Jr Cpl	2226085	C-1-3	1/22/68	20	Paducah, Kentucky

Name	Service #	Unit	Date	Age	Hometown
Tuckwiller, Frank W Capt	087149	B-1-3	9/9/68	29	Lewisburg, Virginia
Tully, James M Maj	078737	Advisor	6/20-21/72	36	Bronx, New York
Tully, Lester A Cpl	2221284	G-2-5	1/31/68	21	Woodville, Florida
Turner, David J 1stLt	0103854	D-2-11	3/19/69	25	Pueblo, Colorado
Turner, Earl J Pfc	2365222	E-2-4	5/17/68	20	Detroit, Michigan
Turner, Lindsay Cpl	2481645	E-2-3	8/10/69*	20	Edgemore, South Carolina
Turner, Willis S LCpl	2335473	D-1-9	5/27/68	20	Dallas, Texas
Tuten, Ernest W Cpl	2513705	K-3-3	5/25/69	20	St. Simons Island, Georgia
Twardowski, John M Sgt	1890804	FLSG	1/30/66	24	Larksville, Pennsylvania
Tweten, Ray G LCpl	2076003	G-2-7	9/24/66	19	Brooklyn, New York
Twilling, Henry M III LCpl	2140441	K-3-27	5/24/68	21	Los Angeles, California
Twohey, Richard B LtCol	054049	CO-1-3	9/2/68	40	New York, New York
Tyler, James H Cpl	2347958	HMM-165	1/12/69	19	Wilmington, California
Tyson, Stuart H Cpl	2412740	M-3-5	6/7/69*	22	Norfolk, Virginia
Uhl, Thomas F Cpl	2244036	B-1-4	10/27/67*	18	New York, New York
Underhill, Herbert SSgt	597136	C-1-7	3/28/66	35	Sioux Falls, South Dakota
Underwood, Billy L Pfc	2340074	M-3-7	2/23/69*	20	Asheboro, North Carolina
Ungar, Thomas D Capt	079467	VMA-121	3/30/68	29	St. Louis, Missouri
Ungerer, William P Sgt	2103251	A-1-7	2/6/68	22	Cleveland, Ohio
Upshaw, Charles R Capt	079074	HMM-364	5/0/67	34	Blythe, California
U'Ren, William F H Jr Maj	076013	H&S-1-7	1/13/68	31	San Bruno, California
Utter, Leon N LtCol	049824	CO-2-7	12/18/65	41	Miami, Oklahoma
Utter, Leon N LtCol	049824	CO-2-7	3/4/66	41	Miami, Oklahoma
Vacca, William P 2dLt	095128	B-1-5	5/24/67	25	Jacksonville, Florida
Valadez, Robert S Pfc	2434714	L-3-3	5/28/68	19	San Antonio, Texas
Valdez, John B Sgt	2388647	I-3-5	6/19/69*	21	Rocky Ford, Colorado
Valle, Guillermo, Cpl	2402392	E-2-7	9/6/70*	26	Brooklyn, New York
Vallerand, Larkin O LCpl	2113431	F-2-4	6/25/66*	21	Tracy, California
Valuzzi, Rocco F Maj	079796	HMM-463	3/31/71	33	Brooklyn, New York
Valvik, Robert A Cpl	2076561	A-1-1	2/1/68	22	New York, New York
VanAntwerp, William M Jr Capt	075400	B-3-Amph	9/16/67*	30	Albany, New York
Vandaveer, James A 2dLt	0107675	K-3-3	6/11/69	23	College Station, Texas
Van Dyke, Gilbert E LCpl	1988072	E-2-5	5/8/66	22	Koats, Indiana
Van Meter, Johnny L LCpl	2278444	I-3-4	3/6/67	21	Scotia, California
Van Riper, James K Capt	089000	Advisor	8/13/66	28	Brownsville, Pennsylvania
Van Riper, Paul K Capt	089001	M-3-7	2/7/66	27	Dormont, Pennsylvania
Van Riper, Paul K Capt	089001	M-3-7	2/23/69	30	Dormont, Pennsylvania
Vanvalkenburgh, Ed J Jr GySgt	1185788	F-2-5	2/7/68	32	Los Angeles, California
Vanzandt, Ray L Cpl	2256470	D-1-5	6/2/67*	21	Austin, Texas
Varelas, Alfred R LCpl	2186264	D-1-26	6/22/67	20	Springfield, Massachusetts
Vargas, Manuel S Jr Capt	083768	G-2-4	3/18/68	29	Winslow, Arizona
Vargas, Pedro R LCpl	2404576	F-2-26	2/11/69	19	Albuquerque, New Mexico
Varney, Ronald T LCpl	2374153	H&S-3-9	2/24/69*	21	Belfry, Kentucky
Vasel, Ralph W LCpl	2515417	CAP-1-2-4	4/12/70	19	Cincinnati, Ohio
Vasterling, Allan C 1stLt	089002	HMM-361	8/19/66	26	Ironton, Missouri
Vaughn, Edward L Cpl	1946372	H-2-4	8/18/65	23	Beaumont, Texas
Vaughn, Joe E Pfc	2412465	K-3-5	5/11/69	19	Los Angeles, California
Vega, Michael C LCpl	2113488	F-2-1	2/28/66	19	San Francisco, California
Veitz, Scott O LCpl	2184674	I-3-26	7/14/68	20	Albuquerque, New Mexico
Vercauteren, Richard F 1stLt	0104377	H-2-9	2/17/69	22	Manchester, New Hampshire
Vermass, Dwight A LCpl	2496489	F-2-3	8/7/69	19	Milwaukee, Wisconsin
Victor, Ralph G LCpl	2131887	C-1-Recon	6/16/66	20	Ogden, Utah
Viera, Marion Jr SgtMaj	310096	H&S-1-2	7/3/69	45	Seekonk, Massachusetts
Villalobos, Arthur G Pfc	2413041	E-2-5	5/15/69*	19	Compton, California
Vivilacqua, Theodore R 2dLt	0106177	H-2-5	5/11/69	22	Long Beach, California
Vogel, Peter J Capt	071964	HMM-163	3/31/65	32	San Miguel, California
Vogelgesang, Donald A 2dLt	092012	H-2-5	10/7/66	28	Canton, Ohio
Vojtisek, James R Jr Pfc	2380642	B-1-27	6/17/68	19	Albany, New York
VonHarten, William R LtCol	067345	XO-1-9	3/4-5/67	34	Beaufort, South Carolina
Voyles, Jerry D 1stLt	087983	C-1-7	3/28/66	24	Bay City, Texas

Name	Service #	Unit	Date	Age	Hometown
Wade, Billy F Cpl	2061433	H-2-1	6/25/66	22	Soperton, Georgia
Wade, Howard W III LCpl	2082177	C-1-26	10/6/66	22	Santa Monica, California
Wade, Nicholas M Pfc	2270139	H-2-5	2/5/68	20	Los Angeles, California
Wade, William G LCpl	2012781	L-3-7	3/21/66*	20	Berea, Ohio
Wadley, Harold E Sgt	1192794	H-2-5	9/10/67	33	Stanley, Idaho
Wagner, John M Capt	092013	VMA-242	10/25/67	24	Chicago, Illinois
Wahlsten, Bruce R Cpl	2313619	B-1-Tanks	2/21/69	21	Minneapolis, Minnesota
Waitulavich, George J Jr Sgt	2244866	G-2-5	12/20/70	24	Wilkes-Barre, Penn
Wajda, Philip J LCpl	2381165	F-2-7	9/19/68*	18	Chicago, Illinois
Waldrop, Roy E Pfc	2382787	G-2-7	5/7/68	21	Dallas, Texas
Walker, Arthur G GySgt	1450549	H&S-1-5	3/19/69	31	New Orleans, Louisiana
Walker, David E 1stLt	088592	HMM-161	4/21/66	25	Abington, Pennsylvania
Walker, Gary W LCpl	2446298	M-3-7	2/23/69	19	Little Rock, Arkansas
Walker, Victor R LCpl	2342545	L-3-9	4/30/68	22	Oxford, Nebraska
Walkley, Robert M LCpl	2365067	B-3-Tanks	3/24/69*	21	Ionia, Michigan
Wallace, Marvin C LCpl	2492237	C-1-7	8/12/69	20	Portland, Oregon
Wallace, Paul H 1stLt	0102365	A-1-9	4/4/68	23	Fresno, California
Wallace, Robert L LCpl	2311134	L-3-9	4/30/68	20	Fargo, North Dakota
Walls, Robert L Pfc	2340399	C-1-7	12/21/67*	18	New Orleans, Louisiana
Walsh, Robert T SSgt	1399730	HMM-361	8/10/66*	30	La Crosse, Wisconsin
Walters, Joseph E 2dLt	0104381	M-3-5	8/9/68	27	Ocean City, New Jersey
Walton, Grover W SSgt	1574776	E-2-1	5/17/68	30	Sutherlin, Virginia
Wandro, James M Pfc	2482623	C-1-5	6/11/69*	19	San Mateo, California
Ward, Harold T Jr Maj	077961	Advisor	9/17/68	26	Asheville, North Carolina
Ward, Joel D Capt	090381	E-2-9	1/20-3/69	26	Mobile, Alabama
Ward, Robert J SSgt	1099309	H-2-4	4/30/68	34	New York, New York
Ward, Robert O SSgt	1549586	B-1-26	6/7/67	28	Kentfield, California
Warmbrodt, Jon F 2dLt	0107564	L-3-26	1/25/69*	22	Santa Monica, California
Warner, James H Capt	092816	POW	67-69	28	Ypisalanti, Michigan
Warshaw, Joel M Capt	087450	VMA-242	10/31/67	28	Hackensack, New Jersey
Washut, Walter J Cpl	2156755	K-3-9	5/20/67	19	Sheridan, Wyoming
Wasko, Michael J Jr Maj	078185	HMM-463	2/23/71	32	Philadelphia, Penn
Watington, Ralph H Jr Cpl	2104550	D-1-4	5/8/67*	21	New York, New York
Watkins, David C Maj	076555	HML-167	1/5/69	32	Cleveland, Ohio
Watkins, Robert W 1stLt	0103873	HMM-263	1/3/70	24	Bloomfield, Missouri
Watson, Albert C Jr LCpl	2260334	F-2-26	10/15/67*	20	Mauston, Wisconsin
Watson, Michael O Cpl	2214943	G-2-7	3/24/68	22	Raleigh, North Carolina
Watters, Kenneth L Cpl	2245676	G-2-9	7/29/67	22	Detroit, Michigan
Wayand, Frederick E Cpl	2206553	HMM-165	10/10/68	20	Greenwich, Connecticut
Weatherholtz, Donald A SSgt	1463029	D-1-4	4/26/68	30	Moorefield, West Virginia
Weaver, Dale L LCpl	2264241	K-3-9	7/17/68*	18	Honey Brook, California
Weaver, Larry H LCpl	2162182	F-2-3	5/3/67	18	Evening Shade, Arkansas
Webb, James H Jr 1stLt	0106180	D-1-5	5/9/69	23	Saint Joseph, Missouri
Webber, Brian L 1stLt	095696	I-3-26	12/8/68*	24	Albuquerque, New Mexico
Webster, Robert E 2dLt	0106396	C-1-9	2/19/69	23	Philadelphia, Pennsylvania
Weede, Richard D Capt	081702	D-1-4	4/26/68	29	Kaneohe, Hawaii
Weeks, Robert W Capt	093814	L-3-1	5/12/69	30	Boston, Massachusetts
Wegener, Joseph B II LCpl	1930274	HMM-165	10/10/68	25	Phoenix, Arizona
Weh, Allen E 2dLt	0100765	A-3-Recon	8/4/67	24	Albuquerque, New Mexico
Weigand, Phillip S Capt	078959	C-1-Tanks	5/26/67	29	San Francisco, California
Weise, William LtCol	057704	CO-2-9	3/18/68	39	Philadelphia, Pennsylvania
Weiss, Peter W 2dLt	0103881	B-1-26	3/30/68	24	Bronx, New York
Weldon, Bucko W Cpl	2356589	G-2-7	3/20/68	19	Oakland, California
Wells, Marshall R Capt	0102378	Advisor	4/8/72	30	St. Louis, Missouri
Wells, Robert D LCpl	2414866	B-1-5	6/9/69	19	Los Angeles, California
Wellmann, Dennis W LCpl	2162886	E-2-4	8/23/66*	21	Hanska, Minnesota
Wenger, Howard W SSgt	1185973	G-2-7	3/4/66	32	Harrisonburg, Virginia
Wenger, Howard W SSgt	1185973	G-2-7	9/24/66	32	Harrisonburg, Virginia
West, Alfred M 1stLt	094212	HMM-362	11/4/68	25	Athens, Alabama
Westbrook, Emmett D Pfc	2124724	F-2-7	3/4/66	20	Dallas, Texas

Westendorf, Gerald C Capt	092018	VMA-533	10/25/67	28	Saginaw, Michigan
Wester, William D Capt	091086	L-3-3	9/13/69	26	Norfolk, Virginia
Whalen, Garland G Cpl	2287396	C-1-3	1/31/69*	20	Denver, Colorado
Wheeler, John B III LCpl	2164264	K-3-1	1/15/67	21	Salt Lake City, Utah
Wheeler, Kenneth W Cpl	2303494	HMM-262	5/10/69*	23	Brownwood, Texas
Wheeler, Thomas C Cpl	2221428	M-3-3	4/30/67	19	Jacksonville, Florida
Whipple, Oliver W Jr Maj	076023	F-2-7	3/16/67	31	Virginia Beach, Virginia
Whisenhunt, James H LCpl	2247981	M-3-3`	4/30/67*	23	Crescent City, California
White, Bobby R Cpl	2290870	H-2-9	2/2/68	19	South Charleston, WV
White, David L Capt	088184	HMM-262	5/25/69	28	Memphis, Tennessee
White, David L Capt	088184	HMM-262	6/7-10/69	28	Memphis, Tennessee
White, Gregory A Cpl	2431766	H-2-9	6/5/69	22	Cleveland, Ohio
White, Harry G Cpl	2043584	G-2-1	6/25/66	21	St. Louis, Missouri
White, John C III 1stLt	0103886	H-2-5	11/1/68*	24	Dayton, Ohio
White, Johnel N Pfc	2396559	B-1-27	6/16/68	22	New Haven, Connecticut
White, Owen Jr Pfc	2439631	M-3-5	9/11/68*	20	Chicago, Illinois
Whitfield, Douglas W Pfc	2232071	H&S-3-1	11/12/66	20	Hillside, Illinois
Whitmer, Maurice P Cpl	2157170	D-1-5	2/15/68	20	Spring Valley, California
Whitted, George L WO	099531	AO	5/5/68	37	Salem, Oregon
Whittingham, Joseph M Pfc	2408757	H-2-3	5/14/68	19	Montgomery, Alabama
Whoolery, Tracy L SSgt	1920881	C-1-1	11/1/67*	26	Baltimore, Maryland
Whorton, William S Capt	090258	Advisor	7/25/68	26	Junction City, Kansas
Whyte, Charles J Sgt	2108109	I-3-27	5/28/68*	24	Olympia, Washington
Wicks, James G Cpl	2347323	E-2-7	8/18/68	20	Philadelphia, Pennsylvania
Wickwire, Peter A LtCol	051969	CO-1-3	7/4/67	39	Mountain Lakes, New Jersey
Wiedhahn, Warren H Jr Maj	060310	XO-3-9	2/0/69	40	Canajoharie, New York
Wielebski, John T Cpl	2201169	A-1-26	5/23/68	20	Milwaukee, Wisconsin
Wigg, Jerry R Cpl	1694359	C-1-7	12/17/68	29	Portland, Oregon
Wiggins, Paul D Jr Pfc	2456227	F-2-5	9/30/68	18	Cincinnati, Ohio
Wildprett, William R Capt	071702	K-3-26	7/31/67	33	Warwick, Rhode Island
Wiley, Joseph F Pfc	2324004	M-3-4	3/13/69	20	Denver, Colorado
Wilhelm, Charles E Capt	090259	Advisor	5/5/69	27	Alexandria, Virginia
Wilke, Edward S Cpl	2053417	HMM-364	9/3/66	22	Waukesha, Wisconsin
Wilker, James A. Cpl	2311095	C-1-Recon	7/15/68	20	Waseca, Minnesota
Wilkerson, Steven D LCpl	2376520	M-3-7	12/26/68*	20	Wakefield, Nebraska
Wilkins, Robert J LCpl	2100350	H-2-9	12/18/65*	20	St. Charles, Missouri
Willcox, Clair E LtCol	059684	CO-1-4	4/3/69	37	St. Louis, Missouri
Williams, Charles E Cpl	2330620	D-1-7	2/16/69	22	San Antonio, Texas
Williams, Clifford D Sgt	2248665	K-3-9	3/27/68*	21	Newberry, Michigan
Williams, Dempsey H III 1stLt	082159	Advisor	3/9/65*	25	Fayetteville, NC
Williams, Freddy R SSgt	1433034	G-2-5	8/18/68*	32	Plains, Georgia
Williams, Gary D Pfc	0000000	F-2-9	2/28/66	00	Lovelock, Nevada
Williams, Howard C LCpl	2384892	A-1-1	7/8/68*	20	Gueydan, Louisiana
Williams, James Sgt	2088210	M-3-26	8/4/67*	20	Oklahoma City, Oklahoma
Williams, James L Capt	081120	H-2-4	3/18/68	30	Winona, Minnesota
Williams, James L Capt	081120	H-2-4	4/30/68	30	Winona, Minnesota
Williams, Johnny B Pfc	2427993	D-1-5	8/29/68*	24	Nacogdoches, Texas
Williams, Ken B Cpl	2378550	D-3-Recon	5/15/69	20	Birmingham, Alabama
Williams, Robert B Capt	086952	B-1-11	2/23/69	35	Kansas City, Missouri
Williams, Terry E Cpl	2325376	C-1-1	3/18/68	20	Atlanta, Georgia
Williams, Theodore J LCpl	2266152	D-3-Recon	1/2/68	22	St. Louis, Missouri
Williams, Thomas E Jr 1stLt	0100873	HMM-262	3/5/69*	25	Pensacola, Florida
Williamson, Curtis C Cpl	2110648	L-3-7	10/17/66	21	Hapeville, Georgia
Williamson, Frederick C Jr 1stLt	090498	E-2-4	12/5-6/65	22	Leonardo, New Jersey
Williamson, Frederick C Jr 1sLt	090498	E-2-4	3/21/66	22	Leonardo, New Jersey
Williamson, Robert M Jr LCpl	2101017	B-3-Tanks	8/17/69	20	Richmond, Virginia
Willis, Robert T 2dLt	093515	F-2-7	9/24/66	25	Evanston, Illinois
Willis, Theodore J LtCol	054059	CO-1-4	4/12/68	38	St. Petersburg, Florida
Willoughby, David H 2dLt	094371	F-2-9	7/29/67	22	Grundy Center, Iowa
Willson, Gordon R 1stLt	089458	B-3-Recon	3/15/66	26	Rapid City, South Dakota
Wilson, Dale E Sgt	2490492	D-1-5	11/17/69	19	Troutman, North Carolina

Name	Service #	Unit	Date	Age	Hometown
Wilson, Donald D Maj	072143	1-MAW	5/9/69	32	Williams, Arizona
Wilson, Douglas E II CWO	088744	VMA-242	10/27/67	29	Los Angeles, California
Wilson, Frederick J III Capt	090444	VMA-164	7/4/67	26	Wakefield, Rhode Island
Wilson, Henry L LCpl	2359352	B-1-1	5/31/68	19	Marion, Alabama
Wilson, Lyndol R Pfc	2344178	H-2-5	2/3/68	18	Okemah, Oklahoma
Winebar, Francis E SSgt	1352529	A-1-4	3/20/66	30	James Island, South Carolina
Winebar, Francis E GySgt	1352529	A-1-4	8/26/66	31	James Island, South Carolina
Winecoff, David F Capt	085492	H-2-9	2/21/68	29	Ellensburg, Washington
Winfrey, James A Cpl	2458900	F-2-3	8/7/69*	23	Webster Groves, Missouri
Winston, Herbert T Maj	080190	Advisor	6/5/69	34	New Orleans, Louisiana
Winston, William O Cpl	2208724	C-1-5	5/10/67	20	Atlanta, Georgia
Withers, Charles R LCpl	2435556	B-1-7	8/12/69	20	Raleigh, North Carolina
Withey, Robert R Cpl	2026895	D-1-3	7/17/65	22	Detroit, Michigan
Witt, James P 2dLt	0106078	D-1-7	2/14/69*	21	Fairview Park, Ohio
Wojcik, Michael F Sgt	2041162	3-Amphib	5/22/66	22	Woronoco, Massachusetts
Wolfendale, Edward J LCpl	2328859	I-3-7	2/24/69*	19	Lawrence, Massachusetts
Womble, William T Jr LCpl	2168383	E-2-3	5/3/67*	18	Norfolk, Virginia
Wood, David R Sgt	2131190	FLSG	2/2/68	23	Clarence, Missouri
Wood, Lester E Sgt	1968049	H-2-5	8/12/66	23	Kingston, New York
Wood, Walter J 1stLt	0105408	F-2-9	4/28/69	22	Chester, Pennsylvania
Woodall, John B 1stLt	093518	K-3-9	4/30/67*	23	East Alton, Illinois
Woodham, Tullis J Jr LtCol	053444	CO-3-27	5/13/68	39	Jacksonville, Florida
Woodring, Willard J Jr Maj	059686	CO-3-9	7/10/67	40	Springfield, Missouri
Woods, Sterling S Cpl	2101086	B-1-3	5/10/67*	21	Virginia Beach, Virginia
Woods, Theodore Cpl	2186943	B-1-5	6/9/69	21	Springfield, Massachusett
Worley, Thomas J Jr LCpl	2386080	G-2-9	4/21/68*	20	Detroit, Michigan
Worrel, Thomas D LCpl	2528075	A-1-Recon	4/23/70*	20	Roanoke, Indiana
Wray, Robert B Sgt	2098565	CAP-1-3-9	9/13/69	22	Winston Salem, North Carolina
Wright, Edward R LCpl	2417283	D-1-9	2/11/69	19	Baltimore, Maryland
Wright, William F Jr Cpl	2040174	F-2-4	8/23-24/66	20	Brooklyn, New York
Wunsch, Michael C Capt	092921	A-3-Tanks	7/27/69*	25	Feasterville, Pennsylvania
Wyman, Michael J LCpl	2404772	D-1-7	2/14/69*	18	Buckner, Illinois
Xavier, Augusto M 1stLt	088544	VMA-311	3/10/66*	24	Beeville, Texas
Yale, Richard S GySgt	1168486	MAG-18	2/23/69	34	Akron, Ohio
Yates, John C Sgt	2022664	A-1-Amp	7/14/67	25	Fergus Falls, Minnesota
Yates, Thurman B LCpl	2130708	H&S-3-3	3/24/67	22	Baltimore, Maryland
Yeddo, Larry J Cpl	2133031	A-1-Amp	12/5/66	21	North Bangor, New York
Yeoman, Richard J 2dLt	094376	C-1-ATanks	3/15/67	22	Sparks, Nevada
Ynacay, Robert S GySgt	1189118	B-1-7	6/4/68	35	Portland, Oregon
Ynda, Benjamin J Cpl	2413651	G-2-7	9/8/70	20	Los Angeles, California
York, Hillous Sgt	2294390	C-1-5	5/10/67	22	Franklin, Kentucky
Young, Gerald V Sgt	2156516	A-1-Recon	10/17/67	21	Detroit, Michigan
Young, James R LtCol	049647	CO-3-1	3/4/66	39	Denver, Colorado
Young, Richard K Capt	078069	I-3-4	9/4/67	29	Greensboro, NC
Yunck, Michael R Col	007484	TAC	12/10/65	46	Northport, New York
Zahn, Leland D SSgt	1084924	D-3-12	4/5/67*	37	Harris, Iowa
Zaptin, Edward R 2dLt	0107579	D-1-7	2/16/69	21	Brooklyn, New York
Zell, Richard M 1stLt	095352	A-1-5	4/11/67	23	Queens, New York
Zende, Floyd W LCpl	2280907	M-3-1	9/18/67	19	Pittsburg, Pennsylvania
Zeno, Stanward Jr LCpl	2147012	F-2-4	8/17/66	21	Detroit, Michigan
Zimmerman, Edward "C" LCpl	2097504	F-2-4	6/25/66*	19	Muncie, Indiana
Zimmerman, Robert E 1stLt	095176	L-3-5	2/22/68	24	New York, New York
Zwicker, Ralph M III Sgt	2027341	L-3-7	6/30/66	22	Wilmington, Massachusette
Zwirchitz, Dennis J Pfc	2381896	H-2-4	.3/16/68*	20	Abbotsford, Wisconsin

The following Marines are believed to have earned the Silver Star but I have not been able to find information to verify this fact: Cpl. James J. Barrett, LCpl. Robert H. Brown, Sgt. John T. Burton, LCpl. Arthur C. Childress, Cpl. Paul K. Christensen, Cpl. Antonoi Crespocruz, LCpl. Michael D. Franks, Cpl. Robert L. Grace, Capt. Alan W. Hinson, LCpl. David M. Hughes, Cpl. James T. Johnson, Sgt. Thomas A. Johnson, Cpl. Timothy P. Johnson, Sgt. William M. Johnson, Cpl. James L. Jones, Jr, Pfc. Jessie Jones,

GySgt. Robert L. Jones, Jr., Cpl. Roger Miller, Cpl.Albert R. Mendoza, Cpl. Raymond L. Powell, LCpl. Donald L. Reek, LCpl. John R. Retrask, LCpl. Gary A. Rice, Sgt. Robert A. Schatz, Pvt. Willie M. Smith, Pfc. Thomas A. Stokes, LCpl. Terry W. Taylor, Pfc. William Eugene Taylor, Pfc. Michael G. Waggoner, Cpl. Charles Walters, Pfc. James R. Ward, Cpl. Kenneth J. Williams, Cpl. Floyd W. Young.

NOTE:
Floyd Book earned an additional Silver Star as a medic with the Army's Company C, 1st Battalion (Airmobile), 8th Calvary Division for action on 9/7/69.

Clyde Bonnelcyke earned two additional Silver Stars as a member of Company D, 2nd Battalion (Airmobile), 8th Army on 8/26/69 and 4/4/70.

Charles G. Blanton and Charles G. Clements is the same Marine, as is Wilton E. Melson, Jr. and Ken B. Williams.

NAVY CORPSMAN

Bardwell, Robert J HMC	6897172	A-1-3	7/17/65	20	Winfield, Kansas
Barnes, Gary L HN	B706764	F-2-7	2/23/69	21	Little Rock, Arkansas
Barraud, Wesley L HM2	5466509	H-2-9	12/18/65	22	Topeka, Kansas
Bates, Gilyard H HMC	327217	BC-1-5	1/7/68	36	Crane, Indiana
Bjishkian, Mark E S HM3	7773427	G-2-3	8/23/66	22	New York, New York
Bollinger, Lawrence C HM3	7784637	A-1-9	5/21/66	18	New York, New York
Bowling, Ronald V HM3	B821789	I-3-26	2/20/68	20	Price, Utah
Bowman, Harry T II HN	B124188	M-3-5	5/9/68*	22	Wood Ridge, New Jersey
Bradford, Richard F III HM2	9975869	C-3-Recon	3/5/69	22	Eastoner, South Carolina
Brasier, John J HM3	8940182	I-3-5	11/16/66	24	Wichita, Kansas
Broad, William R HM3	3532751	M-3-7	1/28/67*	25	Tulsa, Oklahoma
Brown, Bruce E HM3	B811996	CAP-1-3-7	6/3/69*	22	San Francisco, California
Brown, Charles F HM3	5952536	M-3-7	1/28/67	21	Fairbury, Nebraska
Brown, Robert L HMC	9961704	A-3-Recon	1/26/67*	36	East Point, Georgia
Burnley, Earl R Jr HM3	B301618	M-3-4	5/16/68*	18	Jackson, Mississippi
Butlin, Ronald Q HM3	B518888	F-2-7	2/23/69	20	Chicago, Illinois
Byrne, Conal J Jr HM3	7887094	F-2-4	9/21/67*	23	Drexel Hill, Pennsylvania
Campbell, Richard A HM3	B517728	F-2-3	8/7/69	21	Chicago, Illinopis
Campion, Charles G HM3	373910	C-1-3	6/5/65	27	Denver, Colorado
Carper, Loring W Jr HN	6894313	1-FRecon	5/17/66*	19	Winchester, Virginia
Cole, Alonzo P HM3	B889148	H-2-9	2/17/69	20	San Mateo, California
Coles, Alexander Jr HN	8404067	D-1-1	1/23/67*	21	Salem, New Jersey
Cooper, David L HM3	9150584	A-1-1	5/13/67	21	Milwaukee, Oregon
Counce, Donald E HN	6861529	F-2-1	2/28/66	21	Portland, Oregon
Creed, Edward G HM3	6925779	HMM-161	6/22/66*	21	Bossier City, Louisiana
Cress, Kenneth E HM3	7887894	F-2-7	10/28/67	21	Somerset, Kentucky
Davis, Blakely I Jr HM3	B304575	A-1-1	7/29/67*	20	Bradenton, Florida
Dona, Bienvenido C HM3	4680550	L-3-7	1/17/66*	38	Philippines, XP
Donovan, Thomas S HM2	9132355	F-2-5	6/2/67*	22	Natick, Massachusetts
Ellis, Donald R HM3	7973694	H&S-2-3	5/20/67*	21	Florence, Arizona
Elrod, James T HM3	6954121	B-1-5	8/10/66*	20	Moultrie, Georgia
Emanuel, Michael R HN	B137375	K-3-7	10/25/69	20	New Milford, New Jersey
Feldman, Edward M Lt	724224	BS-1-26	1/21/68	26	Forest Hills, New York
Fitzpatrick, Richard HM1	7990183	C-1-Recon	6/16/66	35	Vacaville, California
Fredette, Bradford T HN	6826136	E-2-7	12/10/65	20	Alameda, California
French, James L HM3	6817603	H&S-2-7	3/4/66	22	San Francisco, California
Furman, Richard L HN3	7778661	K-3-9	7/4/66	20	Palmyra, New Jersey
Garnett, Arthur H HN	6980059	A-1-1	10/18/65	20	Decatur, Alabama
Geise, Dell C HM2	9177728	F-2-5	1/26/67*	19	Burlington, Wisconsin
Gibbs, Michael G HM3	7959992	K-3-3	4/25/67*	21	Del Rio, Tennessee
Gibson, James R Jr HM3	6925568	M-3-7	12/25/65	20	Dallas, Texas
Gillies, Robert K HM3	9164494	B-1-5	4/21/69*	21	Mantua, New Jersey
Goldstein, Paul A HN	6881877	M-3-1	3/21/66	20	Minneapolis, Minnesota
Gray, William R HM3	1383786	I-3-9	1/28/69*	22	Fulton, New York

Name	Service #	Unit	Date	Age	Hometown
Greer, Gerald M HN	6951603	D-1-4	3/21/66	20	Atlanta, Georgia
Groshong, Allen E HM3	B204557	F-2-1	4/8/68*	22	Newport News, Virginia
Gunn, Daniel M HM3	6996591	C-1-1	7/23/66*	22	Fredericksburg, Texas
Hamlett, Martin J HM3	6972148	F-2-7	2/28/67	20	Scottsdale, Arizona
Hartigan, Larry A HN	B519786	K-3-7	2/23/69*	21	Maywood, Illinois
Hillhouse, David J HM1	6818625	HMM-262	3/25/69	23	Oakhurst, California
House, Michael A HN	9166335	M-3-9	4/30/67	19	Barnet, Vermont
Hunting, Neil D HM2	819935	H-2-7	8/20/70	21	Idaho Falls, Idaho
Johnson, Charles E HM3	7778277	G-2-3	8/23/67	22	Unknown
Johnson, Lawrence E HM3	5240638	H-2-7	3/4/66	26	Binghamton, New York
Johnson, William D HN	9034183	F-2-4	3/21/66	19	Providence, Rhode Island
Jones, Robert HM3	0000000	HMM-362	2/5/68	22	Rochester, Michigan
Keller, Allen N HM1	6923306	B-1-4	3/22/67	22	Jacksonville, Florida
Kelsey, John F HM2	5930933	L-3-3	3/3/67	24	Germantown, Maryland
Kempel, Michael R HM2	2429248	M-3-5	8/27/70*	22	Cuyahoga Falls, Ohio
Kickham, John V HN	SS#	HMM-364	2/5/71	21	Cleveland, Ohio
Kidder, Ronald W HM3	6866415	K-3-1	3/21/66	22	Casper, Wyoming
Kirkham, Donald A HM3	B506708	G-2-5	2/1/68*	22	Brookfield, Wisconsin
Kovach, Gary S HN	B438456	CAP-1-3-2	5/8/70	20	Toronto, Ohio
Krist, Matthew J HM3	7942936	H&S-2-3	8/21/66*	22	Pueblo, Colorado
Kuklenski, Michael J HN	B727552	A-1-7	5/29/69	20	Independence, Missouri
Kulas, Robert W HM3	6955783	I-3-5	7/24-25/66	21	Minneapolis, Minnesota
Laning, John E HM3	7794774	D-1-4	5/8/67*	20	North Muskegon, Mich
Leitner, Terry L HN	7716728	I-3-1	2/10/66	20	Kansas City, Missouri
Levin, Alan S Lt	689596	HMM-265	9/12/67	25	Skokie, Illinois
Lewandowski, Michael J HM2	9717643	F-2-9	3/18/66	20	Hammond, Indiana
Link, Daniel D HM3	1190687	M-3-5	2/15/67	19	Elmer, New Jersey
Loy, James R HN	B506923	F-2-1	1/11/68*	20	Green Bay, Wisconsin
Luttrell, Lloyd I HM3	B202273	D-1-1	1/24/69*	22	Lexington, Kentucky
Machmer, James A HM3	9982861	A-1-Recon	4/26/68	20	Jacksonville, Florida
Mariskanish, Charles E HN3	B416918	I-3-5	5/9/68*	19	Barnesboro, Pennsylvania
Mathis, James R HM2	5440810	I-3-26	4/9/68	25	Houston, Texas
Matticks, Robert W HM3	6866694	G-2-7	3/4/66	20	Denver, Colorado
Mertlich, Dale E HM3	9182694	M-3-1	1/15/67	20	Salt Lake City, Utah
Mierzwa, Raymond T HM3	6869950	F-2-26	5/16/67	20	Utica, New York
Moffitt, David H HN	B503142	G-2-4	3/10/69	21	Detroit, Michigan
Morris, Shane A HM3	7933529	A-1-9	5/21/66	19	Carmichael, Pennsylvania
Mulhaupt, Richard C HM3	501920	BAS-3-9	3/30/68	22	Chicago, Illinois
Mullen, Thomas A HN	9037446	D-1-3	2/14/66	22	Manchester, New Hampshire
Muller, Daniel S HM3	B586711	I-3-3	6/4/69*	25	Pitsburg, Kansas
Munoz, Pedro HN	6983388	B-1-9	5/12/66*	20	El Paso, Texas
Noah, Dennis L HM2	B509318	H-2-5	9/10/67	20	St. Louis, Missouri
Overmyer, Melvin HM3	1383965	H-2-26	5/25/67	21	Renton, Washington
Pena, Jesse J HN	B584194	B-1-7	2/12/70*	21	Davenport, Iowa
Peterson, Richard A HM2	9188078	H-2-4	5/21/67	20	Minneapolis, Minnesota
Phelps, Huger L HN3	B504280	-1-Amph	2/10/69*	22	Greenfield, Indiana
Porterfield, David E HN	7741866	H&S-3-4	7/18/66*	20	Mamaroneck, New York
Poth, Mansley R Jr HN	B784831	A-1-5	6/17/69	19	Galveston, Texas
Purdin, Patrick L HM3	B845700	B-1-7	11/22/69*	21	Okmulgee, Oklahoma
Rackow, Andrew C HM2	B407475	F-2-5	8/6/68*	20	University Park, Pennsylvania.
Radonovich, Michael F HM2	8400964	M-3-3	9/9/68	20	Cleveland, Ohio
Roach, Richard F HN	6969015	B-3-Recon	6/20/66*	19	East Liverpool, Ohio
Ronchetto, Dominic J Jr HM3	B509270	L-3-3	3/3/68	20	Benton, Illinois
Ross, James C HN	5964486	G-2-1	12/16/65	21	Des Moines, Iowa
Scala, Daniel HM3	B136907	G-2-7	3/15/70	21	Centereach, New York
Scearse, Roger D HN	2904549	F-2-9	6/11/66	20	Louisville, Kentucky
Scheppman, Stanley W HN	1389102	C-1-1	3/13/67	20	Long Beach, California
Schindeler, Theodore K HN	B115636	D-1-4	2/2/68	20	Boston, Massachusetts
Schon, John E HM2	9149203	H&S-3-5	5/26/67*	20	Portland, Oregon
Schultz, Steven O HM3	5990956	H&S-1-1	7/23/66*	25	International Falls, Minnesota

Name	Service #	Unit	Date	Age	Hometown
Seel, Walter P Jr HM2	B411492	G-3-12	2/25/69*	22	Moorestown, New Jersey
Sell, Robert R HM1	6895214	Advisor	2/25/68	22	Seattle, Washington
Sepulveda, Augustin Jr HN	B418645	D-3-Recon	5/16/69	21	Lorain, Ohio
Simmons, Travis A Jr HM3	7715206	F-2-7	3/16/67	23	Midland, Texas
Smith, Robert L HM3	6954300	M-3-4	10/5/66	20	Gadsden, Alabama
Southern, Joe F HM3	7704268	A-1-4	8/26/66	19	Greensboro, N.C
Stanton, Richard E HM3	B604645	E-2-5	2/3/68	20	Kansas City, Missouri
Stauropoulous, George W HN	1394983	C-1-9	7/2/67	20	Los Angeles, California
Stern, Philip L HN	B114699	M-3-5	2/8/68	20	Brunswick, New Jersey
Steward, Larry J HM3	5970591	D-1-1	1/5/66	20	Phoenix, Arizona
Stone, Douglas D HM3	B620093	I-3-3	2/9/69	20	Colorado Springs, Colorado
Strunk, William L HM3	9150696	H-2-4	2/24/67*	32	Portland, Oregon
Sullivan, Caleb J HN	9158021	I-3-12	12/28/66	20	Albuquerque, New Mexico
Szal, Anthony J Jr HM3	7746731	F-2-9	6/11/66	21	Aurora, Illinois
Tam. Michael R HM2	B510314	H-2-7	9/19/68	21	Monroe, Michigan
Tarrance, James C HN	B317129	I-3-26	12/4/68*	20	Jacksonville, Florida
Tarzia, Nicholas C HM3	6941016	E-2-4	8/8/66	20	Westbury, New York
Teague, Michael A HM3	9193333	G-2-4	5/2/68*	23	Brownwood, Texas
Thelen, Robert J Jr HM3	B721526	D-1-7	8/12/69	20	Oklahoma City, Oklahoma
Thirkettle, Michael J HM3	6766414	A-1-1	11/26/67*	20	Whittier, California
Toline, Kenneth D HM3	B629862	E-1-Recon	2/15/70	20	Polk, Nebraska
Trescott, Charles R HM3	5980445	G-2-5	5/3/66*	19	Dearborn, Michigan
Wallace, Clarence E HMC	B310534	A-1-3	11/14/68	20	Raleigh, North Carolina
Warren, Galen E HN	3909539	L-3-9	5/20/67*	21	Seattle, Washington
Watson, Donald P HM3	9987809	C-1-5	5/10/67	20	Chicago, Illinois
Wean, Douglas L HM2	B514066	K-4-12	9/19/69	20	Mt. Morris, Illinois
Wess, Michael A HM2	B502188	I-3-3	9/10/68	20	Winter Springs, Florida
Whinery, Roger L HN	B605148	L-3-9	6/1/67*	22	Fredonia, Kansas
Whitbeck, Eugene HM3	B517024	M-3-5	8/9/68	20	La Crescent, Minnesota
Wiggins, Delmar J HN	7751021	F-2-9	3/18/66	20	Anderson, Indiana
Willeford, Alton W HM3	B704802	M-3-26	1/25/69	21	Corpus Christi, Texas
Williamson, Michael L HM3	7948690	I-3-3	4/3/66	20	Wayne, West Virginia
Wood, Thomas J HM2	B413220	M-3-7	2/23/69	20	Philadelphia, Penn
Youngblood, Roy L HM2	6925243	M-3-5	7/22/66	21	El Centro, California

CHAPLAIN

Name	Service #	Unit	Date	Age	Hometown
Lyons, Richard M Lt	709750	H&S-1-1	1/31/68	24	Baltimore, Maryland

John Abely	Bobby Abshire	James Achterhoff	Henry Banks
Eric Barnes	Wendell Beard	Gary Becham	David Bendorf
Ronald Benigo	Dennis Bergenstein	Larry Berry	R. N. Bloomberg
David Boos	Jack Brandon	William Brown	Rich Buchanan
Chris Buchs	Johnny Burchett	Joseph Calabria	Clifton Canter

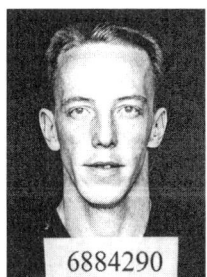

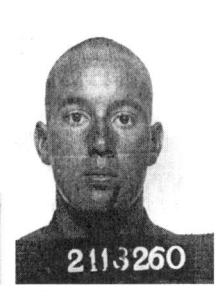

HENRY CASEBOLT	JERRY CAUDLE	JACK CENTERS	 RONALD CENTERS
 WILLIAM CHAPMAN	 HENRY CLAY	 CHARLES CLEMENTS	WILLIAM CONRAD
 JOSEPH CONVERY	CHARLES COURSER	 DAVID DANNER	BUCK DARLING
 FLOYD DARRH	 PHILLIP DENNON	 LELAND DEVINE	 THOMAS DOCKERY
 BILLY DONALDSON	RICHARD DRURY	THOMAS DUBROY	 ANDREW T. DYE

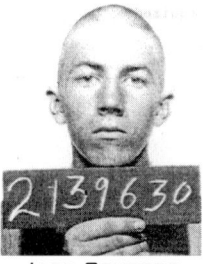

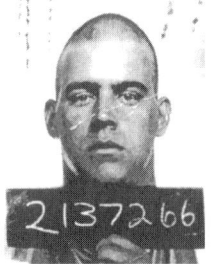

Albert Ellis	 George Emberger	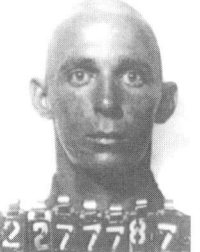 Thomas Foy	Glen Francis
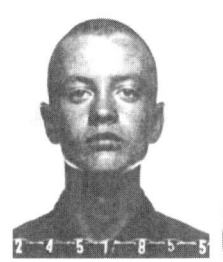 Jerry Frazier	 Duffy DuFriend	 Kenneth Gaffney	 William Gandy
 Conrad Gaywont	Michael Getlin	Richard Gillingham	 Andrew Grignon
William Gum	 Norm Gutzwiler	John Hagler	 Richard Harrelson
 Herbert Hammons	 Mark Herron	 Benjamin Houze	 Jimmy Howard

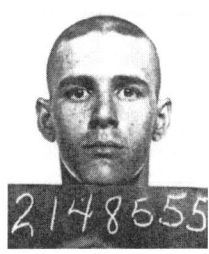

Ralph Howard	Gordon Huckstep	Hubert Hunnicutt	Lawrence Isham
Frank Jandik	Clarence Jones	Larry Jones	Donald Judson
Daniel Kaylor	Billie Kean	Albert Keller	Ira Lackey
Virgil Laizure	Eddie Landry	Dale Lawson	Robert Linn
Roger Lipscomb	Bryce Lockwood	Francis Lovely	John Loweranitis

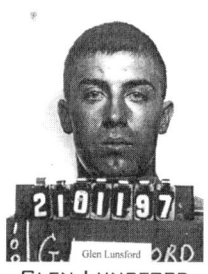

GLEN LUNSFORD	ALFRED LYON	 JACK MARINO	THOMAS MCKEE
TERRANCE MEIER	RICHARD MILLER	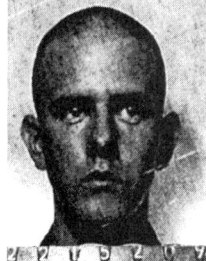 MICHAEL MILLS	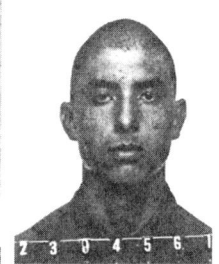 ALBERT MOGUEL
FRED MONOHAN	 NORRIS MONTJOY	DANIEL MORRIS	LONNIE MORRISON
ROBERT MORRISON	 RICHARD MOSS	JOE MUIR	FRANCIS MULDOWNEY
JOHN MULLAN	JAMES MULLOY	WILLIAM MYERS	JOSEPH NESMITH

Gleason Norris	Joseph O'Brien	John O'Donnell	Thurman Owen
Marvin Paxton	Millard Pearson	Elias Pembleton	Anthony Perino
Barnett Person	Jack Phillips	 Woodrow Phillips	Raymond Powell
Harry Ramsburg	Tiago Reis	Oliver Renfro	Earl Richardson
Victor H. Ridlon	Virgilio Rivera	Joseph Roberts	Ray Rogers

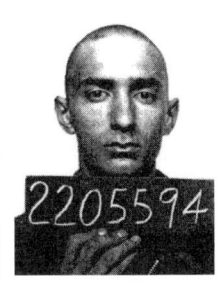

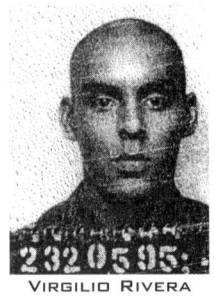

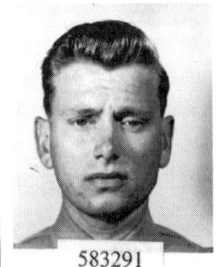

Joseph Roman	Kenneth Ross	Edward Rozanski	Morris Ruddick
David Russell	John Rusth	John Sasser	Kenneth Schauble
Robert Schley	Ronald Schumacher	Fred Scott	Patrick Scully
Raymond Sexton	Edward Sherman	David Siemon	Willie Simpson
Conrad Sipple	Duncan Sleigh	Coy Smith	Ken Stankiewicz

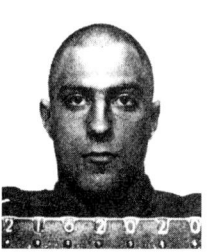

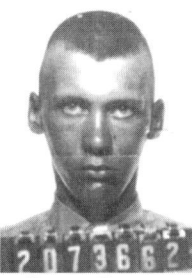

JAMES STUCKEY	MARION STURKEY	JOSEPH SULLIVAN	 ROBERT SYLVESTER
HARVEY THOMPSON	 WAYNE THOMPSON	ROBERT THOMS	RICHARD TILGHMAN
RICHARD TONUCCI	 COREY TRAYLOR	 ELIAS TREVINO	CECILIO VALLEJO
 PAUL VERRETTE	STEVE WALTRIP	BOBBY WHITE	MAURICE WHITMER
 JOE WHITTINGHAM	 PAUL WIGGINS	AUGUSTO XAVIER	JOHN YATES

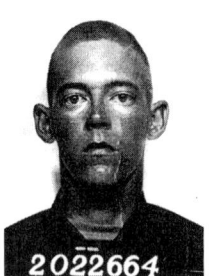

SCOTT BRANK AND JAMES H. TYLER
TWO MARINES WHO
EARNED THE SILVER STAR

UNIDENTIFIED LIEUTENANT, RICHARD
BUCHANAN (NAVY CROSS) AND A
CORPSMAN NAMED WALTER. AT A BASE
CAMP SOUTH OF DANANG IN 1968.
BUCHANAN WOULD BE HOME AND OUT
OF THE CORPS WITHIN 72 HOURS OF
THIS PHOTO.

LEFT TO RIGHT:
KENNETH KEITH, UNIDENTIFIED MARINE,
LEONARD KASSON, UNIDENTIFIED MARINE
KEITH AND KASSON EARNED THE SILVER STAR

Willis Wilson and Thurman Owen

Cpl. Robert Sylvester, 1st Lt. Robert Pitre, HN Mike O'Brien and Pfc. Johnny Haas

BIBLIOGRAPHY

BOOKS
Ballentine, David A. *Gunbird Driver*. Annapolis, MD: Naval Institute Press, 2008.
Buster, R. Dalton. *The Walking Dead*. Edmonton, KY: Self Published, 1987.
Caputo, Phillip. *A Rumor of War*. New York, NY: Holt, Rinehart and Winston, 1977.
Coan, James P. Con Thien: *The Hill of Angels*. Tuscaloosa, AL: The University of Alabama Press, 2004
Conroy, Michael R. *Don't Tell America*. Red Bluff, CA: Eagle Publishing, 1992.
Corbett, John. *West Dickens Avenue: A Marine at Khe Sanh*. Novato, CA: Presidio Press, 2003.
Culver, Richard O. *The Saga of the M-16 in Vietnam*. Internet, 1999.
Fox, Col. Wesley L. *Marine Rifleman*. Washington D.C.: Brassey's Inc., 2002.
Hammel, Eric. *Fire in the Streets: The Battle for Hue*, Tet 1968. Chicago, IL: Contemporary Books, 1991.
Lehrack, Otto J. *No Shining Armor: The Marines in Vietnam*. Lawrence, KS: UniversityPress of Kansas, 1992.
Lehrack, Otto J. *The First Battle*. Havertown, PA: Casemate, 2004.
Murphy, Edward F. *Semper Fi Vietnam*. Novato, CA: Presidio Press, 1997.
Murtha, Gary D. *Timefighter: A Marine in Vietnam*. Kansas City, MO: GDM Publications, 1985
Myers, William L. *Honor the Warrior: The United States Marine Corps in Vietnam*. Maurice, LA: Redoubt Press, 2000.
Nolan, Keith W. *Battle for Hue: Tet, 1968*. Novato, CA: Presidio Press, 1983.
Nolan, Keith W. *Operation Buffalo: USMC Fight for the DMZ*. Novato, CA: Presidio Press, 1991
Norman, Michael. *These Good Men*. New York, NY: Crown Publishers, Inc., 1989.
Prados, John and Ray W. Stubbe. *Valley of Decision*. New York, NY: Dell Publishing, 1991.
Smith, Barry J. *History of Fox 2/7*. Unpublished, 1999.
Spencer, Ernest. *Welcome to Vietnam*, Macho Man. Corps Press, 1987.
Stubbe, Ray W. *Battalion of Kings*. Wauwatosa, WI: Khe Sanh Veterans, Inc., 2005.
Sturkey, Marion F. *Bonnie Sue: A Marine Corps Helicopter Squadron in Vietnam*. Plum Branch, SC: Heritage Press International, 1996.
Warr, Nicholas. *Phase Line Green*. Annapolis, MD: Naval Institute Press, 1997.
West, Francis J. Jr. *Small Unit Action in Vietnam: Summer, 1966*. Washington, D.C. U.S. Marine Corps, 1977.
Winter, Ronald. *Masters of the Art*. New York, NY: Carlton Press, 1989.
Woodruff, Mark W. *Unheralded Victory*. Falls Church, VA; Vandamere Press, 1999.

MAGAZINE ARTICLES
Crumley, B.L. *"The Magnificent Bastards." Hit a Hot LZ."* Leatherneck, 2009.
Keene, R.R. *"The Flaming I' At Getlin's Corner."* Leatherneck, February 2009.
McConnell, Malcom. *"Forever Proud."* Readers Digest, November 1988.

Acknowledgments

A Book of this nature could not have been completed without the help of a large and varied group of people. I would like to express my sincere thanks to those who helped make this project a reality.

I deeply appreciate the assistance of the personnel at the National Personnel Record Center in St. Louis Missouri who have endured the hundreds and maybe thousands of individual military record requests that I have made over the past several years. I must hold the record for the number of individual requests. These records are the glue that holds this work together and the individual combat award citations confirm and provide validation for those whose names appear on the awards lists. Also to be thanked are Mike Fishbaugh and C. Douglas Sterner, Vietnam veterans, who are dedicated to validating those who received combat awards. Both were very helpful in adding names to my extensive but probably incomplete list.

My gratitude also goes to the clerks who work at the National Archives in College Park, Maryland who were graciously and most efficiently helpful on my many visits there. With special thanks to Nathaniel Patch. They provided access to the many Marine Corps Unit Personnel Rosters that were necessary in compiling this work.

I was also inspired by many of the Marines that I served with and some that I have known. Among them are Ed Anderson, R. E. Bair, Leland Devine, Dale Flickinger, Larry Hunt, Kendall Kirksey, Ron Lantz, Curtis Latiolais, Billy Lawless, Sam Meredieth, Danny Murphy, John Mullan, Joe Nesmith, Marvin Paxton, Earl Richardson, John Sommers, Cecilio Vallejo, Paul Verrette, John Warndahl, Franklin West and Laverne Wilson.

Mostly, though, I offer my thanks, appreciation and deepest gratitude to the veterans of the war in Vietnam whom I contacted. Almost all of them agreed to recount their recollections of what they experienced more than forty years ago. They shared their personal stories and in some cases provided personal written accounts of the fights that they had participated in. Some provided photographs and the names of buddies who had fought with them. My gratitude would be impossible to express in words.

Among them are Tobby Baca, Arthur Blades, Scott Brank, James Cannon, Bruce Cruikshank, Richard Culver, Lawson Davis, Pierce Durham, Rudolfo Escotto, Ray Gaul, Ted Golab, Terry Gulch, John Hagler, Richard Harper, Peter Janss, James Kaylor, Tony Kelichner, "Frenchy" Lafountaine, Walter Ledbetter, Gary Linick, Robert Linn, Danny Maguire, Karl Marlantes, Floyd Miller, Richard Neal, Conrad Ortego, Paul Parker, Roy Parr, Ken Pipes, Robert Pitre, Milton Pittman, Richard Porello, Harry Ramsburg, Jim Rider, Victor Ridlon, Jon Rittler, Louis Roman, Lawrence Seavy-Cioffi, Barry Smith, Jim Spence, Bill Stankowski, George Sternisha, Ray Stubbe, Marion Sturkey, Dennis Sykes, Robert Sylvester, Nicholas Warr, Eddie West and Willis Wilson.

NAME INDEX

Aaron, James R. 6, 20
Aaron, Raymond C. 45
Abare, Carl M. 145
Abbott, Lebrun A. 121, 127
Abely, John 325, 369
Abshire, Bobby W. 56, 67, 315-316, 318, 369
Achterhoff, James P. 325, 369
Adams, Charles W. 20
Adams, Glenn E. 134, 145
Adams, James W. 229, 231
Adams Richard G. 14, 20
Adinolfi, John 129, 134-137, 139, 145
Akins, Ronald P. 188
Albano, Paul J. 51, 67
Alexander, Bill 140, 145
Alexander, Donald 104, 106
Aldrich, David A. 196
Alford, Ernest S. 135, 145
Allen, Raymond C. 40
Allen, Yale G. 254-255, 259-260, 264, 318
Alley, Robert 279
Almeida, Richard 134, 138
Almoney, John S. 134, 138
Alvarado, Julio A. 176
Alvarez, Andrew L. 145
Amador, Ernest B. 76, 102
Anderson, David B. 196-197
Anderson, Ed 378
Anderson, George E. 20
Anderson, James 134, 138
Anderson, Jesse R. Jr. 20
Anderson, Larry M. 39, 45
Andrus, Kermit W. 313
Angel, Michael E. 253, 255, 262
Anter, Albert G. 121
Apodaca, Gilbert 173, 176
Arant, George F. 145
Arcand, Paul R. 127
Armenta, Ruben M. 121
Armstrong, Daniel T. 14, 21
Ashby, Patrick D. 245, 248
Atkins, Palmer R. 94, 106
Aungst, Terry L. 145
Avalos, Manuel, Jr. 29, 42, 326
Avent, Roger L. 127
Averitte, William C. 205, 208

Baca, Roger J. 54, 67
Baca, Tobby P. 261-262, 264, 378
Backus, Richard D. 130, 145
Bachta, Thomas E. 48-50, 53, 56, 57,67, 326
Badnek, Samuel J. 11, 15, 318
Baggett, Curtis 40, 45
Bailey, Gordon L. 106

Bailey, Larry J. 60, 67, 106
Bailey, Larry D. 76
Bair, R.E. 378
Baird, John R. Jr. 242, 246, 326
Baker, Allen J. 217, 219, 227, 229
Baker, Daniel P. 215
Baker, Edward M. 89, 107
Baker, Richard L. 157-158, 165
Baker, Sam R. 79, 104, 107, 326
Balkcom, Ronald W. 127
Ballance, Larry D. 269-270, 272, 277
Ballentine, David A. 69, 103-104, 125, 127
Banks, Henry D. 199-202, 204, 209, 215, 369
Banks, Thelbert R. 201, 215
Baptiste, Michael B. 187-188
Barker, Daniel R. 245, 248
Barnard, Ezekial C. 119
Barnes, Eric M. 327, 369
Barnett, Leldon D. 219, 231
Barnett, Robert L. 145
Barrett, John P. 21
Barrow, Robert H. 233
Bartley, Michael 127
Batson, Robert F. 13-14
Bauman, George F. 2
Bauman, Rigor 315
Baylark, Johnnie Jr. 275
Beandette, Thomas 86, 107
Beard, Wendell O. 129, 139, 144-145, 369
Becham, Gary V. 327, 369
Beckner, Duane 275, 277
Belknap, Ronald L. 48-49, 56
Bell, David 40, 45
Bell, Franklin L. 59, 67
Bell, Stanley M. III 253, 255, 264
Bellina, Johnny F. 185-186, 197
Bendorf, David G. 318, 369
Benigo, Ronald 93, 95, 102, 107, 327, 369
Bennett, Ernest L. 45
Benoit, David W. 107
Bergenstein, Dennis 132, 138, 369
Berheide, James R. 28, 42, 45
Bernard, Rod 279
Bernhardt, Webster 34, 45
Berry, Larry J. 327, 369
Best, Earl R. 203, 215
Beyerlein, David A. 236, 242-243, 246, 328
Bianchini, Michael L. 21
Biehl, Gary L. 134, 138
Biehl, Michael C. 254-255, 264, 328
Bielecki, William 79, 99, 107
Bienvenu, Richard 277
Biggers, Archie J. 240, 246, 248, 328

Billingham, Frederick A Jr. 18-189, 191
Binnebose, Paul E. 45
Birch William L. 127
Bish, 292-295, 297
Black, George V. 245, 248
Blackston, James D. 130, 146
Blackston, James L.
Blades, Arthur 71-72, 94, 96-99, 101-102, 105,-106,328, 378
Blakeney, Jane 314
Blanchard, Elmer R. 117-118, 127
Blanchfield, Richard 178, 194, 197
Blair, 296
Bledsoe, Michael A. 40, 45
Blevins, James E. 114-116, 121
Bliss, Robert E. 162-163, 165
Bloom, Allan H. 4, 21
Bloomberg, Richard N. 328, 369
Bloomfield, Harry G. 170, 173
Blundell, Benny 282-283
Bobo, John 109-111, 114-117, 121-127, 317
Boden, Edward E. 127
Bodenweiser, Alec J. 217, 230-231, 328
Boehm, Larry J. 237, 246
Bogard, Robert N. 157-158, 165
Boggia, Richard M. 5, 10-11, 21
Bondarewicz, George S. 60, 67
Boney, William 170-171, 176
Boone, Samuel Jr. 193, 197, 328
Boone William E. 274-275
Boos, David C. 30, 42, 45, 369
Borboa, Oscar F. 235-236, 243, 248
Borjas, Antonio 205, 215
Bornemann, John F. 54, 58, 61-62, 67
Borowitz, John M. 127
Bourgett, Terrance P. 83, 107
Bousquet, Robert G. 10, 15
Bowers, Gene W. 162, 165
Boynton, Charles B. Jr. 208
Bracy, Robert A. 90, 107
Bradley, Christopher M. 119, 127
Brake, Boyd L. 255
Brand, Thomas N. 14
Brandon, Jack A. 329, 369
Brank, Walter S. III 312, 329, 375, 377
Braun, Kenneth R. 109, 117, 120-122,127, 324
Brazzel, Irwin 96, 107
Bredeson, Dale O. 3, 21
Breeding, Fred V. 141, 145
Brees, Marvin F. 127
Brellenthin, Michael J. 188-189, 191-192
Brewczynski, Stanley L. 54, 67
Bright, Calvin E. 179, 183, 197
Brinkman, Clarence "Boo Boo" 278
Britt, Ted D. 192-193, 196-197, 329
Brockman, Russell B. 229, 231
Brogan, Bill 127
Bronson, Paul W. 14-15, 21

Brooks, James R. Jr. 3, 15
Brown, Andrew R. 60, 62-63, 67
Brown, David A.
Broen, David D. Jr. 208
Brown, Glen W. Jr. 145
Brown, James E. 105, 107
Brown, Joseph 86, 107
Brown, William G. 247- 248, 369
Brown, William H. 98, 107
Browne, Chedmond R. 121, 127
Broyer, Clifton L. 251, 255, 263, 329
Brubach, John P. 275
Bruce, Curtis B. 44-45
Bruder, James R. 182, 188-189
Brunk, Ricky 186, 197
Brunson, William 145
Bryant, Alvin 40, 45
Bryant, William P. 186, 197
Buckalew, Raymond W. 40, 45
Buchanan, Richard W. 319, 369, 377
Buchanan, William L. 58, 67
Buchs, Christopher J. 6, 15, 21, 330, 369
Bull, Edward D. 253, 255, 264, 330
Bullard, Cecil E. 173, 176
Burchett, Johnny R. 330, 369
Burgaretta, Carmelo 120, 127
Burghardt, James E. 113, 123, 127, 330
Burke, "Doc" 141
Burke, William 45
Burkes, James 40, 45
Burns, George 159
Burton, Donald R. 36, 41, 43-44
Bush, Richard H. Jr. 107
Buster, Nancy J. 266
Buster, Robert D. 266-272, 277
Butler, Fred III 234, 246
Butler, Gary N. 52-53, 67
Butt, Thomas E. 113-117, 123, 127, 330
Butts, Clifford 97

Cage, Leroy E. Jr. 275
Cahill, John J.H. 202, 214-215
Calabria, Joseph R. 330, 369
Callaway, James M. 45
Callaway, Jerril 222, 227, 231
Callaway, Robert 49, 67
Calogne, Eugene 84-86, 107
Calvin, Nathaniel 53, 67
Calzia, Frank 180-181, 183, 197
Campbell, William R. 225, 229
Campos, Arturo H. 171, 176
Cannon, Edward E. 121-122
Cannon, James R. 129-136, 138-139, 141, 144-145, 161, 330, 378
Canter, Clifton H. Jr. 129-131, 145, 369
Canton, Albert H. 245, 248
Caputo, Phillip J. 24, 45
Cargile, John W. 202, 205, 215

Carroll, Patrick G. 136, 145
Carstens, Thomas J. 130-131, 145
Carter, David O. 203, 209, 215
Carter, Robert L. 275
Carter, Tommie L. 21, 276-277
Casabar, Lorenzo 21
Casebolt, Henry C. 319, 370
Casey, Thomas E. 186, 193, 197
Castillo, Jose 178, 197
Cathie, Harold A. 245, 248
Caudle, Jerry W. 331, 370
Cave, Robert S. 245, 248
Cederholm, Roger W. 14, 21
Centers, Jack 331, 370
Centers, Ronald L. 9-10, 15, 21, 331, 370
Cerillo, Michael E. 38, 52
Chacon, David A. 238, 246, 331
Chamberlain, Gene 117-118, 123, 127, 331
Chapman, Darrell H. 237-238, 246, 248, 332
Chapman, Russell G. 172, 176
Chapman, William W. 332, 370
Cherry, David E. 267, 272-273, 277
Chester, Larry 275
Chervenak, Michael P. 148, 155-157,159, 165
Childers, Lloyd F. 2, 16
Childs, Benny 281
Chiles, Charles R. 21
Chill, John E. 313
Chippero, James A. 215
Chittester, Norman P. 238, 246
Christian, Daniel K. 208
Christman, William J. III 234, 242-244, 246, 319
Christy, Howard A. 56, 67, 332
Cicala, John A. 180-181, 184, 186, 197
Cicio, Robert D. 204, 208
Cimmerman, George 203, 215
Cirrilo, Michael E. 33, 45
Claire, Kenneth 177-178, 180-183, 189-190
Clausen, Raymond M. Jr. 289, 293-295, 298, 307-310,313, 317
Clouse, Jimmie R. 14, 21
Clay, Henry H. 332, 370
Clay, Doyle G. 189
Clements, Charles G. 332, 370
Cloutier, Joseph R. 209, 215
Coan, James P. 148, 217, 231
Cochran, Robert F. 12, 15, 319
Coffman, Harold L. 72-73, 75, 83, 93, 95, 99, 101, 104, 107
Coggins, William F. 250
Cogley, Paul J. 199, 215
Colbert, Robert O. 40, 45
Cole, Elvin H. 138, 146
Collins, Robert H. 208
Collopy, William 127
Compton, Homer D. 201, 215
Conrad, William C. Jr. 168, 176, 370
Conroy, Michael R. 233, 247

Connelly, Edmond J. III 248
Connelly, William C. 205-207, 209, 215, 332
Conner, Simpson Jr. 245, 248
Converse, Richard T. 245, 248
Convery, Joseph F. Jr. 333, 370
Cook, Michael D. 41
Cooney, Charles C. 4-5, 15, 21
Corbett, John A. 177, 197
Corey, Robert A. 127
Coriz, Fidel A. 127
Corso, Charles J. 101, 107
Corson, Richard III 50, 67
Cortez, Joe A, 21
Cosner, Leonard 236-238, 241-242, 248
Costa, Anthony B. 50, 54-55
Costa, James T. 21
Cotton, Ollie R. 14
Courchane, Dale L. 41
Courser, Charles B. 267, 269, 275, 277, 370
Cox, William J. 97, 107
Craig, Bruce K. 208
Craun, Gale E. 208
Crawford, Timothy J. 146
Cretney, Warren C. 288, 313
Crews, Billy D. 162, 165
Cronkite, Walter 166
Cropper, Whitney 278
Crosby, Glen H. 112-113, 127
Cruikshank, Bruce W. 290-298, 300, 302-305, 306, 313, 378
Crumbaker, Larry H. 121
Cruzesquilin, Juan H. 131, 146
Cuellar, Julian C. 41
Culbertson, John J. 160
Culpepper, Jeffrey E. 179, 190-191, 197
Culver, Richard O. 145, 148-161, 164-165, 316, 333, 379
Cummings, William T. Jr. 54, 67
Curry, Melvin R. 203, 209, 215
Curtis, Philip G. 162, 165
Cushing, Barney A. 40, 45
Cushman, Christopher 96, 107
Cutrer, Tyrone 92, 107

Daily, William H. 236, 248
Dalby, John D. 7, 21
Daniel, Stephen A. "Tex" 270, 275, 277
Danner, David J. 319, 370
Daniels, Jesus, Jr. 53, 67
Darling, Marshall "Buck" 70-71, 77, 92-100, 105-107, 370
Darrh, Floyd J. 40, 42, 45, 334, 370
David, Paul 278
Davidson, James G. 170, 176
Davis, Charles C. 130, 134, 138
Davis, James H. 234-236, 241, 248
Davis, Joseph L. 58-59, 67
Davis, Lawson R. 65-67, 378

Davis, Lee R. 40, 45
Davis, Raymond G. 232
Davis, Roger A. 53, 67
Davis, Thomas E. Jr. 201, 215
Decker, James H. 234, 248
Decker, Joseph T. 248
Dedek, John F. 246
Delgado, Raymond R. 253, 255, 262
DeLong, Earl "Pappy" 129, 139, 144, 146, 334
Delmark, Francis J. D. 14
Demaagd, Harvey J. 194, 197
DeMille, Roy W. 139, 143, 146, 334
Dennon, Phillip E. 60, 67, 370
Denton, Charles L. 9, 15, 21, 334
DeRolf, Bruce E. 60, 67
Detrick, Thomas A. 183-184, 186, 197
Devine, Leland W. 379
Dewitt, James P. 15
Dias, Raymond R. 201-202, 209, 215, 334
Dickey, Robert R, III 26, 45
Dickman, Wayne E. 45
Dickson, Harold S. 245, 248
Diez, Larry W. 35, 45
Dillon, John W. 178, 194, 197
Dockery, Charles M. 119-120, 122, 127
Dockery, Thomas J. 95, 100, 107, 370
Dodson, Jerry 181, 188-189
Dolan, William J. 205, 208
Donaldson, Billy M. 47-48, 50, 56, 57, 67, 319, 370
Donaldson, Stephen E. 274-277
Donathan, Richard P. 77-78, 99, 102, 104
Doney, Basil C. 215
Doronzo, Paul F. 298
Dorsett, Herschel D. 4, 21
Doub, Jay A. 15-17, 21
Doucet, James 278
Dowdy, Ralph W. 45
Downing, William K. 135, 138-140, 335
Doyle, Robert W. 174
Drakeford, Jackie L. 146
Drummond, Julius III 245, 248
Drury, Richard L. 57-58, 67, 315, 370
Dubroy, Thomas E. 199, 215, 370
Duerr, Richard 241, 243
DuFriend, James M. "Duffy" 59-60, 64, 67, 371
Duke, Griffith P. 275
Dupont, Stephen P. 146
Duran, Steve G. 15
Durham, Pierce N. 187-188, 197, 379
Dusman, Edward R. 51-52, 67
Dutton, John 43
Duvall, Raymond 88, 104-105, 107
Dye, Andrew T. 335, 370

Easter, Jerry E. 116, 127
Edgar, Gene E. 193, 197-198, 335
Edgell, Joseph C. 21
Edie, Kurt C. 138, 140-141

Edwards, Bill 279
Eernisse, Richard F. 221, 227, 231
Ehlert, Norm 37-38, 45
Eller, Johnnie L. 139, 146
Ellis, Albert J. 80-84, 107, 371
Elrod, James T. 102, 366
Ely, Richard E. 11, 21
Emberger, George J. 336, 371
Emerick, George A. 7-8, 13-15, 18, 21, 336
Engel, Gilbert L. 127
English, Robert T. 79, 101, 107
Enockson, John O. 54, 67, 315-316
Enos, Charles R. 21
Errera, John V. Jr. 127
Escotto, Rudolfo 64, 67, 379
Estep, Michael H. 186, 198
Etheridge, James 121, 127
Eubanks, D. L. 45
Evans, Thomas W. 229, 231

Farrell, Leo J. 51, 67
Faught, David L. 94, 102
Federowicz, Michael J. 203, 209, 215, 336
Feerick, Robert 86, 107
Fennell, Edward S. 130, 146
Ferbos, Stanley 136-137, 146
Ferguson, Marion F. 170, 173
Field, Robert E. 21
Fierro, Alejandro C. 146
Findley, Jon L. 127
Fink, Charles D. 7-8, 21
Fish, Robyn W. 236, 248
Fisher, Joseph R. 2, 5, 15, 18-19, 21, 337
Fitch, Patrick J. 180
Flannery, Dennis S. 229, 231
Flannery, John V. Jr. 146
Flannery, Robert E. 134, 138
Fletcher, Sterling A. 50, 67
Flickinger, Dale 379
Flyte, Forrest J. 229
Folks, Kermit 279
Forbes, Arthur L. 312-313
Ford, David A. 21
Ford, David R. 201, 209, 215
Foreit, Robert L. 140, 146
Foreman, Walter D. 45
Forgacs, Terrance S. 45
Foster, Jerome C. 193, 197-198, 337
Foulkes, Gilmore I. 248
Fox, Wesley L. 233-237, 239-241,
    244-246, 248, 317
Foy, Thomas L. 337, 371
Fracker, Douglas M. 170, 173
Francis, Glen A. 337, 371
Frazier, Pierre R. 337, 371
Frazier, Melvin D. 193, 197-198, 337
Frinzi, Joseph D. 127
Frost, John H. 238, 248

Fry, Daniel E. 245, 248
Fudge, George 81-82, 107
Fuller, Eugene E. 41
Furleigh, James R. 74-80, 83, 99, 101, 104, 107
Futch, Van 86, 107

Gaffney, Kenneth M. 337, 371
Gallegos, William 130, 146
Galloway, Edward R. 21
Galyean, Althea 284
Gandt, Thomas J. 134, 146
Gandy, William G. 338, 371
Gardner, Grady V. 41, 45
Garr, Raymond G. 107
Garrett, Floyd W. Jr. 118, 127
Garrotto, Alfred F. 80, 107
Garza, Franco 264
Garza, Ysidore 40, 45
Gaspard, Robert J. 215
Gatchel, Richard A. 119, 127
Gaul Ray H. Jr. 117-118, 127, 379
Gaywont, Conrad J. 101, 107, 371
Geary, William S. 275-276
Geller, Charles G. 182-183, 186-187, 189
Gessner, William G. 205, 215
Getlin, Michael P. 109-112, 114, 116, 121-122, 319, 371
Gibson, George R. 49, 56, 67, 320
Gigliotti, Pasquale 120, 127
Gilliland, Woody F. 87-90, 102, 107
Gillin, Walter A. 298, 309, 312-313
Gillingham, Richard K. 320, 371
Gillis, "Dobie" 315
Gipson, Billy 127
Glover, Ronald W. 40, 45
Golab, Theodore G. 188, 194, 198, 379
Gomez, David J. 127
Gonzales, Edwardo J. 139, 143-144, 335
Gooding, Robert E. 253, 255
Goodner, Robert 82, 107
Goodnight, David L. 251, 264
Gordon, Irwin 30, 35, 39, 45
Grace, Dennis 305, 307-308, 313
Grace, Ed 308
Greene, Wallace M. Jr. 99, 154, 156
Gregg, Raymond E. 21
Grey, Cameron 229, 231
Guillory, Wendell 203, 208
Guinee, Vincent J. Jr. 90, 102-103, 107, 339
Grignon, Andrew L. 339, 371
Groves, Donald D. 263-264, 339
Gulch, Terrance R. 163-165, 379
Gum, William E. 339, 371
Gutzwiler, Norman P. 339, 371
Guy, Joseph L. 236, 246, 248

Haas, John Adolph 278-284, 376, 377
Haase, Charles T. 201, 215

Haberman, David 202, 208
Haddix, Douglas B. 102
Haga, Gene H. 107, 339
Hagara, Leslie P. 203, 208-209
Haggerty, Michael 65, 67
Hagler, John 26-27, 30-31, 34-43, 45, 371
Haines, Paul A. 41
Hale, William E. 229
Hall, George 21
Hall, Roy 130, 146
Hammel, Eric 166
Hammett, Gary N. 9, 21
Hammons, Herbert D. 170, 173, 340, 371
Hamrick, Benjamin 50, 56
Hanley, Carlos R. 10, 21
Hairston, Clifford O. 107
Harding, Jimmy H. 254, 259, 264
Harman, Milton L. 87, 107
Harper, Richard O. 51-52, 57, 67, 379
Harper, William R. 209, 215
Harrell, Raymond D. 298
Harrelson, Richard C. 60, 67, 371
Harrington, Myron C. 167-173, 176, 320
Harris, Donald L. 11, 21
Hartzell, Charles B. 206-207, 215
Harvey, Richard F. 146
Harwood, Michael H. "Woody" 277
Hasty, William D. 41
Hatcher, Bob 263
Hawley, James G. 133, 139, 146
Hazelbaker, Vincil W. 51, 54-58, 67, 320
Hayden, Farrest 6, 21
Hayden, Michael P. 199, 201-202, 208-212
Hayes, Brien A. 223-226, 231
Hayes, Jerry C. 101, 107
Hayes, Phillip III 189
Hayes, William A. 222-223, 225, 228-229, 340
Hayes, Wayne N. 159, 165
Hays, Larry T. 172, 176
Healey, Edward F. 134, 146
Healy, David Y. 90, 107
Heath, Lloyd L. 131, 138
Heavin, Willis I. 134, 136, 138
Hebert, Teddy 279
Hebron, James 179, 190-191, 198
Hedger, John A. 51-52, 56, 67
Heifner, Thomas E. 312-313
Henderson, James W. 86, 107
Henderson, Jim A. Jr. 175-176
Henrich, Bruce J. 15
Henry, Gwinn A. Jr. 33-34, 36, 42, 45
Herron, James M. 255, 264,
Herron, Lee R. 241-244, 246, 320
Herron, Mark P. 341, 371
Hester, Michael P. 237, 245-246, 248, 341
Hickman, David W. 248
Higbee, Robert D. 102
Hill, Billy Joe 120, 127

384

Himmer, Lawrence 203-208
Hinkle, Jack L. 208
Hinkel, Russell C. 67
Hinman, Donald E. 139, 146
Hodges, Richard P. 239, 246, 342
Hoeck, Richard J. 15, 21, 342
Hoffman, Randall J. 123-124, 127
Hogstad, Edward B. 21
Holdway, David E. 134, 138
Holland, William P. 205, 216
Holliman, William D. 179, 198
Holmes, Lyle A. 275
Homer, Lee E. 21
Hooten, Richard J. 12, 21
Hooyman, Dennis R. 35, 42, 45
Hoppes, Michael L. 21
Horn, Thomas D. 244, 248
Houtz, Larry D. 205, 216
Houze, Benjamin C. Jr. 342, 371
Hovietz, Carl R. 129-131, 146
Howard, Gregory M. 94, 102
Howard, Jimmie E. 105, 107, 317, 371
Howard Joseph 60, 67
Howard, Ralph 342, 372
Huckstep, Gordon W. 342, 372
Hudson, Charles H. 248
Hults, Daniel E. 21
Hunnicutt, Hubert H. 204-207, 209,
    212-214, 216, 321, 372
Hunt, James A. 201, 216
Hunt, Larry 379
Hunt, Richard V. 36, 38, 42, 45
Hunter, Larry E. 40, 45
Hunter, Samuel H. 146
Hurlburt, Roy D. 203, 208
Huskey, James H. 216
Hutson, Charles E. 127
Hutson, Lawrence E. 204, 216
Hutson, William A. 40, 45
Hutto, Ira E. 245, 248
Huwel, Michael F. 134, 138

Ichord, Richard H. 164
Imlah, Jack S. 168, 171-173, 176
Inks, Richard L. 21
Irish, Michael W. 21
Irvin, Ronald H. 130, 146
Isham, Lawrence C. 102-103, 107, 372
Itczak, Richard R. 202, 216
Ito, "Mr." 158
Izenhour, Frank M. Jr. 129, 133, 146
Iverson, Richard L. 50, 67

Jackson James R. 86, 107
Jackson, Lester T. III 245, 248
Jackson, Nathaniel E. 204, 208
Jackson, Walter P. 102
Jacques, Donald 177-180, 184, 189-190

Jager, Arjen S. 127
Jandik, Frank Jr. 168-170, 173, 176, 343, 372
Janss, Peter F. 58-59, 67, 378
Jayne, George W. 185, 198
Jemison, John L. 7-8, 14
Jenkins, Homer K. 2, 4-5, 10-11, 15, 343
Jensen, Peter B. 89, 107
Jensen, Robert R. 243, 246, 248, 343
Jewett, James 186, 198
Jimenez, Steve M. 94, 107
Jmaeff, George V. "Canada" 252-253, 255,
    258-263, 321
Jochem, Gregory T. 116, 128
Johnson, Carrol J. 176
Johnson, Clifford J. 21
Johnson, Anthony L. 238, 246
Johnson, Daryl L. 134, 138
Johnson, Glenn 5, 8-9
Johnson, Ira G. 161, 165
Johnson, Isaac 245, 248
Johnson, James L. Jr. 247-248, 321
Johnson, Leon V. Jr. 146
Johnson, Peter W. 201, 216
Johnston, Dennis N. 208
Johnston, Richard W. 203, 208
Jones, Bruce E. 178, 188-189
Jones Clarence 343, 372
Jones, Homer P. 4, 21
Jones, Jimmie L. 197
Jones, Larry G. 344, 372
Jones, Ronald R. 8, 21
Jordan, Billy G. 245, 248
Jordan, Bobby J. 120, 128
Jordan, Charles G. 45
Jordan, Henry C. 3, 15
Jordan, Willie J. 146
Joy William 97
Joyce, Paris E. 204, 215
Joyce, Walter A. 246
Joyner, Charles D. 51, 67
Judson, Donald H. 344, 372

Kain, Paul J. 120, 128
Kalbhenn, Ronald 227, 231
Kasminoff, Ross M. 205-206, 216
Kasson, Leonard C. 344, 375, 377
Kaus, Harry L. Jr. 4, 15, 344
Kaylor, Daniel R. 32-33, 46, 372
Kaylor, James N. 217-229, 344, 379
Kean, Billie O. 344, 372
Keene, R. R. 109
Kegel, John C. 254-255, 264, 344
Kehn, Alan B. 298, 313
Kehres, Jim 3-5, 22
Keith, Archibald K 344, 375, 377
Kelichner, Anthony K. 31-33, 40, 46, 379
Keller, Albert W. 344, 372
Kelly, John A. 15, 248, 344

Kelly, Paul D. 209, 216
Kelly, 236
Kelley, Joseph P. 107
Kendall, Stuart O. 3, 22
Kendall, Timothy 40, 46
Kennedy, John M. 251, 264
Kepner, Harold E. 129, 146
Kessinger, Charles 281-282
Kettering, Alvah J. 37, 42, 45, 344
Kilgore, Danny A. 208
Kindred, Lawrence J. 76, 102
King, 89
King, Cleveland Jr. 252, 255, 259, 261, 345
Kinnard, Richard E. 40, 46
Kirksey, Kendall 379
Kleinschmidt, Kenneth 136, 146
Kluesner, Robert A. 245, 248
Knight, Roy Jr. 113, 128
Kocsis, John S. 203, 205, 209, 216
Koenig, David B. 138
Kohlbuss, Rodney 81, 83, 107
Kolb, Bruce A. 22
Kolodziej, Ronald E. 133, 139, 146
Kopec, Edward 14
Kopec, Gene R. 146
Korkow, Kenneth A. 196-197-198, 321
Kosaka, Jose A. 204, 216
Kostendt, Kenneth R. 130, 146
Kowalski, 274
Kraner, Marvin R. 130, 146
Kratzer, Everette 40, 46
Kremer, James H. 22
Krick Donald W. Jr. 120-121
Kruger, Robert H. Jr. 134, 138, 140
Krulak, Victor 338
Kuhl, Kenneth S. 248
Kulick, Daniel H. 146
Kumrow, Lawrence L. 46

Lackey, Ira E. 345, 372
Laderoute, Michael J. 189, 191
Lafountaine, Norman "Frenchy" 59-60, 62-64, 67, 379
Laitala, Ellis 88, 90, 107
Laizure, Virgil E. 22, 372
Lalonde, Larry 278
Lamb, Allan W. 13, 15, 22, 345
Landry, Eddie L. 4, 15, 345, 372
Landry, Martin "Frenchy" 121, 128
Landy, Morton S. 295, 310, 313
Lane, Bobby R. 22
Lane, Michael L. 243-244, 246, 248
Lange, John F. 118, 128
Lantry, Thomas H. 107
Lantz, Ron 378
Larock, Robert E. 22
Lasater, Larry E. 312-313
Lassiter, John A. 188-189

Latham, Terry D. 275
Lathrope, Wood W. J. 196, 198
Latiolais, Curtis 379
Lawhorn, Lewis C. 169, 176
Lawless, Billy 279, 378
Lawson, Dale E. 29, 42, 46, 372
Leavell, Richard T. 275
Lebas, Claude G. 29, 42, 46, 321
LeBlanc, Alfred 205, 208
LeBlanc, Donald 130, 146
LeBlanc, Phillip 279
Leary A. 335
Lechowit, John 128
Ledbetter, Walter R. Jr. 288-289, 293-295, 298, 301, 306-313, 321, 379
Lee, Gerals A. 146
Lee, Gregory W. 88, 102, 107, 346
Lee, "Gunny" 210
Lee, Howard V. 47, 50-51, 53, 55-56, 67, 317
Lee, James C. 205, 216
Lego, George M. 108
Lehoullier, 274-276, 346
Lehrack, Otto J. 2, 22, 161
Lemay, George C. 248
Lempa, Joseph S. 113, 128
Lewis, Daniel A. 128
Lewis, Harold 86, 108
Lewis, James H. Jr. 275
Lewis, Richard A. Jr. 108
Lightfoot, Clarence W. 22
Like, Billy Joe 131-132, 139, 140
Linick, Gary P. 60-62, 67, 379
Link, Freddy 8
Linn, Robert A. II "Scar" 263-265, 346, 372, 379
Lipscomb, Roger D. 346, 372
Lira Robert C. 275
Littlefield, Robert H. 201, 208, 214
Lloyd, Raymond W. 128
Lobbezoo, Dennis 227, 229
Lockwood, Bryce F. 347, 372
Logan, Monty R. 216
Logan, William S. 245, 248
Loggins, Raymond B. 146
Lombard, Geoffrey 128
Lomers, A. T. 312
Long, Benjamin L. 178, 194, 197-198
Long, Speedy O. 157
Long, Terry C. 78-79, 99, 104, 108
Longoria, Felipe 40, 46
Louis, Robert Y. Jr. 138
Love, Gregory C. 34, 45
Lovely, Francis B Jr. 199-201, 204, 209, 212, 216, 372
Lowe, Willie Lee 272-275
Loweranitis, John L. 112, 121-123, 321, 372
Lownds, David E. 188, 198
Luke, Wayne A. 128
Lunsford, Glen T. 321, 373

Lyon, Alfred E. 129, 132-133, 136-137, 139, 141-143, 146, 347, 373

Machado, Joseph P. 22
Madden, Jack 120, 128
Madonna, Raymond C. 129, 146
Madsen, Chris 124, 128
Maguire, Danny A. 212-214, 216, 379
Mahalick, Norman J. 37-38, 46
Malone, George M. 234, 239, 243-244, 246, 248, 321
Mangold, Leo 263
Mansir, Paul W. 18, 22
March, Gerald P. 146
Marinkovic, Steve M. 310, 313
Marino, Jack Jr. 12-13, 15-16, 22, 348, 373
Marlantes, Karl A. 165, 250, 253-259, 261, 265, 314, 321, 379
Marquardt, Merlin E. 6, 14
Marshall, A. T. 277
Martel, Norman R. 134, 138
Martin, Bruce R. 276
Martin, James M. 40, 46
Martin, Kenneth E. 27, 29, 40, 42, 46
Martin, Michael C. 54, 67, 119, 128
Martin, Raymond B. 9
Martin, Robert L. 248
Martin, Vernal S. 50, 56
Martinez, Gilbert L. 265
Martinez, Mauro 204-206, 208-209
Martinez, Thomas A. 201, 216
Massey, James 5
Matthews, Don J. 46
Matthews, Robert A. 80-81, 108
Matton, Gregory L. 22
Matzka, Robert E. 178, 183, 198
May, Blaine R. 146
Mayton, James A. 56, 67, 315-316, 324
Mazurak, Peter A. 312-313
McClelland, Aubrey 204, 208
McClelland, George 178, 180-181, 189, 191
McClain, Larry L. 216
McConnell, Malcolm 109
McCoy, Frederick 42-43, 46
McDaniels, Peter W. 146
McDermott, Edward 53, 68
McDonald, Henry III 189
McDonald, Henry James 210
McDonald, Steve 219, 226
McDonald, Walter 96-98
McDougall, Gerold L. 30, 42, 46
McElroy, Ronald L. 170, 173
McGannon, Martin D. 131, 146
McHenry, John K. 275
McHughs, William B. 245, 248
McKay, James E. 54-58, 68
McKee, Thomas E, 349, 373
McKenzie, Richard W. 180, 189

McKeever, Frank G. 312-313
McKnight, James J. 22
McLeod, Bucky 279
McManus, William P. 130, 146
McNamara, Edward M. 171, 174, 176
McNamara, Robert "Strange" 152, 158
McNeer, John K. 134, 146
McWhorter, Samuel T. Jr. 146
Meade, Ronald D. 128
Meadows, Ronald J. 275
Meads, Kim E. 189
Medeiros, William C. 208
Meeks, Joseph R. 228, 231
Meeks, Oscar A. 120, 128
Meharg, Ben A. 54, 56, 68
Mehl, Thomas A. 274-276, 246-248
Meier, Terrance 322, 373
Melanson, James 186, 198
Menninger, Robert P. 253, 255, 262
Menzies, Alexander J. 29-30, 41-42, 349
Mercer, Arthur F. 120, 128
Meredieth, Sam 378
Meuse, John R. 132, 138, 322
Meyer, Ronald W. 105, 108, 349
Michael, Dennis S. 172-173, 176
Militello, Johnny T. 40, 46
Miller, Curtis L. 131, 146
Miller, Floyd E. 66, 68, 378
Miller, Howard G. 2, 7
Miller, Richard E. 350, 373
Miller, Robert E. 108
Miller William E. 97, 108
Millian Carmelo 85-86, 108
Milledge, Warren E. 40, 46
Millis Harlie L. 101, 108
Mills, Michael T. 350, 373
Mills, Zack 279
Miranda, Julio M. 196-198
Mitch, Joseph E. 40, 46
Mitchell, John F. 210, 216
Mitchell, David 128
Mitchell, Thomas E. Jr. 175
Moffett, Dennis 277
Moguel Albert 196-198, 350, 372
Moise, Herve J. 202, 208
Monahan, Frederick G. 139, 146, 160, 165, 322, 373
Monroe, Raymond 263, 265
Monroe, Robert W. 85, 108
Montgomery, Wayne E. 60, 68
Montjoy, Norris D. 350, 373
Moore, Anthony D. 275
Moore, Jack D. 245, 248
Moore, Larry G. 203, 208
Moore, Lloyd W. 185, 189
Moore, Morris J. 275
Moore, Wayne P. 197
Moore, William R. 138

Mora, John H. 312-313
Moreland, Ronald 82, 108
Morgan, Henry L. 140, 146
Morin, Roger L. 131
Morningstar, Robert L. 131-132, 143, 146
Morris, Daniel M. 350, 373
Morris, Gary W. 22
Morris, Marshall L. 101, 108
Morris, Richard A. 175-176
Morrison, Lonnie W. 193-194, 197-198, 350, 373
Morrison, Robert S. 4, 15, 22, 350, 373
Moss, Richard A. 351, 373
Moss, Weldon D. 41
Moser, Richard 88, 108
Mosley, Richard J. 46
Mowan, John E. 108
Muir, Joseph E. 2, 7, 15, 17-18, 22, 322, 373
Muldowney, Francis X. 118, 123, 128, 373
Mullan, John 372, 373
Mulloy, James E. Jr. 13, 15-16, 22, 322, 373
Murff, Herbert S. 51-52, 57, 68
Murphy, Danny 378
Murphy, Edward F. 109, 129, 160
Murray, Ernest M. 161, 165
Murtha, Gary D. 160-161
Myatt, Ramsey D. 3, 22
Myers. Donald F. 235-236, 246, 248, 351
Myers, William L. 373

Nail, Gary D. 36, 39-42
Nasalroad, Lonnie W. 108
Nation, Roy G. 216
Naugle, Robert K. 60, 68
Navadel, George D. 116, 123, 128, 351
Nazarian, David P. 119, 128
Neal, Richard I. 113, 120-121, 123, 128, 351, 378
Neill, Kenneth R. 60, 68
Nerad, Walter J. 121
Nesmith, Joseph Q. Jr. 351, 373
Newell, David G. 60, 67
Nicholas, Warren D, 146
Nick, Chris 312-313
Nickerson, Gilbert R. 7, 14
Nickerson, William W. 15
Nightingale, Charles E. 60, 67
Noakes, David L. 133, 145-146
Noble, Jimmy 60, 64, 68
Noel, Robert L. 22
Noel, Thomas E. 251-252, 255, 322
Nolan Keith W. 109, 129, 166
Noon, Patrick J. Jr. 265
Noonan, Florence 20
Noonan, Thomas P. 20, 22, 317
Norman, Marion Henry 194-195, 197
Norman, Michael 217
Norris, Gleason 352, 374
Northington, William C. 241, 246, 248, 352
Norton, Robert L. "Whimpy" 37-38, 46

Nuncio, Catarino V. Jr. 84-85, 108
Nunez, Victor 3

O'Brien, Joseph J. 352, 374
O'Brien, Mike 284, 376, 377
O'Connell, Terence S. 128
O'Dell, Harry J. 165
O'Donnell, John W. 352, 374
O'Donnell, Phillip F. 20, 22
O'Hara, Michael E. 194-195, 198
Ohanesian, Victor 159, 165
Olson, Charles M. 88, 104-105, 108
O'Malley, Robert E. 5-7, 11-12, 15, 20, 22, 317
Opolski, Frederick 'Ski" 274, 277
Ortego, Conrad L. 52, 68, 378
Ortiz, Eppie 54, 68
Osborne, John E. 147
O'Shields, Charles A. 147
O'Toole, Thomas F. 185-186, 198
Ovalle, Severiano 121, 128
Overstreet, Coy D. 3, 22
Owen, Thurman 26, 32-33, 35-36, 41, 44, 374, 376, 377
Owens, Dave 38
Owens, Robert F. 205, 208
Owens, Timothy E. 208
Ozger, Islam 275

Pace, James 55-56, 68
Pace, Robert L. 50, 68
Pacheco, Jose A. 22
Palma, Francis M. 162, 165
Panykaninec, George 203, 207, 216
Pappas, Ralph B. 116, 120-121, 124-125
Paratore, John M. 40, 46
Parker, David A. 253, 255
Parker, Paul D. II 298, 310-311, 313, 353, 379
Parker, Richard E. 246
Parkerson, Milton J. 60, 68
Parks, Frank E. 91-92, 108
Parnell Randall B. 236, 240, 249
Parr, Roy F. 31, 41-42, 46, 378
Parr, William T. 204, 209, 216, 353
Pass, Richard S. 22
Partridge, James S. 22
Patch, Nathaniel 378
Patrinos, Charles 137, 139, 145, 147, 353
Patten, David D. 27, 34, 42, 353
Paul, Hugh D. 276
Paul, Joe C. 9, 11, 15, 317
Paxton, Marvin L. 25, 46, 374
Pearson, Millard L. 353, 374
Peatross, Oscar F. 1, 15, 18, 353
Peckitt, Arthur E. Jr. 312-313
Peirce, Larry J. 276
Peko, Lolesio 85-86, 102, 108, 353
Pembleton, Elias S. 353, 374
Penney, Thomas E. 239-240, 246, 249

Peretta, Nicholas R. 245, 249
Perdue, Richard W. 275
Perino, Anthony 353, 374
Perry, Howard D. Jr. 40, 46
Perryman, James M. 69, 353
Person, Barnett G. 353, 374
Perzia, Paul A. 275, 277, 247, 249
Peters, George D. 276
Peterson, Dennis A. 26, 46
Petoskey, Regonel L. 22
Petty, Arthur O. 9, 18, 20, 22
Phillips, Jack W. 354, 374
Phillips, Sammy D. 114, 128
Phillips, Woodrow N. 354, 374
Phillips, Willie R. 276
Picanso, Leonard Jr. 138
Picciano, Terrance A. 229-230, 354
Pickett, Bayard "Scotty' 95, 108
Pilcher, William G. 108
Pilson, Darwin R. 84, 102, 108, 354
Pinquet, Jean 8, 22
Piotrowski, Daniel J. 34, 41-42, 354
Pipes, Kenneth W. 180, 183-185, 188, 192, 194-198, 354, 378
Pitre, Henry 284
Pitre, Robert 284-287, 376, 377, 379
Pittman, Milton 35-37, 39-40, 46, 379
Platt, David L. 40, 46
Plumley, James R. 128
Pollard, Richard 246
Pooley, Ronald J. 22
Pope, Gregory A. 85, 108
Porello Richard D. 251-252, 255, 257-261, 265, 354, 378
Porter, Herman P. 276
Porter, John E. Jr. 216
Porter, Mervin B. 2
Post, Gerald W. 204, 209, 216
Pounder, Duane E. 40, 46
Powell, Raymond 374
Powell, Robert C. 133, 135, 138
Prendergast, Edward I. 179, 191, 195, 198
Pressley, Ronald 186, 198
Pritchett, Eddie R. 203, 209, 216, 249
Prickett, John L. 113, 120, 128
Prock, Daniel 227, 229
Proctor, Melvin R. 229, 231
Puhlick, Peter S. 22
Pultz, Daniel 128
Purdy, William R. 295, 313
Purnell, Richard M. 5, 9, 15, 354

Quigley, Thomas N. 194, 198

Rabbitt, Timothy J. 257-258, 265
Radcliffe, Henry J.M. 199-200, 210, 212
Radcliffe, William Martin 210
Raitt, Albert H. 10, 15

Ramirez, Juan J. 138
Ramos, Oscar 139, 147
Rampulla, Terry J. 202, 208
Ramsburg, Harry R. 37, 39-40, 46, 374
Ramsey, David A.15, 17-18, 22, 299-300, 313
Randall, Terry L. 128
Rash, Donald R. 196-197, 322
Rather, Dan 166
Rayburn, Edward C. 181-183, 186, 190-191, 198
Redmond, Randolph 37, 46
Redenius, David G. 203, 206, 208
Reed, Franklin M. 22
Reed, James P. 9, 22
Reed, Kenneth R. 37, 46
Reed, Roy A. 119, 128
Reeff, James M. 22
Reis, Tiago 322, 374
Reising, Sergeant 65
Releford, Isieah Jr. 246, 249
Renfro, Oliver R. 40, 46, 374
Renfrow, Jeffrey R. 5, 22
Reno, Floyd R. 205, 216
Reynolds, William E. 128
Rice, Darwin G. 204, 216
Rice, John C. 134, 138
Rice, Robert 200-201, 208
Rice, Thomas E. 134-135, 147
Richards, Paul 40, 46
Richardson, Clarence A. 147
Richardson, Earl 374, 378
Richey, George C. Jr. 51-52, 68
Rider, James W. 56, 68, 298-301, 307, 313, 315, 355, 378
Ridgeway, Ronald L. 177-178, 181-183, 186-188, 190
Ridlon, Victor H. 64-65, 68, 374
Riley, Walter J. 120, 128
Rimpson, Robert L. 6, 11-12, 22
Rinehart, Benny D. 14-16, 22, 356
Ring, Edwin R. 252, 255, 261, 265, 356
Rittler, Jon B. 59-61, 63-64, 68, 379
Rivera, Arnold J. 189, 191
Rivera, Virgilio 356, 374
Roach, Terence 211
Roberts, Ben 134, 138
Roberts, Joseph T. 51-52, 68, 374
Roberts, Timothy A. 50, 68
Rode, Frederick W. 53, 68
Rodriguez, Damian 173, 176
Rodriguez, Henry Y. 108
Rodrigues, Joe G. Jr. 253, 255
Rogers, Gregory 147
Rogers, Raymond G. 111-116, 119-122, 125-128, 323, 373
Roland, Charles 176
Rolle, Melvin 102
Roman, Joseph G. 356, 375
Roman, Louis 66-68, 379

Rosales, Walter J. 128
Ross, Kenneth G. 59, 68, 375
Roth, Gregory J. 128
Rowell, Peter M. 87, 108
Rozanski, Edward C. 357, 375
Rozell, Edward A. 298
Rudd, Tyrus F. 227, 231, 375
Ruddick, Morris E. Jr. 357, 376
Ruester, Michael 128
Ruff, Willie J. 182, 186-187, 189
Ruiz, Joe A. 128
Ruiz, Jose Jr. 197, 203-204, 208- 209, 357
Rundle, Doc 37
Runick, Paul G. 236, 246, 249
Rush, Marvin G. 205, 208
Russell, David G. 41, 375
Russell, Richard A. 162, 165
Rusth, John E. 323, 375
Ryan, Garry J. 60, 63, 68
Ryan, Thomas R. 245, 249

Salas, Robert M. 40, 46
Sanchez, Javier A. 134, 138-139
Sanders, Walter L. 3, 22
Sanderson, Edward W. 312-313
Sanford, Albert R. 197
Santiago, Anthony 40, 46
Santner, Donald 186, 198
Sapp, Steven A. 253, 265
Sargent, George T. Jr. 257, 265, 323
Sasser, John R. 357, 375
Sawyer, James H. 15
Scaglione, Peter C. 72, 108
Scarborough, David C. 188-189
Schadler, 135
Schaubel, Kenneth 217-221, 227, 229, 375
Schavelin, Hugh E. 205, 208
Schley, Robert J. 323, 375
Schlup, Richard 186, 198
Schirmer, Ronald L. 147
Schmidt, Dennis R. 50, 56
Schneider, John F. Jr. 91
Schrader, Richard C. Jr. 245, 249
Schriner, Ronald L. 131
Schultz, David P. 173, 176
Schumacher, Ronald K. 141, 147, 375
Schwend, Howard L. 7, 14-15, 23
Scott, Fred S. 358, 375
Scullin, Patrick 86, 108
Scully, Patrick R. Jr. 358, 375
Scurlock, Jerry W. 59, 68
Seavy-Cioffi, Lawrence J. 358, 379
Seawright, Walter 227, 229
Sexton, Raymond D. 170, 173, 176, 358, 375
Shafer, Eric D. 276, 358
Shar, Danny A. 308, 313
Shaum, Roy T. 253, 255
Shaw, Franklyn W. 249

Sheehan, James P. 129, 147
Sheehan, Robert J. 89-90, 108
Shepard, Bobby R. 276
Shepheard, Louie D. 245, 249
Shepherd, Allen M. 54, 68
Shepherd, Peter M. 229
Sheppard, Lonnie Jr. 229
Sherman, Andrew W. 48-49, 56, 323
Sherman, Edward J. 358, 375
Shibley, Kamille K. 276, 358
Shier, Ronald J. 253, 255, 263
Shoaff, John W. 119, 128
Short, Mitchell C. 15
Siemon, David A. 118, 121, 375
Siemons, Glenn J. 147
Silvoso, Joseph A. 310, 313
Simonson, Edward H. 100, 108
Simpson, Willie L. 358, 375
Singleton, James K. 312-313
Sipperly, David W. 209, 216
Sipple, Conrad A. 323, 375
Skelton, Roy 139
Skemanski, Jim 31, 41
Skiles, Stanley L. 171, 176
Skinner, Phillip G. 134, 138
Skinner, Richard A. 76, 102
Skinner, Walter F. 189, 191
Slaughter, John E. 10-11, 23
Slaughter, John W. 201, 216
Sleigh, Duncan B. 323, 375
Slowey, Kenneth L. 11, 23
Smith, 203
Smith, Author C. 195, 197
Smith, Barry J. 163-165, 378
Smith, Bill 139
Smith, Clinton 147
Smith, Coy E. 359, 375
Smith, David A. 53, 68
Smith, Douglas W. 189
Smith, George W. 240, 246, 249, 359
Smith, Horace 169, 173, 176
Smith James A. 15
Smith, John D. 76, 102
Smith, Robert N. 60, 68
Smith, Stanley 186, 198
Smith, Steven J. 275
Smith, Sterling M. 176
Smith, Theodore G. 86, 108
Smith, Thomas J. 4, 23
Smith, Walter L. 7, 14
Smith, Walter W. 72, 108
Smith, Willie M. 169, 176
Solliday, Theodore D. 307, 313
Sommers, John 378
Sorenson, Carl M. 83, 108
Sorenson, Merle G. 129, 147
Soto, Octavio Jr. 251, 265
Spanjer, Ralph H. 306, 313

Spahn, Orrin W. 97-98, 106, 108
Spano, Charles D. 147
Sparrow, George R. 128
Spence, James J. 60, 62, 68, 378
Spencer, Ernest E. 198, 250, 265-266, 278
Sperry, Dale S. 245, 249
Spink, Shepard C. 54, 68
Springer, Corporal 300-304
Spurrier, Ronnie 10, 20, 23
Staggs, Daniel A. 229-231, 359
Stankiewicz, Kenneth D. 5, 10-11, 15, 359, 375
Stankowski, William 117, 123, 125-128, 359, 378
Steiner, Lawrence T. 102
Stephen, Luke A. 51-52, 68
Sterner, C. Douglas 314, 378
Sternisha, George 135,136, 139-142, 378
Stevens, Paul D. 109
Stevens, Tommy E. 53, 56, 68
Stevenson, Michael 97, 108
Stigall, Larry S. 238, 246, 249
Stipes, Robert L. 10, 15, 23
Stodelmaier, Frank E. 60, 68
Stone, Daniel M. 170, 173
Stoppiello, Frank 238, 242, 249
Storbeck, William W. 58, 68
Stroble, Thomas W. 40, 46
Strock, Richard A. 53, 67
Stubbe, Ray W. 129, 177, 197-199, 378
Stubbs, Reuben Jr. 23S
Stuckey, James 323, 376
Stueve, Donald E. 312-313
Sturkey, Marion F. 47, 51-52, 68, 109, 376
Sudbury, Paul E. 94, 102
Sullivan, John A. 3
Sullivan, John P.T. 84, 87, 102, 108, 360
Sullivan, Joseph H. 360, 376
Sullivan, Mikal J. 227, 229
Surprise, James M. 276
Sutter, Richard F. 129, 147
Swartz, James J. 50, 68
Swed, Roy F. 170, 173
Sweet, Jerry A. 208
Sykes, Dennis A. 200-202, 207, 209-212, 216, 378
Sylva, Valentin Z. Jr. 46
Sylvester, Harold 282
Sylvester, Robert 281, 284, 374, 376, 377
Szabo, Fernand 40, 46

Talbert, Rudolph H. 60, 68
Taitt, Selwyn H. 176
Tamez, Juan A. 245, 249
Tarzia, Nicholas C. 53, 56, 68, 368
Tasker, Kenneth E. 102
Taylor, Claude E 276
Taylor, James H. 131, 147
Taylor, Philip J. 253, 255
Teague, Frank 161
Terry, James M. 223-236

Tette, John B. 15
Tharp, Jerry D. 2, 10, 15
Theyerl, Clayton 179, 189
Thomas, Allen 246
Thomas, Bobby R. 245, 249
Thomas, Charles W. 255
Thomas, David J. 14
Thomas, Frank H. 121
Thompson, Everette A. 76, 102
Thompson, Harvey E. 174-176, 361, 376
Thompson, Wayne W. 361, 375
Thoms, Robert L. 171, 173, 176, 361, 376
Thrasher, Doss B. 186, 197-198
Thurber, Gary D. 130, 147
Tilghman, Richard A Jr. 255-256, 258, 265, 376
Tino, John F. Jr. 138
Tinsley, Dale 51-52, 68
Tolnay, John J. 26, 42, 46
Tompkins, Rathvon M. 210, 212, 216
Toney, Charles L. 46
Tonucci, Richard L. 3, 9-11, 15, 23, 361, 376
Totten, Kenneth R. 197
Towne, Peter C. 14
Tracey, John N. 191, 198
Tramel, Charles T. 23
Traylor, Corey S. 361, 376
Tressler, Clifford C. 60, 68
Tretiakoff, Alexander 180, 198
Trevino, Elias 361, 376
Trout, Mark A. 141, 147
Trujillo, Arthur, 310, 313
Tuckwiller, David W. 202, 216
Tujaque, John 279
Turner, Eddie H. 147
Turnquist, Roger W. 115, 128
Tyler, James H. 362, 375, 377

Underwood, David 121, 128
Ur, Stanley E. 253, 255

Vallejo, Cecilio 376, 379
Van, Douglas H. 49, 55, 68
Vannatta, Jon D. 14
Vance, Danny R. 52, 68
Vande Velke, Hendrik G. 229, 231
Vargas, Salvador 245, 249
Vasquez, Roy 46
Vaughn, Edward L. 10, 15, 23, 362
Ventus, Hugh G. 36, 46
Verrette, Paul 374, 376
Versimato, Gunny 18
Vicknus, Michael J. 60-61, 68
Vidovich, Matthew 108
Vignere, Joel G. 88, 104, 108
Villa, Feliberto 41
Villamor, Roman 120-121
Vogel, Ronald Lee 84-85, 108
Vroom, James L. 251, 255

Waddell, Lyle K. 23
Wall, Gilbert 178-180, 183-184
Wallace, Ernie 3-4, 15, 19, 23, 324
Walsh, James P. Jr. 172, 176
Walsh, Robert T. 90, 102-103, 363
Walt, Lewis W. 1, 23, 99-100, 301, 315
Walters, Fred M. Jr. 245, 249
Waltrip, Steve W. 313, 376
Wancea, Michael 42, 46
Warr, R. Nicholas 166, 176, 379
Warren, Donald A. 195-197
Warrick, David E. 54, 68
Wasilewski, Leonard P. 130, 147
Wassan, Kenneth L. 60, 68
Watkins, Robert W. 312
Watson, James L. 267, 272, 277
Wayne, John 97
Webb, Bruce D. 2, 5, 7-9, 14-15, 324
Webster, John W. 121, 128
Weiner, Barry K. 130, 141, 147
Weiss, Peter W. 185, 197-198, 363
Welch, Larry 132, 136, 147
Weller, Richard G. 119-120, 123, 128
Wells, James L. 147
Wells, Robert J. Jr. 208
West, Eddie A. 288-300, 302-307, 309, 379
West, Francis J. Jr. 69, 106, 108
West, Franklin 379
West, Seth L. 202, 204, 208-209, 215
Wheeler, Larry G. 128
Whitman, Charles 283
White, Bobby R. 364, 376
White, Donald E. 185-186, 197-198
White, "Gunny" 60
White, James L. 7, 14
White, Levoid 236-237, 249
White, Theodore G. Jr. 102
Whitmer, Maurice P. 170, 172-174, 176, 364, 376
Whittingham, Joseph M. 364, 376
Widdecke, Charles A. 73, 103, 108
Widener, Richard H. 176
Wiggins, Paul D. Jr. 364, 376
Wilcox, Terry L. 37, 46
Wilkerson, Hebert Lloyd 299, 313
Wilkinson, James B. 178
Wilkinson, Kenneth J. 202, 209
Willeumier, Robert C. 90, 108
Williams Danny P. 132, 147
Williams, Joseph R. Jr. 128
Williams, Leroy 53, 68
Williams, Nathaniel M. 204, 208
Williams, Ralph M. 40, 42
Williams, Ronald A. 147
Williams, Thomas C. 40, 46
Williams, Thomas L. 59, 63, 68
Williams, Wallace 121
Williams, Woodrow 134, 138

Williamson, Lynn 128
Williamson, Richard W. 102
Willis, Beasley 54, 68
Wilmot, Donald W. 102-103, 108
Wilson, Bobby G. 40, 46
Wilson, James S. 128
Wilson, Laverne 378
Wilson, Roy R. 40, 46
Wilson, Willis 26-33, 42, 44-46, 324, 376, 377, 379
Wilson, Wilmer D. 208
Winsley, Willie J. 239, 249
Winston, Tom 191
Winter, Gary M. 249
Winter, Ronald 60-61, 68
Winters, Alexander, Jr. 88, 108
Wireman, Charles W. 23
Wolfe, Raymond S. 120, 128
Wolff, Daniel
Womble, William T. 133-134, 138-139, 365
Woodell, Gary B. 186-198
Woodruff, Mark W. 166
Woolman, William H II 179-198
Wright, Carroll L. 312-313
Wright, Curtis D. 276
Wright, Richard H. 42
Wright, Robert 208
Wymer, William W. 208
Wynn, Larry 312-313

Xavier, Augusto M. 365, 376

Yagues, Robert G. 134, 138, 140
Yankey, D. V. 253, 265
Yates, John C. 365, 376
York, Roger D. 23
Young, Carroll W. 40, 46
Young, Clarence W. 132, 139, 147
Young, Donald E. 42
Young, James C. 46
Young, Ralph M. 68
Young, Robert L. 46

Zebrowski, Anthony J. 147
Zertuche, Adolph Jr 128
Zutterman, Joseph A. 119, 128